MATHEMATICS FOR TECHNICIANS

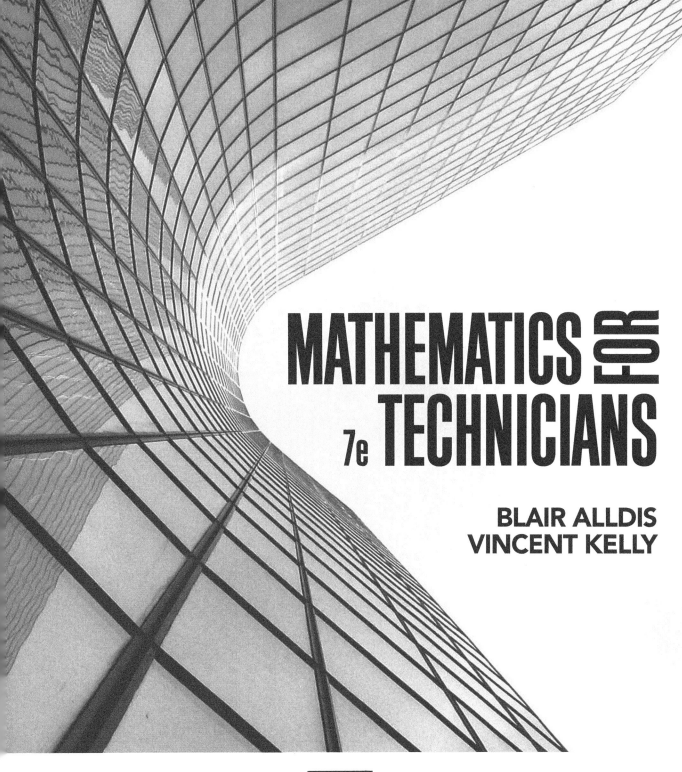

MATHEMATICS FOR TECHNICIANS

7e

BLAIR ALLDIS
VINCENT KELLY

McGraw Hill Education

Reproduction and communication for educational purposes
The Australian *Copyright Act 1968* (the Act) allows a maximum of one chapter or 10% of the pages of this work, whichever is the greater, to be reproduced and/or communicated by any educational institution for its educational purposes provided that the institution (or the body that administers it) has sent a Statutory Educational notice to Copyright Agency Limited (CAL) and been granted a licence. For details of statutory educational and other copyright licences contact: Copyright Agency Limited, Level 15, 233 Castlereagh Street, Sydney NSW 2000. Telephone: (02) 9394 7600. Website: www.copyright.com.au

Reproduction and communication for other purposes
Apart from any fair dealing for the purposes of study, research, criticism or review, as permitted under the Act, no part of this publication may be reproduced, distributed or transmitted in any form or by any means, or stored in a database or retrieval system, without the written permission of McGraw-Hill Australia including, but not limited to, any network or other electronic storage.

Enquiries should be made to the publisher via www.mcgraw-hill.com.au or marked for the attention of the permissions editor at the address below.

National Library of Australia Cataloguing-in-Publication Data
Author: Alldis, Blair K. (Blair Knox)
Title: Mathematics for technicians / Blair Alldis, Vincent Kelly.
Edition: 7th ed.
ISBN: 9781743070772 (pbk.)
Notes: Includes index.
Subjects: Shop mathematics—Problems, exercises, etc.
Engineering mathematics—Problems, exercises, etc.
Other Authors/Contributors:
    Kelly, Vincent Ambrose.
Dewey Number: 510.2462

Published in Australia by
McGraw-Hill Australia Pty Ltd
Level 2, 82 Waterloo Road, North Ryde NSW 2113
Publisher: Norma Angeloni Tomaras
Senior production editor: Yani Silvana
Copyeditor: Ross Blackwood
Proofreader: Pauline O'Carolan
Indexer: Shelley Barons
Cover and internal design: George Creative
Typeset in Berkeley Book 10.5/14.5pt by Laserwords Private Limited,
India Printed in Australia by SOS Print + Media

CONTENTS

PART 1 NUMERACY AND ALGEBRA

PART 2 GEOMETRY AND TRIGONOMETRY

ABOUT THE AUTHORS

Vincent Kelly, BEng (Civil), is currently Senior Civil Engineer at Hydro Tasmania. He is an industry expert with more than 15 years' engineering experience and brings this strong practical knowledge to the text. His particular areas of interest are building design and renewable energy.

Blair Alldis BA, BSc, the original author of the previous six successful editions of *Mathematics for Technicians,* has co-authored this seventh edition with Vincent Kelly. Blair has had extensive experience in teaching mathematics to engineering students at TAFE and was Senior Head Teacher of Mathematics at Sydney Institute of TAFE, Randwick College, prior to retiring.

ACKNOWLEDGMENTS

Many thanks to Blair Alldis for the privilege of working on this book; to my wife Corin for all her support and assistance; and to my sons Zac and Sam for their never-ending distraction and amusement. Thanks also to the teachers who assisted in the development of this new edition with their responses to the review questionnaire: Barry Russell from Skills Tech Eagle Farm, Queensland; Les Smith from South Western Sydney Institute of TAFE; Granville College; and John Hungerford. Finally, thanks to the team at McGraw-Hill who have been involved in the production of this edition, in particular Norma Angeloni Tomaras (publisher), Lea Dawson (development editor), Yani Silvana (senior production editor) and Ross Blackwood (freelance copyeditor).

Vincent Kelly

PREFACE

Mathematics for Technicians was developed to motivate engineering students to see the relevance of mathematics to their future careers. For this seventh edition, Vincent Kelly uses his extensive practical engineering experience to build on the strengths of this essential and inspiring well-known text. Practical topics and examples will help to equip students with a sound working knowledge of the elementary mathematics they require to succeed in the many engineering projects they will work on.

The author has also aligned the text with the requirements of TAFE engineering subject units. To achieve this, the following changes have been made:

- Chapters 17 and 18 have been revised. Chapter 17 now concentrates on circular functions and Chapter 18 focuses on trigonometric functions and phase angles.
- The material on determinants and matrices in Chapter 23 has been extended to more thoroughly cover the solution of simultaneous linear equations using matrices, including the solution of three equations with three unknowns.
- A new chapter (Chapter 24) on statistics and probability has been added.
- Supplementary material previously supplied on a CD is now available on the Online Learning Centre (OLC) and includes additional exercises.

TEXT AT A GLANCE

Setting a clear agenda

Each chapter starts by clearly defining learning objectives.

Learning objectives

On completion of this chapter you should be able to judge whether your knowledge of the following topics is satisfactory as a basis for this course.

- Solve problems involving ratios.
- Solve problems involving direct variation (proportion), inverse variation and joint variation.
- Use percentage to solve problems mentally, manually and using a calculator.
- Perform conversions between fractions, decimals and percentages.
- Apply your knowledge of ratio, proportion and percentage to solve practical problems.

> If you have difficulty with exercises in this chapter, you should consult your teacher before continuing with this course and seek remedial assistance.

Treatment of theory

Rule: When an equation contains fractions, it is usually advisable to clear all fractions first by multiplying both sides by the LCM of the denominators.

Rule: Since in an equation the pronumeral is standing for some definite number (unknown until the equation is solved), we can multiply both sides of an equation by the pronumeral.

Theory is covered clearly throughout each topic. Where appropriate, **mathematical rules**, **advice**, **notes** and **definitions** are highlighted separately for added emphasis.

Worked examples are provided to illustrate the theory and provide a template for students to develop their own skills.

Examples

1 If $A = 3t - 2$ and $B = 5 - t$, express $2A - 3B$ in terms of t.

Solution

$$2A - 3B = 2(3t - 2) - 3(5 - t)$$
$$= 6t - 4 - 15 + 3t$$
$$= 9t - 19$$

2 If $x = \dfrac{3 - k}{2}$ and $y = \dfrac{k - 2}{3}$, express $2x - 6y$ in terms of k.

Solution

$$2x - 6y = 2\left(\frac{3 - k}{2}\right) - 6\left(\frac{k - 2}{3}\right)$$
$$= (3 - k) - 2(k - 2)$$
$$= 3 - k - 2k + 4$$
$$= 7 - 3k$$

Exercises 14.10

1 For each of the following formulae, make the variable in brackets the subject of the formula:

  a $F = k^t$ (t) **b** $y = \log_{10} x$ (x)

  c $y = 3^{-x}$ (x) **d** $x^{\frac{a}{5}} = C$ (a)

  e $m = \log_{10} \dfrac{a}{b - c}$ (b) **f** $\log_{10}\left(\dfrac{x}{y}\right) = n$ (x)

  g $\log x = \log y$ (x) **h** $\log_{10} \dfrac{1}{x} = k$ (x)

+++ **2** Make t the subject of the formula:

  a $y = y_0 \times 10^{-kt}$ **b** $Q = Q_0 \, e^{-kt}$

  c $y = y_0 \times (1 - A^{kt})$ **d** $Q = Q_0 (1 - e^{kt})$

Practice makes perfect

To further develop students' skills and reinforce understanding, several levels of testing are provided throughout the text. A set of exercises follows each theory section. The symbol **+++** denotes more difficult exercises.

To ensure complete understanding of each chapter, self-test questions that address core concepts are provided at the end of most chapters. These tests are an excellent revision tool for the classroom.

Answers to all exercises and self-test questions are provided at the back of the book and online, enabling self-paced learning.

ANSWERS

1 Fractions and decimals

Exercises 1.1

1 a -8 b 4 c -2 d -2 e -4 f 2 2 a -3 b 10 c -3 3 a 0 b 0 c -2 4 a $-2°C$ b $-6°C$ 5 a contraction of 3 mm b contraction of 5 mm 6 a loss of 2 dB b loss of 8 dB c gain of 4 dB 7 a 2 b 14 8 a -6 b -8 c -4 d 4 e 6 f 12 9 a 4 b -2 c 6 d -4 e -8 f -1

Exercises 1.2

1 a 14 b -1 c 26 2 a 12 b 7 c 29 3 a 10 b 5 4 a 40 b 6 c 9 d 2

Exercises 1.3

1 a $2, 3, 5$ b $3, 9$ c $2, 3, 4$ d $3, 5, 9$ e $2, 3, 9$ f $3, 5$ g $2, 4$ h $3, 9$ 2 a $\frac{13}{15}$ b $\frac{16}{17}$ c $\frac{3}{4}$ d $\frac{7}{19}$ e $\frac{3}{4}$ f $\frac{4}{7}$ g $\frac{2}{3}$ h $\frac{3}{5}$ 3 a $\frac{8}{13}$ b $\frac{3}{5}$ c $\frac{997}{7000}$ d 40

d 1.8 e 0.6 f $0.000\,06$ 8 a 0.09 b 0.0004 c 0.36 d 1.44 e 0.0144 f $0.000\,121$ g 0.0009 h 1.21 9 a 0.2 b 0.9 c 0.07 d 1.2 10 a 26.08 b 1.203 c 230.7 d 0.0521 11 a 2.03 b 732.8 c 6.003 d 1.703 e 0.0732 f 0.0039 12 a 0.68 b 5.33 c 62.15 d 234.20 e 0.04 f 0.00 g 0.01 h 0.00 13 a 400 b 0.567 c 0.234 14 a 30 b 0.4 c 0.2 d 0.03 15 a 20 b 20 c 0.3 16 a $3.6\,\Omega$ b 200 m 17 a $0.12\,\Omega$ b $0.16\,\Omega$ c $0.024\,\Omega$ d $0.48\,\Omega$ 18 a 150 b 0.06 s 19 a 1.46 mm b 43.8 mm c 58.4 mm d 7.3 mm 20 3066 mm^2 21 3.4 m/s 22 4000 W 23 19.8 J 24 20 s 25 a 1.5 s b 45 s 26 4.8 min 27 a 7000 kg b 0.002 m^3

Exercises 1.5

1 a 2500 b 2400 c 2500 d 600 e 700 f 800

E-STUDENT / E-INSTRUCTOR

E-student

The Online Learning Centre (OLC) that accompanies this text helps you get the most from your course. It provides a powerful learning experience beyond the printed page.
www.mhhe.com/au/alldis7e

Supplementary material

Supplementary material is now included on the OLC. It provides answers to the questions within the text, as well as additional exercises and theories on a wide range of content. Within the text, the OLC symbol  indicates that supplementary material is available to enhance your learning experience.

E-instructor

Instructors have additional password-protected access to an instructor-specific area, which includes the following resources.

EZ Test Online

EZ Test Online is a powerful and easy-to-use test generator for creating paper or digital tests. It allows easy 'one click' export to course management systems such as Blackboard and straightforward integration with Moodle.

Testbank

A bank of test questions written specifically for this text allows instructors to build examinations and assessments quickly and easily. The testbank is available in a range of flexible formats: in Microsoft Word®, in EZ Test Online or formatted for delivery via Blackboard or other learning-management systems.

PART 1

NUMERACY AND ALGEBRA

CHAPTER 1

FRACTIONS AND DECIMALS

Learning objectives

■ Judge whether your knowledge of basic arithmetic is satisfactory.

 If you have difficulty with any of the exercises in this chapter, you should consult your teacher before continuing with this course and seek remedial assistance.

Do not use a calculator for Exercises 1.1–1.6. It is important that you can perform basic simple arithmetic with understanding and without having to rely on the availability of a calculator.

 ## 1.1 INTEGERS

Addition and subtraction of integers

Exercises 1.1

1   a $-3 - 5$ b $-2 + 6$ c $5 - 7$

    d $-1 - 4 + 3$ e $-3 + 7 + 1 - 9$ f $-5 - 4 + 5 + 2 + 4$

2   Take:

    a   5 from 2   b -3 from 7   c -2 from -5

3   a   Add -3 and 3   b   Take -7 from -7 c   Take -3 from -5

4   In each case below write down the new temperature after the change described:

Original temperature (°C)	Change in temperature (°C)	New temperature (°C)
a 3°C	drop of 5°C	
b −2°C	drop of 4°C	

5   What is the net result of:

    a   an expansion of 2 mm followed by a contraction of 5 mm?

    b   a contraction of 2 mm followed by a contraction of 3 mm?

6   What is the net result of:

    a   a gain of 7 dB followed by a loss of 9 dB?

    b   a loss of 3 dB followed by a loss of 5 dB?

    c   a loss of 3 dB followed by a gain of 7 dB?

Multiplication and division of integers

Exercises 1.1 *continued*

7 a $3 \div 9 \times 12 \div 2$ b $19 \div 13 \times 7 \div 19 \times 26$

8 a Multiply -2 by 3   b Multiply 4 by -2 c Divide 12 by -3

    d Divide -8 by -2 e Divide -12 by -2 f Multiply -2 by -6

9 Write down the missing number in each case below:

    a $-2 \times \ldots = -8$ b $-6 \div \ldots = 3$ c $-3 \times \ldots = -18$

    d $4 \div \ldots = -1$ e $-8 \div \ldots = 1$ f $16 \times \ldots = -16$

1.2 ORDER OF OPERATIONS

Exercises 1.2

1 a $2 + 3 \times 4$ b $8 - 6 - 3$ c $2 \times 3 + 4 \times 5$

2 a $3 \times (2 - 1) \times 4$ b $3 - 4 \times (5 - 6)$ c $(1 + 2) \times 3 + 4 \times 5$

3 a $3 + [2 \times (4 + 1) - 3]$ b $3 + [6 \div (2 + 1)]$

Squares and square roots

Exercises 1.2 *continued*

4 a $(3 + 4)^2 - (7 - 4)^2$ b $\sqrt{8^2 + 6^2} - (8 - 6)^2$

    c $3 + \sqrt{4} + 5 \times 2$ d $\sqrt{16 + 9} - 2^2 \times 3 + 9$

1.3 FRACTIONS

Divisibility rules

The following rules are useful and should be memorised:

- A number is divisible by 5 if its last digit is 5 or 0.
- A number is divisible by 2 if its last digit is divisible by 2 (i.e. it is an even number).
- A number is divisible by 4 if the number formed by its last two digits is divisible by 4.
- A number is divisible by 3 if the sum of its digits is divisible by 3.
- A number is divisible by 9 if the sum of its digits is divisible by 9.

Examples

1 379 516 is divisible by 4 because 16 is divisible by 4.

2 52 764 is divisible by 3 because $5 + 2 + 7 + 6 + 4 = 24$, which is divisible by 3, but this number is not divisible by 9 because 24 is not divisible by 9.

Exercises 1.3

1 Which of the numbers 2, 3, 4, 5, 9 divide into:

   a 390   **b** 513   **c** 2964   **d** 1395

   e 3078   **f** 1365   **g** 2468   **h** 123 456 789

If desired, more practice can be obtained by writing down any number you choose, testing it by the above rules, and then checking with your calculator.

2 Reduce each of the following fractions to lowest terms:

   a $\frac{65}{75}$ **b** $\frac{128}{136}$ **c** $\frac{216}{288}$ **d** $\frac{105}{285}$

   e $\frac{288}{384}$ **f** $\frac{300}{525}$ **g** $\frac{210}{315}$ **h** $\frac{1296}{2160}$

Cancelling

Exercises 1.3 *continued*

3 Cancel first if possible, then express in simplest form:

   a $\dfrac{2 \times 3 \times 4}{3 \times 13}$ **b** $\dfrac{197 \times 3}{5 \times 197}$ **c** $\dfrac{1000 - 3}{1000 \times 7}$ **d** $\dfrac{100 \times 50}{100 + 25}$ **e** $\dfrac{99 - 9}{99 \times 9}$

Mixed numbers

Exercises 1.3 *continued*

4 Convert the following improper fractions to mixed numbers (first reduce to lowest terms if necessary):

   a $\frac{5}{2}$ **b** $\frac{30}{7}$ **c** $\frac{128}{11}$ **d** $\frac{18}{4}$ **e** $\frac{255}{60}$

5 Convert the following mixed numbers to improper fractions:

   a $3\frac{1}{2}$ **b** $2\frac{3}{4}$ **c** $7\frac{2}{5}$ **d** $12\frac{7}{8}$ **e** $11\frac{3}{11}$ **f** $11\frac{5}{12}$

Addition and subtraction of fractions

Exercises 1.3 *continued*

In the following exercises, all working is to be done in fractions and answers should be given as proper fractions or mixed numbers.

6 **a** $\frac{2}{3} + \frac{1}{4}$ **b** $\frac{5}{6} + \frac{3}{4}$ **c** $\frac{3}{200} + \frac{7}{50}$ **d** $\frac{7}{10} + \frac{3}{20}$

   e $\frac{1}{18} + \frac{5}{36}$ **f** $\frac{5}{8} + \frac{13}{20}$ **g** $\frac{8}{9} + \frac{1}{3} + \frac{5}{18}$ **h** $\frac{3}{200} + \frac{7}{100} + \frac{9}{50}$

7 **a** $\frac{3}{4} - \frac{2}{3}$ **b** $\frac{5}{6} - \frac{1}{12}$ **c** $\frac{7}{9} - \frac{1}{6}$ **d** $\frac{21}{100} - \frac{17}{50}$

8 **a** $\frac{2}{3} - \frac{1}{4} + \frac{5}{12}$ **b** $\frac{3}{8} - \frac{1}{12} + \frac{5}{6}$ **c** $\frac{8}{9} - \frac{1}{6} - \frac{2}{3}$

   d $\frac{7}{10} - \frac{1}{3} + \frac{2}{5}$ **e** $\frac{3}{25} + \frac{1}{5} - \frac{3}{10}$ **f** $\frac{11}{12} - \frac{5}{18} + \frac{1}{36}$

Addition and subtraction of mixed numbers

Exercises 1.3 *continued*

9 a $32\frac{3}{5} + 25\frac{2}{5}$ b $64\frac{2}{3} + 31\frac{2}{3}$ c $27\frac{3}{5} + 36\frac{4}{5}$ d $55\frac{5}{8} + 27\frac{7}{8}$

10 a $86\frac{1}{4} - 25\frac{3}{4}$ b $67\frac{2}{5} - 35\frac{3}{5}$ c $30\frac{2}{7} - 16\frac{5}{7}$ d $43\frac{1}{8} - 26\frac{7}{8}$

11 a $37\frac{1}{2} + 29\frac{3}{4}$ b $61\frac{2}{3} - 39\frac{3}{4}$ c $39\frac{1}{2} + 46\frac{4}{5}$ d $82\frac{1}{7} - 52\frac{1}{2}$

Multiplication of fractions

Exercises 1.3 *continued*

12 a $3 \times \frac{1}{4}$ b $3 \times \frac{5}{7}$ c $5\frac{1}{4} \times 6$ d $\frac{4}{7} \times 6$ e $5 \times 20\frac{2}{3}$

13 a $37 \times \frac{19}{37}$ b $\frac{2}{3} \times \frac{5}{7}$ c $\frac{16}{19} \times 19$ d $\frac{11}{12} \times 36$

    e $\frac{5}{7} \times \frac{14}{15}$ f $\frac{3}{8} \times \frac{1}{3}$ g $\frac{13}{15} \times 30$ h $\frac{7}{15} \times \frac{10}{21}$

    i $\frac{2}{3} \times \frac{6}{7} \times \frac{21}{80}$ j $\frac{5}{6} \times 60$ k $\frac{17}{18} \times \frac{3}{5} \times \frac{15}{17}$ l $27 \times \frac{7}{9}$

14 a $1\frac{1}{2} \times \frac{4}{5}$ b $2\frac{2}{3} \times 1\frac{3}{4}$ c $1\frac{1}{3} \times \frac{3}{4}$ d $\frac{2}{3} \times 3\frac{1}{2}$

Division of a fraction by a number

Exercises 1.3 *continued*

15 a $8 \div \frac{2}{3}$ b $\frac{3}{4} \div \frac{2}{3}$ c $\frac{6}{7} \div 2$ d $\frac{5}{6} \div 5$ e $\frac{3}{8} \div \frac{1}{3}$

16 a $4\frac{6}{13} \div 2$ b $3\frac{1}{2} \div \frac{1}{2}$ c $1\frac{2}{3} \div \frac{3}{4}$ d $1\frac{3}{4} \div \frac{3}{4}$ e $2\frac{1}{2} \div 1\frac{1}{3}$

Complex fractions

Exercises 1.3 *continued*

These are most easily simplified by multiplying top and bottom by the same number.

17 a $\dfrac{\frac{7}{8} - \frac{3}{4}}{\frac{5}{8}}$ b $\dfrac{\frac{5}{12}}{\frac{7}{8} - \frac{2}{3}}$ c $\dfrac{\frac{3}{4} + \frac{5}{6}}{\frac{7}{12} - \frac{3}{8}}$ d $\dfrac{\frac{9}{14} + \frac{6}{7}}{\frac{3}{7} - \frac{5}{14}}$ e $\dfrac{1 - \frac{17}{39}}{\frac{5}{13} - \frac{13}{39}}$

Squaring a fraction or a mixed number

Exercises 1.3 *continued*

18 Find the square of each of the following, giving each result as a fraction or a mixed number:

    a $\frac{2}{3}$ b $\frac{1}{7}$ c $\frac{3}{5}$ d $\frac{1}{2}$ e $1\frac{1}{3}$ f $4\frac{1}{2}$ g $2\frac{1}{4}$

19 Evaluate as a fraction or a mixed number:

    a $(\frac{1}{3})^2$ b $(2\frac{1}{3})^2$ c $(1\frac{2}{3})^2$ d $(3\frac{1}{2})^2$ e $(5\frac{1}{2})^2$

continued

continued

20 Find the square root of each of the following, giving each result as a fraction or a mixed number:

    a $\frac{4}{9}$ **b** $\frac{1}{4}$ **c** $\frac{25}{81}$ **d** $\frac{1}{9}$ **e** $\frac{9}{16}$ **f** $\frac{100}{81}$

21 Evaluate as a fraction or a mixed number:

    a $\sqrt{\frac{9}{16}}$ **b** $\sqrt{\frac{1}{25}}$ **c** $\sqrt{2\frac{1}{4}}$ **d** $\sqrt{2\frac{7}{9}}$ **e** $\sqrt{5\frac{4}{9}}$

Miscellaneous

22 Three resistors in parallel carry an electric current. One resistor carries $\frac{2}{5}$ of the current and another carries $\frac{1}{2}$ of the current.

    a What fraction of the total current is carried by these two resistors?

    b What fraction of the total current is carried by the third resistor?

    c If the total current is 5 A, what current is carried by the third resistor?

23 When the wheel of a vehicle makes 2 revolutions, the vehicle moves a distance of 3 m.

    a How many revolutions does the wheel make when the vehicle moves 1 m?

    b How many revolutions does the wheel make when the vehicle moves $2\frac{1}{2}$ m?

    c What distance does the vehicle move when the wheel makes $3\frac{1}{4}$ revolutions?

24 A wire has a resistance of $2\frac{1}{2}$ Ω/m.

    a What will be the resistance of $5\frac{1}{2}$ m of this wire? (Work in fractions.)

    b What length (as a fraction of a metre) has a resistance of $\frac{3}{4}$ Ω?

25 A body moving in a straight line travels 1 km at 6 km/h and then 4 km at 9 km/h. Find the total time taken as a fraction of an hour.

26 If a pump takes 3 h to empty a tank and another pump takes 4 h to empty the same tank, what fraction of the tank is emptied in 1 h when both pumps are working simultaneously?

27 A man travels 5 km at a speed of 3 km/h. How much time has he left (as a fraction of an hour) to complete a journey if the total time allowed is $2\frac{1}{2}$ h?

28 A woman estimates that she walks at a speed of $4\frac{1}{2}$ km/h. How many hours will it take her to walk a distance of 3 km?

Zeros in fractions

$\frac{24}{3}$ represents the number of threes needed to make 24, i.e. the number by which we must multiply 3 in order to obtain 24. We know that this equals 8 because we learnt that $8 \times 3 = 24$. We do not learn division tables; we use our knowledge of the *multiplication* tables in order to do simple divisions.

You know that $\frac{28}{7} = 4$, *not* because you learnt that $28 \div 7 = 4$, but because you learnt that $4 \times 7 = 28$.

    $\frac{12}{4}$ = the number of fours needed to obtain 12 (= 3)

    $\therefore \frac{0}{0}$ = the number of zeros needed to obtain 0

But *any* number of zeros = 0.

    $\therefore \frac{0}{0}$ = any number you like to think of

Since $\frac{0}{0}$ has no *definite* value, we say that it is **undefined.** If the result of a computation is $\frac{0}{0}$, something was wrong with the original problem posed, or with the computation.

$\frac{5}{0}$ is the number of zeros needed to obtain 5, but there is *no* answer to this question, i.e. $\frac{5}{0}$ is *not a number,* and hence has no place in mathematics. We say that, $\frac{N \ (\text{any number})}{0}$, like $\frac{0}{0}$, is undefined—that is, $\frac{7}{0}$, $\frac{-32}{0}$, etc. are mathematically meaningless. So we have the rule: *You cannot divide by zero.*

Note: $\frac{0}{3}$ is the number of threes needed to obtain 0.

Now $0 \times 3 = 0$

$\therefore \frac{0}{3} = 0$

Similarly, $\frac{0}{17}$ is the number of seventeens needed to obtain 0.

$\therefore \frac{0}{17} = 0$, since $0 \times 17 = 0$

Hence: $\dfrac{0}{N} = 0$, where *N* is *any* number except zero (since $\frac{0}{0} \neq 0$, being undefined).

SUMMARY

Learn these two facts:

1 You can't divide by zero.

2 A fraction equals zero if, and only if, the *numerator* is *zero* (and the denominator is *not* zero),

   i.e. $\dfrac{0}{N} = 0$, providing $N \neq 0$.

Exercises 1.3 *continued*

29 For each of the following fractions, either evaluate or write 'U', meaning 'undefined'.

a $\dfrac{3 \times 0}{7}$ **b** $\dfrac{5 \times 0}{0 \times 2}$ **c** $\dfrac{4 + 0}{3 \times 0}$ **d** $\dfrac{3 - 3}{7}$

e $\dfrac{6 - 6}{5 - 5}$ **f** $\dfrac{0}{3\frac{1}{2}}$ **g** $\dfrac{4}{3 - 3}$ **h** $\dfrac{5 - 5}{2 + 5}$

i $\dfrac{7 - 3}{7 - 7}$ **j** $\dfrac{5 \times 0}{3 \times 7}$ **k** $\dfrac{3 \times 0}{4 - 4}$ **l** $\dfrac{6 - 6}{6 - 2}$

1.4 DECIMAL FRACTIONS

You will remember that 0.2 means $\frac{2}{10}$, 0.03 means $\frac{3}{100}$, 0.004 means $\frac{4}{1000}$, and so on.

   0.23 means $\frac{2}{10} + \frac{3}{100}$, i.e. $\frac{23}{100}$

   0.234 means $\frac{2}{3} + \frac{3}{100} + \frac{4}{1000}$, i.e. $\frac{234}{1000}$

   0.067 means $\frac{6}{100} + \frac{7}{1000}$, i.e. $\frac{67}{1000}$

These numbers are called **decimal fractions** (or **decimals**), since they are really fractions written in decimal form.

Note: Decimal fractions less than 1 should be written with a zero before the decimal point, e.g. $\frac{73}{100}$ should be written as 0.73 (*not* as .73).

Addition and subtraction of decimals

The decimal points must be vertically aligned when adding or subtracting decimals.

> **Example**
>
> 3.96 + 426.053 = 430.013 $\downarrow$
>
>   3.96
>   426.053
>   430.013

You should practise adding and subtracting *without* having to write the numbers one under the other.

Exercises 1.4

Remember: no calculator.

1 **a** 4.32 + 16.804   **b** 234.9 + 4.68   **c** 0.068 + 29.94

2 **a** 68.38 − 4.7   **b** 4.03 − 0.98   **c** 11.7 − 0.082

Multiplication and division by powers of 10

Note: When any number has more than four digits before or after the decimal point, these digits should be arranged in groups of three with a space (not a comma) separating them. For example, we write 23 456 (but not 23456 or 23,456), and we write 0.123 45 (but not 0.12345). Note also that a number is never commenced with a decimal point; we write 0.12 (not .12).

Exercises 1.4 *continued*

3 **a** 4.3 ÷ 100   **b** 0.0063 × 100   **c** 2.5 × 1000   **d** 6 ÷ 1000

4 **a** 4.6 ÷ 20   **b** 2.3 × 30   **c** 0.2 × 300   **d** 80 ÷ 4000

Multiplication of decimals

Exercises 1.4 *continued*

5 **a** 3.4 × 6.7   **b** 0.63 × 26   **c** 1.7 × 2.38   **d** 0.456 × 17

Cases such as the following should be computed mentally:

6 **a** 0.2 × 0.03   **b** 2.3 × 0.5   **c** 36 × 0.7
   d 0.2 × 0.5   **e** 0.05 × 0.4   **f** 0.6 × 0.005

7 **a** 0.2 × 0.3 × 0.4   **b** 0.03 × 0.1 × 0.2   **c** 0.12 × 0.12 × 0.2
   d 5 × 1.2 × 0.3   **e** 0.4 × 5 × 0.3   **f** 0.01 × 0.2 × 0.03

8 **a** $(0.3)^2$ **b** $(0.02)^2$ **c** $(0.6)^2$ **d** $(1.2)^2$

  e $(0.12)^2$ **f** $(0.011)^2$ **g** $(0.03)^2$ **h** $(1.1)^2$

9 You should be able to evaluate these mentally:

  a $\sqrt{0.04}$ **b** $\sqrt{0.81}$ **c** $\sqrt{0.0049}$ **d** $\sqrt{1.44}$

Division by an integer

Exercises 1.4 *continued*

Do not use a calculator.

10 **a** $78.24 \div 3$ **b** $8.421 \div 7$ **c** $922.8 \div 4$ **d** $0.4689 \div 9$

11 **a** $69.02 \div 34$ **b** $11\,724.8 \div 16$ **c** $144.072 \div 24$

  d $114.101 \div 67$ **e** $19.1052 \div 261$ **f** $0.3315 \div 85$

Rounding up and down

Exercises 1.4 *continued*

12 Write each of the following correct to two decimal places:

  a 0.6842 **b** 5.3293 **c** 62.1461 **d** 234.1983

  e 0.0432 **f** 0.0039 **g** 0.007 63 **h** 0.004 99

Division by a decimal

Exercises 1.4 *continued*

13 Evaluate the following (without using a calculator) and state each result *exactly*:

  a $14 \div 0.035$ **b** 1.35×0.42 **c** $3.393 \div 14.5$

14 Obtain each answer mentally:

  a $\dfrac{6}{0.2}$ **b** $\dfrac{0.036}{0.09}$ **c** $\dfrac{0.04}{0.2}$ **d** $\dfrac{0.0006}{0.02}$

15 Obtain each answer mentally:

  a $8 \div 0.4$ **b** $30 \div 1.5$ **c** $0.006 \div 0.02$

16 A wire has resistance 0.12 Ω/m.

  a Find the resistance of 30 m of this wire.

  b Find the length of this wire that has resistance 24 Ω.

17 The total resistance of resistors R_1 and R_2 connected in parallel is given by $R = \dfrac{R_1 \times R_2}{R_1 + R_2}$.

  Find the total resistance when the following pairs of resistors are connected in parallel:

  a $R_1 = 0.3\ \Omega, R_2 = 0.2\ \Omega$ **b** $R_1 = 0.8\ \Omega, R_2 = 0.2\ \Omega$

  c $R_1 = 0.06\ \Omega, R_2 = 0.04\ \Omega$ **d** $R_1 = 1.2\ \Omega, R_2 = 0.8\ \Omega$

continued

continued

18   An oscillator is set at a frequency of 5000 Hz (i.e. 5000 cycles per second).

   a   How many oscillations does it make in 0.03 s?

   b   How long does it take to make 300 oscillations? (*Hint*: first write down how long for
          1 oscillation).

19   A load of 50 kg extends a spring by 14.6 mm. What extension would be caused by a load of:

   a   5 kg?   **b**   150 kg?   **c**   200 kg?   **d**   25 kg?

20   A rectangular metal plate has length 70 mm and width 43.8 mm. Find its area.

21   Find the average speed (in metres per second) of an object that travels a distance of 8.5 m
     in 2.5 s.

22   Find the power output of an engine (in watts, i.e. joules per second) if it does 9200 J of work
     in 2.3 s.

23   Find the work done (in joules) by a force F of 16.5 N when it moves its point of application a
     distance S of 1.2 m in the direction of the force ($W = FS$).

24   Find the time taken by a body moving at a speed of 0.03 m/s to travel a distance of 0.6 m.

25   In drilling soft steel, a drill makes 40 rpm with a feed of 0.2 mm (per revolution). Find:

   a   the time taken (in seconds) for one revolution of the drill

   b   the time required for the drill to advance 6 mm

26   A machine operator in a factory finished 25 parts in 2 hours. Find (in minutes) the average time
     taken to finish one part.

27   A volume of 0.04 m^3 of a particular substance has mass 280 kg. Find:

   a   the mass of 1 m^3 of the substance

   b   the volume occupied by 14 kg of the substance

1.5 ROUNDING

We will see in later chapters that it is often desirable to express a numerical result in an *approximate*
form, 'rounding' the number to a certain degree of implied accuracy.

Examples	
The number	2736 rounded to the nearest 10 becomes 2740
	(because 2740 is closer to the original number than is 2730)
	2736 rounded to the nearest 100 becomes 2700
	2736 rounded to the nearest 1000 becomes 3000
	(because 3000 is closer to the original number than is 2000).
Similarly:	45.872 rounded to two decimal places becomes 45.87
	45.872 rounded to one decimal place becomes 45.9
	45.872 rounded to the nearest integer becomes 46

Note: The general rule is that, when digits are discarded, the last of the remaining digits is raised by 1 when the digit that previously followed it was greater than or equal to 5. For example, when rounding to two decimal places, 2.634 becomes 2.63, but 2.635 becomes 2.64.

Exercises 1.5

1 Round the following numbers to the nearest 100:

   a 2460   **b** 2440   **c** 2460   **d** 575

   e 749   **f** 752   **g** 990   **h** 999

2 Round the following numbers to the nearest integer:

   a 73.6   **b** 73.4   **c** 73.5

   d 99.3   **e** 99.5   **f** 999.9

3 Round the following numbers to two decimal places:

   a 1.356   **b** 1.355   **c** 1.352

   d 0.028   **e** 0.0551   **f** 0.007

1.6 SIGNIFICANT FIGURES

In a stated measured value, any digit that must have been read from the scale of a measuring instrument is called a **significant** digit.

Examples

1 345.6 mm: This is a value correct to one decimal place.

   The value may have been read as:

       345.6   on a scale of millimetres, or as

       34.56   on a scales of centimetres, or as

       0.3456   on a scale of metres.

   Clearly, each of these digits must have been read from a measuring scale, and hence each digit is significant. This value is given correct to 4 significant digits.

   All non-zero digits are significant.

2 305.6 mm: As above, each digit, including the zero, must have been read from a measuring scale. Hence each digit is significant. This value is also stated correct to 4 significant digits.

   A zero occurring between non-zero digits is significant.

3 2.30 cm: since the hundredths of a centimetre are stated to be zero, this zero *must* have been read from a scale. (Clearly an instrument with a vernier scale or a micrometer must have been used.)

   If the hundredths of a centimetre had *not* been read (e.g. if an ordinary ruler had been used), the length would have been stated as 2.3 cm (meaning 'correct to the nearest millimetre').

   Zeros at the end of a decimal are significant.

continued

continued

4   0.000 234 km: In this case the zeros may *not* have been read from a scale. The actual scale reading may have been 0.234 m, or 23.4 cm or 234 mm. The only digits that *must* have been read from the scale are the 2, 3 and 4.

Zeros at the beginning of a decimal are not significant.

5   700 km: In this case, we cannot tell whether the zeros were actually read from a scale (and hence are significant) or whether the length is being stated correct to the nearest hundred kilometres, or perhaps to the nearest ten kilometres.

If the reading is correct to the nearest hundred kilometres (i.e. ±50 km), neither of the two zeros is significant, since neither was actually read from a scale.

If the reading is correct to the nearest ten kilometres (i.e. ±5 km), the first zero is significant (since the tens of metres were read from the scale), but the last zero is not significant.

If the reading is correct to the nearest kilometre (i.e. ±0.5 km), both zeros were actually read from a scale, and both are significant.

If you need to rely upon the accuracy of a reading stated as 700 km, it is unwise to *assume* that such a reading is correct to the nearest kilometre. It is safer to assume that the zeros were *not* actually read.

If we were asked: 'How many significant digits are there in the measured quantity 300 N?', the best reply would be: 'Either one, two or three, depending upon how accurate the measurement is. It is safest to assume only *one* significant digit, the 3'.

However, if we were asked: 'How many significant digits are there in the measured quantity 300 N ± 5 N?', the correct reply is 'two' because the value is stated to lie somewhere between 295 N and 305 N.

Zeros at the end of a whole number may or may not be significant. Unless we are told the order of accuracy of the reading, it is safer to assume that these zeros are not significant.

SUMMARY

- Zeros between non-zero digits are significant.
- Zeros at the end of a decimal are significant.
- Zeros at the beginning of a decimal are *not* significant.
- Zeros at the end of a whole number are assumed to be *not* significant, unless information regarding the accuracy of the value shows that they are.

Exercises 1.6

1   State the number of significant digits in each of the following measurements:

   a 30.2 m   **b** 0.0050 kg   **c** 8000 ± 5 km

   d 1.004 mm   **e** 20 ± 0.5 t   **f** 0.004 km

2   Express each of the following measured quantities correct to 3 significant digits, in the same unit of measurement as given:

   a 0.034 56 kg   **b** 34.007 m   **c** 0.049 00 g

   d 600.826 m   **e** 4.007 58 t   **f** 780.68 km

Statement of a result with the appropriate number of significant figures

If we know the measured length, l, and breadth, b, of a particular rectangle, we can then compute its area, A. For example, if $l = 3.2$ m and $b = 2.347$ m, then $A = l \times b = 7.5104$ m^2.

However, our data states that the length lies somewhere between 3.15 m and 3.25 m and that the breadth lies somewhere between 2.3465 m and 2.3475 m. All that we can deduce about the area is that it lies somewhere between 3.15×2.3465 m^2 and 3.25×2.3475 m^2 (i.e. somewhere between 7.391 475 m^2 and 7.629 375 m^2). Since we are not sure of even the *first* decimal place in the value for the area, it would be absurd to state the area as 7.5104 m^2 as originally calculated. The area almost certainly lies between 7.4m^2 and 7.6 m^2 and would be correctly stated as 7.5 ± 0.1 m^2.

In a computation which involves many measured values, it is tedious to find the limits of accuracy by first using the smallest possible values and then using the largest possible values as we have done above. A rule-of-thumb can be used which, although very simple to apply, gives a reasonably good idea of the accuracy.

> **Rule:** State a result only to the number of significant figures present in the least accurately known of the numbers used in its calculation.

> ## Example
>
> Calculate the constant force, F, required to increase the velocity, v, of a 2.357 kg mass, m, from rest to 7.6 m/s over a distance, s, of 13.4 m, given that:
>
> $$F = \frac{mv^2}{2s} \text{ newtons}$$
>
> F computes to 5.079 862 7 newtons (N). The least accurately known number used in the computation is the velocity, which is known to only 2 significant figures. Hence, the *result* should be stated to only 2 significant figures, that is, as 5.1 N.

> **Note:** A constant in a formula is not a measured quantity and hence is an *exact* value. For example, the 2 in the above example and the $\frac{4}{3}$ in the formula $V = \frac{4}{3}\pi r^3$ are exact values.

Exercises 1.6 *continued*

3　Perform the operations on the following measured quantities, expressing each result as a decimal with the appropriate number of significant figures and in the same units as the data:

a　210 g + 307 g	b　5.4 mm + 46.3 mm	c　0.0783 kg + 0.4632 kg
d　9.80 t + 6.28 t	e　380 m + 269.3 m	f　0.0086 m^2 + 0.015 m^2

1.7 USING A SCIENTIFIC CALCULATOR

Exercises 1.7

1 Use a calculator to evaluate the following correct to two decimal places:

a $\dfrac{2.34 + 3.24}{6.26 - 0.23} + \dfrac{1}{13}$

b $2\frac{2}{13} + 4\frac{5}{17}$

c $\dfrac{1}{13} - \dfrac{1}{17}$

d $\dfrac{2.54 + 6.78 \times 2.76}{43.1 - 41.3 \times 0.482}$

e $\dfrac{18.2}{9.36 + 4.23}$

f $4.12 \times (3.54 - 1.76 \times 0.631)$

g $7.23 \times (4.74 - 2.73 \times (0.127 + 0.532))$

h $\dfrac{1}{2.31} + \dfrac{1}{1.75}$

i $\dfrac{6.94 - 2.94}{5.96 - 1.38} + \dfrac{10.96 - 3.76}{1.93 \times 2.95}$

j $2.75 \times 4.67 \div (4.96 - 3.84) + 5.74 \times 1.43$

k $4.85 - 6.94 - 4.36 \times 3.84$

l $7.13 + 7.53 - 6.49 \div (3.65 - 7.84)$

m $\dfrac{13.4}{6.21 \times 4.13} + \dfrac{8.67 + 1.14}{2.19 \times 3.47}$

n $\dfrac{7.64 - 1.28}{1.23 + 5.74 \div (2.54 + 1.65)}$

o $\dfrac{3.87 \times (4.75 - 1.26)}{3.64 - 2.54 \times 1.98} + \dfrac{3.98 + 1.79}{2.74 + 1.18}$

p $\dfrac{1}{24.2 - 7.45 \times (2.54 + 0.675)}$

q $\dfrac{3.62}{1.28 + 2.76} + \dfrac{7.85 - 1.42}{2.94 \times 3.01}$

r $9.21 + 3.21 \times (1.98 - 4.86 \div (1.28 + 2.39))$

s $3.21 \times ((2.65 - 1.29) \times (2.19 + 1.03) - 9.43)$

t $\dfrac{5.19 \times 1.93}{2.46 + 1.21} - \dfrac{6.18 + 15.4}{4.19 \times 3.86}$

u $8.54 + \dfrac{7.94 - 9.45}{1.89 \times 1.96} + \dfrac{4.74}{1.23 + 2.64}$

v $((2.54 - 1.98) \times (1.96 + 2.98 \times (1.75 - 0.983))) \times 1.17$

w $(1.75 + 2.91) \div (3.84 - 1.75) \times 3.71$

x $(2.11 + 3.85) \times 1.17 - 2.18 \div (6.31 + 2.14)$

The fraction key

Exercises 1.7 *continued*

+++ 2 Evaluate the following using the fraction key on your calculator, giving the answer as a mixed number:

a $2\frac{2}{3} + 1\frac{4}{9}$

b $5\frac{3}{7} - 3\frac{2}{5}$

c $4\frac{1}{8} \times 3\frac{2}{5} - 10\frac{3}{20}$

d $7\frac{1}{3} \div 3\frac{2}{3} \div \frac{5}{6}$

e $16\frac{3}{5} + 32\frac{4}{7} \div 3\frac{1}{2}$

f $5\frac{3}{8} \times \frac{7}{3} - \frac{2}{7} \times 5\frac{3}{5}$

1.8 EVALUATING MATHEMATICAL FUNCTIONS USING A CALCULATOR

Exercises 1.8

1 First investigate how to obtain the following results using your calculator:

  a $7^2 = 49$; $\sqrt{9} = 3$; $2^5 = 32$; $(2 + 3)^2 = 25$

  b Using the x^{-1} key: $\frac{1}{5} = 0.2$; $\frac{1}{2} + \frac{1}{4} = 0.75$; $\dfrac{1}{\sqrt{8} \times \sqrt{2}} = 0.25$

  c Using the fraction keys: $\frac{3}{5} = 0.6$; $\frac{19}{20} + \frac{9}{10} + \frac{4}{5} = 2\frac{13}{20} = 2.65$

  d Using the memory and recall keys, but *not* the bracket keys:

$$8^{\frac{2}{3}} = 4; \quad \frac{\sqrt{2.25} + \sqrt{6.25}}{\sqrt{2.25} + \sqrt{12.25}} = 0.8$$

  e Repeat (d) using the bracket keys but *not* the memory key.

Use a calculator to evaluate the following correct to two decimal places:

2 $5.26 \times \sqrt{3.17}$ **3** 2.04×1.77^2 **4** $\dfrac{1}{1.53^3}$

5 $\dfrac{4.17^2 - \sqrt{13.2}}{\sqrt{25.8} - 4.91^2}$ **6** $\dfrac{6.76^3 - 4.38^4}{4.89 \times \sqrt{9.73}}$ **7** $\dfrac{3.27 \times 1.54^3}{8.89 \times 3.61}$

8 $1.12^2 + 1.71^3$ **9** $\sqrt{8.11} + 2.11^2$ **10** $\dfrac{1}{5.19^3 - 9.97^2}$

11 $\dfrac{6.54^3 \times \sqrt{384}}{19.1^2 \times 2.69^3}$ **12** $\dfrac{8.14 \times \sqrt{126}}{5.21 \times 2.33^2}$ **13** $\dfrac{\sqrt{56.9} \times \sqrt{276}}{7.22^2 - 3.31^2}$

SELF-TEST

1 Evaluate:

  a $-7 + 3 + 8 - 2$ **b** $-199 + 67 + 199 - 69$

2 **a** Add -8 and 8 **b** Add -3 and -5 **c** Take -4 from -4

  d Take -1 from 1 **e** Add 2 and -6

3 What is the net result of:

  a a 2°C drop in temperature followed by a 7°C rise?

  b a 3°C rise in temperature followed by a 5°C drop?

4 **a** Multiply -3 by 4, then add 8.   **b** Divide -6 by -2, then subtract 7.

5 **a** $3 - 3 \times 4 + 9$ **b** $3 \times 4 - 6 \div 2$

6 **a** $3 - (2 + 2 \times 3)$ **b** $12 \div (2 - 6)$

7 **a** $1 - 3 - 2 \times (1 - 4)$ **b** $6 \div \{2 \times (5 - 2) - 6 \div (3 - 1)\}$

8 **a** $2 + \sqrt{3^2 + 4^2}$ **b** $\sqrt{25 - 16} + 2 \times 4$

9 **a** $2\frac{5}{6} + 3\frac{2}{3}$ **b** $29\frac{3}{4} - 12\frac{5}{6}$

10 **a** $8 \times 5\frac{3}{7}$ **b** $2\frac{2}{3} \div \frac{3}{8}$

11 **a** $1\frac{3}{4} \times 2\frac{1}{2}$ **b** $\frac{2}{3} \div 1\frac{1}{2}$

12 **a** $\dfrac{\frac{5}{8} + \frac{1}{4}}{1\frac{3}{4} + \frac{1}{8}}$ **b** $\dfrac{\frac{9}{16} - \frac{3}{4} \times \frac{1}{2}}{\frac{3}{8} - \left(\frac{1}{4}\right)^2}$

13 A beam rests on three supports so that one support carries $\frac{2}{3}$ of the load and another carries $\frac{1}{5}$ of the load.

   a What fraction of the load is carried by these two supports?

   b What fraction of the load is carried by the third support?

14 Evaluate mentally:

   a $6.17 + 1.93$ **b** $8.12 - 3.75$ **c** $0.826 + 2.39$ **d** $4.1 - 0.93$

15 Evaluate mentally:

   a 0.3×0.2 **b** 4.2×0.006 **c** 1.1×0.12

   d $0.6 \div 0.02$ **e** $0.96 \div 0.3$ **f** $0.012 \div 0.06$

   g $(0.012)^2$ **h** $\sqrt{0.0121}$ **i** $(1.1)^2$

16 Use a calculator to evaluate the following correct to two decimal places:

   a $\dfrac{8.21}{2.14 + 7.19} + \dfrac{3}{17}$ **b** $\dfrac{9.47}{3.27 \times 2.94}$

   c $\dfrac{4.21 \times 7.31}{2.47 \times 6.87}$ **d** $\dfrac{6.21 - 1.91}{1.74 + 2.70}$

   e $3.21 + \dfrac{1}{2.29}$ **f** $\dfrac{1}{1.09} - \dfrac{1}{1.34}$

   g $1\frac{5}{8} + 5\frac{7}{19} - 2\frac{7}{11}$ **h** $\dfrac{4.81 - 2.14}{3.21 \times 1.04} + \dfrac{1}{1.37}$

   i $\dfrac{1}{4.54 - 1.98 \times 1.56}$ **j** $2\frac{5}{9} \times 1\frac{1}{7} \times 3\frac{3}{20}$

   k $\dfrac{2.38}{1.21 \times 1.62} + \dfrac{1.98}{1.17 \times 1.28}$ **l** $2.17 - \dfrac{1}{1.58} + \dfrac{1}{2.99}$

17 Use a calculator to evaluate the following correct to one decimal place:

   a $17.7 \times \sqrt{12.3}$ **b** 10.2×2.33^2 **c** 3.02^3

   d 4.19^5 **e** $\dfrac{33.8 \times \sqrt{1760}}{12.5 \times 22.2}$ **f** $\dfrac{\sqrt{964}}{4.93^4 - 4.87^4}$

   g $9.03^2 + 1.16^2$ **h** $\sqrt{56.2} + 6.10^2$ **i** $6.03^3 - 4.40^2$

   j $4.49^4 - 6.01^3$ **k** $\dfrac{9.30^2 + 12.4^2}{4.62 \times 8.01^2}$ **l** $\dfrac{34.6 \times \sqrt{854}}{54.2 \times \sqrt{32.7}}$

CHAPTER 2

RATIO, PROPORTION AND PERCENTAGE

Learning objectives

On completion of this chapter you should be able to judge whether your knowledge of the following topics is satisfactory as a basis for this course.

- Solve problems involving ratios.
- Solve problems involving direct variation (proportion), inverse variation and joint variation.
- Use percentage to solve problems mentally, manually and using a calculator.
- Perform conversions between fractions, decimals and percentages.
- Apply your knowledge of ratio, proportion and percentage to solve practical problems.

 If you have difficulty with exercises in this chapter, you should consult your teacher before continuing with this course and seek remedial assistance.

2.1 RATIO

Brass is an alloy usually containing 33 parts of copper to 17 part of zinc, by mass. We say that the **ratio** of copper to zinc is '33 to 17', and we write 'Cu : Zn = 33 : 17 (by mass)'. We are not stating the actual masses of copper and zinc to be found in any particular sample of brass but the *relative* masses in the mix.

We could manufacture brass by alloying, for example:

- 33 g copper with 17 g zinc (giving 50 g brass)
- 66 g copper with 34 g zinc (giving 100 g brass)
- 132 kg copper with 68 kg zinc (giving 200 kg brass.

Since $\dfrac{\text{mass of Cu}}{\text{mass of Zn}} = \dfrac{33}{17}$, ratios are really only fractions written in a different way.

$$33 : 17 = 66 : 34 = 132 : 68 = \ldots$$

$$\tfrac{33}{17} = \tfrac{66}{34} = \tfrac{132}{68} = \ldots$$

Note: To compare the sizes of two quantities, the quantities must be expressed in the *same units*. For example, a time interval T_1 of 3 hours is not less than a time interval T_2 of 240 seconds, even though the number 3 is less than the number 240.

$$T_1 \qquad = 3 \text{ hours} \qquad = 180 \text{ minutes}$$
$$T_2 \qquad = 240 \text{ seconds} = 4 \text{ minutes}$$
$$\therefore T_1 : T_2 = 180 : 4 \qquad = 45 : 1$$

So T_1 is 45 times larger than T_2.

As long as two quantities are expressed in the *same* units, we can compare their sizes without knowing the size of that unit. For example, we do not need to know the size of an electrical capacitance of 1 microfarad (1 μF), or even what electrical capacitance *means*, in order to know that if two capacitances are 300 μF and 200 μF, then their capacitances are in the ratio of 3 : 2 or $1\frac{1}{2}$: 1. The first capacitance is $1\frac{1}{2}$ times larger than the second.

In the exercises in this chapter, quantities to be compared are either stated in the same units or the required conversion rate is given (except for units of time, which are well known by everybody).

Exercises 2.1

Express the answer to each exercise as the ratio of *integers* in simplest form. In questions 1–4, unit conversions are not required.

1   The density of brass is 8.5 t/m³ and the density of cast iron is 7.5 t/m³. What is the ratio of the density of cast iron to that of brass?

2   The coefficients of linear expansion of cast iron and cadmium are 0.000 010 2 K^{-1} and 0.000 028 8 K^{-1} respectively. What is the ratio of the coefficient of cast iron to that of cadmium?

3   The specific heat capacities of mercury and water are 0.14 kJ/kg K and 4.2 kJ/kg K respectively. What is the ratio of the specific heat capacity of water to that of mercury?

4   Young's modulus for cast iron is 120 GN/m² and that for brass is 84 GN/m². Find the ratio of Young's modulus for brass to that for cast iron.

In the following exercises, unit conversions are required.

5   The lengths of two bars are 450 mm and 2 m. What is the ratio of the smaller length to the greater length? (1 m = 1000 mm)

6   The masses of two objects are 3 t and 360 kg. What is the ratio of the smaller mass to the larger mass? (1 t = 1000 kg).

Sometimes a ratio is required in the form $n : 1$ or $1 : n$. In questions 7–9, evaluate n correct to 2 decimal places.

7   Express the ratio 43 : 28 in the form $n : 1$.

8   Express the ratio 1.83 : 5.12 in the form $1 : n$.

9   Resistance $R_1 = 850\ \Omega$ and resistance $R_2 = 1050\ \Omega$.
    a   Express the ratio $R_1 : R_2$ in the form $n : 1$.
    b   Express the ratio $R_1 : R_2$ in the form $1 : n$.

Division into parts

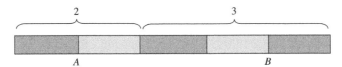

If we mark a bar of any length into five equal parts and then divide it into two sections A and B containing two and three of these parts respectively, we have divided the bar in the ratio 2 : 3. Clearly A is $\frac{2}{5}$ of the total and B is $\frac{3}{5}$ of the total.

   Similarly, if a quantity is divided into two parts A and B in the ratio 4 : 7, $A = \frac{4}{11}$ of the total and $B = \frac{7}{11}$ of the total.

Example

A potential difference of 20 V is divided into two parts, V_1 and V_2, in the ratio 5 : 3. Evaluate V_1 and V_2.

$V_1 = \frac{5}{8}$ of the total

$\quad = \frac{5}{8} \times 20$

$\quad = 12.5\,\text{V}$

$V_2 = \frac{3}{8}$ of the total   or $V_2 = 20 - 12.5$

$\quad = \frac{3}{8} \times 20 \qquad\qquad\quad\; = 7.5\,\text{V}$

$\quad = 7.5\,\text{V}$

Exercises 2.1 *continued*

10 A bar 15 m long is divided into two pieces, the ratio of their lengths being 2 : 3. Find the length of the shorter piece.

11 Fifty litres of liquid is divided into three parts, the volumes of which are in the ratio 4 : 7 : 9. Find the volumes of the three parts.

+++ 12 A beam is divided into three parts, the lengths of which are in the ratio 5 : 3 : 2. If the length of the longest part is 7 m, find:

   a   the length of the original beam

   b   the length of the shortest part

2.2 DIRECT VARIATION: THE UNITARY METHOD

Two quantities are often related in such a way that:
- when one doubles, so does the other
- when one trebles, so does the other
- when one halves, so does the other, and so on.

That is, they always increase and decrease in the same ratio.

Such quantities are said to vary directly as each other, or to be **proportional.**

Examples

1   The number of times a bicycle wheel rotates varies directly as the distance travelled.
2   The cost of buying petrol varies directly as the amount purchased.
3   The current through a resistor varies directly as the voltage across the resistor.
4   The distance travelled at constant speed varies directly as the time travelled.

Warning: It must not be *assumed* that because an increase in one quantity produces an increase in a second quantity, the two quantities will vary *directly.*

Examples

1   The distance you can throw a cricket ball does *not* vary directly with your age. (There is no certainty that when your age doubles, you will be able to throw it exactly twice as far.)
2   The power dissipated by a resistor does *not* vary directly with the current flowing through it.
3   The period of oscillation of a pendulum does *not* vary directly with its length.
4   The speed of a falling body does *not* vary directly with the distance fallen.

In many cases, however, it is obvious that two quantities *do* vary directly, or we know from *observation* that this is the case. In such cases, the **unitary method** is useful for computing changes.

Example

An object travelling at steady speed moves 57 km in 3 h. How far will it move in 5 h?

    3 h gives 57 km

  ∴ 1 h gives 19 km

  ∴ 5 h gives 5 × 19 = 95 km

Exercises 2.2

It may be assumed in each case that the quantities involved do vary directly.

1   If 7 L of a particular fluid costs $2.24, what will be the cost of 5 L?
2   A particular transistor gives a voltage gain of 25 when the load resistance is 1000 Ω. What load resistance is needed to give a voltage gain of 35?
3   A particular coaxial cable of length 84 km is found to have capacitance 6 μF. What length of this cable has capacitance 5 μF?
4   A flywheel rotates 1200 times in 1 min. How many revolutions does it make in 17 s?
5   An object moving at constant speed has a velocity of 180 km/h. How far does it travel in 13 min?

2.3 DIRECT VARIATION: THE ALGEBRAIC METHOD

Although the unitary method is useful for solving simple problems, the **algebraic method** has wider application.

Note: If quantity A varies directly as B, i.e. A is proportional to B, we write $A \propto B$.

This means that $\dfrac{A}{B}$ = a constant, i.e., the ratio $A : B$ is constant. If A doubles, so does B; if A reduces to $\frac{1}{3}$ its present value, so does B, and so on.

$\dfrac{A}{B} = k$, where k is some 'constant', i.e. some definite fixed number.

$\therefore A = k \times B$

Note: We could equally well state that $B \propto A$ and that $\dfrac{B}{A} = K$, and $B = K \times A$, where K is a constant.

Examples

1 When a load of 4 kg is hung on a particular spring, the spring extends by 70 mm. When the load is increased to 6 kg, the extension increases to 105 mm. Does this verify that the extension is proportional to the load?

$$\frac{L_1}{E_1} = \frac{4}{70} = \frac{2}{35}$$

$$\frac{L_2}{E_2} = \frac{6}{105} = \frac{2}{35}$$

Since $\dfrac{L}{E}$ remains constant $\left(= \dfrac{2}{35}\right)$, this *does* verify that $E \propto L$.

Note: The load increased by 50% and so did the extension; the load increased in the ratio $6 : 4 = 3 : 2$ and the extension increased in the ratio $105 : 70 = 21 : 14 = 3 : 2$.

2 When a current of 20 mA flowed through a particular resistor, the power dissipated was 1.2 W. When the current was increased to 50 mA, the power dissipation increased to 7.5 W. Does this verify that the power dissipation is proportional to the current?

$$\frac{I_1}{P_1} = \frac{20}{1.2} = \frac{100}{6} = \frac{50}{3}$$

$$\frac{I_2}{P_2} = \frac{50}{7.5} = \frac{200}{30} = \frac{20}{3}$$

Since $\dfrac{I}{P}$ does not remain constant, P is *not* proportional to I (i.e. $P \not\propto I$).

Note: The current increased in the ratio $20 : 50 = 2 : 5 = 4 : 10$; the power increased in the ratio $1.2 : 7.5 = 12 : 75 = 4 : 25$.

Exercises 2.3

1 The force on the coil in a moving-coil loudspeaker was 20 mN when the current through the coil was 12 mA, and the force was 25 mN when the current was 15 mA. Does this verify that the force is proportional to the current?

2 The open-circuit characteristics of a certain DC generator driven at 1000 rpm show that when the field current is 2.4 A, the generated emf is 140 V, and when the current is 4 A, the emf is 210 V. Does this verify that the generated voltage is proportional to the field current?

3 A simple pendulum whose length (l) is 2.43 m has a period of oscillation (T) of 1.00 s. When its length is 9.72 m, the period is 2.00 s.

   a Is it true that $\dfrac{T_2}{T_1} = \dfrac{l_2}{l_1}$?   b Does this verify that $T \propto l$?

4 When the length (l) of a simple pendulum is 4 m, its period (T) is 4.0 s, and when its length is 9 m, the period is 6.0 s. Does this verify that for a simple pendulum $T \propto \sqrt{l}$?

5 When a certain mass is moving with a speed (v) of 3 m/s, its kinetic energy (E) is 45 J, and when its speed is 4 m/s, the kinetic energy is 80 J. May it be true that:

   a $E \propto v$?   b $E \propto v^2$?   c $v \propto E^2$?   d $v \propto \sqrt{E}$?   e $E \propto \sqrt{v}$?

6 For a mild steel cable the elastic limit load for a cable of diameter 4 mm is 600 kg, and for a cable of diameter 6 mm it is 1350 kg. Does this verify that the elastic limit load of a cable is:

   a proportional to its diameter?

   b proportional to the square of its diameter?

   c proportional to the square root of its diameter?

7 For a certain mass of a gas held at constant temperature, the pressure (p) is 120 kPa when the volume (V) is 150 mm^3 and the pressure is 360 kPa when the volume is 50 mm^3. Does this verify that:

   a $p \propto V$?   b $p \propto \dfrac{1}{V}$?

Computations

If we know that two quantities are proportional, we can immediately write down an equation relating the two variables, involving a constant, k.

1 If we are given a corresponding pair of values for the two quantities, we can then evaluate k.

2 Once we know the value of k, we can find the value of one of the variables corresponding to any given value of the other.

Example

The extension of a cable (e) is proportional to the load (L) suspended from it. Given that a load of 1.25 t on a particular cable produces an extension of 3.57 mm, find the extension produced by a load of 2.18 t.

$$e \propto L \qquad \therefore e = k \times L$$

Now $e = 3.57$ when $L = 1.25$, $\therefore 3.57 = k \times 1.25$

$$\therefore k = \frac{3.57}{1.25}(=2.856)$$

$$\therefore e = \frac{3.57}{1.25} \times L$$

When $L = 2.18$, $e = \dfrac{3.57}{1.25} \times 2.18$

$$\approx 6.23 \text{ mm}$$

Exercises 2.3 *continued*

8   The speed v of a pulley belt is proportional to the rotational speed n of the pulley wheel. Given that when the wheel is making 10.0 revs/s, the belt speed is 9.34 m/s, find the belt speed when the wheel makes 13.0 revs/s.

9   A volume of 425 mL of a certain liquid has mass 386 g. Find the mass of this liquid that has volume 550 mL.

10   The force on the coil of a certain moving-coil loudspeaker is 15.0 mN when the current through the coil is 8.00 mA. Given that the force is proportional to the current, find the force on the coil when the current is 10.0 mA.

11   At a frequency of 500 Hz, the reactance of a certain inductor is 120 Ω. Given that the reactance is proportional to the frequency, find the reactance at a frequency of 350 Hz.

We have learnt that if a quantity A varies directly with another quantity B, this means that they increase and decrease in the same ratio.

Hence $\dfrac{A}{B} = k$ (a constant), so $A = k \times B$.

Similarly, if A varies directly as B^2, then $A = k \times B^2$; if A varies directly as $\sqrt{B}$, then $A = k \times \sqrt{B}$, and so on.

Examples

The period (T) of a simple pendulum varies directly as the square root of its length (L). Given that a pendulum of length 9 m has period 6 s, find:

a   the period of a pendulum whose length is 3 m

b   the length of a pendulum whose period is 1 s

Solution

We have $\qquad\qquad\qquad T = k \times \sqrt{L}$

Now when $L = 9$, $T = 6$

$$\therefore 6 = k \times \sqrt{9}$$

$$\therefore k = 2$$

Hence $T = 2\sqrt{L}$

continued

continued

a   When $L = 3$, $T = 2 \times \sqrt{3}$

≈ 3.46 s

b   When $T = 1$, $1 = 2 \times \sqrt{L}$

$\therefore \sqrt{L} = \frac{1}{2}$

$\therefore L = \frac{1}{4}$ m $= 0.25$ m

Exercises 2.3 *continued*

12 **a**   If x varies directly as y^2, and given that $x = 18$ when $y = 3$, evaluate:

   i x when $y = 4$ **ii** y when $x = 50$

   b   If t varies directly as $\sqrt{m}$, and given that $t = 17.0$ when $m = 321$, evaluate:

   i t when $m = 832$ **ii** m when $t = 13.4$

   c   If A varies directly as $\dfrac{1}{B}$, and given that $A = 117$ when $B = 83.4$, evaluate:

   i A when $B = 146$ **ii** B when $A = 63.2$

13 The maximum safe load (L) that can be suspended from a steel cable varies directly as the square of the diameter of the cable. For a certain mild steel cable, the maximum safe load is 0.25 t when the diameter is 5.0 mm. Find:

   a   the maximum safe load for a cable made of the same material with diameter 7.5 mm

   b   the diameter of the cable needed if the maximum safe load is to be 4.0 t

14 When a constant resultant force F is applied to any particular body originally at rest, over a distance of 2 m, the final velocity of the body varies directly as the square root of the force. Given that for a particular body, a constant force of 12.0 N produces a final velocity of 15.7 m/s, find:

   a   the final velocity produced by a force of 17.0 N

   b   the force required to produce a final velocity of 20.0 m/s

2.4 INVERSE VARIATION

The statement 'A varies inversely as B' means that A varies directly as $\dfrac{1}{B}$.

Hence $A = k \times \dfrac{1}{B}$ and $A \times B = k$

When one of the quantities *increases*, the other quantity *decreases in the same ratio,* so the product $A \times B$ remains constant. For example, for a car travelling at constant speed (V), the time (t) it will take to travel a distance of 120 km varies inversely as the speed.

When $V = 20$ km/h, $t = 6$ h
When $V = 40$ km/h, $t = 3$ h
When $V = 60$ km/h, $t = 2$ h
When $V = 80$ km/h, $t = 1\frac{1}{2}$ h

The product $V \times t$ remains constant ($=$ the distance travelled).

$20 \times 6 = 40 \times 3 = 60 \times 2 = 80 \times 1\frac{1}{2}$

SUMMARY

- If A varies inversely as B, then $A = k \times \dfrac{1}{B}$, $\therefore k = A \times B$

- If P varies inversely as Q^2, then $P = k \times \dfrac{1}{Q^2}$, $\therefore k = P \times Q^2$

- If S varies inversely as $\sqrt{t}$, then $S = k \times \dfrac{1}{\sqrt{t}}$, $\therefore k = S \times \sqrt{t}$

Examples

The time taken to raise the temperature of 1 L of water in an insulated container from 20°C to 100°C by means of a heating coil varies inversely with the square of the current, I, passed through the coil. Given that for a particular heating coil, a current of 2.0 A boils the water in 3.2 min, find:

a   the time taken to boil the water with a current of 3.0 A
b   the current required to boil the water in 2.5 min

Solution

$t = k \times \dfrac{1}{I^2}$, $\therefore k = t \times I^2$

When $I = 2.0$, $t = 3.2$, $k = 3.2 \times 2^2$

$\qquad\qquad\qquad\quad = 12.8$

$\qquad\quad \therefore t \times I^2 = 12.8$

a   When $I = 3$, $t \times 3^2 = 12.8$ b   When $t = 2.5$, $2.5 \times I^2 = 12.8$

$\qquad\qquad\qquad t \approx 1.4$ min $I \approx 2.3$ A

Exercises 2.4

1 a   If x varies inversely as y, evaluate x when $y = 31.6$, given that $x = 7.62$ when $y = 128$.
   b   Given that P varies inversely as the square of Q and that $P = 0.621$ when $Q = 17.3$, evaluate P when $Q = 2.47$.
   c   Given that A varies inversely as the square root of B and that $A = 13.9$ when $B = 2.07$, evaluate A when $B = 62.5$.

2 The time taken to raise the temperature of 1 L of water in an insulated container from 20°C to 30°C by means of a given heating coil varies inversely as the square of the current, I, passed through the coil. For a given heating coil it is found that the time taken when the current passed is 4.0 A is 33 s. Find (correct to 2 significant figures):
   a   the time taken when the current is 5.0 A
   b   the current required if the time taken is to be 1 min

continued

continued

+++ 3 The frequency of revolution, n (i.e. the number of orbits per unit time), of a satellite above the earth varies inversely as $\sqrt{r^3}$, where r is the radius of orbit. Given that a satellite whose radius of orbit is 5080 km orbits the earth each hour, find:

  a the frequency of revolution of a satellite whose orbital radius is 8000 km (in revolutions per hour correct to 3 significant figures)

  b the time taken for one orbit of the satellite in part (a) above (in minutes correct to the nearest minute)

  c the radius of orbit for a satellite that orbits the earth once per day (correct to the nearest 100 km)

  d the radius of orbit of the moon, given that it orbits the earth each 27.4 days (answer correct to the nearest 1000 km)

2.5 JOINT VARIATION

If the value of a quantity Q depends upon the values of *several* variables $x, y, z, \ldots$, such that:

$$Q = k \times x \times y \times z \times \ldots$$

then $Q = k_1 \times x$ when the other variables are constant

(i.e. Q varies directly as x),

and $Q = k_2 \times y$ when the other variables are constant

(i.e. Q varies directly as y), . . . and so on.

Hence, if Q varies directly as x, and also Q varies directly as y, and also Q varies directly as z, etc., then $Q = k \times x \times y \times z \times \ldots$

Example

If Q varies as t^2 and also inversely as v and also inversely as d, it follows that:

$$Q = k \times t^2 \times \frac{1}{v} \times \frac{1}{d}, \text{ i.e. } Q = k \times \frac{t^2}{vd}$$

If one set of corresponding values is known for all the variables, k can be evaluated and hence the formula for Q obtained.

For example, it is known that the force of gravitational attraction, F, between two masses m_1 and m_2 varies directly as m_1, varies directly as m_2 and varies inversely as d^2, where d is the distance separating the centres of gravity of the masses. Hence $F = k \times m_1 \times m_2 \times \dfrac{1}{d^2}$, i.e. $F = k \times \dfrac{m_1 m_2}{d^2}$.

By laboratory measurements of the force for known values of the masses and the separating distance, this constant is known to an accuracy of 0.01%. It is called the Universal Gravitation Constant and is usually denoted by G.

Exercises 2.5

1 If Q varies directly as x and inversely as y^2, and given that $Q = 0.731$ when $x = 7.23$ and
   $y = 13.6$, evaluate:
   a Q when $x = 3.86$ and $y = 17.2$ b y when $Q = 0.429$ and $x = 13.4$

2 When a load is suspended from a cable, the extension produced varies directly as the length of
   the cable, directly as the load and inversely as the square of the diameter of the cable. Given that
   for a mild steel cable of length 9.8 m and diameter 7.5 mm, the extension produced by a load of
   0.43 t is 4.7 mm, find the extension produced in a mild steel cable of length 14 m and diameter
   12 mm by a load of 1.3 t.

+++ 3 The number of oscillations made in a given time period t by a simple pendulum varies directly as
   t and inversely as the square root of the length of the pendulum. Given that a pendulum of length
   248.5 mm makes 10.0 oscillations in 10.0 seconds, determine:
   a the number of oscillations made in 14.0 seconds by a pendulum of length 3.98 m
   b the length of a pendulum that makes 50.0 oscillations in 1.00 minutes
   c how long it would take a pendulum of length 95.0 cm to make 15.0 oscillations.

2.6 PERCENTAGES

$\dfrac{3}{100}$ is often written as 3% and is then called '3 per cent'.

Similarly:

$7\% = \frac{7}{100} = 0.07; \quad 13\% = \frac{13}{100} = 0.13; \quad 0.08\% = \frac{0.08}{100} = 0.0008$

Exercises 2.6

For questions 1–4, find the answers mentally.

1 Write each of the following as a decimal:
   a 38%   b 2%   c 1.63%   d 0.014%

2 Find:
   a 3% of 16.7   b 41% of 6.00   c 12% of 81.5
   d 0.38% of 40   e 71% of 600

3 In each of these cases, express the percentage as a fraction and evaluate the result as a decimal:
   a 25% of $\frac{9}{25}$ b 90% of $\frac{2}{3}$ c 28% of $\frac{3}{4}$ d 60% of $\frac{1}{3}$

continued

continued

4 In each case, express the percentage as a fraction and evaluate:
  a 50% of 7.60   b 75% of 120   c 25% of 8.12
  d 20% of 0.250   e 10% of 2.73

5 Find:
  a 7.5% of $4.63 (to the nearest cent)   b 15.3% of 327 Ω
  c 3.81% of 2.47 h (in minutes and seconds to the nearest second)
  d 63.2% of 5.86 t   e 18.7% of 253 m

Expressing any number as a percentage

Exercises 2.6 *continued*

6 Express as percentages:
  a 0.431   b $\frac{5}{9}$ c 2.36   d $\frac{5}{6}$ e 4.00
  f $\frac{2}{3}$ g 0.003 18   h $\frac{3}{8}$ i 26.8   j $\frac{11}{12}$

Miscellaneous

7 The overall efficiency of an engine is given by $\dfrac{\text{output power}}{\text{input power}}$. In each of the following cases, find the efficiency as a percentage:
  a a power-generating station with input power 450 000 MW and output power 95 000 MW
  b a four-stroke diesel engine with input power 80 kW (from fuel) and output power 22 kW (brake power)

8 Type metal contains 18% antimony. What mass of antimony is there in 780 kg of type metal?

9 Out of a batch of 640 castings, 16 were defective. What percentage of the castings was defective?

10 The rotational speed of a shaft is specified to be 350 rpm, but measurement shows a speed of 335 rpm. What is the percentage error?

Increasing or decreasing a number by a given percentage

Example

$\frac{107}{100} \times 13 = (\frac{100}{100} + \frac{7}{100}) \times 13 = 100\%$ of 13 + 7% of 13

So:

- To increase a number by 7%, find 107% of it (i.e. multiply by 1.07).
- To increase a number by 31%, find 131% of it (i.e. multiply by 1.31).
- To increase a number by 0.9%, find 100.9% of it (i.e. multiply by 1.009).

Similarly:

- To decrease a number by 7%, find 93% of it (i.e. multiply by 0.93).
- To decrease a number by 31%, find 69% of it (i.e. multiply by 0.69).
- To decrease a number by 0.9%, find 99.1% of it (i.e. multiply by 0.991).

Exercises 2.6 *continued*

Do question 11 mentally.

11 **a** Increase 7.00 by 3%   **b** Increase 9.00 by 29%   **c** Increase 12.00 by 71%

    d Decrease 3.00 by 7%   **e** Decrease 11.00 by 53%   **f** Decrease 17.00 by 20%

12 **a** Increase 17.6 by 13%   **b** Decrease 8.97 by 29%

    c Increase 97.6 by 3.5%   **d** Decrease 0.871 by 16.2%

    Note: If a number receives a succession of multiplications by some number, it will make successive *percentage* increases in its value. For example, successive multiplications of a number by 1.07 will result in successive increases of 7%.

SELF-TEST

1 $272 is divided between two people so that the amounts they receive are in the ratio 3 : 5. How much does each receive?

2 In a test, a student scores 67 marks out of a maximum possible mark of 82. What is the approximate percentage mark?

3 The frequency of vibration of a stretched wire when plucked varies inversely as its length and directly as the square root of its tension. If a wire that is 1.40 m long vibrates with a frequency of 376 per second (376 Hz), when its tension is 125 N:

    a what will be its frequency if its length is 2.20 m and its tension is 234 N?

    b what will be the approximate percentage increase or decrease in the frequency?

4 Evaluate mentally:

    a 37% of 8 mL   **b** 12% of $23.50

5 A 50 m cable extends by 25 mm under a load of 1 t. What is the percentage increase in its length? (1000 mm = 1 m)

6 The introduction of an extra machine into a workshop increases the production rate from 170 to 204 articles per week. What is the percentage increase in the production rate?

7 Clock brass consists of $64\frac{1}{4}$% copper, 34% zinc and $1\frac{3}{4}$% lead.

    a What mass of copper is needed to make 200 kg of clock brass?

    b What mass of lead is present in 40 kg of clock brass?

    c A manufacturer has an unlimited supply of copper and lead, but has only 50 kg of zinc. How much clock brass can she make? (Answer to the nearest kilogram.)

CHAPTER 3

MEASUREMENT AND MENSURATION

Learning objectives

- Use SI units to express lengths, areas, volumes, masses and densities and convert between different SI units of the same quantity.
- State the absolute and/or percentage errors in quantities obtained from measurement.
- State results using the appropriate number of significant figures.
- Solve problems involving Pythagoras' theorem.
- Solve mensuration problems using SI units:
  - to calculate perimeters and areas of right-angled triangles, rectangles, circles and combinations of these shapes
  - to calculate volumes of prisms and pyramids with triangular, rectangular and circular cross-sections, and combinations of these shapes.

3.1 SI UNITS (SYSTÈME INTERNATIONALE D'UNITÉS)

In order to overcome the inconvenience of the great variety of units that existed at the time, an international conference in 1960 agreed to adopt the SI system of units.

Basic units

Any system of units must be based upon a small number of **basic** units, all other units being derived from these. The choice of the basic quantities and their basic units is a matter of convenience, and the following were chosen for the SI system and their sizes defined:

Quantity	Basic unit
length	metre (m)
mass	kilogram (kg)
time	second (s)
electric current	ampere (A)
temperature	kelvin (K)
luminous intensity	candela (cd)
amount of substance	mole (mol)

Prefixes are used to create larger and smaller units.

Prefix	Abbreviation	Meaning	Prefix	Abbreviation	Meaning
deca-	da	10	deci-	d	$\frac{1}{10}$
hecto-	h	100	centi-	c	$\frac{1}{100}$
kilo-	k	1000	milli-	m	$\frac{1}{1000}$
mega-	M	1 000 000	micro-	μ	$\frac{1}{1\,000\,000}$
giga-	G	1 000 000 000	nano-	n	$\frac{1}{1\,000\,000\,000}$
tera	T	1 000 000 000 000	pico-	p	$\frac{1}{1\,000\,000\,000\,000}$

The above table is for reference purposes only. In this chapter the only prefixes we will use are the prefixes:

- kilo- (k) = 1000
  mega- (M) = 1 000 000
  milli- (m) = 0.001
  micro- (μ) = 0.000 001　　These are prefixes that increase or decrease by factors of 1000 and are called 'preferred' prefixes.
- centi- (c) = 0.01

Although centi- is not a *preferred* prefix, it is a convenient one and commonly used.

Note: The kilogram is the *only* basic unit that has a prefix. The kilogram is the basic unit for mass, *not* the gram. Kilometre is pronounced with the accent on the first syllable. *All* units with prefixes are pronounced with the main accent on the first syllable.

Examples

1 kg = 1000 g (because 'kilo-' means 1000)

1 Mg = 1 000 000 g = 1000 kg

$1 \text{ mg} = \frac{1}{1000} \text{ g}$

$1 \text{ } \mu\text{g} = \frac{1}{1\,000\,000} \text{ g} = \frac{1}{1000} \text{ mg}$

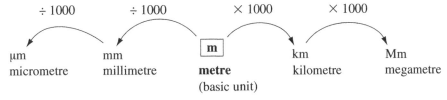

length:

| ÷ 1000 | ÷ 1000 | × 1000 | × 1000 |

μm — micrometre　　mm — millimetre　　**m** — **metre** (basic unit)　　km — kilometre　　Mm — megametre

and remember that 1 cm = 10 mm and 100 cm = 1 m

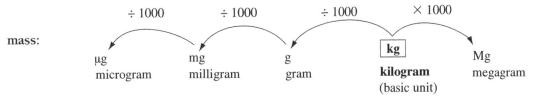

mass:

| ÷ 1000 | ÷ 1000 | ÷ 1000 | × 1000 |

μg — microgram　　mg — milligram　　g — gram　　**kg** — **kilogram** (basic unit)　　Mg — megagram

Examples

From the above diagrams, it can be seen for example that:

1000 km = 1 Mm, 1 000 000 mm = 1 km, 1 000 000 000 mg = 1 Mg

Note: 1 megagram is usually called 1 tonne (1 t = 1000 kg = 1 000 000 g)

Exercises 3.1

1   Express each of the following in terms of the basic unit, as either an integer or a decimal:

    a 1 g   **b** 1 t   **c** 1 mm   **d** 1 μm   **e** 1 cm   **f** 1 mg

2   Express:

    a 2630 mg in grams   **b** 0.0754 Mm in kilometres

    c 500 000 g in tonnes   **d** 0.00 35 km in millimetres

    e 0.008 16 cm in micrometres   **f** 0.0008 t in grams

3.2 ESTIMATIONS

When an estimate is made without any actual measurements taken (sometimes colloquially called a 'guesstimate'), caution must be exercised in relying on its accuracy. As a general rule, the estimate may be regarded as probably accurate to the number of significant figures given. For example, we may regard:

    'about 3 km' as meaning 'correct to the nearest km' (i.e. 3 ± 0.5)

    'about 30 km' as meaning 'correct to the nearest 10 km' (i.e. 30 ± 5)

    'about 300 km' as meaning 'correct to the nearest 100 km' (i.e. 300 ± 50)

    'about 0.3 km' as meaning 'correct to the nearest 0.1 km' (i.e. 0.3 ± 0.05)

Numbers that are estimated are seldom given to more than 1 significant figure except for numbers greater than about 20, where a 5 is often used to indicate that the estimator is unsure as to which of two figures to give. For example, 'about 25 years old' probably means 'between 20 and 30 and most likely about halfway between' (i.e. 25 ± 5).

Exercises 3.2

(The answers given are only approximate. Any answer of about the same magnitude will satisfy.)

1   A person in a concert hall estimates that there are about 30 seats per row, about 70 rows of seats and about $\frac{3}{4}$ of the seats are occupied. What is her estimate of the number of occupied seats?

2   A person estimates that he walks for about 2 hours per day, taking an average of two steps per second, with each step being an average length of half a metre. What is his estimate of the distance he walks in one week?

3.3 APPROXIMATIONS

Before finding a value using a calculator, it is a good practice to obtain a quick approximation of the value *without* using the calculator. It may be that a rough idea of the value is all that is required and, in any case, it provides a check that the result you obtain on the calculator is a *reasonable* one. Pressing one wrong key on a calculator can give a very false result and this is more likely to be detected if an approximate result has been estimated beforehand.

An approximation of the result is obtained quickly by 'rounding off' numbers to approximate values that are simpler to handle.

It is useful to note that $\pi \approx 3.14 \approx 3$ and $\pi^2 \approx 10$.

We try to round off numbers in such a way that the various increases and decreases tend to compensate for each other.

Examples

1 $9.62 \times 0.773 \times 66.4 \approx 10 \times 0.8 \times 60$
$$\approx 480$$
We round the third number *down* to help compensate for the rounding *up* of the first two numbers.

2 $\dfrac{8.41 \times \sqrt{13.6} \times 17.2}{326 \times 61.4} \approx \dfrac{8 \times 4 \times 20}{300 \times 60}$
$$\approx \tfrac{32}{900} \approx \tfrac{1}{30} \approx 0.03$$

Note: A *constant* in a formula is not a measured quantity and hence is an *exact* value. For example, m and v are the only measured quantities in the formula $\frac{1}{2}mv^2$ and r is the only measured quantity in the formula $\frac{4}{3}\pi r^3$.

Exercises 3.3

In each case below, approximate the value, writing down the steps you used as in the examples above. Then use your calculator to obtain the value correct to the appropriate number of significant figures. Check that the actual value, when expressed correct to one significant figure, agrees fairly well with your approximation. If it does not do so, examine the method you used to obtain your approximation to see how it could have been improved.

1 $\frac{4}{3}\pi \times 1.87^3$ **2** $8.32 \times \pi^2 \times \sqrt{42.1} + 7.5$

More practice at approximating can be obtained by using Exercises 1.7 (the calculator exercises at the end of Chapter 1).

3.4 ACCURACY OF MEASUREMENT

Question: If the price of an article increases by $1, is this a 'large' increase or not?

Reply: It is impossible to give a 'yes' or 'no' answer to this question because we are given only the *absolute* (i.e. actual) value of the increase, whereas the importance of a change depends upon the fractional (or percentage) change. If the $1 increase is in the price of a $40 000 vehicle, this is an increase of $\frac{1}{40\,000}$ ($= 0.0025\%$), a change of little consequence, but if this same $1 increase was in the price of a $2.50 loaf of bread, this would be an increase of $\frac{1}{2.50}$ ($= 40\%$) and would result in public outcry and newspaper headlines.

No quantity can be measured *exactly*. For example, even when using a high-precision instrument, the *exact* length of an object cannot be determined. It is even doubtful whether any object *has* an exact length because its length will be continuously changing with alterations in its temperature, the atmospheric pressure, the evaporation of its molecules, and so on. We never know the *actual* error made in making any measurement but we can usually make an *estimate* of the *maximum* probable error (MPE). For example, most people would agree that if a length is measured (with care) using a steel tape measure, the MPE would be about 2 mm.

An error can be expressed as an **absolute** error, a **fractional** error or as a **percentage** error (MP%E).

- If an error of 4 mm is made in measuring a length of 5 m, we have:
  absolute error = 4 mm
  fractional error $= \frac{4}{5000} = \frac{1}{1250}$
  percentage error $= \frac{1}{1250} \times 100\% = 0.08\%$
- But if the same error of 4 mm is made in measuring a length of 20 cm, we have:
  absolute error = 4 mm
  fractional error $= \frac{4}{200} = \frac{1}{50}$
  percentage error $= \frac{1}{50} \times 100\% = 2\%$

The same absolute error of 4 mm is much more 'serious' in the second case above, and this is indicated by a consideration of the percentage errors.

At first sight it might seem that a length measurement correct to the nearest millimetre is quite an accurate measurement (MPE $= \pm 0.5$ mm) *but this is not necessarily so*. If the value was 1 m correct to the nearest millimetre, this would be an accurate measurement (MP%E $= \frac{0.5}{1000} \times 100 = \pm 0.05\%$). However, if the length was, say, 2 mm $\pm$ 0.5 mm, this would *not* be very accurate (MP%E $= \frac{0.5}{2} \times 100 = \pm 25\%$).

Similarly, if the mass of the earth was measured correct to the nearest tonne, this would be an extremely accurate measurement. However, if the mass of a vehicle was determined correct to the nearest tonne, this would be a very crude measurement.

An estimated MPE or an estimated MP%E may be expressed as a *plus-or-minus* value.

SUMMARY
When we speak of a measurement as being 'accurate', we mean that it has a *small percentage error*, not necessarily a *small absolute error*.

Example
If a mass is measured to be 500 g with an MPE of 2 g, this may be stated to be 500 $\pm$ 2 g or 500 g $\pm$ 0.4%. This statement means that the mass is somewhere between 498 g and 502 g, most probably close to 500 g. The '± 2' states the *limits of accuracy*, 498 g being the *least probable value* and 502 g being the *greatest probable value*.

Exercises 3.4

1 State the MPE, correct to one significant figure, in each of the following measurements:

   a 5.3 kg ± 2%   **b** 820 mg ± 5%

2 State the MP%E, correct to one significant figure, in each of the following measurements:

   a 190 ± 5 mm   **b** 7.60 ± 0.05 kg   **c** 8650 ± 100 km

3.5 STATED ACCURACY

Towns have survey marks from which distances are accurately measured. If I am told that the distance from town A to town B is 120.312 km, and I wish to know how long it would take me to drive between these towns at an average speed of about 40 km/h, the answer would be approximately $\frac{120}{40} = 3$ h. I would not calculate time required $= \frac{120.312}{40} = 3.0078$ h, since such accuracy would be neither desired nor justified (as the average speed is not known accurately).

An accurately measured quantity can be stated correct to various numbers of decimal places, depending upon the accuracy required.

Example

A length of 6.382 54 m becomes 6.4 m when stated correct to *one* decimal place.

   It becomes 6.38 m when stated correct to *two* decimal places.

   It becomes 6.383 m when stated correct to *three* decimal places.

In each case the value is stated as accurately as possible using the number of decimal places required.

Note: The last digit is raised by 1 if the following digit is 5 or greater.

3.6 IMPLIED ACCURACY IN A STATED MEASUREMENT

When the value of a measured quantity is stated, it is assumed that it is correct to the number of decimal places given.

Example

If a length is stated to be 6.3 km, we do *not* assume that this means 6.300 000 km, but rather that this length is stated correct to one decimal place, i.e. we assume that the length is somewhere between 6.25 and 6.35 km. This means assuming that the length is 6.3 ± 0.05 km. Hence, the MPE is ±0.05 km (= 50 m) and the MP%E $= \frac{0.05}{6.3} \times 100 \approx 0.8\%$.

Example

A measured value of 23 units means 23 ± 0.5 (MPE $= \pm 0.5$). A measured value of 6.74 units means 6.74 ± 0.005 (MPE $= \pm 0.005$). A measured value of 0.000 049 units has an MPE of 0.000 000 5.

Examples

1   52.3 g has MPE 0.05 g.

    $\therefore$ MP%E $= \frac{0.05}{52.3} \times 100\%$

              $\approx 0.1\%$

2   157 W has MPE 0.5 W.

    $\therefore$ MP%E $= \frac{0.5}{157} \times 100\%$

              $\approx 0.3\%$

Note: The MPE is 5 in the next decimal place after the last one stated. 2.300 N states a more accurate value than 2.3 N, since it implies an MPE of 0.0005 N, whereas 2.3 N implies an MPE of 0.05 N.

Maximum probable percentage errors are quoted only *approximately*.

Example

For the stated value 14.8 mm, the:

- MPE $= 0.05$ mm
- MP%E $= \frac{0.05}{14.8} \times 100\% \approx 0.34\%$
- limits of accuracy are from 14.75 to 14.85 mm

Exercises 3.5

1   For each of the following stated measurements, find: (i) the MPE, (ii) the MP%E and (iii) the limits of accuracy of the measurement:

    a 78 km   **b** 0.083 kg   **c** 4.000 m

    d 0.000 063 t   **e** 0.47 t   **f** 654 km

2   Write each of the following numbers correct to two decimal places:

    a 354.114 99   **b** 0.006 03   **c** 20.001 98

    d 0.146   **e** 0.000 345   **f** 6.299

3.7 ADDITIONAL CALCULATOR EXERCISES INVOLVING SQUARES AND SQUARE ROOTS

Exercises 3.6

Use a calculator to evaluate the following correct to 3 significant figures:

1    $\sqrt{5.62^2 + 7.21^2}$ 2 $\sqrt{9.42^2 - 6.32^2}$

3    $\sqrt{6.39^2 - 4.17^2}$ 4 $\sqrt{19.8^2 - 11.9^2}$

5    $\sqrt{4.55^2 + 5.19^2}$ 6 $\sqrt{1.33^2 + 7.71^2}$

7    $\sqrt{12.1^2 - 10.3^2}$ 8 $\sqrt{2.97^2 - 1.68^2}$

9    $(\sqrt{4.93} + \sqrt{7.62})^2$ 10 $(1.27^2 + 3.85^2)^2$

11    $\dfrac{\sqrt{50.7}}{3.28 \times 2.15}$ 12 $(\sqrt{1.11} + \sqrt{2.13} + \sqrt{1.36})^2$

3.8 PYTHAGORAS' THEOREM

This is probably the most famous and the most used theorem in the whole of mathematics. It is named after its discoverer, Pythagoras, the Greek philosopher and mathematician (*c.* 580-500 BC).

In a right-angled triangle, the side opposite the right angle is called the **hypotenuse**. Pythagoras discovered that 'the square on the hypotenuse equals the sum of the squares on the other two sides'.

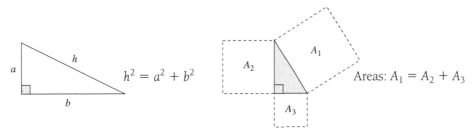

In particular, note that: $3^2 + 4^2 = 5^2$ and $5^2 + 12^2 = 13^2$

and any multiples of these: $6^2 + 8^2 = 10^2$ $10^2 + 24^2 = 26^2$

                                 $9^2 + 12^2 = 15^2$ $15^2 + 36^2 = 39^2$

Hence, any triangle having these lengths of sides must be a *right-angled* triangle (the hypotenuse being the longest side).

Examples

1    $x^2 = 6^2 + 4^2 = 52$

     $\therefore x = \sqrt{52}$

     ≈ 7.21 m

2    $(7.32)^2 = (5.14)^2 + d^2$

     $\therefore d^2 = (7.32)^2 - (5.14)^2$

     $\therefore d = \sqrt{(7.32)^2 - (5.14)^2}$

     ≈ 5.21 m

Exercises 3.7

Find the lengths of the third sides of the following triangles:

1 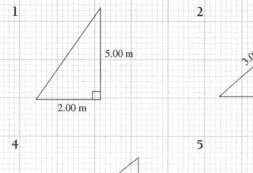   2   3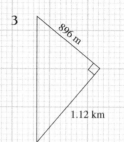

5.00 m

2.00 m

3.00 m

4.00 m

896 m

1.12 km

4    5 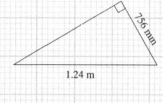   6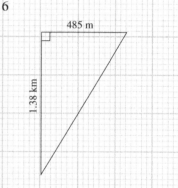

4.200 km

3.700 km

756 mm

1.24 m

485 m

1.38 km

7   A rectangle has one side of length 27.3 mm and a diagonal length of 32.4 mm. Find the length of the other side.

8   Find the length of the diagonal of a square which has sides of length 837 mm.

9   Chord *AB* of a circle has a length of 38.4 mm, and its distance from the centre, *CK*, is 15.2 mm. Find the radius of this circle.

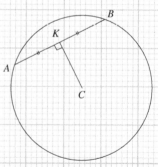

10   A vertical pole 15.3 m tall is supported by guy wires running from a point 3.50 m from the top to points on the ground that are 5.60 m from the foot of the pole. What is the length of each of the wires?

11   A person wishes to use a ladder that is 9.0 m long to climb from the ground to the gutter of a building. If the gutter is 7.8 m above the ground and the person requires at least 1 m of the ladder to extend above the gutter, what is the maximum distance of the foot of the ladder from the base of the building?

3.9 'IDEAL' FIGURES

When we say that in this triangle, the length of the hypotenuse must be 13 m, we are regarding the triangle as being an 'ideal'. In *practice*, nothing has a length of *exactly* 5 m or 12 m and

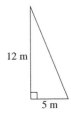

12 m

5 m

no angle is *exactly* 90°. If these were *measured* values, they would be regarded as 5 ± 0.5 m, 12 ± 0.5 m and 90° ± 0.5°. It would then be possible that, for example, the sides are 5.4 m and 12.4 m, and the marked angle 90.4°, in which case the hypotenuse would have a length of approximately 13.6 m.

Although 'ideal' figures cannot exist in practice, it is true, for example, that in a triangle whose adjacent sides have lengths that are extremely close to being 5 m and 12 m, then the hypotenuse will have a length that is extremely close to 13 m. So if you are asked for the 'exact' value of its length, you are being asked to consider an ideal situation—you are being asked for the length of the hypotenuse *if* the given measurements were exact.

Exercises 3.8

1 Write down the values of the pronumerals in the following table, which refers in each case to a right-angled triangle:

Sides (m)	Hypotenuse (m)
3, 4	(a)
12, (b)	20
0.6, 0.8	(c)
5, 12	(d)
15, (e)	39

2 What is the ratio of the length of the diagonal of a square to the length of one of the sides?

3 In the (ideal) figures below, lengths are labelled in metres. Find the exact values of the pronumerals. (The marked angles are right angles.)

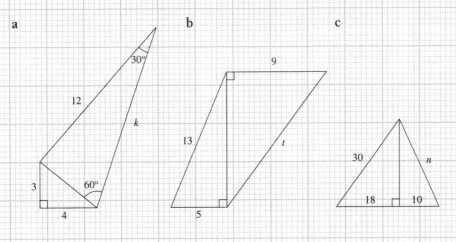

a

12

3   60°

4

30°

k

b

13

5

9

t

c

30   n

18   10

3.10 PERIMETERS AND AREAS OF RIGHT-ANGLED TRIANGLES, QUADRILATERALS AND POLYGONS

A plane rectilinear figure is a two-dimensional figure whose sides are straight line segments:

- A **triangle** has three sides.
- A **quadrilateral** has four sides.
- A **polygon** has more than four sides (a **regular polygon** has all its sides equal in length).
- The **perimeter** of a figure is the distance around its boundary.

Rectangle

Opposite sides equal.
All angles right angles.

Square

All sides equal.
All angles right angles.

3.11 THE AREA OF A RECTANGLE

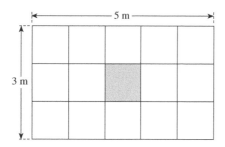

$\therefore$ area of rectangle $= l \times b$ m^2

The small shaded area in this diagram is a square
1 m $\times$ 1 m. This area is called 1 square metre (1 m^2).

In this rectangle, there are 3 rows, each containing 5 of these areas.

$\therefore$ area of rectangle $= 3 \times 5 = 15$ m^2

In the rectangle below there would be b rows, each containing l areas of 1 m^2 each.

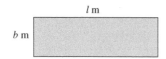

Rule: The area of a rectangle $= l \times b$

Units

As with all formulae, consistent units must be used:

- metres $\times$ metres gives square metres (m^2) of area
- millimetres $\times$ millimetres gives square millimetres (mm^2) of area
- kilometres $\times$ kilometres gives square kilometres (km^2) of area.

Note: 10^2 is an abbreviation for 100

10^3 is an abbreviation for 1000

10^4 is an abbreviation for 10 000

The index indicates the number of zeros.

$10^2 \times 10^3 = 100 \times 1000 = 100\,000 = 10^5$

When multiplying, the sum of the indices gives the number of zeros in the result.

$(10^2)^2 = 10^2 \times 10^2 = 10^4 = 10\,000$

$(10^3)^2 = 10^3 \times 10^3 = 10^6 = 1\,000\,000$

$(10^2)^3 = 10^2 \times 10^2 \times 10^2 = 10^6 = 1\,000\,000$

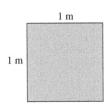

In this diagram:

$1\,m^2 = 1\,m \times 1\,m$

$\qquad = 1000\,mm \times 1000\,mm$

$\qquad = 1\,000\,000\,mm^2$

$\qquad = 10^6\,mm^2$

Note: $1\,m = 10^3\,mm$ $1\,km = 10^3\,m$ $1\,km = 10^6\,mm$

$1\,m^2 = 10^6\,mm^2$ $1\,km^2 = 10^6\,m^2$ $1\,km^2 = 10^{12}\,mm^2$

Exercises 3.9

1   A rectangular piece of wood has length 3.45 m, breadth 23.0 mm and thickness 7.50 mm. Find its total surface area:

  a expressed in square metres   **b** expressed in square millimetres

2   A spool contains a length of 1.35 km of recording tape. Given that the tape is 7.25 mm wide, find the cost of applying a chemical coating to one side of this tape, at 16.3 cents per square metre.

3.12 THE AREA OF A TRIANGLE

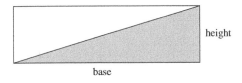

area of right-angled triangle $= \frac{1}{2}$ area of rectangle

$\qquad\qquad\qquad\qquad = \frac{1}{2}$ base $\times$ height

Examples

It is assumed that you know that $a(x + y) = ax + ay$. This is revised in section 4.5.

1 area of $\triangle ABC$

  $=$ area of $\triangle ABD +$ area of $\triangle ACD$

continued

continued

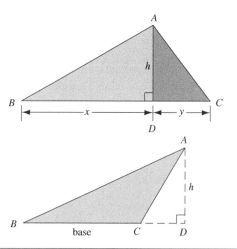

$$= \tfrac{1}{2}xh + \tfrac{1}{2}yh$$
$$= \tfrac{1}{2}h(x + y)$$
$$= \tfrac{1}{2}BC \times h$$
$$= \tfrac{1}{2}\text{base} \times \text{height}$$

2   area of $\triangle ABC$

$$= \text{area of } \triangle ABD - \text{area of } \triangle ACD$$
$$= (\tfrac{1}{2}BD \times h) - (\tfrac{1}{2}CD \times h)$$
$$= \tfrac{1}{2}h(BD - CD)$$
$$= \tfrac{1}{2}BC \times h$$
$$= \tfrac{1}{2}\text{base} \times \text{height}$$

Rule: The area of a triangle $= \tfrac{1}{2}b \times h$

Note: Since the area of each of these triangles is $\tfrac{1}{2}BC \times h$, they all have the same area. (Triangles on the same base and between the same parallels have equal areas.)

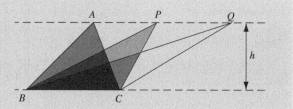

By turning a triangle around, it can be seen that any of the three sides can be regarded as being the base. In the formula, area $= \tfrac{1}{2}b \times h$, the height is the perpendicular distance to the base line from the opposite vertex.

Examples

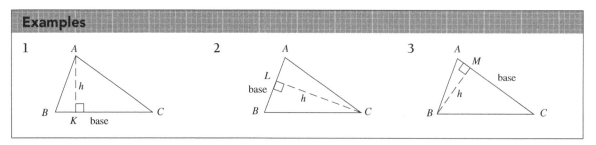

Exercises 3.10

All figures are 'ideal'. State all results exactly.

1   Find the area of $\triangle ABC$.

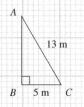

2   Find the area of $\triangle DEF$.

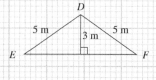

3 Find the area of ΔHJK.

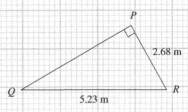

4 Find the area of ΔPQR.

5 a Find the area of ΔLMK.

b Find the length of LT.

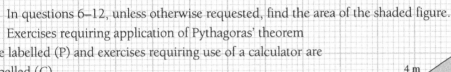

In questions 6–12, unless otherwise requested, find the area of the shaded figure.

Exercises requiring application of Pythagoras' theorem are labelled (P) and exercises requiring use of a calculator are labelled (C).

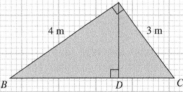

6 (P) Find:

a the area of the shaded triangle

b the length of the interval AD

7 (P)

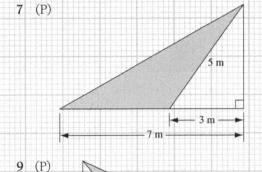

8 (P)

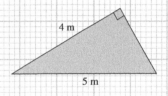

10 (P) C is the centre of the circle. Find the area of quadrilateral PQRS.

9 (P)

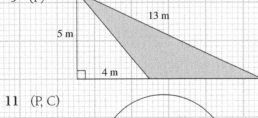

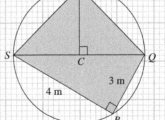

11 (P, C)

12 (P)

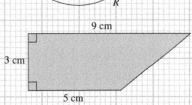

3.13 THE CIRCLE: CIRCUMFERENCE AND AREA

The circumference of any circle is a little more than three times its diameter. The exact ratio cannot be written down since it is a non-terminating, non-repeating decimal, 3.141 592 65 . . . (By coincidence, this is *approximately* $3\frac{1}{7}$, i.e. $\frac{22}{7}$.) This ratio is represented by the Greek letter π(pi). Hence, if a circle has diameter D units long and circumference C units long, $\frac{C}{D} = \pi$ (where $\pi = 3.141\ 592\ 65\ldots$).

$$\therefore C = \pi D$$

$$C = 2\pi r \text{ (where } r \text{ is the radius of the circle)}$$

The area of a circle is given by $A = \pi r^2$ (a formula we can prove after studying integration in a later year of this course).

$$\text{Now } r = \frac{D}{2}, \therefore A = \pi\left(\frac{D}{2}\right)^2 = \frac{\pi D^2}{4}$$

When the diameter of a circle is known, it is easier to use $A = \frac{\pi D^2}{4}$ than to calculate the radius $\left(=\frac{D}{2}\right)$ and then use $A = \pi r^2$. In practice, it is usually the diameter that is stated, since this is easier to measure than the radius (which requires accurate location of the centre).

SUMMARY

Remember these formulae:

- Circumference: $C = \pi D (= 2\pi r)$

- Area: $A = \frac{\pi D^2}{4} (=\pi r^2)$

Exercises 3.11

State answers correct to 3 significant figures.

1   Find the circumference and the area of a circle, given that:

    a   the radius measures 20 mm   b   the diameter measures 2.4 m

    c   the radius measures 63.7 mm   d   the diameter measures 826 mm

    e   the radius measures 163 mm

2   Find the area of the annulus between two concentric circles, given that the radii are:

    a   2 m and 3 m   b   7.4 mm and 9.2 mm   c   62 mm and 83 mm

    d   28.4 mm and 31.1 mm   e   1.08 m and 1.73 m

3   Find the radius of a circle whose circumference measures 3.45 m.

4   Find the diameter of a circle whose circumference measures 517 mm.

5   Find the radius of a circle whose area is 6.08 m^2.

6   Find the diameter of a circle whose area is 237 cm^2.

7   Find the area of a circle whose circumference measures 1.62 m.

+++ 8   Find the areas of the shaded regions. (*C* marks the centre of a circular arc.)

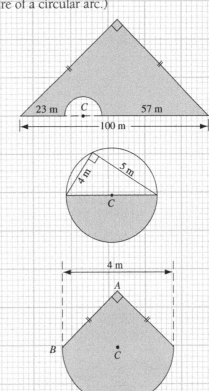

a

11 m   11 m

C

3 m   3 m

b

23 m   *C*   57 m

|← 100 m →|

c

4 m   *C*   *C*   6 m

10 m

d

4 m   5 m

C

e

4 m   4 m

C

f

4 m

A

B   *C*

3.14 VOLUMES OF PRISMS

A cube 1 m × 1 m × 1 m is said to have a volume of one cubic metre (1 m³).

If we build a stack of these unit cubes *l* m × *b* m × *h* m, as shown, there will be *h* layers each containing *l* × *b* cubes.

Hence the volume of the stack

= number of unit cubes

= *l* × *b* × *h* cubic metres

= area of base × height

Changing the shape of the base (without changing its area) will not affect the volume.

h m

b m

l m

A **prism** is a solid figure in which there are two identical parallel faces (the ends) with the surface(s) joining these faces perpendicular to them.

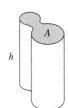

A

h

> **Rule:** volume of prism = area of base × height

Prisms are named according to the shape of the ends (i.e. the shape of the base).

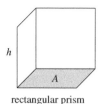

rectangular prism

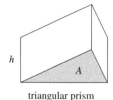

triangular prism

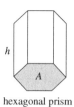

hexagonal prism

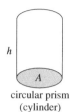

circular prism
(cylinder)

Units

An area is a *length* × a *length*: units are mm², cm², m², etc.

A volume is a *length* × a *length* × a *length*: units are mm³, cm³, m³, etc.

Length: 1m = 1000 mm   100 cm = 1 m   1m = 0.001 km

Area: 1m² = (1000)² mm²   (100)² cm² = 1 m²   1m² = (0.001)² km²

Volume: 1m³ = (1000)³ mm³   (100)³ cm3 = 1 m³   1m³ = (0.001)³ km³

In any formula, *consistent* units must be used. For example, if a rectangular metal plate has a length of 1.3 m, a breadth of 74 cm and a thickness of 0.26 mm, to find its volume we must first express all its dimensions in consistent units. If we decide to use centimetres, we have l = 130 cm, b = 74 cm and h = 0.026 cm. Hence, the volume is 130 × 74 × 0.026 cm³ ≈ 250 cm³. (Working in millimetres we would have: volume = 1300 × 740 × 0.26 ≈ 250 000 mm³.)

The relationship between the SI units for volume is shown below:

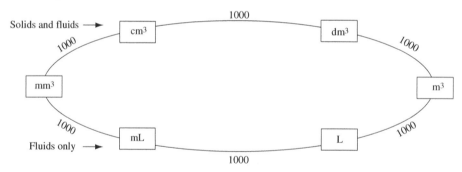

Note:	■ **For fluids:**	1000 mm³ = 1 millilitre	(mL)	(N.B. 1 mL = 1 cm³
		1000 mL = 1 litre	(L)	
		1000 L = 1 cubic metre	(m³)	

■ A teaspoon holds approximately 5 mL (= 5000 mm³).

Exercises 3.12

1 Convert:

   a 8630 mm³ to cubic metres   **b** 0.069 L to millilitres

   c 4.5 mL to cubic centimetres   **d** 27 000 cm³ to cubic metres

   e 0.023 L to cubic millimetres

2 A rectangular tank has length 1.03 m, breadth 4.87 cm and depth 7.50 mm. Find the volume of the tank:

   a in cubic millimetres   **b** in millilitres   **c** in cubic metres

   d in litres   **e** in cubic centimetres

3 A rectangular prism measures 68.3 mm × 52.7 mm × 43.3 mm. Find its volume in:

   a cubic centimetres   **b** millilitres

4 A rectangular prism measures 8.47 m × 3.62 m × 57.0 mm. Find its volume in cubic metres.

5 Find the volume of this triangular prism in cubic centimetres:

6 Find the volume of timber in a telegraph pole of diameter 0.40 m and length 15 m.

7 Find the total volume of liquid metal required to make 2000 cylindrical roller bearings, each of diameter 4.00 mm and length 12.0 mm. State the answer in millilitres.

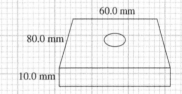

+++ 8 A cylindrical hole of diameter 10.0 mm is drilled through a metal block 80.0 mm × 60.0 mm × 10.0 mm, as shown. Find the volume of metal remaining in the block in cubic centimetres.

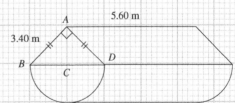

+++ 9 A cylindrical water pipe of length 1.00 km has external diameter 50.0 mm and internal diameter 40.0 mm. Find:

   a the volume of water required to fill the pipe, in kilolitres

   b the volume of metal present in the pipe, in cubic metres

+++ 10 The cross-section of this right prism consists of a semicircle surmounted by an isosceles right-angled triangle. Find:

   a the diameter of the semicircle

   b the area of the semicircle

   c the area of $\triangle ABD$

   d the volume of the prism

3.15 SURFACE AREAS OF PRISMS

If a hollow circular cylinder is cut parallel to its axis and then opened out, its surface becomes a rectangle.

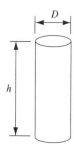 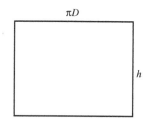

The external surface area of a *hollow* cylinder is the area of this rectangle:

   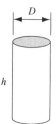

$$S = \pi Dh (= 2\pi rh)$$

For an *open* cylindrical can, the external surface area = πDh + area of the circular base:

$$S = \pi Dh + \frac{\pi D^2}{4}$$

For a *solid* cylinder, the external surface area = πDh + area of top + area of base:

$$A = \pi Dh + \frac{\pi D^2}{4} + \frac{\pi D^2}{4} = \pi Dh + \frac{\pi D^2}{2}$$

For a *pipe*, the total surface area = external curved surface area + internal curved surface area + surface area of the two ends:

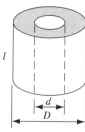

$$S = \pi Dl + \pi dl + 2\left(\frac{\pi D^2}{4} - \frac{\pi d^2}{4}\right)$$

Exercises 3.13

Find the surface areas of the following bodies described in Exercises 3.14. For the purposes of these exercises, assume that all dimensions given are correct to 0.1%. State all answers correct to 3 significant figures.

1 The solid prism in question 3, in m².

2 The solid prism in question 4, in m².

3 The surface area of *one* of the roller bearings in question 7, in mm².

+++ 4 The *total* surface area of the metal block in question 8 (including the surface area in the hole), in cm².

+++ 5 The *total* surface area (internal and external) of the water pipe in question 9 (ignore the two ends), in m².

+++ 6 The solid prism in question 10 in m².

SELF-TEST

1   In each of the following, state the result to only the number of significant figures justified by the given measured quantity:

Express:

a   56 700 mm in kilometres
b   0.002 700 kg in grams
c   5400 mm² in square centimetres
d   120 mL in cubic centimetres
e   86.00 kg in tonnes
f   4830 mm³ in millilitres

2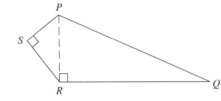

In the quadrilateral *PQRS*, side lengths are:

$QR = 24.0$ cm, $RS = 8.00$ cm, $SP = 6.00$ cm.

Find: **a**   the perimeter   **b**   the area.

+++   3   The cross-section of this right prism consists of a semicircle surmounted by an isosceles right-angled triangle. Expressing the results correct to 3 significant figures, find:

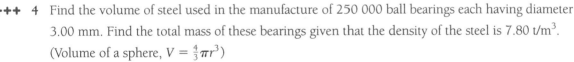

a   the length of *AB* in millimetres
b   the area of Δ*ABD* in square millimetres
c   the area of the semicircle in square millimetres
d   the volume of the prism in cubic centimetres
e   the exterior surface area of the prism

+++   4   Find the volume of steel used in the manufacture of 250 000 ball bearings each having diameter 3.00 mm. Find the total mass of these bearings given that the density of the steel is 7.80 t/m³. (Volume of a sphere, $V = \frac{4}{3}\pi r^3$)

5   Find the diameter and the surface area of a sphere ($4\pi r^2$) that has volume 887 mm³.

6   Use a calculator to evaluate the following correct to 3 significant figures:

a $\sqrt{6.44^2 + 3.28^2}$
b $\sqrt{8.49^2 - 7.11^2}$
c $\sqrt{1.09^2 + 1.57^2}$
d $(\sqrt{9.17} + \sqrt{5.69})^2$
e $(2.10^2 + 3.09^2)^2$
f $(\sqrt{1.50} + \sqrt{1.61} + \sqrt{2.66})^2$
g $1.13^2 + \sqrt{4.70^2 - 3.08^2}$
h $\sqrt{\sqrt{9.06} + \sqrt{7.21}}$
i $(2.02^2 + 1.33^2 - 2.31^2)^2$
j $\sqrt{6.37^2 - 4.08^2}$

CHAPTER 4

INTRODUCTION TO ALGEBRA

Learning objectives

- Substitute values for pronumerals in a simple algebraic expression and evaluate the expression using basic algebraic techniques.
- Perform the four basic operations on algebraic expressions.
- Solve simple linear algebraic equations.
- Create a linear equation to solve a verbal problem and hence solve the problem.
- Solve simultaneous linear equation systems in two unknowns.
- Create simultaneous equations to solve a verbal problem and hence solve the problem.

4.1 SUBSTITUTIONS

A **pronumeral** is a symbol (usually a letter of the Roman or Greek alphabet) that stands in place of a number. For example:

If $x = 2$, then $x + 3 = 5$, but if $x = 6$, then $x + 3 = 9$

If $\theta - 2 = 4$, it follows that θ must equal 6

When substituting values for pronumerals, the following points should be noted carefully:

- When no sign is placed between two pronumerals (or between a numeral and a pronumeral), *multiplication* is implied:

    $2k$ means $2 \times k$ $5a^2b$ means $5 \times a^2 \times b$

- The negative of a negative number is *positive*:

    $2 - -3 = 2 + 3$ If $a = 7$ and $b = -2$: $a - b = 7 - -2$
    $\qquad\quad = 5$ $= 7 + 2$
    $\qquad\qquad\qquad\qquad\qquad\qquad\qquad\qquad\qquad\qquad\qquad\quad = 9$

- When two numbers are multiplied or divided, the result is positive if the two numbers have the same sign and negative if the two numbers have different signs.

    $2 \times (-3) = -6;$ $(-2) \times 3 = -6$
    $(-2) \times (-3) = 6;$ $(-2) \times (-3) \times (-4) = -24;$ $(-2) \times (-3) \times (-4) \times (-5) = 120$

■ An index operates only on the numeral or pronumeral above which it is placed:

$ab^2 = a \times b^2 = a \times b \times b$

$a^2b = a^2 \times b = a \times a \times b$

Note: $(ab)^2 = ab \times ab = a \times b \times a \times b = a^2b^2$.

Examples

If $k = -2$, $n = -3$ and $t = -16$:

1 $kn^2 - t$ $= k \times n \times n - t$

$\qquad\qquad = (-2) \times (-3) \times (-3) - (-16)$

$\qquad\qquad = -18 + 16$

$\qquad\qquad = -2$

2 $2kt - 3n^2 = 2 \times k \times t - 3 \times n \times n$

$\qquad\qquad = 2 \times (-2) \times (-16) - 3 \times (-3) \times (-3)$

$\qquad\qquad = 64 - 27$

$\qquad\qquad = 37$

Note: The use of brackets will help to avoid errors with negative signs.

Exercises 4.1

1 Given that $a = 2$ and $b = 3$, evaluate:

    a $3a$ **b** $3a + 4b$ **c** $5b - 2a$ **d** $a - b$ **e** $2a - 3b$

2 Given that $p = 2$ and $q = 3$, evaluate:

    a q^2 **b** $p^2 - q^2$ **c** pq^2 **d** $(pq)^2$ **e** p^2q

3 Given that $x = 3$ and $y = 5$, evaluate:

    a $(x + y)^2$ **b** $(2y - x)^2$ **c** $(x - y)^2$ **d** $(y - 3x)^2$

4 Given that $m = 2$ and $t = 3$, evaluate:

    a t^2 **b** $(-t)^2$ **c** $-t^2$ **d** $(m + t)^2$

    e $(t - m)^2$ **f** $m^2 - t^2$ **g** $2t + m^2$ **h** $3m - t^2$

    i mt^2 **j** m^2t **k** $(mt)^2$ **l** m^2t^2

5 Given that $a = 2$, $b = -3$ and $c = 0$, evaluate:

    a $abc + b^2c + ac^2$ **b** $2a + 3b + 4c$ **c** $a - b$ **d** $3a - 2b$

continued

continued

6   Given that $a = 2$ and $b = 3$, evaluate:

   a $\dfrac{a}{b} + \dfrac{1}{3}$ **b** $\dfrac{a}{b} + \dfrac{b}{a}$ **c** $\dfrac{2a - b}{ab} + \dfrac{a - b}{b}$

7   Given that $m = -2$, $k = -1$ and $t = -3$, evaluate:

   a $3m + 2k$ **b** $2t - 4k$ **c** $mk + t$

   d mkt **e** $kt - mk$ **f** $\sqrt{m^2 + k^2}$

4.2 ADDITION OF LIKE TERMS

Only 'like' terms (terms that are identical and hence stand for the same number) can be added and subtracted.

Examples

1   $2a + 3a = 5a$ but $2a + 3b$ *cannot* be written as a single term.
2   $3m^2t + 5m^2t = 8m^2t$ but $3m^2t + 5mt^2$ *cannot* be written as a single term, because m^2t and mt^2 do not stand for the same number.

Exercises 4.2

1   Write down the numbers missing from the spaces indicated by (i), (ii), (iii) and (iv).

   a   3 nines + 1 nine + 2 nines + 2 elevens + 3 elevens + 2 elevens

        = (i) nines + (ii) elevens

        $3a + a + 2a + 2b + 3b + 2b =$ (iii) $a +$ (iv) b

   b   2 sevens + 3 tens + 4 sevens + 2 tens + 3 sevens − 2 tens

        = (i) sevens + (ii) tens

        $2x + 3y + 4x + 2y + 3x - 2y =$ (iii) $x +$ (iv) y

   c $\dfrac{3}{14} - \dfrac{2}{14} + \dfrac{2}{14} + \dfrac{5}{13} + \dfrac{2}{13} + \dfrac{1}{13} = \dfrac{\text{(i)}}{14} + \dfrac{\text{(ii)}}{13}$

        $3p - 2p + 2p + 5q + 2q + q =$ (iii) $p +$ (iv) q

   d $\dfrac{2}{7} - \dfrac{1}{9} + \dfrac{1}{7} + \dfrac{5}{9} + \dfrac{3}{7} - \dfrac{2}{9} = \dfrac{\text{(i)}}{7} + \dfrac{\text{(ii)}}{9}$

        $2k - m + k + 5m + 3k - 2m =$ (iii) $k +$ (iv) m

2   Simplify:

   a $5a + 3b - 2a - 5b$ **b** $2x - 3y - 4y + x$

   c $2f - 5f + 3g - f - g$ **d** $3ab^2 - 3a^2b - 5ab^2 - 2a^2b$

   e $6xy - 6x - 6y + 2x + 7xy$ **f** $2ab - 2ab^2 + 2a^2b + 2ab^2 - 2ab$

4.3 REMOVAL OF BRACKETS

A positive sign before brackets renders the brackets unnecessary:

$$a + (b + c) = a + b + c$$

A negative sign before brackets negates *each* of the terms within the brackets. It thus give the *opposite of each* of these terms, i.e. changes the sign of each:

$$-(a + b - c) = -a - b + c$$

Exercises 4.3

1 Express without brackets:

   a $a + (b + c)$ **b** $(a + b) + c$ **c** $a - (b + c)$

   d $a - (b - c)$ **e** $a - (-b + c)$ **f** $b - (b - a)$

2 Express without brackets, then simplify:

   a $2 + (x - 1)$ **b** $2m - (m - 3)$

   c $3k - (3 + k) - (3 - k)$ **d** $(a - 2) - (a + 2)$

   e $(3 - 2b) - (3b - 5)$ **f** $(5t - 3x) + (x - 2t) - (t - 2x)$

3 Express without brackets, then simplify:

   a $a^2 - (b^2 - a^2)$ **b** $x^2 - (x + x^2)$ **c** $(p^2 - q^2) - (p^2 + q^2)$

4.4 MULTIPLICATION AND DIVISION OF TERMS

As we have already seen, numbers may be multiplied in *any order*. Using this fact, we can often write algebraic products more simply.

Examples

1 $2m \times 3t = 2 \times m \times 3 \times t$

           $= 2 \times 3 \times m \times t$

           $= 6mt$

2 $2a - 3ab = 2 \times a \times 3 \times a \times b$

           $= 2 \times 3 \times a \times a \times b$

           $= 6a^2b$

When we *divide* 24 by 8, we look for a number by which we can *multiply* 8 in order to obtain 24.

$$24 \div 8 = 3 \text{ because } 3 \times 8 = 24$$

Example

When we divide $6ab^2$ by $3b$, we look for the expression by which we can multiply $3b$ in order to obtain $6ab^2$.

$3b \times \ldots = 6ab^2$

The answer is $2ab$, since $3b \times 2ab = 6ab^2$.

Exercises 4.4

1 Write down the following products in simplest form:

a $2p \times 6q$ b $14l \times 3m$ c $-8a \times 4$

d $3x \times -5t$ e $5x \times x$ f $2 \times x \times 4 \times x$

g $3x \times 2x$ h $a \times ab \times 3$ i $2ab \times 4a$

j $4x \times \frac{1}{2}a \times x$ k $2\frac{1}{2}k \times 3k \times 2m$ l $3 \times \frac{1}{2}x \times 2y \times 5x$

2 Simplify:

a $12xy \div 12y$ b $6mx^2 \div 2mx$ c $18kt^2 \div 6k$

d $6m^2t^2 \div \frac{1}{3}m^2t$ e $2msy^2 \div \frac{1}{2}my$ f $5(a + b) \div 5$

g $12(x + y) \div 4$ h $18a(2b - c) \div 18$ i $18a(2b - c) \div 6(2b - c)$

4.5 THE DISTRIBUTIVE LAW

To evaluate $2 \times (3 + 4)$, the 'order of operations' convention requires that we evaluate the operation inside the brackets, before doing the multiplication:

$2 \times (3 + 4) = 2 \times (7)$

$\qquad = 14$

There is, however, another method by which this expression may be evaluated:

$2 \times (3 + 4) = (2 \times 3) + (2 \times 4)$

$\qquad = 6 + 8$

$\qquad = 14$

Here we are using what is called the **distributive law.**

> **Note:** The distributive law applies only for *multiplication.*
>
> We *cannot* evaluate $\sqrt{(9 + 4)}$ by $\sqrt{9} + \sqrt{4}$, nor can we evaluate $(2 + 3)^2$ by $2^2 + 3^2$.

Examples

1 $3(x + 2) = 3x + 6$ 2 $2a(3a + b) = 6a^2 + 2ab$

3 $3mt (4 + 2t) = 12mt + 6mt^2$

Exercises 4.5

1 Multiply out so that the following are expressed without brackets:

a $3(x + 2y)$ b $2(a - b)$ c $3(2x + 4y)$

d $a(2b + 3x)$ e $a(a + 2b)$ f $m(3m - 2)$

g $3(a - b) - 5(b - a)$ h $m(2p - q) - q(3p - m)$ i $3(2x - 3) + 2(2 - 3x)$

Extension of the distributive law

$$\overbrace{(a + b)}(x + y) = (a + b)x + (a + b)y$$
$$= ax + bx + ay + by$$

Hence, $(a + b)(x + y)$ can be expanded by forming the products as shown by the arrows:

$$(a + b)(x + y) = ax + ay + bx + by$$

Examples

1 $(2 + 3)(4 + 5) = 8 + 10 + 12 + 15$
$$= 45$$

Note: $(2 + 3)(4 + 5) = (5)(9) = 45$

2 $(m - 3)(2m - 4) = 2m^2 - 4m - 6m + 12$
$$= 2m^2 - 10m + 12$$

Exercises 4.5 *continued*

2 Multiply out to express without brackets:

a $(a + b)(x + y)$	**b** $(a - 3)(t + 2)$	**c** $(m - k)(t - f)$
d $(x - 3)(x - 2)$	**e** $(2x + 3)(3x + 2)$	**f** $(2x - 1)(x - 3)$
g $(m + 2)(k + 3)$	**h** $(a - x)(b - x)$	**i** $(x - a)(x + a)$

Standard expressions

There are three *most important* results, which should be committed to memory. Failure to memorise these or recognise them when encountered in later work will seriously hinder your progress.

$$(a + b)^2 = (a + b)(a + b) = a^2 + ab + ab + b^2 = a^2 + 2ab + b^2$$
$$(a - b)^2 = (a - b)(a - b) = a^2 - ab - ab + b^2 = a^2 - 2ab + b^2$$
$$(a + b)(a - b) = a^2 - ab + ab - b^2 = a^2 - b^2$$

Memorise:
$$(a + b)^2 = a^2 + 2ab + b^2$$
$$(a - b)^2 = a^2 - 2ab + b^2$$
$$(a + b)(a - b) = a^2 - b^2$$

Note: a and b above stand for *any* numerical or algebraic expressions.

Examples

1 $(3k + t)^2 = (3k)^2 + 2(3k)(t) + (t)^2 = 9k^2 + 6kt + t^2$

2 $(1 - 5ab)^2 = (1)^2 - 2(1)(5ab) + (5ab)^2 = 1 - 10ab + 25a^2b^2$

3 $(4k + 1)(4k - 1) = (4k)^2 - (1)^2 = 16k^2 - 1$

Exercises 4.5 *continued*

3 Multiply out to express without brackets:

   a $(x + 3)^2$ b $(2a + 4)^2$ c $(k - 3)^2$ d $(x + 3)(x - 3)$

   e $(3 + 2x)^2$ f $(h + l)(h - l)$ g $(2 - 3t)^2$ h $(5b - 1)^2$

4.6 HCF AND LCM

$$3 \text{ divides into } 12 \quad \begin{cases} 3 \text{ is called a } \textbf{factor} \text{ of } 12 \\ 12 \text{ is called a } \textbf{multiple} \text{ of } 3 \end{cases}$$

HCF

The numbers that will divide into both 12 and 18 are 2, 3 and 6. Hence, 2, 3 and 6 are **common factors** of 12 and 18. The **highest** common factor (HCF) of 12 and 18 is 6.

> ### Example
>
> The common factors of $12a^2b$ and $8abc$ are 2, 4, 2a, 4a, 2b, 4b, 2ab and 4ab. Hence, the HCF is 4ab.

Exercises 4.6

1 Write down the HCF of:

   a $6x$ and $3x$ b $6x$ and $3y$ c $6x$ and $2xy$

   d xy^2 and x^2y e $6p^2q^2$ and $8pq^2$ f $4ab^3$ and $6a^2b^2$

2 Write down, in factored form, the HCF of:

   a $2(a + b)$ and $3(a + b)$ b $6(x^2 - y)$ and $9(x^2 - y)$

   c $6ab(p - 3q)$ and $4a^2b(p - 3q)$ d $p(m - 2k)$ and $3p(m - 2k)$

   e $3a(b - 2x)$ and $2a(b - 2x)$ f $a(p + q)(a + b)$ and $a^2(a + b)$

3 Reduce the following fractions to lowest terms by dividing the numerator and denominator by their HCF:

   a $\dfrac{6ax}{2ay}$ b $\dfrac{x^2y}{xy^2}$ c $\dfrac{12ax}{8a^3bx}$

   d $\dfrac{3ax(p - q)}{6a(p + q)}$ e $\dfrac{(a + b)(c - d)}{3(a + b)}$ f $\dfrac{12k^2(x + y)^2}{16k^3(x + y)}$

LCM

Multiples of 4 are 8, 12, 16, 20, 24, and so on.

Multiples of 6 are 12, 18, 24, 30, 36, and so on.

Common multiples of 4 and 6 are 12, 24, 36, and so on.

The **least** common multiple (LCM) of 4 and 6 is 12.

> **Example**
>
> The LCM of 2*a* and 4*b* is 4*ab*, since this is the *smallest* expression they will both divide into.

Exercises 4.6 *continued*

4 Write down the LCM of each pair of terms in question 1.

5 Write down, in factored form, the LCM of each pair of expressions in question 2.

4.7 ALGEBRAIC FRACTIONS

Algebraic fractions are added, subtracted, multiplied and divided in exactly the same way as arithmetical fractions. The following should be noted:

$\frac{8 + 3}{4} = \frac{8}{4} + \frac{3}{4} = 2 + \frac{3}{4} = 2\frac{3}{4}$ The 4 is divided into *each* of the numbers in the numerator.

$\frac{8 \times 3}{4} = \frac{8}{4} \times 3 = 2 \times 3 = 6$ The 4 is divided into only *one* of the numbers in the numerator.

Similarly:

$\dfrac{6a + 3b}{2} = \dfrac{6a}{2} + \dfrac{3b}{2} = 3a + 1\frac{1}{2}b$

$\dfrac{6a \times 3b}{2} = \dfrac{6a}{2} \times 3b = 3a \times 3b = 9ab$

Exercises 4.7

1 Express in simplest form:

   a $\dfrac{5 \times 15x}{5}$ **b** $\dfrac{5 + 15x}{5}$ **c** $\dfrac{x + 3ax}{x}$ **d** $\dfrac{x \times 3ax}{x}$

2 Reduce to lowest terms:

   a $\dfrac{2x + 6xy}{8xy}$ **b** $\dfrac{2x \times 6xy}{8xy}$ **c** $\dfrac{9k \times 6k^2}{6kt}$ **d** $\dfrac{9k - 6k^2}{6kt}$

3 Simplify:

   a $\dfrac{12x - 3}{3}$ **b** $\dfrac{6ab + 2a}{2a}$ **c** $\dfrac{14t - 7}{7t}$

> **Examples**
>
> **1** $\dfrac{2x}{3} + \dfrac{3x}{4} = \dfrac{8x + 9x}{12}$ **2** $\dfrac{3}{2a} - \dfrac{4}{ab} = \dfrac{3b - 8}{2ab}$
>
> $= \dfrac{17x}{12}$
>
> **3** $\dfrac{2x}{3} \div \dfrac{3x^2}{2} = \dfrac{2x}{3} \times \dfrac{2}{3x^2}$ Divide top and bottom by *x*.
>
> $= \dfrac{4}{9x}$

Exercises 4.7 *continued*

4 Simplify:

a $\dfrac{2}{7} + \dfrac{3}{7}$ **b** $\dfrac{2}{x} + \dfrac{3}{x}$ **c** $\dfrac{5}{m} - \dfrac{2}{m}$

d $\dfrac{a}{m} - \dfrac{b}{m}$ **e** $\dfrac{3}{x} + \dfrac{2}{x} - \dfrac{1}{x}$ **f** $\dfrac{2}{a+b} + \dfrac{5}{a+b}$

5 Express as a single fraction:

a $\dfrac{x}{2} + \dfrac{2x}{3}$ **b** $\dfrac{2R}{3} - \dfrac{R}{4}$ **c** $\dfrac{3C}{4} - \dfrac{C}{8}$

6 Express as a single fraction:

a $\dfrac{3}{x} + \dfrac{4}{m}$ **b** $\dfrac{2}{C} - \dfrac{C}{2}$ **c** $\dfrac{1}{3M} + \dfrac{3}{2M^2}$

7 Express as a single fraction:

a $\dfrac{m+1}{2} + \dfrac{m-1}{3}$ **b** $\dfrac{4k+1}{6} + \dfrac{1-2k}{3}$

c $\dfrac{3x+11}{6} + \dfrac{2-3x}{6}$ **d** $\dfrac{2+x}{x^2} + \dfrac{x-3}{3x}$

8 **a** $\dfrac{a}{b} \times \dfrac{x}{y}$ **b** $\dfrac{a^2b}{3x} \times \dfrac{x}{ab}$ **c** $\dfrac{k}{3t} \times \dfrac{6t^2}{3km}$

9 **a** $666 \times \dfrac{2132}{222}$ **b** $\dfrac{3t}{14ab} \times 7a$ **c** $\dfrac{5a^2}{12mt^2} \times 4mt$

10 **a** $\dfrac{x}{y} \div \dfrac{3x}{k}$ **b** $\dfrac{5}{ab} \div \dfrac{b}{3a}$ **c** $\dfrac{3t^2}{5mt} \div \dfrac{2mt}{3m}$

11 **a** $\dfrac{n}{k} \div k$ **b** $\dfrac{3a^2}{2b} \div 3a$ **c** $\dfrac{3k^2}{2t} \div 5t$

12 **a** $\dfrac{m+t}{b} \times \dfrac{a}{3(m+t)}$ **b** $\dfrac{3m}{a-b} \times \dfrac{5(a-b)}{6m^2}$

c $\dfrac{a+b}{a-b} \div \dfrac{2a+3b}{a-b}$ **d** $\dfrac{x-m}{abc} \div \dfrac{3(x-m)}{ak}$

Note: $1 - \dfrac{b}{a+b} = \dfrac{a+b}{a+b} - \dfrac{b}{a+b}$ $2t - \dfrac{2t^2}{t+1} = \dfrac{2t(t+1)}{t+1} - \dfrac{2t^2}{t+1}$

$\qquad\qquad\qquad\quad = \dfrac{a+b-b}{a+b}$ $= \dfrac{2t^2+2t-2t^2}{t+1}$

$\qquad\qquad\qquad\quad = \dfrac{a}{a+b}$ $= \dfrac{2t}{t+1}$

Exercises 4.7 *continued*

+++ **13** Express as single fractions:

    a $\dfrac{x+y}{y} - 1$ **b** $1 + \dfrac{a+b}{a-b}$ **c** $t - \dfrac{t}{t+1}$

Complex fractions

As with arithmetical complex fractions, algebraic complex fractions are usually most easily solved by multiplying above and below by the same term, preferably the lowest common denominator of all the fractions. (The multiplier is given in brackets after each example below.)

Examples

1 $\dfrac{1 - \dfrac{1}{x}}{1 + \dfrac{1}{x}} = \dfrac{x-1}{x+1}$, $(\times x)$ **2** $\dfrac{\dfrac{a}{b}}{\dfrac{1}{2a} - \dfrac{1}{b}} = \dfrac{2a^2}{b - 2a}$, $(\times 2ab)$

Exercises 4.7 *continued*

14 Express as single fractions in simplest form:

    a $\dfrac{\frac{1}{2} - \frac{1}{x}}{\frac{1}{2x} - 1}$ **b** $\dfrac{\frac{2}{t} - \frac{t}{2}}{\frac{3}{4t}}$ **c** $\dfrac{\frac{1}{x} - \frac{1}{y}}{\frac{1}{x} + \frac{1}{y}}$

4.8 ALGEBRAIC FRACTIONS OF THE FORM $\dfrac{a \pm b}{c}$

Online Learning Centre

$$\dfrac{9x - 6}{3} = 3x - 2$$

$$-\dfrac{9x - 6}{3} = -(3x - 2) \qquad \text{The negative sign applies to the whole fraction.}$$

$$= 3x + 2$$

    It is recommended that with all algebraic fractions containing more than one term in the numerator, you put *brackets* around these *numerators*.

Example

$$\frac{R + 1}{2} - \frac{2 - 5R}{3} = \frac{(R + 1)}{2} - \frac{(2 - 5R)}{3}$$

$$= \frac{3(R + 1) - 2(2 - 5R)}{6}$$

$$= \frac{3R + 3 - 4 + 10R}{6}$$

$$= \frac{13R - 1}{6}$$

Exercises 4.8

1 Simplify:

a $5x - \dfrac{6x - 4}{2}$ b $T - \dfrac{6T - 9}{3}$ c $1 - \dfrac{8 + 6R}{2}$

d $R + 3 - \dfrac{6R - 8}{2}$ e $3V - \dfrac{2V - 6}{2}$ f $1 - 4t - \dfrac{8t + 4}{4}$

2 Express as single fractions:

a $\dfrac{x}{2} - \dfrac{x + 1}{3}$ b $\dfrac{3}{4} - \dfrac{2 - 4E}{3}$ c $\dfrac{2}{3} - \dfrac{3C - 1}{6}$

d $\dfrac{3V}{4} - \dfrac{2 + 3V}{6}$ e $\dfrac{5}{6} - \dfrac{3L + 5}{4}$ f $\dfrac{4k}{9} - \dfrac{k - 5}{6}$

3 Express as single fractions:

a $\dfrac{x - 2}{3} - \dfrac{x - 3}{2}$ b $\dfrac{1 - x}{2} - \dfrac{x + 1}{3}$ c $\dfrac{5R - 2}{3} - \dfrac{2R + 1}{6}$

d $\dfrac{2n + 1}{4} - \dfrac{1 - n}{3}$ e $\dfrac{7 - 2G}{8} - \dfrac{5G - 1}{6}$ f $\dfrac{5R + 1}{6} - \dfrac{7 + 5R}{9}$

+++ 4 Express as single fractions:

a $1 - \dfrac{\frac{1}{x} - 1}{\frac{1}{x} + 1}$ b $1 - \dfrac{\frac{1}{2} - \frac{1}{m}}{\frac{1}{m} - 1}$

4.9 TWO IMPORTANT POINTS

■ The negative of $a - b$ is $b - a$ since $-(a - b) = -a + b = b - a$.

Examples

1 $-(x - 3) = 3 - x$ 2 $-(2k - t) = t - 2k$

■ The negative sign in a fraction may be placed before the numerator, before the denominator, or before the whole fraction.

Example

$$\frac{-6}{3} = -2 \qquad\qquad \frac{6}{-3} = -2 \qquad\qquad -\frac{6}{3} = -2$$

Note: $-\dfrac{k-3}{2} = \dfrac{-(k-3)}{2} = \dfrac{3-k}{2}$

$-\dfrac{3}{R-4} = \dfrac{3}{-(R-4)} = \dfrac{3}{4-R}$

$-\dfrac{x-3}{x-5}$ may be written as $= \dfrac{3-x}{x-5}$ or $\dfrac{x-3}{5-x}$

Exercises 4.9

1 Simplify:

a $\dfrac{a+b}{b+a}$

b $\dfrac{-(a+b)}{a+b}$

c $\dfrac{a-b}{b-a}$

d $\dfrac{t-m}{-m+t}$

e $\dfrac{3-x}{3-x}$

f $\dfrac{ax^2-ax}{ax-ax^2}$

2 Write as a positive fraction in simplest form:

a $-\dfrac{1}{x-y}$

b $-\dfrac{3+k}{3-k}$

c $-\dfrac{2x-3}{x+2}$

3 Simplify:

a $\dfrac{x-t}{3} \times \dfrac{a-b}{t-x}$

b $\dfrac{R}{2R-3} \times \dfrac{3-2R}{R-6}$

c $\dfrac{m-k}{k-t} \times \dfrac{t-k}{k-m}$

d $\dfrac{C-2}{3k} \div \dfrac{2-C}{6k}$

Examples

1 If $A = 3t - 2$ and $B = 5 - t$, express $2A - 3B$ in terms of t.

Solution

$$\begin{aligned}
2A - 3B &= 2(3t - 2) - 3(5 - t) \\
&= 6t - 4 - 15 + 3t \\
&= 9t - 19
\end{aligned}$$

2 If $x = \dfrac{3-k}{2}$ and $y = \dfrac{k-2}{3}$, express $2x - 6y$ in terms of k.

Solution

$$\begin{aligned}
2x - 6y &= 2\left(\frac{3-k}{2}\right) - 6\left(\frac{k-2}{3}\right) \\
&= (3 - k) - 2(k - 2) \\
&= 3 - k - 2k + 4 \\
&= 7 - 3k
\end{aligned}$$

Exercises 4.9 *continued*

4 If $P = 3 - 2n$ and $Q = 2 - 5n$, express the following in terms of n:

  a $P - Q$ **b** $2P + Q$ **c** $Q - 3P$

  d $3Q - P$ **e** $\frac{1}{2}P + \frac{1}{3}Q$ **f** $\frac{1}{3}Q - \frac{2}{3}P$

5 If $A = \dfrac{2t - 1}{3}$ and $B = \dfrac{3 - t}{2}$, express the following in terms of t:

  a $3A + 2B$ **b** $4B + 9A$ **c** $6B - 9A$

  d $3B - 2A$ **e** $6A + \frac{1}{2}B$ **f** $\frac{2}{3}B - \frac{4}{5}A$

4.10 SOLVING LINEAR EQUATIONS

If two numbers are equal and we operate on them both in exactly the same way, this will clearly result in two new numbers that are equal. We solve linear equations by performing the same operation on both sides of the equation.

Examples

1 If $4 \times -3 = 7$

  $\qquad 4x = 10$ (adding 3 to both sides)

  $\qquad \therefore x = 2.5$ (dividing both sides by 4)

2 If $\dfrac{5x - 4}{2} = 5x - 8$

  $\qquad 5x - 4 = 10x - 16$ (multiplying both sides by 2)

  $\qquad -4 = 5x - 16$ (subtracting 5x from both sides)

  $\qquad 12 = 5x$ (adding 16 to both sides)

  $\qquad 2.4 = x$ (dividing both sides by 5)

  $\qquad \therefore x = 2.4$

Exercises 4.10

1 Solve:

  a $2x + 3 = 11$ **b** $3x + 2 = 8$ **c** $-2x = 10 - 3x$

  d $5 - 3x = x + 1$ **e** $4 + 3x = 8$ **f** $2W = 3W - 1$

  g $5x + 1 = 1$ **h** $1 - 5R = 3 - 4R$ **i** $7 = 1 - 4t$

  j $2 - 3x = 2x + 1$ **k** $5 + 7x = 3$ **l** $1 - 3x = 10$

  m $5E - 3 = 1 - 2E$ **n** $1 - l = 4 + 2l$

Rule: When an equation contains brackets, clear the brackets first.

Example

If $3(2 - x) + 1 = 4 - 2(2x - 3)$

$6 - 3x + 1 = 4 - 4x + 6$

$7 - 3x = 10 - 4x$

$7 + x = 10$

$x = 3$

Exercises 4.10 *continued*

2 Solve:

a $2x = x - (3 - 2x)$

b $3(2 - x) = 2(x - 1)$

c $5(2x - 3) = 3(1 + x)$

d $4(2 - 3x) = 3(3 - x)$

e $5(2x - 1) = 2(2 + 3x)$

f $3 - 2(d - 3) + d = 0$

g $1 - (W - 1) = W - 1$

h $2(x + 2) = 3(4 - 2x)$

i $1 - 3(L - 1) - (L + 1) = 0$

j $3(x - 3) = 2(3 - x)$

Rule: When an equation contains fractions, it is usually advisable to clear all fractions first by multiplying both sides by the LCM of the denominators.

Example

If $\dfrac{x}{3} = \dfrac{1}{2} - \dfrac{3x}{4}$

$4x = 6 - 9x$ (multiplying both sides by 12)

$13x = 6$

$\therefore x = \dfrac{6}{13}$

Exercises 4.10 *continued*

3 Solve:

a $\dfrac{x}{2} - \dfrac{x}{3} = 1$

b $2 + \dfrac{x}{2} = 1 - \dfrac{x}{3}$

c $\dfrac{a + 1}{2} = 3 + \dfrac{2a + 1}{2}$

d $\frac{1}{2}k + \frac{1}{4} = \frac{1}{3}k + 1$

e $\dfrac{m - 2}{3} = \dfrac{m - 1}{2} + 1$

f $\frac{2}{3}x = \frac{1}{2}$

g $\dfrac{2n}{3} = \dfrac{n - 2}{3}$

h $\dfrac{1 + 2W}{3} = \dfrac{W}{6}$

i $\frac{1}{2}y + 2 = \frac{2}{3}y + 1$

j $\frac{3}{4}x = \frac{1}{2} + x$

Rule: Since in an equation the pronumeral is standing for some definite number (unknown until the equation is solved), we can multiply both sides of an equation by the pronumeral.

Example

If $\dfrac{2x + 3}{3x} = 1 - \dfrac{5}{6x}$

$\qquad 4x + 6 = 6x - 5 \qquad$ (multiplying both sides by 6x)

$\qquad\qquad 11 = 2x$

$\qquad\quad \therefore x = 5\tfrac{1}{2}$

Exercises 4.10 *continued*

4 Solve:

a $\dfrac{1}{x} = 3$

b $\dfrac{1}{x} - 3 = 2$

c $\dfrac{2}{a} = \dfrac{1}{2} + \dfrac{3}{a}$

d $\dfrac{3}{a} - 2 = \dfrac{1}{a}$

e $\dfrac{x - 2}{3x} = 5$

f $\dfrac{3 - m}{2m} = \dfrac{2}{3m} + 4$

Cross–multiplication

When we have an equation consisting of 'a single fraction = a single fraction', we can 'cross–multiply'.

Examples

1 If $\dfrac{x + 2}{3} \diagdown\!\!\!\diagup \dfrac{x - 2}{2}$

$\qquad 3x - 6 = 2x + 4 \qquad$ (this is equivalent to multiplying both sides by 6)

$\qquad\quad \therefore x = 10$

2 If $\dfrac{2x - 3}{x + 4} = \dfrac{3}{5}$

$\qquad 10x - 15 = 3x + 12 \qquad$ (this is equivalent to multiplying both sides by 5(x + 4))

$\qquad\quad 7x = 27$

$\qquad\quad \therefore x = 3\tfrac{6}{7}$

Cross–multiplication may *not* be used in cases where there are other terms as well as the single fraction on each side.

Example

$\dfrac{x + 1}{5} = \dfrac{x}{3} + 5$

We cannot cross–multiply because of the '+ 5' term. In this case, multiply both sides by 15.

Exercises 4.10 *continued*

5 Solve for the pronumeral:

a $\dfrac{x+1}{2} = \dfrac{x+2}{3}$

b $\dfrac{2x+3}{x} = \dfrac{2}{3}$

c $\dfrac{3}{x} = \dfrac{2}{1-x}$

d $\dfrac{2-x}{3} = \dfrac{2x-1}{5}$

e $\dfrac{2x-3}{4} = \dfrac{2}{3}x$

f $\dfrac{x-1}{2} = \dfrac{2-x}{3}$

Fractions equal to zero

We have learnt in our study of arithmetical fractions that a fraction equals zero if, and only if, the *numerator* = 0 (and the *denominator* ≠ 0).

Examples

1 If $\dfrac{x}{17.3} = 0$, then $x = 0$.

2 If $\dfrac{x}{3x-6} = 0$, then $x = 0$.

3 If $\dfrac{7-3x}{2x-8} = 0$, then $7 - 3x = 0$.

These results can, of course, also be obtained by multiplying both sides by the denominator.

It is a common elementary error to argue that if, for example, $\dfrac{38}{x-2} = 0$, then $x - 2 = 0$. This is *wrong* because the *denominator* can *never* be zero. In fact, $\dfrac{38}{x-2}$ can *never* equal 0, no matter what value is given to x.

Note: If $\dfrac{a}{b} = 0$, then $a = 0$ and $b \neq 0$.

Exercises 4.10 *continued*

6 Solve:

a $\dfrac{x-2}{x-3} = 0$

b $\dfrac{x+5}{x-2} = 0$

c $5\left(\dfrac{3x}{x-2}\right) = 0$

4.11 SOLVING MORE DIFFICULT LINEAR EQUATIONS

When an equation contains fractions of the form $\dfrac{a \pm b}{c}$, clear the fractions in the usual way by multiplying both sides by the same number. It is recommended that brackets be placed around the numerator of any fraction when it consists of more than one term. Exercise care with minus signs when removing brackets.

Examples

1 If
$$\frac{x}{2} - \frac{3x - 1}{4} = 7$$
$$\frac{x}{2} - \frac{(3x - 1)}{4} = 7$$
$$2x - (3x - 1) = 28$$
$$2x - 3x + 1 = 28$$
$$-x + 1 = 28$$
$$-x = 27$$
$$\therefore x = -27$$

2 If
$$\frac{3}{4} - \frac{3 - 5C}{2} = \frac{3C}{8}$$
$$\frac{3}{4} - \frac{(3 - 5C)}{2} = \frac{3C}{8}$$
$$6 - 4(3 - 5C) = 3C$$
$$6 - 12 + 20C = 3C$$
$$-6 + 20C = 3C$$
$$17C = 6$$
$$\therefore C = \frac{6}{17}$$

Exercises 4.11

1 Solve:

a $\dfrac{x}{2} - \dfrac{2x + 1}{3} = \dfrac{5}{6}$

b $\dfrac{2}{3} - \dfrac{n - 1}{4} = \dfrac{7n}{12}$

c $3 - \dfrac{2 + 3R}{5} = \dfrac{7R}{10}$

d $2\frac{1}{2} - \dfrac{3L - 5}{4} = \dfrac{L}{2}$

e $2R + 3 - \dfrac{3R + 2}{4} = \dfrac{R}{3}$

f $\dfrac{1 - M}{2} - \dfrac{1 - M}{3} = 1 - M$

2 Solve:

a $2x - \dfrac{3x - 5}{2} = 1$

b $4 - \dfrac{1 + 2L}{2} = \dfrac{L}{6}$

c $\dfrac{2V}{3} = 3 - \dfrac{V - 10}{4}$

d $\dfrac{n}{2} - \dfrac{5 + 2n}{3} = 0$

+++ 3 Solve:

a $x - \dfrac{x - 2}{5} = 5 - \dfrac{x}{5}$

b $\dfrac{2(t - 3)}{3} - \dfrac{4(2t - 3)}{6} = \dfrac{1}{5}$

c $\dfrac{4(1 - k)}{3} - 2 = \dfrac{3(k - 2)}{2} - 4k$

d $1 - \dfrac{3 - 8E}{4} = \dfrac{1 - 3E}{2} - 4(E + 1)$

e $\dfrac{1 - x}{x} = 3 - \dfrac{1 + 2x}{5x}$

f $\dfrac{1}{2} - \dfrac{2(1 - 4W)}{3} = \dfrac{W}{2} - 2(W - 7)$

4.12 USING LINEAR EQUATIONS TO SOLVE PRACTICAL PROBLEMS

If we are trying to find an unknown number (e.g. a number of kilograms, a number of amperes, a number of seconds), and we have enough information about that number to enable us to find its value, algebra provides a very powerful and convenient tool for doing the job. We let x (or any other pronumeral) represent the unknown number and then form equations containing x using the information we have about this number.

Examples

1 I am thinking of a number and I give you the following information about this number: if I double it, then add 3 to the result, and then double the result, the final result is 36. What is the number?

  Now the number *could* be found by 'trial and error' (trying various possibilities in turn), or by other methods, but algebra provides us with a quick and simple method for finding the number.

Solution

Let n represent the number to be found.
I double the number, obtaining $2n$.
I then add 3, obtaining $2n + 3$.
I then double this result, obtaining $4n + 6$.
But I am told that this result is 36.
Hence $4n + 6 = 36$
$$4n = 30$$
$$n = 7.5$$

The solution would be set out thus:
Let the number be n.
$$[(n \times 2) + 3] \times 2 = 36$$
$$(2n + 3)2 = 36$$
$$2n + 3 = 18$$
$$2n = 15$$
$$n = 7.5$$

2 A man says to his brother. 'I am now three times your age. In eight years' time I will be twice your age.' What is the difference between their ages?

Solution

Let the present age of the younger brother be x years; the present age of the elder brother is $3x$ years. In eight years' time, the younger brother's age will be $x + 8$ and the elder brother's age will be $3x + 8$.

$$3x + 8 = 2(x + 8)$$
$$3x + 8 = 2x + 16$$
$$x = 8$$

The present age of the younger brother is 8 and the present age of the elder brother is 24. The difference between their ages is 16 years.

It is easy to invent such puzzles, but only those with a little knowledge of algebra can solve them quickly.

Here is another one for you to try (but only students who know how to remove brackets will be able to do this one quickly).

Exercises 4.12

1 I am thinking of a number. If I subtract 5 from this number and then subtract the result from three times the number, the result is 28. Find the number.

Example

The length of a certain rectangle is 7 m more than its breadth. Given that the perimeter of this rectangle is 40 m, find its length.

Solution

Let the breadth be b metres.
The length is $b + 7$ m.
The perimeter is $2b + 2(b + 7)$

$$= 2b + 2b + 14$$
$$= 4b + 14 \ m$$

Hence $4b + 14 = 40$
$$4b = 26$$
$$\therefore b = 6.5$$

Hence the breadth is 6.5 m
$\therefore$ the length is $6.5 + 7 = 13.5$ m

Exercises 14.12 *continued*

2   A rectangle has a perimeter of 20.0 m. Given that its length is 3.00 times its breadth, find (a) its breadth, (b) its area.

3   The height of a tower was increased by 4.50 m. A few years later its height (at that time) was doubled. If its final height was 50.0 m, what was its original height?

4   One wire, having resistance R_1, is 3.50 times as long as another piece of the same wire whose resistance is R_2. Hence, the first wire has 3.50 times the resistance of the second wire. The total resistance of the two wires in series, $(R_1 + R_2)$, is measured to be 2.64 Ω. Find the resistance of the longer wire.

5   In triangle ABC, the length of side BC is three times the length of side AB. The length of side CA is two-thirds of the total length of the other two sides. Given that the perimeter measures 60 mm, find the length of the longest side.

6   If two opposite sides of a certain square have their lengths each increased by 1 m, and the other two sides each decreased by 3 m, a rectangle is formed whose perimeter is 16 m. Find:
    a   the measure of the sides of the original square
    b   the area of the rectangle formed by the changes

+++ 7   At noon, two cars leave towns separated by a distance of 240 km, travelling towards one another. If the average speed of one car is 8 km/h greater than that of the other, find the average speed of each car, given that they meet at 3.00 pm.

+++ 8   A motor boat whose speed in still water is 8.5 km/h travelled 6.0 km in a river against the current, then back again *with* the current. Given that the total time taken was 1 h 42 min, find the speed of flow of the river. (*Hint:* write down the times taken in each direction in terms of v, the speed of the river, and put their sum equal to the total time.)

+++ 9 An electricity supplier has two tariff rates for charging, and a consumer is free to choose whichever rate is desired:

- Rate A: 10c/unit for lighting and 2c/unit for power.
- Rate B: a fixed charge of $10 per quarter plus 3c/unit for each unit used in the quarter, for either lighting or power.

a In one quarter, a consumer uses 100 units for lighting and 400 units for power. Find the amount he or she would have to pay under each of the two tariff rates.

b If a consumer uses 1000 units for power each quarter, find the least number of units he or she would need to use for lighting in order to make rate B the cheaper rate.

4.13 SIMULTANEOUS LINEAR EQUATIONS

Consider the equation $y = x + 2$. There is no limit to the number of pairs of values for x and y that satisfy this equation:

$$\begin{cases} x = -3 \\ y = -1 \end{cases} \quad \begin{cases} x = -1 \\ y = 1 \end{cases} \quad \begin{cases} x = 0 \\ y = 2 \end{cases} \quad \begin{cases} x = 2 \\ y = 4 \end{cases} \quad \begin{cases} x = 1.63 \\ y = 3.63 \end{cases} \quad \text{etc., ad infinitum}$$

Similarly, for the equation $y = 3x - 2$, there is no limit to the number of solutions:

$$\begin{cases} x = -3 \\ y = -11 \end{cases} \quad \begin{cases} x = -1.28 \\ y = -5.84 \end{cases} \quad \begin{cases} x = 0 \\ y = -2 \end{cases} \quad \begin{cases} x = 2 \\ y = 4 \end{cases} \quad \begin{cases} x = 7 \\ y = 19 \end{cases} \quad \text{etc.}$$

But there is only *one* pair of values for x and y that satisfies *both* equations (simultaneously),

namely: $\begin{cases} x = 2 \\ y = 4 \end{cases}$

That is, the solution to the simultaneous equations $\begin{cases} y = x + 2 \\ y = 3x - 2 \end{cases}$ is $\begin{cases} x = 2 \\ y = 4 \end{cases}$

Exercises 4.13

1 a Does $x = 4, y = 5$ satisfy the equation $y = 2x - 3$?

b Does $x = 4, y = 5$ satisfy the equation $y = 3x - 7$?

c Is $x = 4, y = 5$ the solution to the simultaneous equations $\begin{cases} y = 2x - 3 \\ y = 3x - 7 \end{cases}$?

continued

continued

2 a Is it true that $\begin{cases} x = -3 \\ y = 2 \end{cases}$ satisfies the equation $y = x + 5$?

   b Is it true that $\begin{cases} x = -3 \\ y = 2 \end{cases}$ satisfies the equation $y = 2x + 7$?

   c Is it true that $\begin{cases} x = -3 \\ y = 2 \end{cases}$ is the solution to the simultaneous equations $\begin{cases} y = x + 5 \\ y = 3x - 7 \end{cases}$?

Simultaneous equations can be solved by trial and error, but this can be a long and tedious method.

   We will study two logical methods that give the solutions directly:

1 the substitution method

2 the elimination method.

Chapter 23 describes 2×2 determinants and how these can be used to solve simultaneous equations.

4.14 THE SUBSTITUTION METHOD

This method involves using *one* of the equations to make a substitution in the *other* equation, which gives us a single equation with only one variable.

Example

Solve: $\begin{cases} \dfrac{3y + 14}{2x + 3} = 5 \\ y = 2x + 3 \end{cases}$

The second equation tells us that wherever we see a *y*, we may replace it with $2x + 3$ if we wish to do so, or wherever we see a $2x + 3$, we may replace it with a *y*.

Method A

Now there is a *y* in the first equation.
Replacing this by $2x + 3$, we obtain:

$$\frac{3(2x + 3) + 14}{2x + 3} = 5$$

$$3(2x + 3) + 14 = 5(2x + 3)$$

$$6x + 9 + 14 = 10x + 15$$

$$6x + 23 = 10x + 15$$

$$4x = 8$$

$$\therefore x = 2$$

Method B

There is a $2x + 3$ in the first equation.
Replacing this by *y*, we obtain:

$$\frac{3y + 14}{y} = 5$$

$$3y + 14 = 5y$$

$$2y = 14$$

$$\therefore y = 7$$

We then evaluate *x* using either of the two given equations.

We then evaluate y using either of the

two given equations.

$$y = 2x + 3$$
$$\quad = 2(2) + 3$$
$$\quad = 7$$

Solution: $\begin{cases} x = 2 \\ y = 7 \end{cases}$

$$y = 2x + 3$$
$$7 = 2x + 3$$
$$\therefore x = 2$$

Solution: $\begin{cases} x = 2 \\ y = 7 \end{cases}$

Exercises 4.14

Use the substitution method to solve the following pairs of simultaneous equations:

1 a $\begin{cases} x = 2 \\ 3x + 2y = 12 \end{cases}$ b $\begin{cases} R = -3 \\ R - 4r = 1 \end{cases}$ c $\begin{cases} 6F + 3W = -3 \\ F = -\frac{2}{3} \end{cases}$

2 a $\begin{cases} y = 4x + 1 \\ y = 3x + 2 \end{cases}$ b $\begin{cases} E = 7 - 2V \\ E = 4V - 5 \end{cases}$ c $\begin{cases} 3C_1 = C_2 + 4 \\ 3C_1 = 3C_2 - 12 \end{cases}$

3 a $\begin{cases} x + 2y = 12 \\ 2x + 3y = 19 \end{cases}$ b $\begin{cases} 3T + l = -1 \\ 6T - 5l = 12 \end{cases}$ c $\begin{cases} 5V - E = -13 \\ 2V + 3E = -12 \end{cases}$

4.15 THE ELIMINATION METHOD

We can eliminate one of the variables by subtracting the equations when the coefficients of that variable are the same, or by adding when the coefficients are opposites.

Note: If we subtract equals, the result is zero: $(6x) - (6x) = 0$ and $(-3y) - (-3y) = 0$

If we add opposites, the result is zero: $(6x) + (-6x) = 0$ and $(-3y) + (3y) = 0$

Examples

1 $\begin{cases} 7x - 5y = 7 \\ 3x + 5y = -47 \end{cases}$ (different signs, add)

Adding: $\underline{10x = -40}$

$\therefore x = -4$

$y = -7$ (found by substituting $x = -4$ into either of the two given equations)

Solution: $x = -4$, $y = -7$

continued

continued

2    $\begin{cases} 7E - 4V = 1 \\ 2E - 4V = -14 \end{cases}$ (same signs, subtract)

Subtracting: $\underline{5E = 15}$

$\therefore E = 3$

$V = 5$ (found by substituting $E = 3$ into either of the two given equations)

Solution: $E = 3, V = 5$

Note: When the signs are *different*, *add* (as above with $-5y$ and $+5y$).

When the signs are the *same*, *subtract* (as above with $-4V$ and $-4V$).

Exercises 4.15

1   Solve for the pronumerals using the elimination method (giving answers as integers, fractions or mixed numbers):

a $\begin{cases} 2x + 5y = 19 \\ 2x + 3y = 13 \end{cases}$

b $\begin{cases} 3W - 2d = 16 \\ 2W + 2d = 4 \end{cases}$

c $\begin{cases} 4I_1 - 3I_2 = 3 \\ I_1 - 3I_2 = 12 \end{cases}$

d $\begin{cases} 2t + 3x - 2 = 0 \\ 5t + 3x - 8 = 0 \end{cases}$

e $\begin{cases} 3L_1 - 7L_2 = 4 \\ 2L_1 - 7L_2 = 1 \end{cases}$

f $\begin{cases} 3i + 5V = 10 \\ 4i - 5V = 11 \end{cases}$

Sometimes it is necessary to multiply both sides of one (or both) of the equations by the same number so that a variable will vanish when we add or subtract the equations.

Examples

1   $\begin{cases} 6x + 2y = 40 & —① \\ 3x - 8y = 2 & —② \end{cases}$

Multiplying both sides of equation ② by 2, we obtain:

$\begin{cases} 6x + 2y = 40 & —① \\ 6x - 16y = 4 & —② \times 2 \end{cases}$

Subtracting: $\underline{18y = 36}$

$\therefore y = 2$

Now $3x - 8y = 2$

$3x - 16 = 2$

$3x = 18$

$\therefore x = 6$

Solution: $x = 6, y = 2$

2   $\begin{cases} 7x - 4y = 18 & —① \\ 12x + 6y = 18 & —② \end{cases}$

$\begin{cases} 21x - 12y = 54 & —① \times 3 \\ 24x + 12y = 36 & —② \times 2 \end{cases}$

Adding: $\underline{45x = 90}$

$\therefore x = 2$

Now $7x - 4y = 18$

$4y = 14 - 18$

$\therefore y = -1$

Solution: $x = 2, y = -1$

Exercises 4.15 *continued*

2   Solve for the pronumerals, using the elimination method (giving answers as integers, fractions or mixed numbers):

a $\begin{cases} 4b + 5a = 15 \\ 2b + 3a = 10 \end{cases}$ b $\begin{cases} 5R - E = 11 \\ 2R + 3E = 1 \end{cases}$ c $\begin{cases} 3d - 2v - 4 = 0 \\ 7d - 6v - 14 = 0 \end{cases}$

3   Solve for the pronumerals, using the elimination method (giving answers as integers, fractions or mixed numbers):

a $\begin{cases} 3x + 5b = 18 \\ 2x + 7b = 23 \end{cases}$ b $\begin{cases} 4V_1 - 3V_2 = 17 \\ 13V_1 - 4V_2 = 38 \end{cases}$ e $\begin{cases} 5v - 4S = -10 \\ \frac{2}{3}v + 6S = -34 \end{cases}$

+++   4   Solve for the pronumerals, correct to 3 significant figures, using the elimination method, giving all values as decimal fractions:

a $\begin{cases} 2a + 3b = 4 \\ 3a - 7b = 1 \end{cases}$ b $\begin{cases} 15x - 4y = 2 \\ 13x - 3y = 3 \end{cases}$

4.16 PRACTICAL PROBLEMS

Exercises 4.16

State all answers correct to 3 significant figures.

1   The diameter of a tapered rod at a distance l mm from the smaller end is given by $D = d_e + kl$ mm, where k is a constant and d_e mm is the diameter at the smaller end. From measurements taken, it was found that when $l = 100$, $D = 7.50$; when $l = 150$, $D = 8.10$.

a   Form a pair of simultaneous equations and solve them to evaluate d_e and k.

b   Find the diameter at a distance of 130 mm from the smaller end.

2   For a lifting machine it is known that $E = kL + b$, where E kN is the effort force applied, L kN is the load lifted, and k and b are constants for the machine. For a particular machine, it is known that when the load is 4 kN, the effort force required is 1.2 kN, and when the load is 6 kN, the effort force is 1.7 kN.

Form a pair of simultaneous equations and solve them to find:

a   the values of k and b for this machine

b   the effort force required to raise a load of 10 kN

3   If a body of mass m kg rests on a horizontal plane, the force parallel to the plane, P N, required to give the body an acceleration of a m/s^2, is given by $P = F + ma$, where F is the limiting friction force. It is found that when $P = 48$ N, $a = 2$ m/s^2, and when $P = 78$ N, $a = 3.5$ m/s^2.

continued

continued

Form a pair of simultaneous equations and solve them to find the mass of the body and the friction force.

4 The length of a spring in tensile stress, supporting a load L kg, is given by $l = kL + C$, where k is a constant for the spring and C is the original length of the unstressed spring. For a particular spring, it is known that when $L = 4$ kg, $l = 232$ mm and when $L = 6$ kg, $l = 248$ mm.

Form a pair of simultaneous equations and solve them to find:

  a the values of k and C for the spring

  b the length of the spring when supporting a load of 9 kg

  c the load that gives the spring a length of 300 mm

5 For a body travelling with constant acceleration of A m/s², its velocity at a time t s after passing a certain fixed point P is given by $v = u + At$ m/s, where u m/s is its velocity when passing P. For a particular body, the velocity was 27 m/s 3 seconds after passing P, and its velocity was 42 m/s 2 seconds later.

Form a pair of simultaneous equations and solve them to find:

  a the acceleration   b the velocity when passing P

  c the velocity 7 s after passing P

6 A voltage divider is required consisting of two resistors R_1 and R_2 in series, such that the sum of their values is 20 000 Ω and the ratio $R_1 : R_2 = 7 : 1$.

Form two equations in R_1 and R_2 and solve them simultaneously to evaluate R_1 and R_2.

7 For this circuit, $E + V = IR$.

  a Given that when $V = 2$ V, $I = 5$ A, and when $V = 5$ V, $I = 7$ A, evaluate R and E.

  b For *different* values of R and E it is found that $I = 591$ mA when $V = 86.0$ V, and $I = 710$ mA when $V = 114$ V. Find the new values of R and E.

8 Equations obtained by applying Kirchhoff's laws to this circuit are:

$$I_1R + (2I_1 - I_2)r = 34$$
$$I_2R + (I_2 - I_1)r = 42$$

  a Given that if $I_1 = 3$ A then $I_2 = 4$ A, evaluate R and r.

  b Given that if $I_1 = 7$ A then $I_2 = 9$ A, evaluate R and r correct to 3 significant figures.

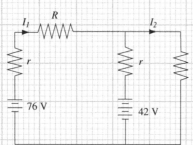

The following exercises require you to define pronumerals to stand for the two unknown values and then to form two simultaneous equations to solve for their values.

+++ 9 One bottle of copper sulphate solution contains 30 g of copper sulphate per litre and another contains 70 g/L. What volume of each must be mixed in order to obtain 400 mL of solution with a concentration of 45 g/L of copper sulphate?

+++ 10 A motor boat moving at its maximum speed travels in a river a distance of 3.6 km upstream against the current and then back again with the current to its starting point. The passage upstream took 48 minutes and the return trip took 18 minutes. Determine:

  a the maximum speed of the boat in still water, and

  b the speed of the current.

11 A manufacturer wishes to cut a rectangular piece of copper sheet that has a perimeter of 24 cm and an area of 2975 mm². What must the length and the breadth of the rectangle be?

+++ 12 Two blocks of metal each has a volume of 1000 cm³. One block is an alloy of 80% iron and 20% nickel by volume and a mass of 8.078 kg. The other block consists of 50% iron and 50% nickel and has a mass of 8.390 kg. Deduce the densities of iron and nickel.

+++ 13 A mixture of two powdered substances contains 20% of substance A by mass and 10% of substance B. How much of each substance must be added to 50 kg of this mixture in order to produce a new mixture containing 25% of substance A and 15% of substance B?

SELF-TEST

1 Given that $m = 2$, $k = -3$ and $t = -4$, evaluate:

  a $m - kt$

  b $\dfrac{t}{m} - \dfrac{2k}{t}$

  c $2m - (k - 2)$

  d $m^2 + 2k^2 - t^2$

  e $(m - 2t) - (k - 2)$

  f $(m - k)^2 - 3kt$

2 Multiply out to express without brackets:

  a $(2x - 3)(2x + 3)$

  b $(2t - 3)^2$

  c $(x + 3t)^2$

  d $(2a - 3b)(2a + 3b)$

  e $(5k - 2m)^2$

3 Express as a single fraction:

  a $\dfrac{2}{ab} + \dfrac{1}{2a^2b}$

  b $\dfrac{V}{R} - \dfrac{2}{3RL}$

  c $\dfrac{3}{2xy} + \dfrac{1}{6x^2}$

  d $\dfrac{x}{3p^2t} - \dfrac{1}{4pt^2}$

4 Simplify:

  a $\dfrac{\frac{2}{t}}{3 - \frac{1}{t}}$

  b $\dfrac{\frac{1}{x - 1}}{1 - \frac{1}{1 - x}}$

5 Express as single fractions:

  a $\dfrac{1 - 7L}{12} - \dfrac{3 - 7L}{8}$

  b $\dfrac{4V + 5}{6} - \dfrac{3V + 5}{8}$

6 Solve:

a $8 - 2(1 - 4a) = 6a$

b $\dfrac{3 - 2b}{2b - 9} = 5$

c $\dfrac{k - \frac{2}{3}}{\frac{1}{2} - 6k} = 1$

d $\dfrac{1}{2} - \dfrac{E - 1}{3} = \dfrac{E - 1}{6}$

e $3 - \dfrac{2(3 - 2t)}{3} = t$

f $\dfrac{3}{2 - 3x} = \dfrac{2}{1 - x}$

g $\dfrac{13(57 - a)}{29(a - 2)} = 0$

+++ h $\dfrac{\frac{2}{3}(1 - \frac{W}{2})}{1 - \frac{W}{3}} = \dfrac{10}{13}$

+++ i $\dfrac{\frac{2}{3}(\frac{3x}{4} - \frac{1}{2})}{x - \frac{1}{5}} = 0$

+++ j $\dfrac{1}{2 - \frac{3 - t}{4 - t}} = \dfrac{1}{3}$

7 Glycerine has a density (mass per unit volume) that is 1.5 times that of the density of methylated spirits. Given that 4 L of glycerine weighs 1.7 kg more than 4 L of methylated spirits, find:

a the density of methylated spirits (in kg/m³),

b the mass of 4 L of glycerine.

(*Hint:* let the density of methylated spirits be *m* kg/L.)

8 In one week it costs a firm $2\frac{1}{2}$ times as much for wages as for all other expenses. Given that the total expenditure for this week is $87 500, how much is spent on wages?

9 Solve, using the substitution method:

a $\begin{cases} 6k - 4m = 3 \\ 3k - m = 1 \end{cases}$

b $\begin{cases} I_1 = \frac{1}{2}I_2 + \frac{1}{3} \\ 2I_1 = \frac{2}{3}I_2 - \frac{1}{2} \end{cases}$

c $\begin{cases} \frac{3}{5}V - t + 5 = 0 \\ 2V + 5t = 0 \end{cases}$

10 Solve, using the elimination method:

a $\begin{cases} 7F - 5m + 39 = 0 \\ 14F + 3m + 13 = 0 \end{cases}$

b $\begin{cases} 3a - 5k + 6 = 0 \\ 5a - 20k + 10 = 0 \end{cases}$

11 Solve, using the elimination method:

a $\begin{cases} 7a - 5t + 3 = 0 \\ 13a - 2t - 9 = 0 \end{cases}$

b $\begin{cases} 5S - 13t - 29 = 0 \\ 4S - 17t - 11 = 0 \end{cases}$

+++ 12 *Full–cream* milk contains 3.8% fat and *Slim* milk contains 1.6% fat. How many litres of each would have to be mixed in order to obtain 100 L of milk containing 2.5% fat?

+++ 13 When working simultaneously, two conveyors fill a certain container in 25 minutes, delivering 875 kg of material. If one of the conveyors breaks down after 15 minutes, the other conveyor must continue delivering for a further 25 minutes in order to fill the container. Find the rates of delivery of each of the two conveyors.

CHAPTER 5

FORMULAE: EVALUATION AND TRANSPOSITION

Learning objectives

- Evaluate a non-linear algebraic expression by substitution of given values.
- Evaluate formulae for physical quantities:
    - by direct substitution of values given in basic SI units
    - by conversion of values given in derived units into basic SI units before substitution.
- Make transpositions in formulae so as to make a particular pronumeral the subject of a formula.

5.1 EVALUATION OF THE SUBJECT OF A FORMULA

By now you should be familiar with your calculator and you should have had some experience in using it. These exercises afford further practice.

In working these exercises, you are advised to do the following:

1 First plan the *order* in which you will do the computation. It is best to use the *shortest* method, not so much because it is the quickest but because the fewer the keys you have to press, the less likely you are to make errors. However, *obtaining the correct result* is the only really important consideration, and so sometimes you may choose not to use the shortest method because you find it too involved.

2 Exercise constant vigilance. Each time you key in a number, *look at the display* to check that the correct number has been registered. Do not work too quickly; make quite sure that you do press the correct key.

3 As you work through a computation, record the results of the steps as you proceed. This provides you with a better opportunity for checking your work and also provides an examiner with an opportunity to award some marks (if he or she wishes to do so) for a computation that is *mostly* correct but gives the wrong result (e.g. because you forgot to take the square root or reciprocal in the last step).

4 Let your calculator do the *whole* computation. If you perform only *part* of a calculation, write down this answer and then re-enter it at a later stage, you will increase the chance of error. If

the intermediate result is not retained to a sufficient degree of accuracy, approximation here can lead to significant variation in the final answer. The memory can be used to temporarily store an intermediate result.

Exercises 5.1

Use a calculator to evaluate the following correct to 3 significant figures:

1 $\dfrac{ab}{c^2 + d^2}$ when $a = 1.68, b = 7.55, c = 3.84, d = 4.26$

2 $\dfrac{1}{k^2} - \dfrac{1}{\sqrt{m}}$ when $k = 0.832, m = 0.913$

3 $x\sqrt{x^2 + y^2}$ when $x = 3.27, y = 2.81$

4 $\dfrac{1}{n} + \dfrac{1}{k^2} - \dfrac{1}{\sqrt{t}}$ when $n = 1.28, k = 1.16, t = 3.47$

5 $\dfrac{a^2 - b^2 + c^2}{\sqrt{ab - ck}}$ when $a = 38.7, b = 19.3, c = 24.9, k = 16.1$

6 $\dfrac{\sqrt{s}}{tu}$ where $s = 50.7, t = 3.28, u = -2.15$

7 $-\sqrt{x^2 - y^2}$ where $x = 19.8, y = 11.9$

8 $\sqrt{m^2 + n^2}$ where $m = 4.55, n = -5.19$

9 $\dfrac{1}{\sqrt{u^2 + v^2}}$ where $u = -1.33, v = -7.71$

10 $\sqrt{f^3 + g^3}$ where $f = 12.1, g = -10.3$

11 $(\sqrt{a} + \sqrt{b} + \sqrt{c})^2$ where $a = 1.11, b = 2.13, c = 1.36$

12 $-f^2 + \sqrt{g^2 - h^2}$ where $f = 5.18, g = -3.17, h = 2.98$

13 $\sqrt{\dfrac{k^2 + q^3}{dh^2}}$ where $k = 8.21, q = 5.14, d = 1.26, h = -2.99$

14 $\dfrac{1}{\sqrt{\sqrt{r} - \sqrt{s}}}$ where $r = 95.9, s = 73.4$

15 $u^3 + (v^2 - w^2)^3$ where $u = -5.01, v = 3.17, w = 2.27$

16 $\dfrac{\sqrt{s - t}}{(u - v)^3}$ where $s = 65.2, t = 18.7, u = 6.22, v = 7.88$

5.2 MORE SI UNITS

Units are important when substituting values into formulae. Every measurable quantity must have units of measurement whose sizes are defined. In Chapter 3 we learnt and used the SI basic units for length, mass and time and the list below gives the basic SI units of some other quantities that you will meet in this and later chapters of this book. **You do not need to know the *sizes* of these units but merely to recognise that the given symbols do represent basic SI units:**

Quantity	Basic unit	Abbreviation symbol
Mass	kilogram	kg
Length	metre	m
Time	second	s
Force	newton	N
Pressure	pascal	Pa (1 Pa = 1 N/m^2)
Work, energy	joule	J
Power	watt	W (1 W = 1 J/s)
Frequency	hertz	Hz (1 Hz = 1 oscillation/s)
Electric current	ampere	A
Electric charge	coulomb	C
Electric potential difference	volt	V
Electric resistance	ohm	Ω
Electric capacitance	farad	F
Electric inductance	henry	H

It is important to remember that the only basic SI unit that bears a prefix is the first one in the above list—the unit of mass, the *kilogram* (kg). Whenever the value of a mass is substituted into a formula, it must be expressed in *kilograms* and not in grams. This anomaly arose because the gram is inconveniently small for the measure of most commonly encountered masses.

It is useful to have a rough idea of the size of the units you use:

Mass:　1 g is roughly the mass of a paperclip.

　　　　1 kg is the mass of 1 L of water.

　　　　1 Mg (= 1 t) is the mass of 1 m^3 of water and roughly the mass of a small motor car.

Force:　1 N is roughly the force required to lift a small apple (of mass 102 g).

　　　　1 kN is roughly the force required to lift a very heavy person (of mass 102 kg).

Power: 1 kW is roughly the maximum human power output (for a short period).

Some other useful facts

■ The unit in which electrical energy is sold is the kilowatt hour (kWh), which is the energy delivered in 1 h at a rate (power) of 1 kW (= 1000 J/s).

Hence 1 kWh = 1000 × 3600 J = 3.6 × 10^6 J = 3.6 MJ.

- To raise the temperature of 1 kg (1 L) of water by 1°C requires 4.18 kJ of energy (the **specific heat capacity** of water is 4.18×10^3 J/kg/°C).
- The density of water is 1 g/mL = 1 kg/L = 1 t/m³.
- 1 m/s = 3.6 km/h. Check this yourself. This result is worth memorising.

Weight

The weight of a body is the *force* of gravitational attraction upon it and hence must be measured in *newtons* (or newtons with a prefix). The weight of a body *cannot* be measured in a unit of mass (e.g. kilograms).

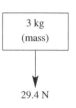

The weight of a 3 kg mass near the earth's surface where $g = 9.80$ m/s² is approximately 29.4 N. This weight *cannot* be expressed as '3 kg' or '3 kg wt'.

(The term 'load' may refer to a *mass,* in which case it is expressed in a unit of mass, or it may refer to a *weight,* in which case it is expressed in a unit of force. For more information about 'weight', see Appendix B.)

A practical household exercise

If you are interested, you may like to answer the following questions using information above:

A typical domestic hot-water tank holds 250 L and uses a 3.6 kW heating element.

1 How much energy is required to raise the temperature of all the water in this tank from 20°C to 70°C?

2 How much will this cost if the price of the electricity is 5.8c per unit? (Answer to the nearest cent.)

3 How long would it take to perform this task? (Answer to the nearest minute.)

Answers: **1** 52.25 MJ (14.5 kWh)   **2** 84c   **3** 4 h 2 min

5.3 EVALUATION OF A FORMULA

When evaluating a quantity by substitution into a formula, it must be remembered that the formula has been derived on the understanding that *consistent* units are to be used for all quantities.

Example

If a body travels at 40 km/h for 3 hours, then the distance travelled,

$s = v \times t$

$\quad = 40 \times 3$

$\quad = 120$ km

We have used the *kilometre* for distance and the *hour* for time, so the velocity must be expressed in *km/h.*

Example

In calculating the area of a rectangle, if we use 7 cm for the length, then the breadth must also be expressed in *centimetres* and the area in square *centimetres*.

In the simple example above, it was obvious that it was not necessary to use the basic SI units (the *metre*, the *second* and the *metre per second*), and other (but *consistent*) units were used. **However, in general it is advisable to use the basic SI units for every quantity.**

Example

The constant force that is required to increase the velocity of a mass m by an amount v during the time interval t, is given by $F = \dfrac{m \times v}{t}$ N.

Find the constant force required to increase the velocity of a mass of 5.80 t from 15.0 km/h to 125 km/h during a time interval of 1.30 minutes.

Solution

First express the quantities in basic SI units. (It is useful to remember that to convert km/h into the basic unit of m/s, divide by 3.6 because 1 km/h $= \dfrac{1000}{60 \times 60}$ m/s $= \dfrac{1}{3.6}$ m/s.)

$$m = 5.80 \text{ t} = 5800 \text{ kg}$$

$$v = 125 \text{ km/h} - 15 \text{ km/h} = 110 \text{ km/h} = \frac{110}{3.6} \text{ m/s}$$

$$t = 1.3 \text{ mins} = 78 \text{ s}$$

$$\therefore F = \frac{m \times v}{t}$$

$$= \frac{5800 \times \dfrac{110}{3.6}}{78} \text{ N}$$

$$\approx 2270 \text{ N}$$

Answer: 2.27 kN

Note: The final result is more neatly expressed in a prefixed unit, kN.

In all the following exercises:

- every value substituted into a formula should be expressed in the basic SI unit for that quantity
- every result of an evaluation should be expressed as a number between 1 and 1000, using if necessary an appropriate prefix.

This is called 'engineering notation', which will be explained more fully in section 11.7.

Examples

A result of 6840 mm would be expressed as 6.84 m.
A result of 73 600 N would be expressed as 73.6 kN.
A result of 0.000 413 F would be expressed as 413 μF.
A result of 2 830 000 g would be expressed as 2.83 Mg (or 2.83 t).

Exercises 5.2

In questions 1–7:

- part **a** requires direct substitution of the given values
- part **b** requires conversion of values into basic SI units before substitution.

1 The kinetic energy of a body, in joules (J), is given by the formula $E = \frac{1}{2}mv^2$. Find the kinetic energy of (a) an 81.3 kg mass that has a velocity of 37.2 m/s, (b) a car of mass 1.36 t and a velocity of 93.5 km/h.

2 The constant force (in newtons, N) required to increase the velocity of a body of mass m, from u to v, while it travels a distance s in a straight line is given by:

$$F = \frac{m(v^2 - u^2)}{2s}$$

Find the force required in the following cases:

a $m = 5.23$ kg, $u = 19.7$ m/s, $v = 36.2$ m/s, $s = 234$ m

b $m = 1.28$ t, $u = 6.37$ km/h, $v = 58.3$ km/h, $s = 3.86$ km

3 The radius of gyration of a rectangular prism about an axis through its centre and perpendicular to a face having a length l and breadth b is given by $k = \sqrt{\dfrac{l^2 + b^2}{12}}$. Find the radius of gyration in the following cases:

a $l = 0.863$ m, $b = 0.241$ m   b $l = 1.68$ m, $b = 972$ mm

4 The velocity of a projectile of mass m in free flight at height h above its launching point is given by: $v = \sqrt{\dfrac{2(W - mgh)}{m}}$ where W is the energy expended in its projection and $g \approx 9.81$ m/s^2.

Find the velocity in each of the following cases:

a $m = 1.63$ kg, $h = 9.72$ m, $W = 381$ J   b $m = 26.4$ g, $h = 3.18$ km, $W = 1.28$ kJ

5 When capacitances C_1, C_2 and C_3 are connected in series, the total capacitance of the combination is given by: $C = \dfrac{1}{\dfrac{1}{C_1} + \dfrac{1}{C_2} + \dfrac{1}{C_3}}$.

Find the total capacitance in the following cases:

a $C_1 = 2.62$ F, $C_2 = 1.83$ F, $C_3 = 3.09$ F

b $C_1 = 987$ μF, $C_2 = 1.26$ mF, $C_3 = 1.08$ mF

6 For a circuit comprising an inductance L in series with a capacitance of C, the resonance frequency is given by: $f = \dfrac{1}{2\pi\sqrt{LC}}$.

Find the resonance frequency in the following cases:

a $L = 2.46$ H, $C = 3.87$ F   b $L = 2.17$ mH, $C = 8.72$ μF

7 When a coil having an inductance L and a resistance R is connected in parallel with a capacitance C, the resonance frequency is given by: $f = \dfrac{1}{2\pi L}\sqrt{\dfrac{L}{C} - R^2}$ Hz.

Find the resonance frequency in the following cases:

a $L = 2.55$ H, $C = 0.001\ 32$ F, $R = 32.6\ \Omega$

b $L = 6.86$ mH, $C = 8.41\ \mu$F, $R = 19.8\ \Omega$

8 $T = 2\pi\sqrt{\dfrac{m + \frac{1}{3}m_1}{F_1}}$ gives the period (time of oscillation) of a mass m suspended from a spiral spring of mass m_1, where F_1 is the force required for unit extension.

Evaluate T when $m = 2.38$ kg, $m_1 = 58.6$ g and $F_1 = 5.26$ N.

5.4 TRANSPOSITION

When one variable is expressed in terms of other variables, this variable is called the **subject** of the formula.

Example

In the formula $t = 2a + b - c^2$, t is expressed in terms of a, b and c:

t is the subject of this formula.

Note: In the formula $x = 8k - 3x - 4m$, x is not expressed in terms of the other variables, since x also appears on the right-hand side. However:

if $x = 8k - 3x - 4m$

$\therefore 4x = 8k - 4m$

$\therefore x = 2k - m$ and *now* x is the subject of the formula.

SUMMARY

To make x the subject of a formula, we must obtain an equation of the form $x =$ (terms not containing x).

To make a particular variable the subject of a formula, we proceed in exactly the same manner as when solving an equation for x.

Examples

Compare these two examples:

1 Solve for x: $3(x + 2) = 8$

$$3(x + 2) = 8$$
$$3x + 6 = 8$$
$$3x = 2$$
$$\therefore x = \tfrac{2}{3}$$

2 Make x the subject of the formula:

$$a(x + t) = m$$
$$a(x + t) = m$$
$$ax + at = m$$
$$ax = m - at$$
$$\therefore x = \frac{m - at}{a}$$

Exercises 5.3

1 **a** Solve for x: $\dfrac{2(x - 3)}{5} = 4$

   b Make x the subject of the formula: $\dfrac{a(x - y)}{n} = 3k$

2 **a** Solve for x: $\dfrac{2x}{3} - 4 = 6$

   b Make x the subject of the formula: $\dfrac{ax}{b} - t = m$

3 **a** Solve for x: $\dfrac{x}{2} + \dfrac{3}{4} = 5$

   b Make x the subject of the formula: $\dfrac{x}{n} + \dfrac{a}{b} = t$

4 **a** Solve for x: $\dfrac{2}{x - 3} = 4$

   b Make x the subject of the formula: $\dfrac{h}{x - m} = t$

Transpositions involving squares and square roots

If the pronumeral required as subject lies in a term that is squared or appears under a square root sign, this term must first be isolated.

Examples

Make x the subject of the following:

1 $m = k - ax^2$

$$ax^2 = k - m$$
$$x^2 = \frac{k - m}{a}$$
$$\therefore x = \pm\sqrt{\frac{k - m}{a}} \quad \text{(isolating the term containing } x)$$

2 $t = 2\sqrt{\dfrac{x - a}{b}} + 3$

$$\frac{t - 3}{2} = \sqrt{\frac{x - a}{b}}$$
$$\left(\frac{t - 3}{2}\right)^2 = \frac{x - a}{b}$$
$$x = b\left(\frac{t - 3}{2}\right)^2 + a \quad \text{(isolating the term containing } x)$$

Exercises 5.3 *continued*

Make x the subject of the following formulae:

5 $m - \sqrt{x} = t$

6 $3a - kx^2 = m$

7 $\dfrac{2a}{kx^2} = 3y$

8 $\left(\dfrac{x-a}{3}\right)^2 = t$

9 $\left(\dfrac{3-2x}{5}\right)^2 = t^2$

10 $k - \sqrt{\dfrac{a-2b}{x}} = t$

Miscellaneous

In each of the following exercises, for students of mechanical engineering, a formula is given followed by a pronumeral in parentheses. In each case, make this pronumeral the subject of the formula (i.e. express the given pronumeral in terms of the other variables and constants).

11 $\dfrac{n_a}{n_b} = \dfrac{T_b}{T_a}$ **a** (T_b) **b** (n_b)

12 $V = \dfrac{2d_1}{d_1 - d_2}$ (d_2)

13 $P = \dfrac{(T_1 - T_2)v}{1000}$ **a** (T_1) **b** (T_2)

14 $F \times S = \frac{1}{2}mv^2 - \frac{1}{2}mu^2$ (v)

15 $E = \dfrac{W}{V} + \dfrac{F}{V}$ (F)

16 $q = \dfrac{m_s}{m + m_s}$ (m)

These exercises, intended for students of electrical engineering, also provide additional practice for other students.

17 $V = IR$ (R)

18 $P = \dfrac{W}{t}$ (W)

19 $C = \dfrac{Q}{V}$ (V)

20 $R = B.\,I.\,l$ (I)

21 $H = \dfrac{IN}{l}$ (l)

22 $F = C \times \dfrac{I_1 I_2 l}{d}$ (l)

Questions 23–26 involve taking a square or square root.

23 $W = \frac{1}{2}\dfrac{Q^2}{C}$ (Q)

24 $M = k\sqrt{L_1 L_2}$ (L_1)

25 $W = \frac{1}{2}\mu H^2$ (H)

26 $f = \dfrac{1}{2\pi\sqrt{LC}}$ (L)

Questions 27–31 are more difficult.

+++ 27 $G = \dfrac{1}{\frac{1}{a} \times \rho}$ (a)

+++ 28 $m = \dfrac{2w}{v^2 + 2gh}$ (h)

+++ 29 $I_1 = \dfrac{R_2}{R_1 + R_2} I$ (R_1)

+++ 30 $\dfrac{1}{R} = \dfrac{1}{R_1} + \dfrac{1}{R_2}$ (R)

+++ 31 The total inductance of this circuit is L, where:

$$\dfrac{1}{L} = \dfrac{1}{L_1} + \dfrac{1}{L_2 + L_3}$$

Make L_3 the subject of this formula.

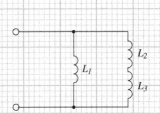

5.5 TRANSPOSITIONS IN WHICH GROUPING IS REQUIRED

When the variable required as subject occurs in more than one term, these terms must be collected on one side of the equation.

Examples

1 $\dfrac{x - a}{b} = x + t$

$x - a = bx + bt$

$x - bx = a + bt$ (collecting the x terms on one side)

$x(1 - b) = a + bt$ (taking out the common factor, x)t

$x = \dfrac{a + bt}{1 - b}$

2 $b(x + t) = a(m - 2x)$

$bx + bt = am - 2ax$

$bx + 2ax = am - bt$

$x(2a + b) = am - bt$

$x = \dfrac{am - bt}{2a + b}$

3 $\dfrac{x - a}{b} = \dfrac{3m - x}{k}$

$kx - ka = 3bm - bx$

$kx + bx = ka + 3bm$

$x(k + b) = ka + 3bm$

$x = \dfrac{ka + 3bm}{k + b}$

Exercises 5.4

In each exercise below, a formula is given followed by a pronumeral in parentheses.

In each case make the pronumeral in parentheses the subject of the formula:

1   $x(a + 4) = 5$ (x)
2   $x(3t - m) = k$ (x)
3   $ax + mx = t$ (x)
4   $kn = 5 - bn$ (n)
5   $\dfrac{k - w}{m} = 3w$ (w)
6   $a - \dfrac{b - t}{m} = t$ (t)

Exercises 5.5

Mechanical

In each case, make the pronumeral in parentheses the subject of the formula (express the given pronumeral in terms of the other variables and constants). SI units are used unless otherwise stated.

1   $v = u + at$ (t) Velocity of a body moving in a straight line after t seconds, where u is the initial velocity and a is the acceleration.

2   $PV = \dfrac{m}{M}RT$ The Universal Gas Equation.

a (m) **b** (M)

3 $\eta = \dfrac{Q_1 - Q_2}{Q_1}$

Efficiency of a perfect heat engine where Q_1 is the rate of heat input and Q_2 is the rate of heat rejection.

  a (Q_2) **b** (Q_1)

4 $P = \dfrac{(T_1 - T_2)v}{1000}$

Power (kW) transmitted by a belt rotating a pulley, where T_1, T_2 are the tensions on the tight and slack sides respectively, and v is the speed of the belt.

  a (T_1) **b** (T_2)

5 $F \times S = \frac{1}{2}mv^2 - \frac{1}{2}mu^2$ (m)

Work done by a force in accelerating a body on a smooth horizontal plane = the increase in the kinetic energy of the body.

6 $E = \dfrac{W}{V} + \dfrac{F}{V}$

Effort used on a lifting machine where W = the load, V = the velocity ratio and F = the 'friction load'.

  a (F) **b** (V)

7 $F = \dfrac{mv - mu}{t}$

Resultant force applied to a body = its rate of change of momentum.

  a (v) **b** (m)

8 $E = aW + b$ (W)

The 'law of a machine'.

9 $P = 2q(l + b)t$

Force on a punch required to punch a rectangular piece of metal of length l metres and width b metres from a piece of metal of thickness t metres. q = the ultimate shear stress of the metal.

  a (q) **b** (b)

10 $Q = mc(t_2 - t_1)$

Heat energy required to raise temperature of a body of mass m from $t_1°C$ to $t_2°C$. c = the specific heat capacity of the substance.

  a (c) **b** (t_1)

11 $V = V_0\left(1 + \dfrac{t}{273}\right)$ (t)

Volume of a fixed mass of gas at $t°C$, where V_0 = volume at $0°C$, the pressure remaining constant.

12 $\eta = \dfrac{W}{(aW + b) \times V}$

Efficiency of a machine where W = the load, V = the velocity ratio and a, b are constants for the machine.

  a (V) **b** (a) **c** (W)

Electrical

In each exercise below, a formula is given followed by a pronumeral in parentheses. In each case, make the pronumeral in parentheses the subject of the formula. SI units are used unless otherwise stated.

13 $R_s = \dfrac{I_F \times r}{I - I_F}$ (I_F) Shunt resistance needed for an ammeter conversion.

14 $R_x = \dfrac{E - V}{V} \times R_i$ (V) Measurement of resistance using battery and voltmeter only.

15 $R_a = \dfrac{R_1 R_2}{R_1 + R_2 + R_3}$ (R_1) Star network equivalent to a delta network.

16 $\dfrac{E}{e} = \dfrac{R + r}{r}$ (r) E is the total voltage drop across two resistors R, r connected in series, where e is the voltage drop across r.

17 $I_2 = \dfrac{R_1 I}{R_1 + R_2}$ (R_1) Current through resistor R_2 when resistors R_1, R_2 are connected in parallel.

18 $R = R_1 + \dfrac{R_2 R_3}{R_2 + R_3}$ (R_2) Resistance of circuit with R_1 in series with parallel combination of R_2 and R_3.

19 $E = I_1 R_a + (I_1 + I_2) R_b$ (I_1) Equation arising from application of Kirchhoff's laws.

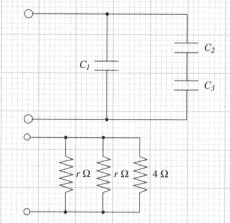

+++ 20 The total capacitance of this circuit is given by:

$$C = C_1 + \dfrac{C_2 C_3}{C_2 + C_3}$$

Make C_3 the subject of this formula.

+++ 21 The total resistance of this circuit is given by:

$$R = \dfrac{1}{\dfrac{2}{r} + \dfrac{1}{4}}$$

Make r the subject of this formula.

5.6 MORE TRANSPOSITION AND EVALUATION

As explained earlier, during a computation you should show the results of the steps as you proceed. Although these partial results need be shown only approximately, do not make unnecessary approximations during the computation. You should work with at least 5 significant figures (preferably with the full capacity of your calculator) and only approximate when the final result is obtained.

In Exercises 5.6 a formula is given.

a   Evaluate the subject using the data provided.

b   Change the subject to the pronumeral named.

c   Evaluate the new subject using the data provided.

Example

Distance, $s = d + vt$ **a** $d = 840$ m, $v = 53$ km/h, $t = 6.5$ min   **b** t

c $s = 1.24$ m, $d = 765$ mm, $v = 2.49$ km/h

Solution

a $d = 840$ m

$v = \frac{53}{3.6}$ m/s

$t = 6.5 \times 60$ s

$s = d + vt$

$= 840 + (\frac{53}{3.6} \times 6.5 \times 60)$

≈ 6582 m

≈ 6.58 km

b $s = d + vt$

$s - d = vt$

$t = \dfrac{s - d}{v}$

c $s = 1.24$ m

$d = 0.765$ m

$v = \frac{2.49}{3.6}$ m/s

$t = \dfrac{s - d}{v}$

$= (1.24 - 0.765) \div \frac{2.49}{3.6}$

≈ 0.687 s

≈ 687 ms

Answers: (a) 6.58 km   (b) $t = \dfrac{s - d}{v}$ (c) 687 ms

Exercises 5.6

1 Power, $P = \dfrac{W}{t}$

    a $W = 1.40$ kJ, $t = 1.50$ h

    b W

    c $P = 1.60$ kW, $t = 850$ ms

2 Work, $W = Fs$

    a $F = 1.45$ kN, $s = 25.0$ mm

    b s

    c $F = 850$ mN, $W = 2.38$ kJ

3 Potential difference, $E = IR$

    a $I = 450$ μA, $R = 2.70$ MΩ

    b I

    c $E = 870$ mV, $R = 1.20$ kΩ

4 Current, $I_1 = \dfrac{R_2 I}{R_1 + R_2}$

    a $I = 735$ mA, $R_1 = 1.50$ kΩ, $R_2 = 820$ Ω

    b R_1

    c $I = 1.35$ mA, $I_1 = 430$ μA, $R_2 = 1.20$ kΩ

Squares and square roots

5 Force, $F = \dfrac{M(v^2 - u^2)}{2s}$

    a $M = 1.28$ t, $u = 6.37$ km/h,

       $v = 58.3$ km/h, $s = 3.86$ km

    b v

    c $F = 230$ mN, $s = 1.38$ km, $M = 631$ g,

       $u = 85.0$ km/h

6 Velocity, $v = \sqrt{\dfrac{2(W - mgh)}{m}}$

    a $g = 9.81 \text{ m/s}^2$, $m = 26.4$ g,

      $h = 3.18$ km, $W = 1.28$ kJ

    b h

    c $W = 410$ mJ, $m = 350$ g,

      $v = 645$ mm/s, $g = 9.81 \text{ m/s}^2$

7 Power, $P = \dfrac{V^2 R_1}{(R_1 + R_2)^2}$

    a $R_1 = 1.50 \text{ m}\Omega$, $R_2 = 2.20 \text{ m}\Omega$,

      $V = 84.0$ mV

    b V

    c $P = 75.5$ mW, $R_1 = 4.70 \ \Omega$, $R_2 = 2.20 \ \Omega$

8 Frequency, $f = \dfrac{1}{2\pi \sqrt{LC}}$

    a $L = 2.50$ mH, $C = 8.50 \ \mu\text{F}$

    b C

    c $L = 150 \ \mu\text{H}$, $f = 3.25$ kHz

Grouping of like terms

9 Work, $W = Fs + Ps$

    a $F = 1.32$ N, $P = 940$ mN, $s = 374$ mm

    b s

    c $F = 865$ N, $P = 1.08$ kN, $W = 1.62$ kJ

10 Friction force, $F = \dfrac{W - mgh}{h}$

    a $g = 9.80 \text{ m/s}^2$, $m = 25.4$ g,

      $W = 1.38$ kJ, $h = 1.25$ km

    b h

    c $g = 9.80 \text{ m/s}^2$, $m = 1.25$ t,

      $F = 850$ N, $W = 12.5$ kJ

11 Current, $I_1 = \dfrac{I_2 R}{R + r}$

    a $I_2 = 1.40$ mA, $R = 1.20 \text{ M}\Omega$, $r = 820 \text{ k}\Omega$

    b R

    c $r = 1.20 \ \Omega$, $I_1 = 850$ mA, $I_2 = 2.50$ A

+++ 12 Capacitance, $C = C_1 + \dfrac{C_2 C_3}{C_2 + C_3}$

    a $C_1 = 1.35$ mF, $C_2 = 1.80$ mF, $C_3 = 850 \ \mu\text{F}$

    b C_3

    c $C = 2.50$ mF, $C_1 = 850 \ \mu\text{F}$, $C_2 = 3.50$ mF

SELF-TEST

Evaluation

1 When a potential difference of e volts exists across a resistor of R ohms in series with a capacitance of C farads, then $e = iR + \dfrac{Q}{C}$ volts where the current flowing is i amps and the charge on the capacitor is Q coulombs.

Evaluate e for the case where the current is 28.6 mA, the resistance is 1.63 kΩ, the charge is 2.44 mC and the capacitance is 43.8 μF.

2 At a frequency of f hertz (Hz), the impedance of a circuit that consists of a resistance of R ohms in series with an inductance of L henries (H), is given by: $Z = \sqrt{R^2 + (2\pi f L)^2}$ ohms.

   a Calculate the impedance of a series circuit at a frequency of 50 Hz, given that the resistance is 10 Ω and the inductance is 200 mH, stating the result correct to 3 significant figures.

   b Calculate the impedance of a series circuit at a frequency of 350 Hz, given that the resistance is 23.5 Ω and the inductance is 6.45 mH.

3 For a certain smoothing circuit of a full-wave rectifier: $\dfrac{V_o}{V_i} = \dfrac{1}{\omega^2 LC - 1}$ where V_o the output ripple voltage, V_i is the input ripple voltage, $\omega = 2\pi f$, where f is the ripple frequency in hertz (Hz), L is the inductance in henries (H), C is the capacitance in farads (F).

Evaluate the ratio $\dfrac{V_o}{V_i}$ when the ripple frequency is 100 Hz, the inductance is 15.0 mH and the capacitance is 2.75 mF.

4 Make the given pronumeral the subject of the formula:

   a $V = V_o\left(1 + \dfrac{t}{273}\right)$ (t)

   b $E = \dfrac{V^2 t}{R}$ (R)

   c $\rho = \dfrac{R \times a}{l}$ (l)

   d $M = \dfrac{L_A - L_s}{4}$ (L_s)

   e $\dfrac{R_x}{R_s} = \dfrac{R_1}{R_2}$ (R_2)

   f $\dfrac{1}{C} = \dfrac{1}{C_1} + \dfrac{1}{C_2}$ (C_2)

   g $V = i_1 R_1 - i_2 R_2$ (i_2)

   h $I = \dfrac{E}{r + \dfrac{R}{n}}$ (n)

In Exercises 5–9, all data is given to 3 significant figures.

5 Given that force $F = \dfrac{mv}{t}$:

   a evaluate F when $m = 4.25$ g, $v = 560$ km/h and $t = 500$ μs

   b make t the subject of this formula

   c evaluate t when $m = 435$ g, $v = 5.62$ m/ms and $F = 4.52$ kN

6 Given that the electric charge $Q = CV$ coulombs:

  a evaluate Q when $C = 650\ \mu F$ and $V = 800$ mV

  b make C the subject of this formula

  c evaluate C when $Q = 450\ \mu C$ and $V = 950$ mV

7 Given that velocity $v = \sqrt{u^2 + 2as}$:

  a evaluate v when $u = 20$ km/h, $a = 8$ m/s^2 and $s = 685$ mm

  b make s the subject of this formula

  c evaluate s when $u = 50$ km/h, $v = 51$ km/h and $a = 10$ m/s^2

+++ 8 Given that time interval $t = \dfrac{\sqrt{u^2 + 2as} - u}{a}$:

  a evaluate t when $u = 28.4$ km/h, $a = 8.13$ m/s^2 and $s = 860$ mm

  b make s the subject of this formula

  c evaluate s when $t = 1.50$ min, $u = 22.0$ m/s and $a = 95.0$ mm/s^2

9 In your electrical theory you will learn that when a capacitor is in series with a resistor, the total impedance, Z, is given by the diagonal of the rectangle which has sides of lengths R and X_c, where the reactance of the capacitor is given by $X_c = \dfrac{1}{2\pi f C}$ where f is the frequency (in hertz) and C is the capacitance (in farads).

Evaluate Z, in ohms (Ω), correct to 3 significant figures when:

  a $R = 2000\ \Omega$ and $X_c = 5000\ \Omega$

  b $R = 3.2$ kΩ and $X_c = 800\ \Omega$

  c $R = 30$ kΩ, $C = 0.01\ \mu F$ and $f = 400$ Hz

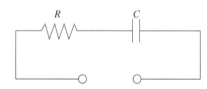
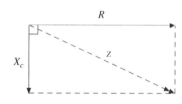

PART 2

GEOMETRY AND TRIGONOMETRY

CHAPTER 6

INTRODUCTION TO GEOMETRY

Learning objectives

- Use geometrical instruments to make simple constructions.
- Solve numerical problems using the relationships between adjacent angles, vertically opposite angles and angles formed by parallel straight lines and a transversal.

6.1 POINTS, LINES, RAYS AND ANGLES

A **point** is a position in space. We cannot actually draw a point because it has no area, but we *represent* a point by drawing a small but clearly visible dot or cross.

A **line** is a continuous extent of length, straight or curved. We cannot actually draw a line but we *represent* a line by a thin but clearly visible drawing.

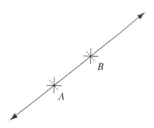

A **straight** line extends indefinitely in both directions. The diagram represents a straight line *AB*. It passes through the points *A* and *B* but does not terminate at *A* or *B* (or anywhere else).

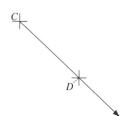

The diagram represents the **ray** *CD*. It starts from point *C*, passes through *D* and then extends indefinitely.

The diagram represents a **line segment** or **interval** *EF*. It extends from point *E* to point *F* and has a definite length.

An **angle** is formed at the point of intersection of two lines, rays or segments. The angle in the diagram is called the angle *θ* or the angle *Q* (∠*Q*), or the angle *PQR* (∠*PQR*), or the angle *RQP* (∠*RQP*).

6.2 ANGLES

360° = 1 revolution

1° is divided into 60 parts called **minutes**, i.e. 60′ = 1°

1′ is divided into 60 parts called **seconds**, i.e. 60″ = 1°

You may like to investigate why the same terms 'minutes' and 'seconds' are used for units of both time and angles.

Addition

Examples

1   43°47′ + 38°28′ = 81°75′
    = 82°15′

2   23°52′43″ + 8°49′68″ = 31°101′111″
    = 31°102′51″
    = 32°42′51″

Subtraction

Examples

73°16′ + 28°49′ = 72°76′ − 28°49′
        = 44°27′

If necessary it is advisable to change the notation for the first angle so that no 'borrowing' or 'paying back' is needed.

Types of angle

1   An angle less than 90° is an **acute angle**.

2   An angle of 90° is a **right angle**.

3   An angle between 90° and 180° is an **obtuse angle**.

4   An angle of 180° is a **straight angle**.

5   An angle greater than 180° is a **reflex angle**.

Note: Throughout the remainder of this book any angle marked thus

is a right angle and any angle marked thus    may or may not be a right angle.

Exercises 6.1

1   In each case find the measure of angle A and state which type of angle it is (acute, right, obtuse, straight or reflex). You should practise obtaining these results with and without a calculator.

a $A = 77°53' + 81°58'$ b $A = 340°24' - 263°46'$

c $A = 68°47' + 19°58'$ d $A = 263°19' - 73°38'$

e $A = 27°27' + 32°44'$ f $A = 67°52' + 22°08'$

g $A = 261°41' - 88°53'$ h $A = 168°44' + 11°56'$

i $A = 103°47' + 76°13'$ j $A = 39°28' + 38°53'$

   # 6.3 COMPLEMENTS AND SUPPLEMENTS

Two angles whose sum is 90° (1 right angle) are called **complementary** angles.

Examples

1   70° and 20° are complementary angles.
    70° is the complement of 20°; 20° is the complement of 70°.
2   38°24′ is the complement of 51°36′.

Two angles whose sum is 180° (2 right angles) are called **supplementary** angles.

Examples

1 120° and 60° are **supplementary** angles. 120° is the supplement of 60°; 60° is the supplement of 120°.
2 72°26′ and 107°34′ are supplementary.

To find the complement of an angle, subtract it from 90° (or 89°60′).

Examples

1 The complement of $3t°$ is $(90 - 3t)°$.
2 The complement of $(5 - k)°$ is $90° - (5 - k)° = 90° - 5° + k° = (85 + k)°$.
3 The complement of 26°29′ is 89°60′ − 26°29′ = 63°31′ Check, using your calculator.

To find the supplement of an angle, subtract it from 180° (or 179°60′).

Examples

1 The supplement of $\theta°$ is $(180 - \theta)°$.
2 The supplement of $(x - 23)°$ is $180° - (x - 23)° = 180° - x° + 23° = (203 - x)°$
3 The supplement of 123°45′ is 179°60′ − 123°45′ = 56°15′ Check, using your calculator.

Exercises 6.2

1 Find the:

a complement of 28°46′
b supplement of 28°46′
c complement of 73°28′
d supplement of 9°24′
e supplement of 138°17′
f complement of 48°25′

2 Find the:

a complement of $\theta°$
b supplement of $x°$
c complement of $90° - 38°26′$
d supplement of $180° - 67°18′$
e complement of $(t - 43)°$
f supplement of $(86 - \theta)°$

3 In each of the following cases, evaluate the pronumeral:

a   b   c

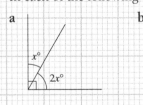 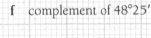

d   e   f

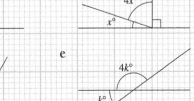   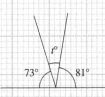

6.4 VERTICALLY OPPOSITE ANGLES

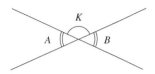

Angle *A* is the supplement of angle *K*. Angle *B* is the supplement of angle *K*.

$$\therefore \angle A = \angle B$$

A and *B* are called **vertically opposite** angles.

Note: Vertically opposite angles are equal.

6.5 PARALLEL LINES AND A TRANSVERSAL

We place arrows on lines to indicate that they are parallel.
A line that cuts two parallel lines is called a **transversal.**

Angles and parallel lines

Rule: Alternate angles (i.e. angles between the parallel lines on opposite sides of the transversal) are equal.

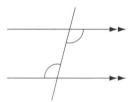   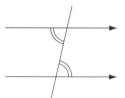

Rule: Corresponding angles (i.e. angles on the same side of the transversal and on the same side of the parallel lines) are equal.

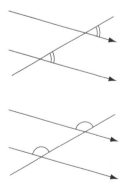   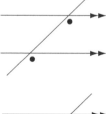

Rule: Cointerior angles (i.e. angles between the parallel lines and on the same side of the transversal) are supplementary.

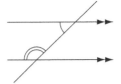

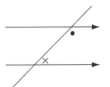

SUMMARY

- Alternate angles–Z-shaped, equal.

- Corresponding angles–F-shaped, equal.

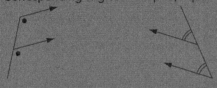

- Cointerior angles–U-shaped, supplementary.

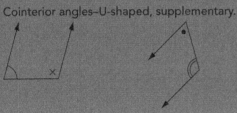

Exercises 6.3

In each case below, evaluate the pronumeral. (Parallel lines are indicated by arrows.)

1

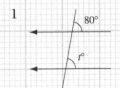

2

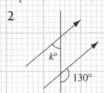

3

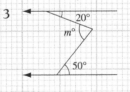

6.6 THE ANGLES OF A TRIANGLE

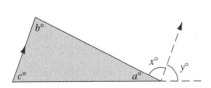

In the diagram:

$$a° + x° + y° = 180°$$ (a straight angle)

Now $x° = b°$ (alternate angles)

and $y° = c°$ (corresponding angles)

so:

$$x° + y° = b° + c°$$

and $a° + b° + c° = 180°$

We have established two very important results:

- An exterior angle of a triangle equals the sum of the two interior opposite angles. (An exterior angle of a triangle is the angle formed by extending one of the sides.)
- The sum of the three angles of a triangle is 180° (2 right angles).

Exercises 6.4

Evaluate the pronumerals in the following:

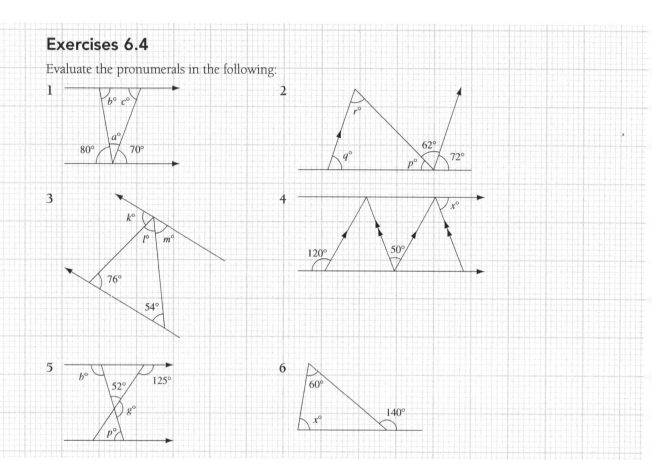

6.7 THE USE OF THE RULE, PROTRACTOR AND SET SQUARE

- **The rule** is used for three purposes:
  (i) to draw a straight line
  (ii) to measure the length of a given interval (The rule should always be held so that the graduation marks are *in contact with* the object being measured. If the rule has appreciable thickness, it would need to be stood on edge.)
  (iii) to draw an interval of a given length. (First draw a *long* straight line; mark a point on the line, and then mark a second point on the line at the required distance from the first point.)

- **The protractor** is used:
  - (i) to measure a given angle (the arms of the angle may need to be extended to reach the graduation marks on the protractor)
  - (ii) to draw a line making a given angle with a given line.
- **The set square**, which is usually a triangular laminar made of plastic, has 'set' angles. These are usually 90°, 30°, 60° or 90°, 45°, 45°. The set square is used for two main purposes:
  - (i) to draw a line parallel to a given line and passing through a given point
  - (ii) to draw a line making one of the set angles with a given line and passing through a given point, in particular to draw a line *perpendicular* to a given line and passing through a given point.

To construct a line parallel to a given line and passing through a given point:

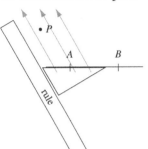

- place one side of the set square against the given line *AB*
- place the straight edge (rule) against one of the other sides of the set square
- holding the rule stationary, slide the set square along the rule until the edge that was along the given line now passes through the given point *P*
- draw the line through *P*, parallel to *AB*.

To construct a line perpendicular to a given line and passing through a given point:

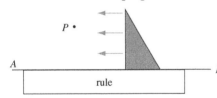

- place the straight edge along the given line *AB*
- slide the set square along the rule until one edge passes through the given point *P*
- draw the required line.

Note: In geometrical language:

$AB \parallel CD$ means 'line *AB* is parallel to line *CD*'.

$AB \perp CD$ means 'line *AB* is perpendicular to line *CD*'.

Example

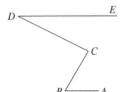

Using only a rule and a set square, construct the figure *ABCDE* (not drawn to scale here), where $AB = 36$ mm, $\angle ABC = 77°$, $BC = 47$ mm, $CD \perp BC$, $CD = 82$ mm, $DE \parallel BA$ and $DE = 95$ mm.

This construction is shown on below.

Note: In the construction below, *as in all constructions:*

■ A rough sketch is drawn first so that the figure can be positioned conveniently on the page. (In this case, the sketch shows that we must place the starting point *A* near the right-hand margin of the page.)

■ All lines are drawn *through* points and are drawn longer than the lengths of the given intervals. This allows more accurate use of the protractor and the set square.

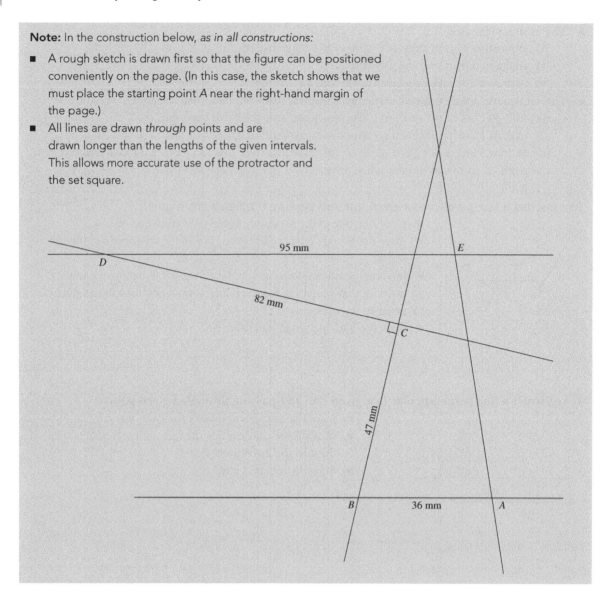

Exercises 6.5

1 Carry out the construction described above and then measure:
   a the length of the interval *AE*   **b** the size of ∠*BAE*.
2 Draw *any* triangle *ABC*. Draw *any* straight line parallel to *BC* and cutting the sides *AB* and *AC* at *P* and *Q*. Measure the intervals *AP*, *PB*, *AQ*, *QC*. Hence evaluate the ratios *AP* : *PB* and *AQ* : *QC*.

   Repeat this construction using a different triangle. State any likely conclusion you can draw from these ratios.

3 Using a rule and set square construct any quadrilateral whose opposite sides are parallel. (This figure is called a *parallelogram*.) Measure the lengths of the four sides and mark them on your figure.

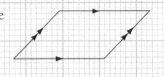

   Repeat this construction drawing a *different* parallelogram.
   State any likely conclusion you can draw from these measurements.

6.8 GEOMETRICAL CONSTRUCTIONS USING A COMPASS

To construct the perpendicular bisector of a given interval AB:

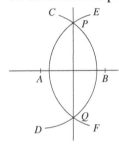

- With centres at *A* and *B*, and keeping the radius constant, draw arcs *CD* and *EF*, intersecting at *P* and *Q*.
- The line *PQ* is the perpendicular bisector of the interval *AB*.

Note: The point where *PQ* cuts *AB* is the midpoint of the interval *AB*.

To construct a line perpendicular to a given line and passing through a given point:

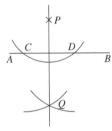

- With the given point *P* as centre and a convenient fixed radius, draw an arc cutting the given line *AB* at points *C* and *D*.
- With *C* and *D* as centres and any convenient fixed radius, draw the arcs shown in the diagram intersecting at *Q*.
- Draw the required line *PQ*.

Note: The same construction is used if *P* is a point *on* the given line.

To bisect a given angle:

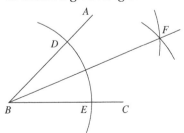

- Draw an arc cutting the arms of the given angle (∠*ABC*) at points *D* and *E*.
- With centres at *D* and *E* and any convenient fixed radius, draw arcs intersecting at *F*.
- Draw the line *BF*, the bisector of ∠*ABC*.

To construct an angle equal in size to a given angle:

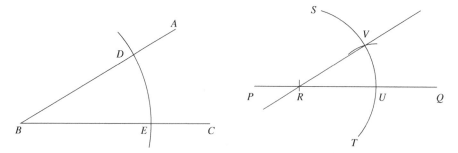

To copy the angle ABC:

- Draw a straight line PQ and mark on it a point R.
- With B as centre, draw an arc cutting the arms of the given angle at D and E. Then using the *same* radius and with R as centre, draw the arc ST, cutting PQ at U.
- Adjust the radius of the compass so that with E as centre, an arc drawn with the compass passes through D. Then with U as centre and the same radius, draw an arc cutting the arc ST at V.
- Draw the line RV. ∠VRQ has been constructed equal to ∠ABC.

Exercises 6.6

All the following constructions should be as large as your paper allows.

1 Draw a line AB and mark on it two points P and Q.
   Draw a line CD through P. At Q construct the angle BQE
   equal to ∠QPD.
   a Check with a rule and set square that QE
      is parallel to PD.
   b Check with a protractor that ∠BQE = ∠QPD.

2 Draw a line AB and on it mark two points P and Q.
      Using a compass and with P and Q as centres, construct
   the triangle PQR such that PQ = QR = RP. Using a compass,
   construct the line PD, perpendicular to QR:
   a measure the angle PRQ
   b measure the angle QPD
   c measure the lengths of PQ and PD and hence determine
      the ratio PQ:PD correct to 3 significant figures.

3 Mark a point P and, with P as centre, draw a circular arc
   with a chord AB. Construct the perpendicular bisector of the
   chord and check that this bisector passes through the centre
   of the arc, P.

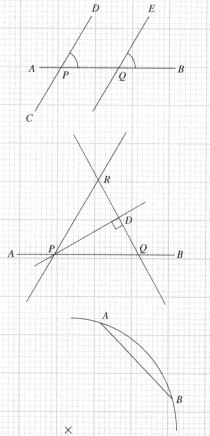

4   Draw a straight line *AB* and on it mark two points *P* and *Q*.
Draw a line *PC*. From *Q* draw the line *QD* such that the angle *AQD*
is constructed with a compass to equal the angle *BPC*.
Bisect each of the three angles of the triangle *PQR* and check that
these three bisectors intersect at a common point inside the triangle.
Label this point, *K*.
Check that the line *RK* is the perpendicular bisector of the
interval *PQ*.

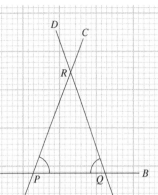

SELF-TEST

1   Write down:

   a   the supplement of $57°28'$

   b   the complement of $43°51'$

   c   the supplement of $(k - 20)°$

   d   the complement of $(180 - t)°$

   e   the supplement of $(2x - 90)°$

2   Evaluate the pronumerals:

   a

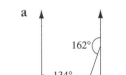

   b

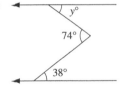

   c

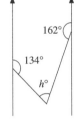

   d

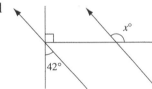

3   Evaluate the pronumerals:

   a   **+++**   **b**

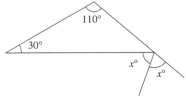

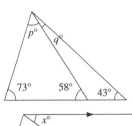

+++   **c**

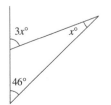

   d

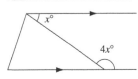

4   Carry out the following construction using only a *rule* and *compass*. Draw an angle *ABC* and then construct another angle *PQR* of the same size. Mark a point *S* such that *QS* bisects ∠*PQR*. Then use a protractor to check that:

    a ∠*PQR* = ∠*ABC*   **b** ∠*PQS* = ∠*RQS*

5   **a** Use the following method to divide a given interval into any number of equal intervals:

    ■ On a straight line, mark off an interval *PQ*.

    ■ Draw a straight line *PR* inclined to the line *PQ* at any convenient angle.

    ■ On *PR*, use a compass to mark off three equal intervals *PA*, *AB* and *BC*.

    ■ Draw the line *CQ*.

    ■ Using a rule and set square, draw lines through *A* and *B*, parallel to *CQ*, intersecting the interval *PQ*.

    ■ Check that the interval *PQ* is now divided into three equal parts.

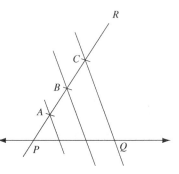

    b Repeat the above construction on an interval of different length, and divide this new interval into a *different* number of equal parts (e.g. 4 or 5).

6   Using only a straight edge, a set square, and a compass, construct a *rhombus* (a quadrilateral whose opposite sides are parallel and whose sides all have the same length). Do not measure the length of any of the sides.

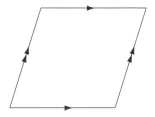

    Using a rule and a *compass*, verify that **a** the bisectors of the angles of a rhombus are the diagonals, and **b** the diagonals of a rhombus bisect each other at right angles.

CHAPTER 7

GEOMETRY OF TRIANGLES AND QUADRILATERALS

Learning objectives

- Identify and classify triangles and quadrilaterals.
- Identify and locate by construction the incentre, circumcentre, centroid and orthocentre of a triangle.
- Identify congruent and similar triangles.
- Use the properties of similar triangles to solve practical problems.
- Identify special quadrilaterals and know some of their important properties.
- Find the area of a triangle when given the lengths of the three sides.
- Find the areas of special quadrilaterals.

7.1 CLASSIFICATION OF TRIANGLES

A triangle with no two sides equal is a **scalene** triangle. A triangle that is not scalene has either two sides equal or all three sides equal. A **vertex** is a point of intersection of two sides.

Isosceles triangles

This triangle has *two* sides equal. It is an **isosceles** triangle.
(Greek 'isos' means equal, as in 'isobars', 'skeles' means 'legs'.)

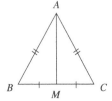

Let ABC be any isosceles triangle with $AB = AC$. Let M be the midpoint of the base BC.
It is clear from symmetry that the following pairs of angles are equal:

$$\angle ABC = \angle ACB$$
$$\angle BAM = \angle CAM$$
$$\angle AMB = \angle AMC \ (= 90°)$$

(These results will be proved later in the chapter.)

Remember these diagrams.

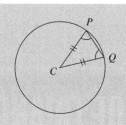

Note: Intervals that are the same length always have the same mark, angles that are equal always have the same mark.

Note: Triangle CPQ (above) is isosceles because $CP = CQ$ (radii of the circle).

Equilateral triangles

A triangle with all **three** sides equal is an **equilateral** triangle.

It is obvious, by symmetry, that all three angles must also be equal, so that each angle equals 60°.

Right-angled triangles

If you have not already studied Chapter 3 of this book, you should now study section 3.8 and Exercises 3.7. An understanding of **Pythagoras' theorem** is assumed during the remainder of this course.

Exercises 7.1

Evaluate the pronumerals:

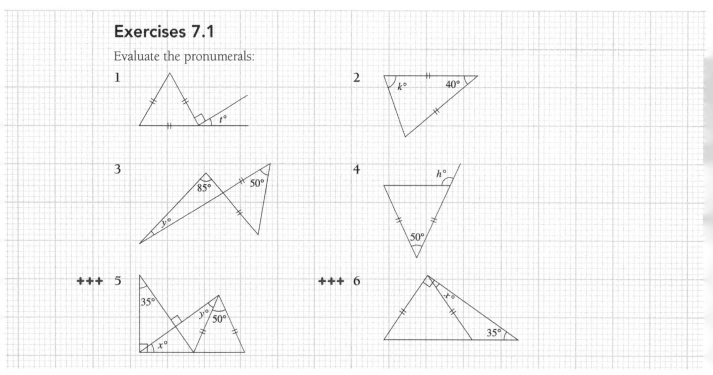

7.2 DETECTING EQUAL ANGLES

It is important that you can quickly discover equal angles in figures. This will be required, for example, when we study similar triangles.

> **SUMMARY**
> Two angles may be equal because they are:
> - vertically opposite angles
> - angles opposite equal sides of a triangle
> - alternate angles (Z-shaped) with parallel lines
> - corresponding angles (F-shaped) with parallel lines
> - complements of the same angle or of equal angles
> - supplements of the same angles or of equal angles
> - the third angles of two triangles, the other two angles being respectively equal.

If the above summary is not understood, you should now revise Chapter 6, sections 6.3–6.6.

Exercises 7.2

For each of the following figures, state which angles must be equal to each other. It is suggested that in each case you make a quick copy of the diagram you are working on.

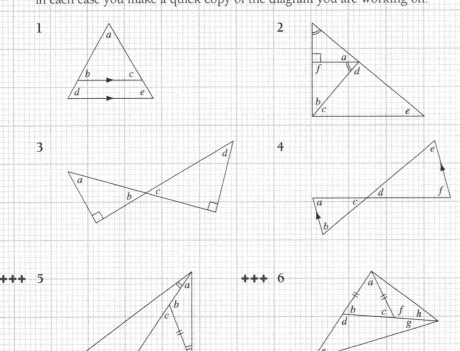

7.3 THE INCENTRE, THE CIRCUMCENTRE, THE CENTROID AND THE ORTHOCENTRE

Three or more straight lines are **concurrent** if they intersect at a common point. There are four important points of concurrency associated with a triangle.

1 **The bisectors of the three angles are concurrent**

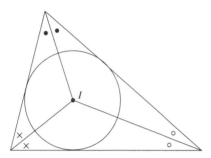 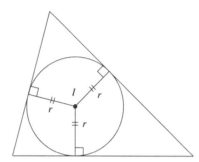

The point of intersection of the three bisectors is called the **incentre,** which is the centre of the **incircle** of the triangle, the circle inside the triangle that touches each of the three sides.

This circle is said to be **inscribed** in the triangle, the triangle being **circumscribed** about this circle. Note that the radius of the incircle is the perpendicular distance from the incentre to a side of the triangle, the incentre being equidistant from the three sides.

To inscribe a circle inside a given triangle, the incentre must first be located by construction.

You should now draw a large triangle, bisect the three angles, locate the incentre and draw the incircle.

2 **The perpendicular bisectors of the three sides are concurrent.**

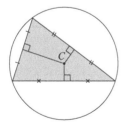 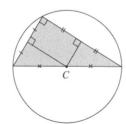 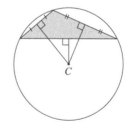

The point of intersection of the perpendicular bisectors of the three sides is called the **circumcentre,** which is the centre of the **circumcircle** of the triangle. The circumcircle is the circle that is circumscribed about the triangle. A particular triangle has only *one* circumcircle, but, of course, there is no limit to the number of different triangles that you could inscribe in a given circle. As can be seen from the three diagrams above:

- if the triangle is acute, the circumcentre lies inside the triangle
- if the triangle is right-angled, the circumcentre lies on the hypotenuse
- if the triangle is obtuse, the circumcentre lies outside the triangle.

The radius of the circumcircle is the distance from the circumcentre to a vertex of the triangle. Note that the circumcentre is equidistant from the three vertices.

You should now draw a large triangle, construct the perpendicular bisectors of the three sides, locate the circumcentre and draw the circumcircle.

3   The three medians of a triangles are concurrent.

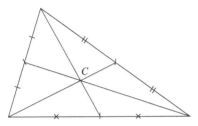

A **median** of a triangle is an interval joining a vertex to the midpoint of the opposite side. The point of intersection of the three medians is called the **centroid** of the triangle.

For a triangular sheet of material of uniform thickness and density, the centroid of one of its triangular faces gives the **centre of gravity** of the object, this being the point midway between the centroids of the two triangular faces.

Note: The centre of gravity of *three equal spherical masses* is the centroid of the triangle formed by joining their centres.

You should now cut a large triangular piece of cardboard, plywood or sheet metal, draw the three medians on one face of the object and locate the centroid. By placing a pin or nail through a small hole at the centroid and spinning the object about this as a horizontal axis, you should check that you have located the centre of gravity of the object.

4   The three altitudes of a triangle are concurrent.

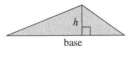 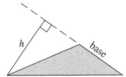 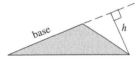

An **altitude** of a triangle is an interval drawn from a vertex perpendicular to the opposite side, produced if necessary. With respect to this altitude, the opposite side is called the *base* of the triangle. In each of the three diagrams above the altitude is labelled *h*.

The point of intersection of the three altitudes of a triangle is called the **orthocentre** of the triangle.

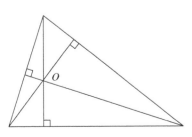

You should now draw a large triangle, construct the three altitudes and locate the position of the orthocentre. You should be able to verify that:
- for an acute triangle, the orthocentre lies inside the triangle
- for a right-angled triangle, the orthocentre lies at the right angle vertex
- for an obtuse triangle, the orthocentre lies outside the triangle.

Construction of triangles and quadrilaterals

To construct a given triangle or quadrilateral, all the techniques learnt in the previous chapter concerning the uses of the rule, set square, compass and protractor are available. *To construct two intervals of equal length, it is usually more accurate to use a compass to span the intervals rather than to measure the lengths of the intervals using a rule.*

Exercises 7.3

1 Practise drawing a triangle and then locating its incentre, circumcentre, centroid or orthocentre. When you locate an incentre or a circumcentre, check by drawing the incircle or circumcircle.

+++ 2 In the triangle *ABC*, *AB* = *CA* = 156 mm and *BC* = 294 mm. By construction, locate the incentre and the orthocentre of this triangle and measure the distance between them. (*Hint:* a preliminary rough sketch will show that in order to draw as large a figure as possible, it is only necessary to construct *half* of the triangle *ABC*, i.e. the triangle *ABM* where *M* is the midpoint of side *BC*.)

3 Using only a straight edge and a compass, construct a large triangle whose angles are 90°, 45° and 45°. Check that the perpendicular bisector of the hypotenuse passes through the opposite vertex.

4 Draw a large triangle and, by constructing the three medians, locate the centroid. Check that the centroid is a point of trisection of each of the medians.

7.4 CONGRUENT TRIANGLES

Two triangles are said to be **congruent** if they have exactly the *same size* and *shape,* as are the triangles below. Each angle in one triangle has an equal angle in the other, and each side in one has an equal side in the other. A tracing of one would fit exactly over the other.

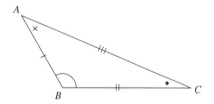
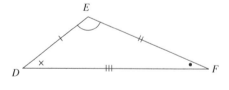

Note: The sides that 'correspond' in the two triangles are opposite the angles that correspond. For example, ∠*B* = ∠*E* and the sides that are opposite these equal angles are equal, i.e. *AC* = *DF*.

Proving that two triangles are congruent

It is not necessary to prove the six equalities (the three angles equal and the corresponding three sides equal) to show that two triangles are congruent. It can be shown that it is sufficient to prove one of the following (and then it follows that the other three equalities are also true):

- three sides equal (the SSS test)
- two angles equal and a corresponding side equal (the AAS test)
- two sides equal and the angles contained by these sides equal (the SAS test)
- a right angle in each, the hypotenuses equal and one other side equal (the RHS test).

Example

Given any isosceles triangle *ABC* in which *AB* = *AC*, join *A* to *M*, the midpoint of *BC*.

In △*ABM* and △*ACM*:

$$\begin{cases} AB = AC \text{ (given)} \\ MB = MC \text{ (since } M \text{ is the midpoint of } BC) \\ AM = AM \text{ (common side)} \end{cases}$$

∴ △*ABM* is congruent to △*ACM* (SSS test)

(This is written △*ABM* ≡ △*ACM*.)

Since the triangles *ABM* and *ACM* are congruent (i.e. identical), one triangle would fit exactly on the other. Therefore:

∠*BAM* = ∠*CAM*

∠*ABC* = ∠*ACB*

∠*AMB* = ∠*AMC*   (and hence each = 90°)

Naming the angles and sides of a triangle

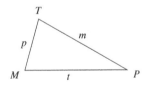

In this triangle, the three angles are called ∠*T*, ∠*M* and ∠*P*. The three sides *opposite* these angles are called *t*, *m* and *p* respectively.

Example

In triangle *BWN*,

∠*B* = 32°

∠*N* = 19°

w = 7 m

b = 5 m

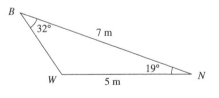

Exercises 7.4

1   For each pair of triangles below, state whether the triangles must be congruent (answer yes or no). If they must be congruent, state the test used to determine this congruency (use abbreviations, e.g. SAS). (Equal sides and angles are marked in the same way.)

continued

continued

a 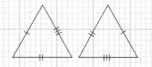   b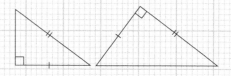

c 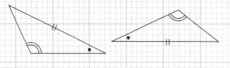   d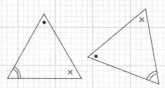

e 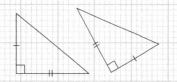   f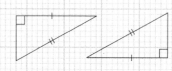

g 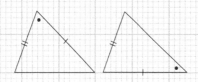   h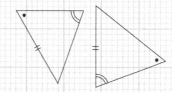

2   In each case below, sketch the two triangles and label the given equal sides and angles. Then state whether the two triangles must be congruent and if so state the test that proves the congruency:

   a $\triangle ABC$ and $\triangle DEF$, given $a = e$, $b = f$, $c = d$

   b $\triangle PQR$ and $\triangle GHT$, given $P = H$, $Q = G$, $R = T$

   c $\triangle MLT$ and $\triangle ACF$, given $M = C$, $t = f$, $m = c$

   d $\triangle GHK$ and $\triangle AFN$, given $G = A$, $K = N$, $g = a$

 # 7.5 SIMILAR TRIANGLES

If we enlarge a photograph, we can still recognise the people in the picture. This is because all *shapes* have remained the same, even though the *lengths* have changed.

- All angles have remained unchanged.
- All lengths have increased proportionately (i.e. in the same ratio).

> **Rule:** Two triangles are called **similar** if one is an enlargement of the other, that is:
>
> - if they have the same sized angles
>
> - if the lengths of the corresponding sides (i.e. sides opposite the equal angles) are in the same ratio.

It can be proved that if either of these conditions is satisfied, the other is also, so we can recognise similar triangles by *either* of these properties.

Examples

1 The triangles below are similar because they are equiangular. Hence the corresponding sides must be in the same ratio.

Note: In similar triangles, 'corresponding' sides are those opposite equal angles.

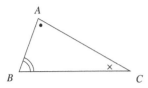 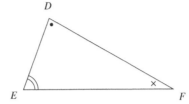

If $DE = 1.8 \times AB$, then $EF = 1.8 \times BC$ and $FD = 1.8 \times CA$, i.e.

$$\frac{DE}{AB} = \frac{EF}{BC} = \frac{FD}{CA} = (1.8)$$

The **enlargement** (or **magnification**) **factor** is 1.8.

2

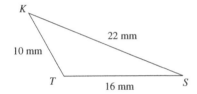 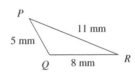

These triangles are similar, since:

$$\frac{KT}{PQ} = \frac{TS}{QR} = \frac{SK}{RP} = (2)$$

Hence the corresponding angles are equal, i.e.

$$\angle P = \angle K, \angle Q = \angle T, \angle R = \angle S$$

When finding the lengths of unknown sides in pairs of similar triangles, it is suggested that you set up equal ratios of corresponding sides (i.e. **magnification factors**).

Example

The triangles below are similar since they are equiangular.

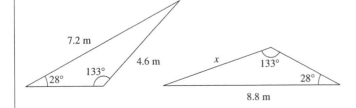

$$\therefore \frac{x}{4.6} = \frac{8.8}{7.2}$$

$$\therefore x = \frac{4.6 \times 8.8}{7.2}$$

$$\approx 5.6 \text{ m}$$

Exercises 7.5

Evaluate each pronumeral in questions 1–14. All interval lengths are in metres. It is recommended that, in each case, a proportion be first set up and then solved. Angles and sides marked in the same way are equal. Sides with arrows are parallel.

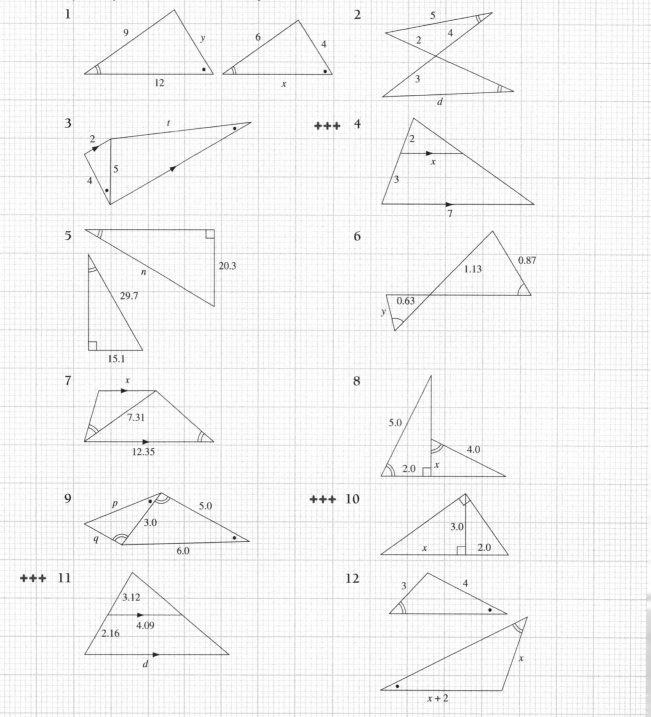

+++ 13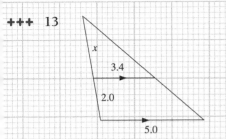

+++ 14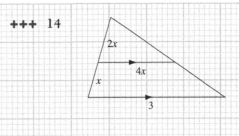

+++ 15 A boy 1.50 m tall is standing 20.0 m from a point directly below a street lamp, which is 5.00 m above the ground. Find the length of his shadow.

+++ 16 In the triangle ABC: $\angle A = 90°$, $AB = 4.0$ m and $CA = 3.0$ m. D is the point on BC such that $AD \perp BC$. Mark three pairs of equal angles in the figure and using similar triangles find the length of AD.

7.6 SPECIAL QUADRILATERALS AND THEIR PROPERTIES

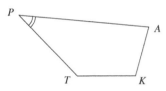

A **quadrilateral** is a plane figure bounded by four straight lines. This diagram shows the quadrilateral *PAKT*. The vertices (the corners) must be named in either clockwise or anticlockwise order. We could, for example, call this quadrilateral *APTK* but not *PATK*.

The intervals *PK* and *AT* are the two **diagonals.** The angle marked in the figure may be called 'the angle *P*' ($\angle P$), 'the angle *APT*' ($\angle APT$) or 'the angle *TPA*' ($\angle TPA$). **Opposite angles** are at the two ends of a diagonal (e.g. angles *A* and *T*). The **perimeter** is the total length of the boundary (i.e. the sum of the lengths of the four sides). By drawing a diagonal, any quadrilateral can be divided into two triangles. The sum of the angles in each triangle is 180°, hence the sum of the angles in any quadrilateral is 360°.

Special quadrilaterals

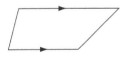

A **trapezium** is a quadrilateral that has two of its sides parallel.

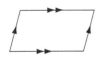

A **parallelogram** is a quadrilateral that has *both* pairs of opposite sides parallel (hence the name *paralle*logram).

Note the following properties of the parallelogram:
- opposite sides parallel
- opposite sides equal
- opposite angles equal
- diagonals bisect each other.

If any *one* of these is true, it can be shown that they are *all* true and hence the figure is a parallelogram. It is also sufficient to show that any one pair of opposite sides is both equal and parallel.

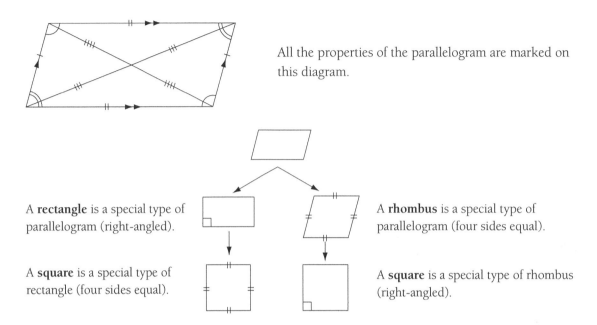

All the properties of the parallelogram are marked on this diagram.

A **rectangle** is a special type of parallelogram (right-angled).

A **rhombus** is a special type of parallelogram (four sides equal).

A **square** is a special type of rectangle (four sides equal).

A **square** is a special type of rhombus (right-angled).

Since the rectangle, rhombus and square are all special parallelograms, they all have the four properties of the parallelogram (opposite sides parallel, opposite sides equal, opposite angles equal, diagonals bisect each other).

Note the following properties carefully:

	Parallelogram	**Rectangle**	**Rhombus**	**Square**
Sides	Opposite sides equal	Opposite sides equal	Four sides equal	Four sides equal
Diagonals	Diagonals bisect each other but are *not* equal in length	Diagonals bisect each other and *are* equal in length	Diagonals bisect each other *at right angles,* but are *not* equal in length	Diagonals bisect each other *at right angles* and *are* equal in length

Exercises 7.6

1 Draw a parallelogram and draw one of its diagonals. Use the definition of a parallelogram to mark two pairs of equal alternate angles.

   a Which test is used to prove that the two triangles in your figure are congruent?

   b What deductions can you make about the properties of a parallelogram?

2 Draw a quadrilateral that has its opposite sides equal. Mark the equal opposite sides and draw one of the diagonals.

   a Which test is used to prove that the two triangles in your figure are congruent?

   b Which test is used to prove that the quadrilateral is a parallelogram?

3 A quadrilateral $ABCD$ has its opposite angles equal, $\angle A = \angle C$ and $\angle B = \angle D$.

   a What can you deduce about the relationship between the pairs of angles A and D, B and C, A and B, C and D?

   b What can you deduce about the quadrilateral? State your reasoning.

4 In each case below, if a quadrilateral has the property given, state what (if any) special type of quadrilateral it *must* be:

   a Both pairs of opposite sides are equal.

   b Its diagonals bisect at right angles.

   c Its diagonals are equal in length.

   d Two opposite sides are parallel.

   e Two opposite sides are equal in length.

   f Its diagonals are equal in length and bisect each other at right angles.

   g Its opposite angles are equal.

   h Its four sides are equal in length.

   i Its diagonals bisect each other and are equal in length.

   j Two opposite sides are parallel and equal in length.

   k Its four angles are equal.

   l Its opposite sides are parallel.

5 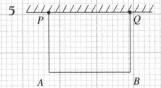 A beam AB hangs horizontally from two equal vertical ropes PA and QB.

   a If the beam is set swinging in any direction, will it always remain horizontal? If so, how do we know this?

   b If each rope has a length 26 m and the beam is displaced 10 m to the right, through what vertical distance will the beam rise?

   c If Q were much higher than P so that the rope lengths were $PA = 26$ m and QB very large (hundreds of metres), what could you say about the vertical distance through which B would rise when A is displaced 10 m to the right?

continued

continued

>   **d** If the two ropes are not equal in length, would the beam remain horizontal during the
>   above displacement?
>
> **6** *PQRS* is a quadrilateral in which sides *QR* and *SP* are each 2 m long and are parallel. The diagonal
>   *QS* has length 4 m. $\angle PSQ = 40°$.
>   **a** What type of quadrilateral is this?
>   **b** Do the diagonals bisect each other?
>   **c** Do the diagonals intersect at right angles?
>   **d** Find the size of $\angle RPS$.

7.7 HERON'S FORMULA

If you have not studied Chapter 3 of this book, you should first study sections 3.10–3.12 before
proceeding further.

Area of a triangle given the lengths of the three sides

When we are given the lengths of the three sides of a triangle, we use Heron's formula to find the area:

$$\text{area} = \sqrt{s(s - a)(s - b)(s - c)}$$

where $\begin{cases} a, b, c \text{ are the lengths of the sides} \\ s = \dfrac{a + b + c}{2}, \text{ the semiperimeter} \end{cases}$

Example

$s = \dfrac{5 + 6 + 9}{2} = 10 \text{ m}$

$\therefore \text{area} = \sqrt{10(5)(4)(1)}$

$= \sqrt{200}$

$\approx 14.1 \text{ m}^2$

6.00 m   9.00 m

5.00 m

General method: As usual, let your calculator do the work. Heron's formula provides practice in
the use of the bracket keys and the memory keys.

For a triangle whose sides have lengths *a, b,* and *c* units, using M for 'put into memory' and R for
'recall the contents of the memory', use the keys

$a + b + c = \div 2 = \text{M}$

then $\text{R} \times (\text{R} - a) \times (\text{R} - b) \times (\text{R} - c) = \sqrt{} =$

Try this method when $a = 3, b = 4$ and $c = 5$. The correct result is 6.

Try this method when $a = 3.17, b = 4.32$ and $c = 5.25$. The correct result is 6.84.

Exercises 7.7

1 In each case below, the lengths of the three sides of a triangle are given. Find the areas of these triangles:

   a 4.00 m, 5.00 m, 7.00 m   **b** 4.00 m, 7.00 m, 9.00 m

   c 497 mm, 536 mm, 824 mm   **d** 43.2 mm, 64.8 mm, 83.5 mm

2 Find the area of a parallelogram whose adjacent sides measure 13.0 cm and 14.0 cm and whose shorter diagonal measures 15.0 cm.

3 Find the area of the quadrilateral $PQRS$ given that $PQ = 3.00$ m, $QR = 5.00$ m, $RS = 8.00$ m, $SP = 4.00$ m and $PR = 6.00$ m.

7.8 AREAS OF THE PARALLELOGRAM AND THE RHOMBUS

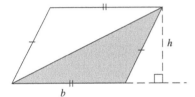

$$\begin{aligned}\text{area of parallelogram} &= 2 \times \text{area of triangle} \\ &= 2 \times (\tfrac{1}{2}bh) \\ &= b \times h\end{aligned}$$

Rule: area of parallelogram = length of one of two parallel sides × perpendicular distance between them

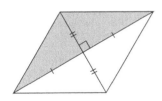

In addition to its properties as a parallelogram, a rhombus also has the property that its diagonals bisect at *right angles*.

If the lengths of the diagonals are D and d units,

$$\begin{aligned}\text{area of rhombus} &= 2 \times \text{area of the shaded triangle} \\ &= 2 \times (\tfrac{1}{2}bh) \\ &= 2 \times \tfrac{1}{2}D(\tfrac{1}{2}d) \\ &= \tfrac{1}{2}Dd\end{aligned}$$

Rule: area of rhombus = half the product of its diagonals

(This formula can, of course, be used for a square, since a square is a special type of rhombus.)

Examples

1 ABCD is a parallelogram in which $AB = 15$ m, $BC = 5$ m and the distance between AB and CD is 4 m. Find the lengths of the diagonals correct to 3 significant figures.

It can be deduced that:

$k = 3$ (Pythagoras)

$t = 3$ (symmetry)

$\therefore n = 12$

$x = 4$ (data)

$AC^2 = 18^2 + 4^2$ (Pythagoras)

$\therefore AC \approx 18.4$ m

$BD^2 = 12^2 + 4^2$ (Pythagoras)

$\therefore BD \approx 12.6$ m

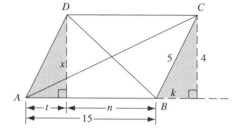

2 A rhombus has diagonals of lengths 172 mm and 246 mm. Find the lengths of the sides.

The diagonals bisect at right angles. Hence we have a right-angled triangle with sides of lengths 86 mm and 123 mm.

$L = \sqrt{123^2 + 86^2}$

≈ 150 mm

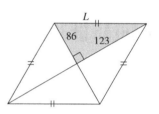

Exercises 7.8

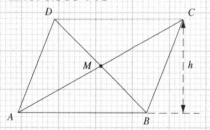

In each of the following exercises, M is the point of intersection of the diagonals and h is the distance between the sides AB and CD.

All answers are to be given correct to 3 significant figures.

Figure	Data	Find
1 rhombus *ABCD*	$BC = 6.820$ m $h = 4.230$ m	area
2 parallelogram *ABCD*	$BD = 13.00$ m $AC = 15.00$ m $h = 12.00$ m	**a** *AB* **b** *BC* **c** area
3 square *ABCD*	$AM = 5.000$ m	**a** *BD* **b** *AB* **c** area
4 rhombus *ABCD*	$BC = 13.00$ m $h = 12.00$ m	**a** *AC* **b** *BD* **c** area
5 parallelogram *ABCD*	$AB = 140$ mm $BC = 100$ mm area $= 11\ 200$ mm^2	**a** *h* **b** *AC* **c** *BD*
6 rhombus *ABCD*	$AB = 8.620$ m area $= 55.60$ m^2	**a** *h* **b** *AC* **c** *BD*
7 rhombus *ABCD*	$AB = 2.000$ m $\angle DAB = 60°$	**a** *h* **b** area **c** *AC*

8 In the triangle *ABC*: $AB = CA = 13$ m and $BC = 10$ m Find:

   a the distance from *A* to the midpoint of *BC*,

   b the area of the triangle using the formula $A = \dfrac{1}{2}bh$,

   c the area of the triangle using the three-sides (Heron's) formula.

9 A rhombus has sides of length 5 m and its longer diagonal has a length of 8 m. Find:

   a the area of the rhombus,

   b the length of the shorter diagonal.

10 In a quadrilateral *ABCD*: $AB \parallel CD$, $AB = 20.00$ cm, $BC = 15.00$ cm, $CD = 34.00$ cm, $DA = 13.00$ cm, $BD = 27.73$ cm.

   a What type of quadrilateral is this?

   b Find the area of the quadrilateral correct to 3 significant figures.

SELF-TEST

1   Evaluate the pronumerals

a 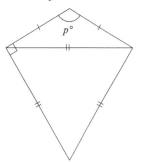   b 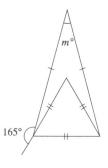   c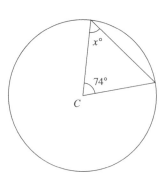

+++   2   For the following figures, state which angles must be equal to each other.

a 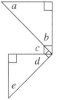   b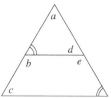

c 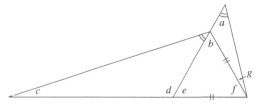   d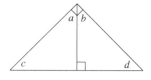

3   For each pair of triangles below, state whether they must be congruent and if so which congruency test applies:

a 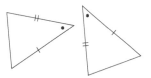   b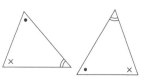

c    d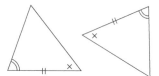

e   *PTK* and *MBW*, given $\angle P = \angle B$, $k = w$, $t = b$

f   *ABC* and *DEF*, given $\angle A = \angle F = 90°$, $b = e$, $c = d$

4   *KSTV* is a parallelogram in which $\angle KST = 90°$, $ST = 6$ m and $SV = 10$ m.

   a   What type of quadrilateral is this?

   b   Do the diagonals bisect each other?

   c   Do the diagonals intersect at right angles?

   d   Find the length of the side *KS*.

   e   Find the perimeter of this quadrilateral.

5   *AB* is a vertical windscreen-wiper blade, being one side of a rigid
    rectangular frame *ABCD*. Arms *PD* and *QA* are hinged at *D* and *A*.
    PQ is horizontal and equals *AD*.

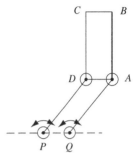

   a   If the arms oscillate as shown by the arrows, will the blade always
       remain vertical? If so, how do we know this?

   b   If each arm is 300 mm long and oscillates through an angle of 90°
       (45° to each side of the vertical), through what horizontal distance will the blade move?

   c   Through what vertical distance will the blade rise as the arms oscillate through 90°?

+++ 6   In the parallelogram *ABCD*, $AB = 8.000$ m, $BC = 5.000$ m and $h = 4.000$ m. Find:

   a   the area of the parallelogram

   b   the length of the diagonal *AC*

   c   the length of the diagonal *BD*

+++ 7   In the rhombus *ABCD*, $BD = 862$ mm and $AB = 537$ mm. Find:

   a   the length of the diagonal *AC*

   b   the area of the rhombus

8   Using Heron's formula, find the area of a triangle whose sides have lengths:

   a   4.00 m, 6.00 m and 8.00 m.

   b   23.4 mm, 38.7 mm and 59.1 mm.

CHAPTER 8

GEOMETRY OF THE CIRCLE

Learning objectives

- Name the various terms associated with a circle.
- Solve problems that involve:
  - angles associated with a circle and the relationships between them
  - chords of a circle
  - tangents to a circle
  - intersecting circles.
- Apply your knowledge about all the above relationships to practical situations.

8.1 THE CIRCLE: SOME TERMS USED

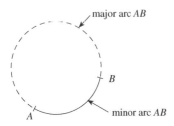

AB is an **arc** of the circle, i.e. a portion of the circumference. The solid curve is the **minor arc** *AB*; the broken curve is the **major arc** *AB*.

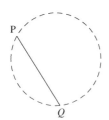

The interval *PQ* is a **chord** of the circle, i.e. the join between two points on the circumference. If a chord passes through the centre of the circle, it is called a **diameter**.

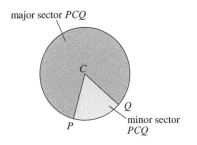

PCQ is called a **sector** of the circle, i.e. an area enclosed between two radii and an arc.

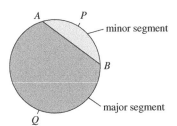

APB and AQB are **segments** of the circle, i.e. areas cut off by a chord.

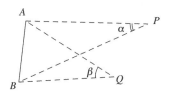

Interval AB **subtends** the angle α at point P, and it subtends the angle β at point Q.

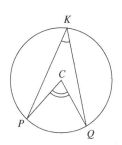

Angle PCQ is the angle at the centre 'standing on' (or 'subtended by') the minor arc PQ. Angle PKQ is the angle at K on the circumference 'standing on' (or 'subtended by') the minor arc PQ.

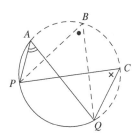

The angles marked at points A, B and C are angles at the circumference standing on the same arc PQ, and in the same segment (the major segment cut off by chord PQ).

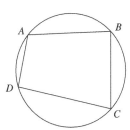
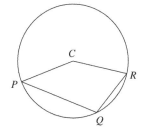

A quadrilateral whose vertices all lie on the circumference of a circle is called a **cyclic quadrilateral.** ABCD is a cyclic quadrilateral. PQRC is *not* a cyclic quadrilateral (since C does *not* lie on the circle).

Exercises 8.1

1   What is the length of a chord that subtends an angle of 90° at the centre of a circle whose diameter is 24.2 m?

2   What is the angle subtended at the centre of a circle by a chord whose length equals the radius of the circle?

8.2 THE CIRCLE: PROPORTIONAL RELATIONSHIPS

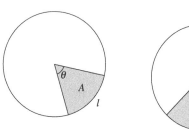 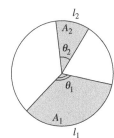

If a sector of a circle contains angle θ and arc length l and has area A, then θ, l and A are all proportional to each other, i.e. if one of these is increased or decreased by a certain ratio, the other two are also increased or decreased in the same ratio.

$$\frac{\theta_1}{\theta_2} = \frac{l_1}{l_2} = \frac{A_1}{A_2}$$

Example

If θ_1 is 2.73 times θ_2, then l_1 is 2.73 times l_2 and A_1 is 2.73 times A_2.

Exercises 8.2

1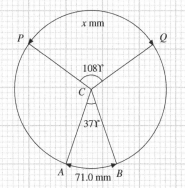

Arc lengths x mm and 71.0 mm subtend angles at the centre of 108° and 37.0° respectively.

a   Evaluate x.

b   If the area of sector CPQ is 236 mm^2, find the area of sector CAB.

2   Given that the areas of sectors CTK and CKM are, respectively, 1.34 m^2 and 3.82 m^2, and that the length of arc KM is 0.675 m, find:

a   the measure of $\angle KCM$

b   the arc length TK

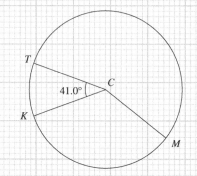

8.3 ANGLES ASSOCIATED WITH THE CIRCLE

Questions 1–6 below should be worked through in sequence; this establishes four important results.

Exercises 8.3

1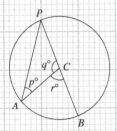

    a If $\angle APB = 40°$, evaluate p, q and r.

    b If $\angle APB = 25°$, evaluate p, q and r.

    c If $\angle APB = x°$, find p, q and r in terms of x.

2 **a** If $\angle APC = 20°$, and $\angle BPC = 30°$, find the measure of:

      i $\angle ACD$ **ii** $\angle BCD$ **iii** $\angle ACB$

  b If $\angle APC = x°$ and $\angle BPC = y°$, find the measure of:

      i $\angle ACD$ **ii** $\angle BCD$ **iii** $\angle ACB$

  c State the general result you have established.

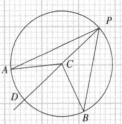

3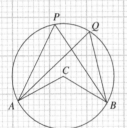

    a If $\angle ACB = 100°$, what is the measure of:

      i $\angle APB$?   **ii** $\angle AQB$?

    b If $\angle ACB = x°$, what is the measure (in terms of x) of:

      i $\angle APB$?   **ii** $\angle AQB$?

    c State the general result you have established.

4 **i** 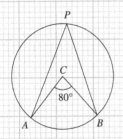   **ii**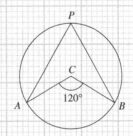

    iii 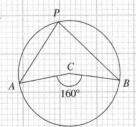   **iv**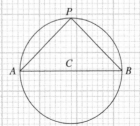

  a In each case above, write down the measure of $\angle APB$.

  b State the general result you have established concerning a semicircle.

continued

continued

5 a

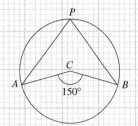

b

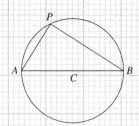

c

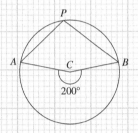

d

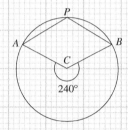

In each case above, write down the measure of ∠APB.

6

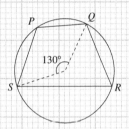

a Write down the measures of:

   i ∠R

   ii ∠P

   iii ∠P + ∠R

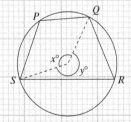

b Write down the measures of:

   i ∠P, in terms of y

   ii ∠R, in terms of x

   iii ∠P + ∠R

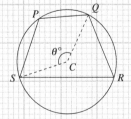

c Write down the measures of:

   i ∠R in terms of θ

   ii reflex ∠SCQ in terms of θ

   iii ∠P in terms of θ

   iv ∠P + ∠R

d State the general result you have established.

Note: These four important results should be memorised:

- The angle at the centre is double an angle at the circumference standing on the same arc.

   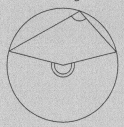

- Angles in the same segment are equal (i.e. angles at the circumference standing on the same arc are equal).

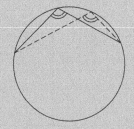   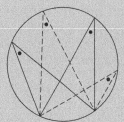

- An angle in a semicircle is a right angle.

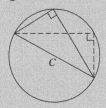

- The opposite angles in a cyclic quadrilateral are supplementary, for example:

  $a + b = 180°$

  $c + d = 180°$

  $x + y = 180°$

  $t + n = 180°$

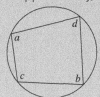

Exercises 8.3 *continued*

Evaluate the pronumerals in the figures below. Every point labelled C is the centre of the circle.

7

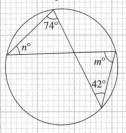

8

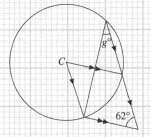

9

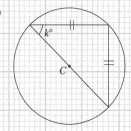

continued

continued

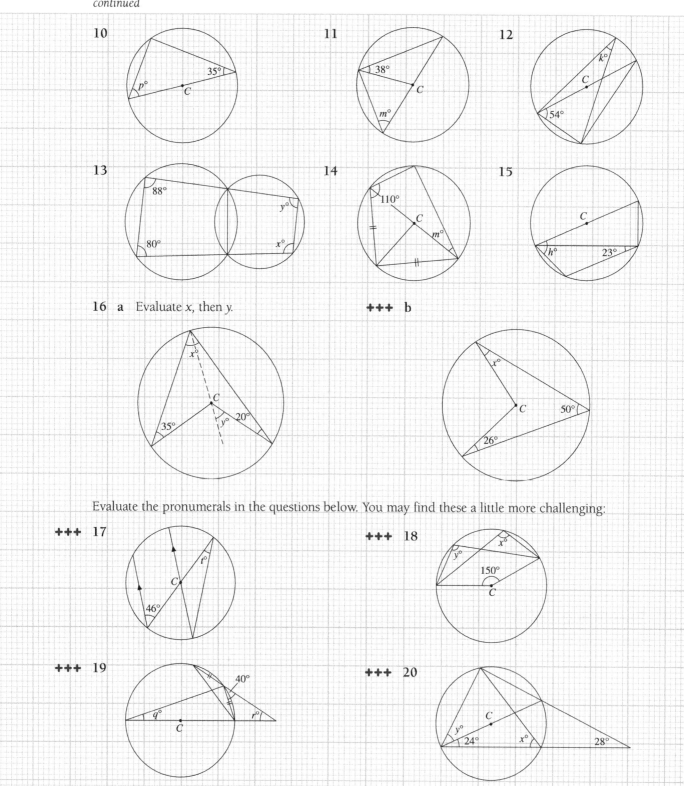

10

35°

p°

C

11

38°

C

m°

12

k°

C

54°

13

88°

80°

y°

x°

14

110°

C

m°

15

C

h°   23°

16 a   Evaluate *x*, then *y*.

x°

C

35°   *y*°   20°

+++ b

x°

C   50°

26°

Evaluate the pronumerals in the questions below. You may find these a little more challenging:

+++ 17

C

r°

46°

+++ 18

x°

y°

150°

C

+++ 19

40°

q°   *r*°

C

+++ 20

C

y°   24°   *x*°   28°

If you managed those, try the following:

+++ 21

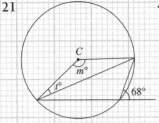

+++ 22

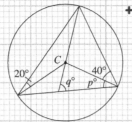

+++ 23

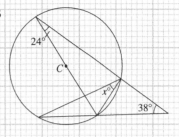

8.4 CHORDS OF A CIRCLE

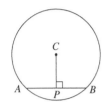

- The perpendicular from the centre of a circle to a chord bisects the chord.
- The line joining the centre of a circle to the midpoint of a chord is perpendicular to the chord.

These results are proved by showing triangles ACP and BCP to be congruent.

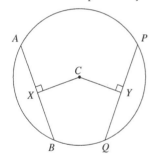

- Equal chords of a circle are equidistant from the centre.
- If two chords are equidistant from the centre, they are equal chords.

Thus, if $AB = PQ$, then $CX = CY$,
and if $CX = CY$, then $AB = PQ$.

These results are proved by showing that triangles ACX and PCY are congruent.

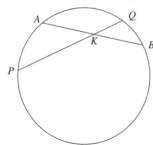

- If two chords of a circle intersect, the products of their segments are equal. If chords AB and PQ intersect at K:
  $AK \times KB = PK \times KQ$

This result is proved by showing that triangles APK and QBK are similar (see Exercises 8.3, question 7). Hence:

$$\frac{AK}{QK} = \frac{PK}{BK}$$

$$\therefore AK \times BK = QK \times PK$$

Exercises 8.4

1

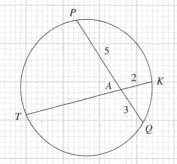

Given that $AQ = 3.0$ m, $AP = 5.0$ m and $AK = 2.0$ m, find the length of AT.

2 Find the lengths of the two chords.
(All measurements are in metres.)

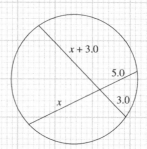

8.5 TANGENTS TO CIRCLES

Line AT touches the circle at only one point. It is called a **tangent** to the circle from point A. **The tangent to a circle is perpendicular to the radius drawn to the point of contact.** (This result, which we will not prove, should be remembered.)

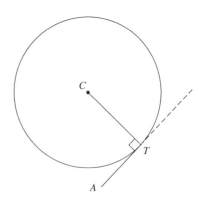

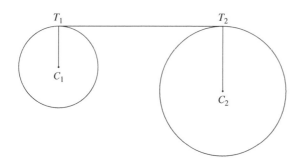

T_1T_2 is a tangent to *both* circles, and is called a **direct common tangent.**

Note: $C_1T_1 \parallel C_2T_2$ (since $\angle C_1T_1T_2$ and $\angle C_2T_2T_1$ are each 90°).

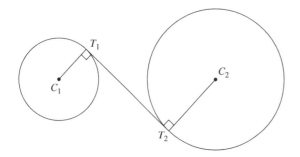

T_1T_2 is called an **indirect common tangent.**

Again, $C_1T_1 \parallel C_2T_2$

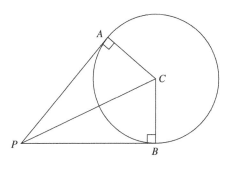

PA and PB are tangents to a circle whose centre is C.
Triangles PCA and PCB are congruent (RHS test).

$\therefore PA = PB$

That is, the two tangents to a circle from an external point
are equal in length.

Hence also $\angle APC = \angle BPC$, i.e. PC bisects $\angle APB$.

Exercises 8.5

In questions 1–4, AT, AP and AQ are tangents

1 Find the measure of:

  a $\angle CAT$

  b $\angle ACT$

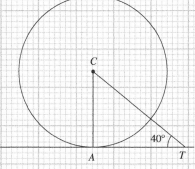

2

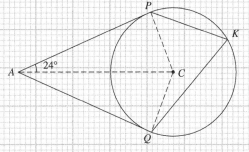

3

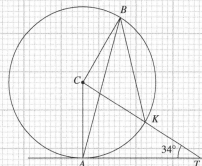

Find the measure of:

  a $\angle CAQ$

  b $\angle PCQ$

  c $\angle PKQ$

Find the measure of $\angle ABK$.

4 Find the measure of:

  a $\angle PCQ$

  b $\angle PBQ$

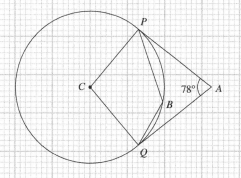

continued

continued

5   A circle has a diameter of 24 m and a tangent is drawn from point *P*, 16 m from its centre, to contact the circle at *T*. Find the length of interval *PT* correct to 4 significant figures.

6   Calculate the length of the interval *PQ* on the common tangent drawn to two circles of radii 7 m and 2 m, given that their centres are 13 m apart.

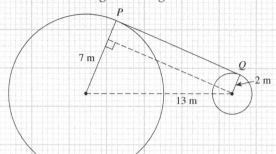

+++ 7

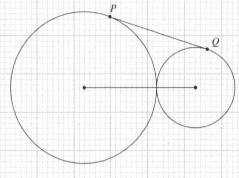

Two circles having radii of 9 m and 4 m touch externally. Find the length of the interval *PQ* on the common tangent.

+++ 8

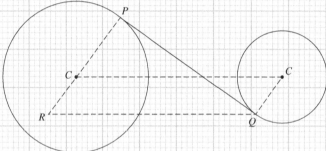

Given that the radii of these circles are 3 m and 2 m, and that their centres are 7 m apart, find the length of the indirect common tangent *PQ*. State the length to 3 significant figures.
(*Hint:* Draw broken lines as indicated and consider triangle *PQR*.)

8.6 THE ANGLE IN THE ALTERNATE SEGMENT

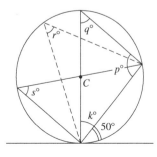

Write down the values of *k* and *p*, and hence the values of *q*, *r* and *s*.

If $\angle PTA = x°$, what can you deduce about the size of $\angle TBA$? Hence, what can you deduce about the size of $\angle TMA$?

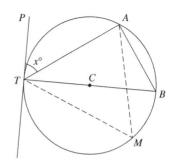

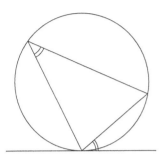

Note: The angle between a tangent and a chord drawn to the point of contact of the circle and the tangent is equal to the angle subtended by that chord in the other segment of the circle. That is, the angle between a tangent and a chord equals the angle in the alternate segment.

Exercises 8.6

In each case below, find the value of the pronumeral or pronumerals. In each case, C is the centre of the circle.

1

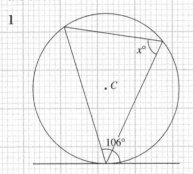

+++ 2

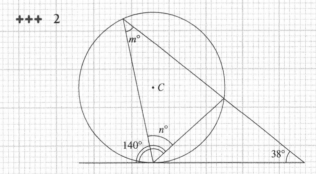

8.7 INTERSECTING CIRCLES

Consider two circles with centres C_1, C_2 and radii R_1, R_2.

    If two circles 'touch' at point P:

- the distance between the centres $= R_1 + R_2$
- points C_1, C_2 and P are collinear
- the common tangent is perpendicular to the line of centres, $TT' \perp C_1C_2$.

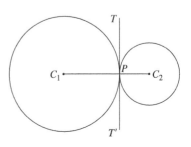

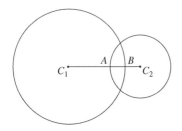

If the smaller circle is moved distance x to the left:

- the distance between the centres $= R_1 + R_2 - x$
- the 'overlap' $(AB) = x$.

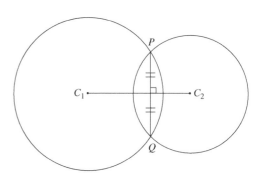

You should be able to prove that, when two circles intersect, cutting at P and Q:

- the common chord is perpendicular to the line of centres
- the common chord is bisected by the line of centres.

Exercises 8.7

Two circles having radii 5.00 m and 3.00 m intersect so that the length of the common chord is 2.00 m.

+++ 1 Find the distance between the centres.

+++ 2 Find the overlap.

8.8 PRACTICAL SITUATIONS

Exercises 8.8

In these exercises the data is correct to 3 significant figures.

+++ 1 A flat section, 26 mm wide, is to be machined on a bar 58 mm in diameter. Calculate the maximum depth of cut h.

+++ 2 Calculate the depth h of the keyway cut in a shaft of diameter 100 mm, as shown.

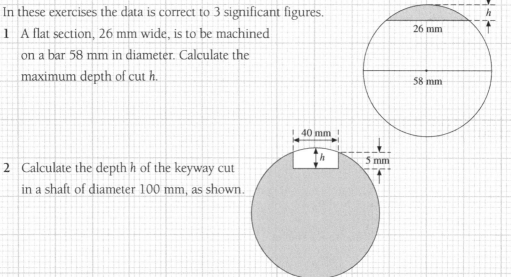

+++ 3 In order to check the accuracy of this template (radius 87 mm), two plugs of diameter 10 mm are placed under it as shown. Calculate the checking distance *x*.

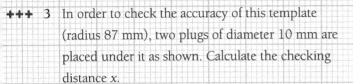

4 Two ball bearings of diameters 8 mm and 12 mm rest on a plane touching one another. How far apart are their contact points with the plane, *A* and *B*?

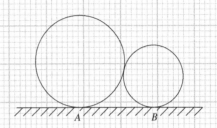

+++ 5 A ball bearing is to be placed in a conical container so that the top of the ball is level with the top of the cone. Given that the cone has vertical angle 40° and height 30 mm, find the diameter of the ball.

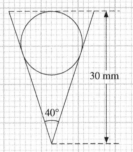

+++ 6 a Two pulley wheels with diameters 600 mm and 100 mm are connected by a flat drive belt. Given that their centres *A* and *B* are 500 mm apart and that *P* and *Q* are contact points as shown in the diagram, find:

   i the distance *PQ*

   ii ∠*PAB*

   iii the total length of the belt

b If the above pulleys are connected with a crossed belt as shown in the diagram, find the total length of the belt.

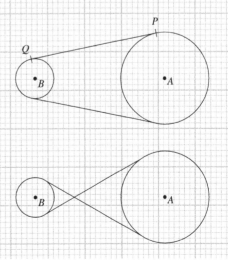

1   Given that $\angle PCR = 117°$, $\angle QCR = 26.0°$,
    arc length $QR = 23.9$ mm, and area of
    sector PCR is 153 mm, find:

    a   the arc length PQ

    b   the area of sector PCQ

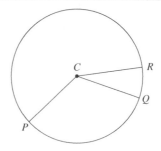

2   **a**   Express p in terms of x.

    b   Express q in terms of x.

    c   Hence evaluate x.

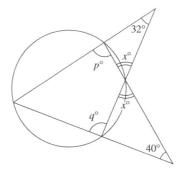

3

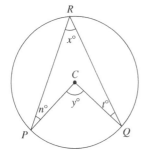

Evaluate:

a x, when $y = 80$

b y, when $x = 32$

c x, when $x + y = 150$

d x, when $y - x = 40$

e y, when $y - x = 50$

4   Given that AB is a diameter of
    the circle, evaluate:

    a m, when $k = 25$

    b t, when $k = 36$

    c k, when $x = 43$

    d t, when $m + k = 69$

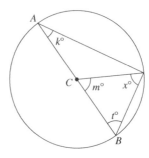

+++ 5

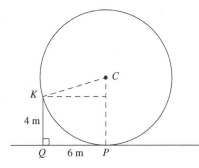

QP is a tangent to the circle and *KQ* ⊥ *QP.*
Given that *KQ* = 4.0 m and *PQ* = 6.0 m, find
the radius of the circle.
(*Hint:* Draw broken lines as indicated.)

+++ 6 Two circles, of radii 32 mm and 12 mm,
have their centres 52 mm apart. *PQ* is a
common tangent touching the circles at
P and *Q.* Find the length of *PQ.*

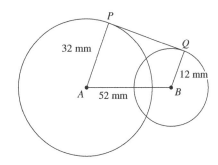

+++ 7 Evaluate the pronumerals:

a

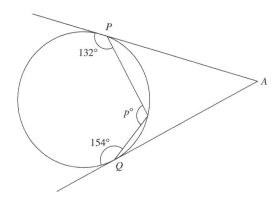

b

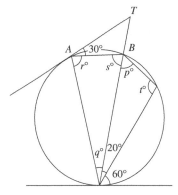

AP and *AQ* are tangents to the circle.
(*Hint:* Join *PQ.*)

AT is a tangent to the circle.

CHAPTER 9

STRAIGHT LINE COORDINATE GEOMETRY

Learning objectives

■ Use the given coordinates of points on a straight line to:
  - find the gradient of the line
  - find the equation for the line.
■ From given empirical data:
  - plot points to represent the data and to determine the straight 'line of best fit'
  - use the 'line of best fit' to interpolate values and to obtain the equation for the line.
■ Plot a straight line graph from a given equation.
■ Use straight line graphs:
  - to solve linear equations
  - to solve systems of two linear simultaneous equations.

9.1 THE NUMBER PLANE

As you will have learnt previously, points are plotted on the **number plane** by drawing two **axes** at right angles and marking scales on these axes. The axes are said to be horizontal and vertical and are called the **x-axis** and the **y-axis** respectively. The axes divide the number plane into four **quadrants,** and the point of intersection of the axes is called the **origin.**

In the diagram, the origin is the point O and the four quadrants (called the 'first', 'second', 'third' and 'fourth' quadrants) are labelled ①, ②, ③, ④. Note that the scales on the two axes are not the same and that this is often, but not always, the case.

We specify the position of a point by stating its x-value (or **abscissa**) and its y-value (or **ordinate**). Point A in this diagram is the point $x = 8, y = 5$. It is more convenient, however, to specify these coordinates by two numbers in parentheses. This is called an **ordered pair** because the order is important, the x-value always being the first number stated. Hence, A is the point $(8, 5)$. The other points shown are $B(-4, 10)$, $C(-6, -5)$ and $D(2, -15)$. The origin is always the point $(0, 0)$.

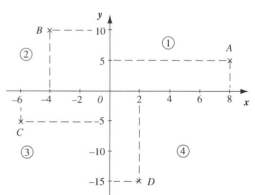

Exercises 9.1

1   *AB* is a line segment, that is, a part of a straight line.
    The coordinates of point *B* are (x_2, y_2).

    a   What are the coordinates of point *A*?

    b   What is the length of *AK*?

    c   What is the length of *BK*?

    d   Using Pythagoras' theorem, write down the length
        of *AB*.

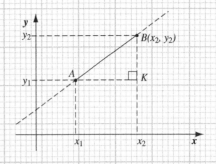

Rule: The distance between the points $A(x_1, y_1)$ and $B(x_2, y_2)$ (i.e. the length of the line segment *AB*) is given by:

$$d = \sqrt{(x_2 - x_1)^2 + (y_2 - y_1)^2}.$$

Note: $(x_2 - x_1)^2 = (x_1 - x_2)^2$, e.g. $(5 - 3)^2 = (3 - 5)^2$.

Example

The distance between the points $(2, 8)$ and $(7, 5)$ is

$\sqrt{(7 - 2)^2 + (8 - 5)^2}$

$= \sqrt{5^2 + 3^2}$

$= \sqrt{34}$ (≈ 5.38 units)

The units in which the length is expressed are the units on the two scales when the same scale is used on each axis.

Exercises 9.1 *continued*

2   Let *M* be the midpoint between *A* and *B*
    (i.e. *AM* = *BM*).

    a   Which test shows that the triangles *AQM* and
        MPB are congruent? (The congruency tests
        were treated in section 7.4.)

    b   In these triangles state the corresponding
        sides which are equal.

    c   How do we know that *MP* = *QK* and
        MQ = *PK*?

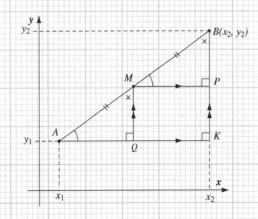

continued

continued

   d What conclusion can you draw about the points Q and P?

   e State the coordinates of the points Q and P. (Note that the number that is halfway between two numbers a and b is the 'average' of those two numbers, i.e. $\dfrac{a+b}{2}$.)

   f State the coordinates of the point M.

Rule: If two points have coordinates (x_1, y_1) and (x_2, y_2), the coordinates of the midpoint are
$$\left(\frac{x_1 + x_2}{2}, \frac{y_1 + y_2}{2}\right).$$

Exercises 9.1 *continued*

3 In each case below the coordinates of two points are given. State (i) the distance between the two points, and (ii) the coordinates of the midpoint.

   a $A(2, 3)$ and $B(5, 7)$ **b** $P(3, 4)$ and $Q(8, 16)$

   c $K(4, 9)$ and $L(-2, 1)$ **d** $C(-1, 2)$ and $D(3, -1)$

   e $G(-3, -13)$ and $H(-15, -8)$ **f** $M(-7, -3)$ and $P(2, -5)$

4 M is the midpoint of the line segment PQ.

   a If P is the point $(1, 3)$ and M is the point $(2, 6)$, what are the coordinates of point Q? (*Hint*: Let (a, b) be the required coordinates and use the fact that the coordinates of a midpoint are $\left(\dfrac{x_1 + x_2}{2}, \dfrac{y_1 + y_2}{2}\right)$).

   b If Q is the point $(-3, -2)$ and M is the point $(4, -1)$, what are the coordinates of the point P?

9.2 THE GRADIENT OF A LINE

If P and Q in the diagram are two points on a straight incline (e.g. a road or driveway), the incline is said to have a gradient (or slope) of 1 in 5 (or 1 : 5, or $\frac{1}{5}$, or 0.2).

   The **rise** from P to Q is 1 m.

   The **run** (horizontal distance to the right) from P to Q is 5 m.

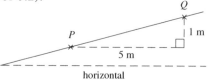

Remember: Gradient between two points $(m) = \dfrac{\text{rise} \uparrow}{\text{run} \rightarrow}$

Examples

1 Here, the gradient of line *PQ* is given by:

$$m = \frac{\text{rise} \uparrow}{\text{run} \rightarrow} \text{(between } P \text{ and } Q\text{)}$$

$$= \frac{10 - 6}{3 - 1}$$

$$= 2$$

The gradient of line *KQ* is:

$$m = \frac{\text{rise} \uparrow}{\text{run} \rightarrow} \text{(between } K \text{ and } Q\text{)}$$

$$= \frac{14}{5}$$

$$= 2.8$$

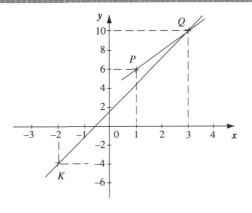

Note: A line that slopes 'downhill' from left to right has a *negative* gradient.

2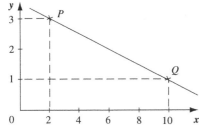

The gradient of line *PQ* is:

$$m = \frac{\text{rise} \uparrow}{\text{run} \rightarrow} \text{(between } P \text{ and } Q\text{)}$$

$$= \frac{-2}{8}$$

$$= -\frac{1}{4}$$

$$= -0.25$$

The rise from *P* to *Q* is *negative*, since there is actually a *fall* from *P* to *Q*.

Exercises 9.2

For each of the graphs sketched below, find:

   a the (horizontal) run from *P* to *Q*
   b the (vertical) rise from *P* to *Q*
   c the gradient of the line.

1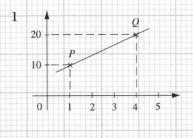

2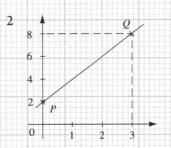

continued

continued

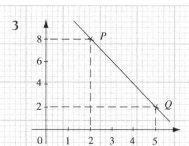

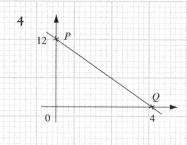

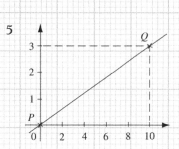

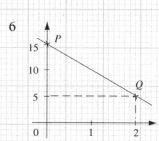

Formula for the gradient

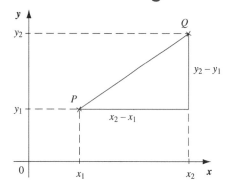

If P is the point (x_1, y_1) and Q is (x_2, y_2):

From P to Q: $\begin{cases} \text{rise} = y_2 - y_1 \\ \text{run} = x_2 - x_1 \end{cases}$

Gradient of the line through P and $Q = \dfrac{y_2 - y_1}{x_2 - x_1}$

(which also $= \dfrac{y_1 - y_2}{x_1 - x_2}$)

Example

If P and Q are the points $(3, 8)$ and $(7, 6)$,

$x_1 = 3 \qquad x_2 = 7$

$y_1 = 8 \qquad y_2 = 6$

$\therefore$ gradient of line $PQ = \dfrac{y_2 - y_1}{x_2 - x_1} \qquad or \qquad = \dfrac{y_1 - y_2}{x_1 - x_2}$

$\qquad\qquad\qquad = \dfrac{6 - 8}{7 - 3} \qquad\qquad\qquad = \dfrac{8 - 6}{3 - 7}$

$\qquad\qquad\qquad = \dfrac{-2}{4} \qquad\qquad\qquad\quad = \dfrac{2}{-4}$

$\qquad\qquad\qquad = -0.5 \qquad\qquad\qquad\;\; = -0.5$

We can summarise the gradient formula thus:

> **Rule:** $m = \dfrac{\Delta y}{\Delta x}$ i.e. $\dfrac{\text{(difference between the y-values)}}{\text{(difference between the x-values)}}$
>
> where the subtractions are made in the same 'directions'.

Examples

1 $P(3, 8)$ and $Q(7, 6)$:

$$m = \frac{6 - 8}{7 - 3} = -0.5$$

or $P(3, 8)$ and $Q(7, 6)$:

$$m = \frac{8 - 6}{3 - 7} = -0.5$$

2 The gradient of the line that passes through the points $K(3, 1)$ and $M(2, 5)$ is given by:

$$m = \frac{1 - 5}{3 - 2} = \frac{-4}{1} = -4 \text{ or } m = \frac{5 - 1}{2 - 3} = \frac{4}{-1} = -4$$

3 The gradient of the line through $T(-3, -9)$ and $W(1, -2)$ is given by:

$$m = \frac{(-9) - (-2)}{(-3) - (1)} = \frac{-9 + 2}{-3 - 1} = \frac{-7}{-4} = 1.75$$

$$\text{or } m = \frac{(-2) - (-9)}{(1) - (-3)} = \frac{-2 + 9}{1 + 3} = \frac{7}{4} = 1.75$$

Exercises 9.2 *continued*

7 In each case below, find the gradient of the line that passes through the given pair of points:

a $(2, 3)$ and $(5, 12)$ **b** $(0, 1)$ and $(2, 13)$

c $(-1, 1)$ and $(2, 13)$ **d** $(-3, 1)$ and $(-1, 5)$

e $(1, 3)$ and $(3, -1)$ **f** $(2, -7)$ and $(3, 0)$

9.3 INTERCEPTS

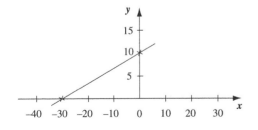

The value on an axis at the point where a straight line cuts that axis is called the **intercept** made by the line on that axis.

For the line shown, the x-intercept is -30 and the y-intercept is 10.

Note: The x-intercept is the value of x when $y = 0$: the y-intercept is the value of y when $x = 0$.

From the graph we cannot read off the value of x when $y = 0$, or the value of y when $x = 0$, since these points are not shown.

The x-intercept is *not* 12 ($x = 12$ when $y = \mathbf{20}$).

The y-intercept is *not* 24 ($y = 24$ when $x = \mathbf{10}$).

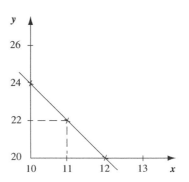

9.4 THE STRAIGHT LINE EQUATION $y = mx + b$

Exercises 9.3

1 a

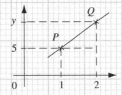

Given that the gradient of this line is 3, evaluate y (the ordinate of Q) using the definition $\dfrac{\text{rise}}{\text{run}} = \text{gradient}$.

b

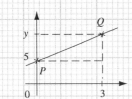

Given that the gradient of this line is 2, evaluate y.

c

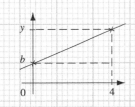

Given that the gradient of this line is 5, find y (in terms of b).

d

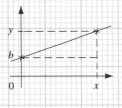

Given that the gradient of this line is 4, find y (in terms of x and b).

e

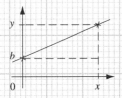

Given that the gradient of this line is m, find y (in terms of x, m and b).

Once the equation for a straight line is written in the form $y = mx + b$ (i.e. with y as the subject of the formula), the gradient and the y-intercept can be read from the equation.

> **Rule:** The equation for a straight line is:
>
> $$y = mx + b$$
>
> where m = the gradient
>
> b = the y-intercept (the value of y when $x = 0$)

Examples

1 For the straight line $2y = 4 + 8x$

$$y = 4x + 2$$

$$\therefore \text{gradient} = 4$$

$$y\text{-intercept} = 2$$

2 For the straight line $3x + 4y - 5 = 0$

$$4y = -3x + 5$$

$$\therefore y = -\tfrac{3}{4}x + \tfrac{5}{4}$$

$$\therefore \text{gradient} = -\tfrac{3}{4} = -0.75$$

$$y\text{-lintercept} = \tfrac{5}{4} = 1.25$$

3 For the straight line $$\frac{x + 3}{y - 1} = 5$$

$$x + 3 = 5y - 5$$

$$5y = x + 8$$

$$y = \tfrac{1}{5}x + \tfrac{8}{5}$$

$$\therefore \text{gradient} = \tfrac{1}{5} = 0.2$$

$$y\text{-intercept} = \tfrac{8}{5} = 1.6$$

In order to find the gradient, it is advisable to first express the equation in the form $y = mx + b$ (i.e. with y as the subject), but with the equation written in any form, we can deduce the coordinates of any point on the line, since every point on the line must have coordinates that satisfy the equation.

Examples

1 Does the point (4, 10) lie on the line $y = 3x - 2$?

When $x = 4$, $y = 3(4) - 2$

$$= 10$$

Hence the coordinates (4, 10) satisfy the equation, and (4, 10) is a point on this line.

continued

continued

2 Does the point (8, 2) lie on the line $2y - \dfrac{x-2}{3} = x - 5$?

When $x = 8$, $2y - \dfrac{8-2}{3} = 8 - 5$

$$2y - 2 = 3$$
$$2y = 5$$
$$\therefore y = 2.5$$

Hence the coordinates (8, 2) do *not* satisfy the equation and (8, 2) is *not* a point on this line.

3 Given that the point $(k, -3)$ lies on the line $\dfrac{x+1}{2} - \dfrac{y-3}{4} = x + y$, evaluate k.

When $y = -3, \dfrac{x+1}{2} - \dfrac{-3-3}{4} = x - 3$ or, substituting the coordinated:

$$\frac{x+1}{2} + \frac{3}{2} = x - 3$$

$$x + 1 + 3 = 2x - 6$$

$$\therefore x = 10$$

$$\therefore k = 10$$

$$\frac{k+1}{2} - \frac{-3-3}{4} = k - 3$$

$$\therefore \frac{k+1}{2} + \frac{6}{4} = k - 3$$

$$\therefore 2k + 2 + 6 = 4k - 12$$

$$20 = 2k$$

$$\therefore k = 10$$

Finding intercepts

To find the *x*-intercept of a line, find the value of *x* when $y = 0$.
To find the *y*-intercept of a line, find the value of *y* when $x = 0$.

Example

For the line $4x - 3y + 6 = 0$:

When $x = 0$,   When $y = 0$,

$$-3y + 6 = 0$$ $$4x + 6 = 0$$

$$3y = 6$$ $$4x = -6$$

$$\therefore y = 2$$ $$\therefore x = -1.5$$

$$\therefore y\text{-intercept} = 2$$ $$\therefore x\text{-intercept} = -1.5$$

Exercises 9.3 *continued*

2 For each of the lines given by the following equations, write down (i) the gradient, (ii) the *y*-intercept, (iii) the *x*-intercept.

a $y = 3x + 5$ **b** $y = 2x - 7$ **c** $y = -5x - 3$

d $y - 2x = 4$ **e** $3x - y + 6 = 0$ **f** $2x - 3y - 9 = 0$

3 a Which of the following points lie on the line $y = 2x + 3$?

$P(0, 3)$ $Q(1, 5)$ $R(2, 8)$ $S(-1, 1)$

$T(-2, 2)$ $U(3, 12)$ $V(-3, -3)$ $W(-5, 7)$

b Which of the following points line on the line $x - 2y - 3 = 0$?

$J(4, 1)$ $K(6, 0)$ $L(0, -3)$ $M(3, 0)$

$N(-2, -4)$ $P(-1, -2)$ $Q(-3, -3)$ $R(1, 5)$

c Given that the point $(2, 3)$ lies on the line $y = 2x + b$, find the y-intercept.

d Given that the point $(3, k)$ lies on the line $y = 2x - 5$, evaluate k.

e Given that the point $(2, -3)$ lies on the line $y = mx + 3$, find the gradient of the line.

f Given that the point $(a, -3)$ lies on the line $2x + 3y + 4 = 0$, evaluate a.

9.5 FINDING THE EQUATION FOR A PARTICULAR STRAIGHT LINE

When you are given the gradient and the coordinates of one point on the line:

1 substitute the gradient for m in the equation $y = mx + b$, then

2 substitute the coordinates of the given point into the equation.

Examples

1 Find the equation for the straight line that passes through the point $(2, -5)$ and has gradient -3.

$y = mx + b$

$m = -3, \therefore y = -3x + b$

Substitute $(2, -5)$: $-5 = -3(2) + b$

$-5 = -6 + b$

$\therefore b = 1$

Hence the equation is: $y = -3x + 1$

2 Find the equation for the line that has gradient 2.5 and x-intercept 2.

$y = mx + b$

$m = 2.5, \therefore y = 2.5x + b$

An x-intercept of 2 means that $x = 2$ when $y = 0$, i.e. $(2, 0)$ is a point on the line.

Substitute $(2, 0)$: $0 = (2.5)2 + b$

$0 = 5 + b$

$\therefore b = -5$

Hence the equation is: $y = 2.5x - 5$

When you are given the coordinates of two points on the line:

1 evaluate the gradient $\left(m = \dfrac{\Delta y}{\Delta x} \right)$

2 substitute the coordinates of one of the given points.

Example

Find the equation for the line that passes through the points $(1, -3)$ and $(-1, -9)$.

$$m = \frac{y_2 - y_1}{x_2 - x_1} = \frac{(-9) - (-3)}{(-1) - (1)} = \frac{-9 + 3}{-2} = \frac{-6}{-2} = 3$$

$\therefore y = 3x + b$

Substitute $(1, -3)$: $-3 = 3(1) + b$

$-3 = 3 + b$

$\therefore b = -6$

Hence the equation is: $y = 3x - 6$

Exercises 9.4

1   By substituting into the equation $y = mx + b$, find the equation of the line in each case below, given that:

   a   the gradient is 2 and the y-intercept is 3

   b   the y-intercept is 5 and $(1, 2)$ lies on the line

   c   the y-intercept is 3 and the x-intercept is 1

   d   the y-intercept is -2 and the x-intercept is 3

   e   the gradient is -3 and $(-2, 1)$ lies on the line

2   Find the equation of the line that passes through the points:

   a $(2, 5)$ and $(4, 11)$ **b** $(2, 1)$ and $(-4, 16)$ **c** $(-1, -2)$ and $(-3, 4)$

3.   Find the equation for each of the lines in Exercises 9.2, questions 1–6, assuming in each case that y is plotted against x.

9.6 DEPENDENT AND INDEPENDENT VARIABLES

Before commencing this course, you learnt to plot pairs of corresponding values (ordered pairs) on graph paper.

Example

Suppose that when various loads (M kg) were suspended from a spiral spring and the length (L mm) of the spring was measured for each load, the following results were obtained:

   When $M = 6$ kg, $L = 220$ mm.

   When $M = 8$ kg, $L = 260$ mm.

   When $M = 10$ kg, $L = 300$ mm.

We can show this information in a convenient visual form on a graph:

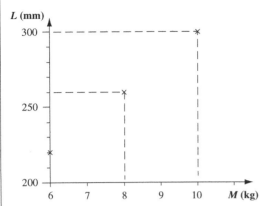

The given values are usually shown in a table:

M (kg)	6	8	10
L (mm)	220	260	300

Technology is greatly concerned with the changes in the magnitude of one quantity that are produced by changes in the magnitude of another quantity; for example, how a change in the wing length of an aircraft affects the take-off run required, how a change in an electric current affects the magnetic field produced by the current, or how the speed of a turbine affects the efficiency of the machine.

When the value of one quantity, D, depends upon the value of another quantity, I, the former is called the **dependent** variable and the latter is called the **independent** variable. The independent variable is usually assigned equally spaced values over the range of values required.

When drawing up a table of values, it is customary to state the prearranged values of the independent variable first, and then to fill in, beneath these, the corresponding values of the dependent variable.

Independent variable	I	10	12	14	16	18	20
Dependent variable	D						

When graphing these results, the independent variable is plotted on the *horizontal* axis and the corresponding values of the dependent variable are plotted against these on the *vertical* axis.

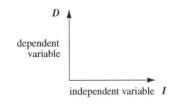

If we are interested in the time of oscillation (T) of a simple pendulum for various lengths (L) of the pendulum, we would say that the time of oscillation depends upon the length of the pendulum. Hence, T would be the dependent variable.

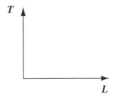

If, however, we are interested in the length of pendulum required to obtain various times of oscillation, we would say that the required length depends upon the required time of oscillation. In this case, L would be the dependent variable.

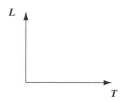

Note: In any case, if you were *instructed* (e.g. in an examination question) to plot the values of a particular variable on a particular axis, then of course you would do so, even though the instruction might appear to contravene accepted practice.

9.7 PLOTTING AND INTERPOLATION

We have revised the plotting of pairs of corresponding values (ordered pairs) on graph paper. Our main concern now is to emphasise the importance of several procedures and to extend your knowledge to drawing straight lines and curves of best fit.

1 *Plot all points clearly* but mark as small an area as possible. A dot is not satisfactory, since it is likely to be obscured later by a line or curve that passes through the point. Every point plotted should be clearly visible, even after a curve has been drawn through it. A large blob is not accurate and is also untidy. It is suggested that ✕ be used to mark the position of a point.

2 *Each axis must be labelled with:*
   a the name of the quantity plotted or its symbol (e.g. current or I);
   b the unit used on the scale or its symbol (e.g. milliamperes or mA).

3 *Each axis must be marked clearly with the scale divisions.* There should be enough of these marks for the scale to be easily read along its length.

4 *A convenient scale should be chosen for each axis.* The scale should be one that can be easily and quickly read at all points along its length and that spreads the points plotted over as large an area of the graph paper as possible.

Note: a It is often advisable to rotate the sheet of graph paper through 90°.
     b The scale on either axis does not have to commence from zero.

5 *The straight line or curve of best fit should be drawn after plotting the points.* This should, in general, be a *smooth* curve (or a ruled straight line if the relationship appears to be linear or if requested to draw the 'line of best fit' (see p. 155).

6 Deductions and calculations from the graph should be shown on the graph sheet when possible. If there is not sufficient space on the graph sheet, use a separate sheet of paper. *Never* use the back of the graph sheet for this purpose.

Choice of scale

As mentioned above, there are two considerations when deciding upon the scale to use on an axis. The most important requirement is that the scale should be easy to read quickly at all points along its length.

Examples

1

This shows a very unsatisfactory scale, since it is difficult to read off values quickly. The value to which the arrow is pointing is approximately 131.5, but it is a tedious task to determine this fact, since each small scale division represents 2.6 units.

2
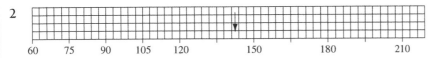

This is a better scale than in example 1, since each small scale division represents 3 units. The reading to which the arrow is pointing is approximately 142.5.

3
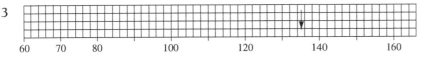

The scale shown here is even more convenient to read since each small scale division represents 2 units. The reading to which the arrow is pointing is approximately 135.

In general it is best to have each small scale division representing 1, 2 or 5 units, multiplied or divided by powers of 10, for example:

1, 2, 3, 4, . . . or . . . 10, 20, 30, 40, . . . or . . .
2, 4, 6, 8, . . . or . . . 20, 40, 60, 80, . . . or . . .
5, 10, 15, 20, . . . or . . . 50, 100, 150, 200, . . . or . . .

Line or curve of best fit

When instrument readings (e.g. readings of current from an ammeter) are being plotted, it must be remembered that no reading from an instrument is *exact* (except in the case of a digital instrument, which is merely *counting*). There are always instrument errors and human errors; no instrument is perfect, nor is any pair of eyes. The scale of an instrument can always be read to a certain accuracy and no more. If, for example, a current is read as 23.7 mA, it is quite certain that the current is *not exactly* 23.7 mA.

When a series of readings is taken, it usually happens that some of the readings are too high and some are too low. It may happen that, due to an error in the instrument (e.g. a 'zero error'), all the readings are too high or they are all too low, but in general it is safe to assume that some are too high and some too low.

Using the readings plotted, we draw a straight line or curve of 'best fit' to show the most likely pairs of corresponding values at all points between those we have plotted. If, as usually happens, the points plotted do not fall along a straight line or a smooth curve, we assume that some of the points are too high and some are too low and we use the plotted points as a guide in predicting the most likely line or curve of best fit.

The line or curve that we draw may not pass exactly through *any* of the plotted points. We draw a *smooth* curve (or straight line) that passes as close to as many of the plotted points as possible. Most relationships between variables are such that the graph is a *smooth* curve (i.e. a curve with no sudden bumps or wobbles), and unless we suspect that the relationship is otherwise, we draw a *smooth* curve (or a straight line).

The three diagrams below show the same six points plotted on three different graphs with three attempts to draw the line of best fit.

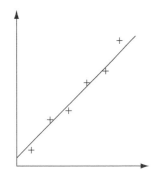 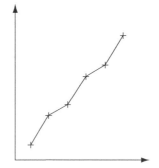 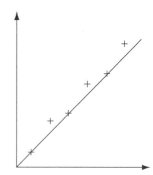

The relationship here appears to be linear. This is the line of 'best fit', a good 'average', assuming some readings are too high and some are too low.

This is a very poor graph. If the relationship really *is* a 'zigzag' one, there is no reason to assume that the 'zigs' and the 'zags' occur at the points that we happened to read. This would be a remarkable coincidence!

This does *not* show a good line. It is most unlikely that you would have three readings that are perfect and three others that are *all* too high.

The gradient of a straight line graph $\left(\dfrac{\Delta y}{\Delta x}\right)$ gives the increase in the quantity represented on the y-axis per unit increase in the quantity represented on the x-axis. This is often referred to as the 'rate of increase of y with respect to x'.

Example

In the graph on page 153 the gradient of the straight line drawn through the three points,

$\dfrac{\Delta L}{\Delta M}$, $= \dfrac{80}{4} = 20$ mm/kg. The spring extends by 20 mm for each additional kilogram load. A gradient $\dfrac{\Delta y}{\Delta x}$

indicates the rate at which y increases with respect to x. In this case, the gradient $\dfrac{\Delta L}{\Delta M}$ gives the rate at

which the length of the spring (L) increases with the load (M), which equals 20 millimetres per kilogram (20 mm/kg).

Since the gradient specifies a rate of increase of one quantity with respect to another, the *units* of the quantities must be stated.

Note: When only the first quantity is mentioned, 'time' is implied for the second quantity. A 'rate of increase of distance' implies 'with respect to time', and hence means 'velocity'. A 'rate of increase of velocity' means 'acceleration'.

Exercises 9.5

1 Given the following table of corresponding values for the variables x and y, plot y (vertical axis) against x, and draw the line of best fit.

x	0.7	2.2	3.6	5.1	7.0
y	5.0	9.5	14.0	17.5	23.5

From your graph, find:

a the value of y when x = 3.1

b the value of x when y = 21.5

c the gradient of the graph

2 From the following table of values, plot v (vertical axis) against t and draw the line of best fit. (*Note:* In this case it would not be satisfactory to begin either scale at zero.)

t (s)	1.35	1.40	1.45	1.50	1.55	1.60	1.65	1.70	1.75	1.80
v (m/s)	50.24	50.38	50.48	50.60	50.72	50.84	50.94	51.06	51.18	51.30

From your graph, find:

a the value of v when t = 1.62 s

b the value of t when v = 51.22 m/s

c the rate at which the velocity increases with time (i.e. the acceleration)

continued

continued

3 The diameter *d* of a tapered rod was measured at several places along its length, at distances *s* from one end. The readings were:

s (mm)	0	50	100	150	200
d (mm)	15.04	15.38	15.68	16.04	16.32

Plot *d* (vertical axis) against *s*, and from the line of best fit, find:

a   the diameter at the position 75 mm from the thin end of the rod

b   the distance from the thin end of the rod to the place where the diameter is 16.14 mm

c   the rate at which the diameter increases along the length of the rod

4 A mass of a certain gas was heated at constant pressure, its volume at various temperatures being read:

Temperature (K)	300	350	400	450	500	550
Volume (L)	1.00	1.16	1.34	1.52	1.67	1.83

Plot the volume (vertical axis) against the temperature, and from the line of best fit find:

a   the volume of the gas when the temperature was 430 K

b   at what temperature the volume of the gas was 1.6 L

c   the rate at which the volume increases with the temperature

5 For a certain resistor, readings were taken of the current (*I*) for various applied potential differences (*PD*).

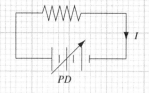

PD (V)	0	40	80	120	160
I (mA)	0	13.0	23.5	37.5	48.0

Plot the potential difference (vertical axis) against the current and draw the line of best fit. From your graph determine:

a   the current produced by a potential difference of 100 V

b   the potential difference required to produce a current of 25 mA

c   the rate at which *I* increases with the *PD*

9.8 PLOTTING GRAPHS FROM EQUATIONS

An equation (formula) generates an *infinite* set of ordered pairs. For example, if the length (*l* mm) of a spring when a load of *W* kg is suspended from it is given by the equation $l = 20W + 100$, we can compute the value of *l* for *any* load *W* (e.g. when $W = 2.36, l = 147.2$; when $W = 2.37, l = 147.4$, etc.).

When the equation is of the form $y = mx + b$, where m and b are constants, the points will fall on a straight line. Hence such an equation is called a **linear equation**.

Examples

1 $y = 3x + 7$	**2** $I = 5.3W + 9.1$
3 $v = 6.2t - 4.6$	**4** $s = -8t + 5$

Note: The *subject* of a formula is always regarded as being the *dependent* variable. If $v = 6.2t - 4.1$, we say that the value of v depends upon the value of t, rather than vice versa.

Exercises 9.6

1 Given that $y = 2x + 100$, copy and complete the following table of values:

x	0	1	2	3	4
y					

   a Plot the ordered pairs in the table of values and then draw the straight line given by the equation $y = 2x + 100$.

   b From your graph, determine the value of y when $x = 3.25$.

   c From your graph, determine the value of x when $y = 106.7$.

   d Check your answers to parts **b** and **c** above by substituting into the equation.

2 The current (I) in a certain conductor at various times (t) is given by the formula $I = 10t - 100$. Copy and complete the following table of values and hence draw the graph of I against t from 20 to 36.

t(s)	20	26	30	36
I(A)				

   a From your graph, read off the value of t when $I = 194$ A.

   b Check your answer to part **a** above by substituting into the equation.

3 Plot the following graphs accurately (y against x). Label each axis and show its scale clearly.

   a $y = 2x + 3$ $\quad\quad\quad\quad (0 \leq x \leq 4)$

   b $2x + 3y + 4 = 0$ $\quad\quad (-6 \leq x \leq 3)$

   c $\dfrac{x}{y} + 2 = 0$ $\quad\quad\quad\quad (-400 \leq x \leq 200)$

continued

continued

4 The power output of the engine that is required to move a certain locomotive at a speed of 25 km/h up a steady incline having a gradient G is given by $P = 2520G + 17$ kW. Plot the graph showing the power required (vertical axis) for gradients from 0 to 0.025. From your graph, determine the power required when the gradient is 1 in 50, and check your reading by substitution into the given equation.

5 A manufacturer makes a profit of \$157 on each article produced, but out of this profit has to pay overhead charges of \$5000 per year. Draw a graph showing the net income, \$$I$, corresponding to the production of N articles per year for $0 \le N \le 150$. From your graph, find:

   a the net income for a production of 90 articles a year

   b the minimum number of articles that must be produced in order to cover overhead expenses

   c the annual output necessary to achieve a net annual income of \$10 000.

9.9 GRAPHICAL SOLUTION OF LINEAR EQUATIONS AND SYSTEMS OF TWO (SIMULTANEOUS) LINEAR EQUATIONS

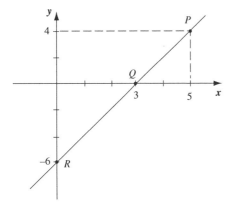

This graph is a straight line, so we know that its equation is of the form $y = Cx + K$ where C and K are constants. For points on the line, the y-values (*ordinates*) give the values of $Cx + K$ for different values of x. For example:

- point P shows that $y = 4$ when $x = 5$,
  i.e. $Cx + K = 4$ when $x = 5$
  i.e. $x = 5$ is the solution of the equation $Cx + K = 4$
- point Q shows that $x = 3$ is the solution of the equation $Cx + K = 0$
- point R shows that $x = 0$ is the solution of the equation $Cx + K = -6$

  It is easy to see that the equation for this line is $y = 2x - 6$ and hence that the above solutions do satisfy the equation.

Exercises 9.7

1

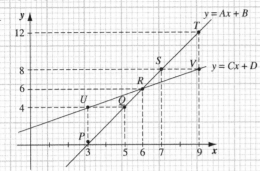

In the above graphs state which point, (P, Q, . . . V), gives the solution to the equation:

a $Ax + B = 12$

b $Ax + B = 0$

c $Ax + B = 4$

d $Cx + D = 4$

e $Ax + B = 8$

f $Cx + D = 8$

g $Ax + B = Cx + D$

h $(A - C)x = D - B$

2

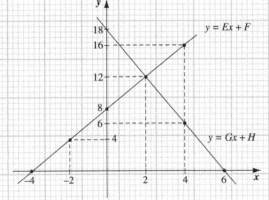

From the above graphs find the solutions of the equations:

a $Ex + F = 16$ **b** $Gx + H = 0$ **c** $Ex + F = 0$

d $Gx + H = 6$ **e** $Ex + F = Gx + H$ **f** $(E - G)x = H - F$

3

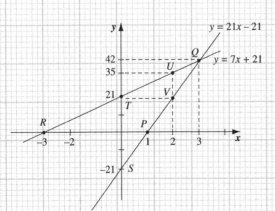

continued

continued

On the above graphs there are 7 points, labelled P, Q, R, S, T, U and V.

State which of these points gives the solution to the equation:

a **i** $7x + 21 = 0$ **ii** $21x - 21 = 42$ **iii** $7x + 21 = 21$

     iv $21x - 21 = -21$ **v** $7x + 21 = 35$ **vi** $21x - 21 = 7x + 21$

b Which of the points shows that the ordered pair (2, 35) satisfies the equation $y = 7x + 21$ but does *not* satisfy the equation $y = 21x - 21$?

c Which of the points shows that the ordered pair (2, 21) satisfies the equation $y = 21x - 21$ but does *not* satisfy the equation $y = 7x + 21$?

d Which of the points shows that there does exist an ordered pair that satisfies *both* equations $y = 21x - 21$ and $y = 7x + 21$?

e Which of the points gives the solution to the simultaneous equations $\begin{cases} y = 21x - 21 \\ y = 7x + 21 \end{cases}$?

f From the graphs determine the ordered pair that is the solution to the simultaneous equations $\begin{cases} y = 21x - 21 \\ y = 7x + 21 \end{cases}$.

4 On the same set of axes, plot the graphs of $y = x + 3$ and $y = 2x + 1$ for $0 \le x \le 4$.

  a From your graph find and write down five ordered pairs (integral values only) that satisfy the equation $y = x + 3$.

  b From your graph find and write down five ordered pairs (integral values only) that satisfy the equation $y = 2x + 1$.

  c Write down all the ordered pairs that satisfy both equations.

  d Write down the solution to the simultaneous equations $\begin{cases} y = x + 3 \\ y = 2x + 1 \end{cases}$.

  e Complete the statement: 'To solve a pair of simultaneous equations graphically, plot the graphs of both equations on the same set of axes and . . .'

5 Solve the following pairs of simultaneous equations by plotting the graphs of the two equations on the same set of axes for the values of x given. (Evaluate y for the extreme values of x before drawing the axes and affixing scales to them.)

  a $\begin{cases} y = 4x - 3 \\ y = 6 - 2x \end{cases}$ **b** $\begin{cases} y = 2x - 6 \\ y = -3x - 9 \end{cases}$

     Plot for $0 \le x \le 4$.   Plot for $-2 \le x \le 2$.

6 Two cars A and B are travelling along a long straight road, each with a constant velocity. The distances of the cars from a fixed point P on the road were noted at some time $t = 0$ and then each hour later for 3 hours, the readings being given in the table below:

Time (h)	0	1	2	3
Car A: Distance from P, D_A (km)	0	68	128	188
Car B: Distance from P, D_B (km)	144	100	44	0

a On the same axes plot for each car the distances from P against time, and draw the straight lines of best fit.

b From the graphs find the time and the distance from point P at the instant when the cars pass one another.

c Write down the equation connecting D with t for car A and the equation connecting D with t for car B.

d Solve the two simultaneous equations obtained in part c above, using the method of elimination.

e Compare your answers to b and d above.

SELF-TEST

1 For each of the following straight lines, find (i) the gradient, (ii) the x-intercept, (iii) the y-intercept:

a $y = 3x - 6$ b $5x + 2y + 4 = 0$

c $\dfrac{x - 1}{2} = \dfrac{y + 1}{5}$ d $\dfrac{x + 1}{y - 1} = 5$

2 Find the equation for the straight line in each case below, given that:

a the gradient is 2 and the line contains the point $(-1, 3)$

b the x-intercept is 4 and the gradient is -3

c the line contains the point $(2, -5)$ and the y-intercept is -2

d the line passes through the points $(2, -1)$ and $(-2, 3)$

e the gradient is -3 and the line contains the point $(-1, -3)$

f the x-intercept is -2 and the y-intercept is 5

3 The following readings were obtained for the common emitter static characteristic of a certain transistor, in which the collector current (I_c) was measured for various values of the collector voltage (V_c), the base current being held constant at 150 mA.

V_c (V)	2	4	6	8	10	12	14	16
I_c (mA)	4.10	4.14	4.19	4.27	4.31	4.36	4.42	4.49

Plot I_c (vertical axis) against V_c and draw the line of best fit. From your graph:

a determine the collector current corresponding to a collector voltage of 7 V

b estimate the percentage error made in the 4.27 mA reading, assuming that the collector voltage was measured accurately

c determine the rate at which I_c increases with V_c

4 The resistance of a certain coil of copper wire was measured at different temperatures:

Temperature (°C)	0	25	50	75	100	125	150
Resistance (Ω)	120	135	145	155	170	183	195

Plot the resistance (vertical axis) against the temperature *using a temperature scale from* $-250°C$ *to* $150°C$ *and a resistance scale from 0 to 200* Ω, and draw the line of best fit.

a By continuing the line, find the temperature at which the resistance would be zero *if* this linear relationship continued to hold down to this low temperature (i.e. find the inferred zero-resistance temperature for this material).

b From your graph determine the increase in resistance per degree Celsius rise in temperature for this coil.

5 Plot the following graphs accurately (*y* against *x*):

a $y = -3x + 1$ $(-2 \le x \le 5)$

b $\dfrac{x - 2}{y} - 3 = 0$ $(-4 \le x \le 5)$

6 An endless conveyor belt transports loads up an incline. The power (in watts) required to drive the belt is given by $P = 0.014W + 45$ where W kg is the total load transported per hour. Plot P (vertical axis) against W for values of W from 0 to 35 000. From your graph, determine the power required to transport 17 000 kg in 1 hour and check your reading by substitution into the given equation.

7 On the same axes, for $0 \le x \le 5$, plot the graphs of the simultaneous equations
$$\begin{cases} y = 12x + 20 \\ y = 80 - 16x \end{cases}$$ and hence solve these equations.

8 One mass A was heated while another mass B was cooling. Their temperatures θ_A and θ_B were measured first at some time $t = 0$ and then each hour thereafter for 5 hours, the readings being shown in the table below:

Time (h)	0	1	2	3	4	5
θ_A (°C)	19.9	32.2	36.9	50.1	61.9	65.5
θ_B (°C)	80.1	62.0	53.4	34.9	12.1	7.2

a On the same axes, plot the temperatures against time and draw the straight lines of best fit.

b From the graphs find the time when the two bodies have the same temperature and the value of this common temperature.

c Write down the equations connecting θ with t for each of the two masses.

d Solve the simultaneous equations obtained in **c** above.

e Compare your answers to **b** and **d** above.

9 (This graph is a curve.)

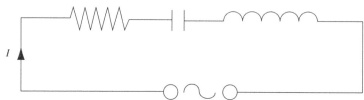

For a particular series circuit containing resistance, inductance and capacitance, the current was measured for various frequencies of a constant alternating potential difference supply:

Frequency, f (Hz)	200	210	220	230	240	250	260	270
Current, I (mA)	32.2	35.9	43.4	49.0	48.0	39.8	34.0	31.2

Plot the current (vertical axis) against the frequency and draw the curve of best fit. From your graph, determine:

a   at what frequency the current is maximum

b   the maximum value of the current

CHAPTER 10

INTRODUCTION TO TRIGONOMETRY

Learning objectives

- Convert angles between sexagesimal measure and decimal degrees.
- Use the trigonometrical ratios of sine, cosine and tangent to find the dimensions of sides and angles of right-angled triangles.
- Apply trigonometry using the above ratios to solve practical problems.

10.1 CONVERSION OF ANGLES BETWEEN SEXAGESIMAL MEASURE (DEGREES, MINUTES AND SECONDS) AND DECIMAL DEGREES

(You may wish to do some research to find out why angles are measured in units that have the same name as the units of time–minutes and seconds.)

$$1 \text{ minute of angle} = \frac{1}{60} \text{ of a degree} \qquad \left(1' = \frac{1°}{60}\right)$$

$$1 \text{ second of angle} = \frac{1}{60} \text{ of a minute} \qquad \left(1'' = \frac{1'}{60}\right)$$

$$= \frac{1}{3600} \text{ of a degree} \qquad \left(1'' = \frac{1°}{3600} \approx 0.0003°\right)$$

Examples

1 $48°23'37'' = \left(48° + \frac{23°}{60} + \frac{37°}{3600}\right) \approx 48.3936°$

2 $1° = 60'$ and $1' = 60''$

To convert 48.3936° into sexagesimal measure using your calculator:

Write down 48°.

Then press $0.3936 \; \boxed{\times} \; 60 \; \boxed{=}$ and write down 23′.

Then press $\boxed{-} \; 23 \; \boxed{=} \; \boxed{\times} \; 60 \; \boxed{=}$ and write down 37″.

The result you have written down is 48° 23′37″.

Your calculator probably provides a special key for these conversions, in which case you should use it in preference to the above laborious calculations. This key is usually labelled $\boxed{°\; ' \; ''}$ or $\boxed{DMS}$.

Exercises 10.1

1 Use your calculator to verify the following conversions:
   a $65° 41' 27'' \approx 65.6908°$ b $65.6908° \approx 65° 41' 27''$
   c $28.6108° \approx 28° 36' 39''$ d $28° 36' 39'' \approx 28.6108°$

10.2 THE TANGENT RATIO: CONSTRUCTION AND DEFINITION

The three triangles below are equiangular and, hence, similar (see section 7.5). They are different magnifications (enlargements) of the same triangle. Hence, corresponding sides are in the same ratio:

$$\frac{a}{b} = \frac{h}{m} = \frac{t}{x}$$

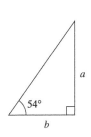 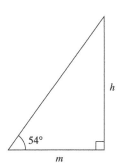 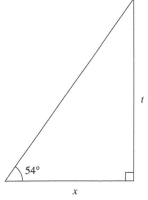

If we draw *any* such triangle (90°, 54°, 36°), the above ratio will be the same.

By inspection of the above triangles, it would seem that the ratio is roughly 1.5 (i.e. side a appears to be about $1\frac{1}{2}$ times as long as side b, side h about $1\frac{1}{2}$ times as long as side m and side t about $1\frac{1}{2}$ times as long as side x).

Let us evaluate this ratio more accurately.

Exercises 10.2

1 On a sheet of graph paper or ordinary paper, using a protractor, draw *accurately* a *large* triangle
   DEF, similar to those above (i.e. a 90°, 54°, 36° triangle).

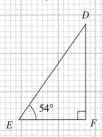

Measure accurately the lengths of the sides *DF* and *EF,* and hence find the value of the ratio $\dfrac{DF}{EF}$ correct to 3 significant figures.

continued

continued

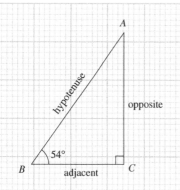

The side opposite the right angle (the *largest* side) in a right-angled triangle is always called the **hypotenuse.** With respect to the 54° angle, side *AC* is called the **opposite** side (being opposite the 54° angle), and the third side, *BC*, is called the **adjacent** side. Which is the 'opposite' side and which is the 'adjacent' side depends upon which angle we are concerned with.

In the exercise above we found an approximate value for the ratio $\dfrac{\text{opposite}}{\text{adjacent}}$ for a 54° angle. This ratio is called the **tangent** of 54° or tan 54°.

> **Rule:** For any angle θ in a right-angled triangle:
>
> $$\tan \theta = \frac{\text{length of the opposite side}}{\text{length of the adjacent side}} \qquad \textit{Learn: } \tan \theta = \frac{\text{opposite}}{\text{adjacent}}$$

There are three common systems of units used to measure the size of an angle: degrees, radians and grads. Scientific calculators use all three systems. You *must* be in the correct trigonometric mode (the correct angle unit system) when you are performing calculations involving trigonometric functions.

To evaluate tan 54° using your calculator first ensure your calculator is in *degree* mode.

Exercises 10.2 *continued*

2 Verify on your calculator that tan 54° ≈ 1.3764. (Your calculator must be in *degree* mode (i.e. ready to accept angles measured in *degrees*).)

   Compare the value tan 54° obtained with a calculator with the value you obtained by construction and measurement in the example above. State the approximate percentage error in your previous result.

3 By construction and measurement (as in question 1), find an approximate value for tan 62°. Compare this value with that obtained by the calculator and state your percentage error.

4 Use your calculator to evaluate the required ratio in each of the triangles below. Give answers correct to 4 significant figures.

a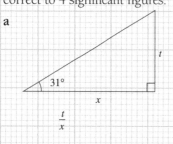

$$\frac{t}{x}$$

b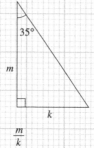

$$\frac{m}{k}$$

c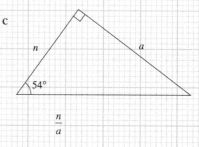

$$\frac{n}{a}$$

5　Use your calculator to verify the following results:

a　$\tan 54°25' \approx 1.398$　　　　　　**b**　$\tan 54.8637° \approx 1.4209$

c　$\tan 38°20' \approx 0.7907$　　　　　　**d**　$\tan 38°20'41'' \approx 0.7910$

6　Use your calculator to evaluate the following, correct to 4 significant figures:

a　$\tan 26°47'$　　　　**b**　$\tan 72°23'$　　　　**c**　$\tan 15°55'$

7　Use your calculator to evaluate the following, correct to 3 significant figures:

a　$3.08 \tan 41°17'$　　　　　　**b**　$\dfrac{\tan 63°26'}{1.39}$

c　$\dfrac{4.86}{\tan 13°43'}$　　　　　　**d**　$\dfrac{\tan 32°41'}{\tan 27°53'}$

10.3 THE TANGENT RATIO: FINDING THE LENGTH OF A SIDE OF A RIGHT-ANGLED TRIANGLE

$\tan 74° = \dfrac{51.3}{d}$

$\therefore d = \dfrac{51.3}{\tan 74°}$

$\approx 14.7 \text{ mm}$

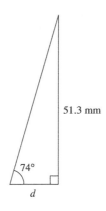

51.3 mm

74°

d

Exercises 10.3

1　In each of the following triangles, find the length of the side marked with a pronumeral:

a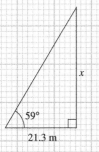

x

59°

21.3 m

b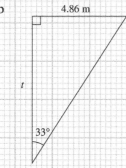

4.86 m

t

33°

c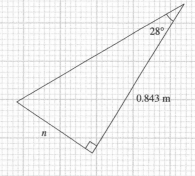

28°

0.843 m

n

continued

continued

2 For each of the following triangles, find, correct to 3 significant figures, the length of the side marked by a pronumeral:

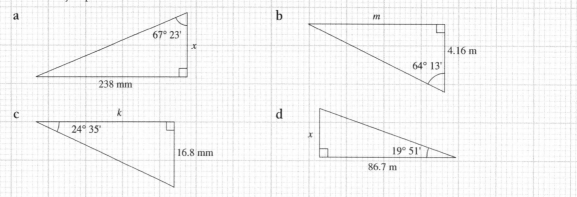

a

67° 23'

x

238 mm

b

m

4.16 m

64° 13'

c

k

24° 35'

16.8 mm

d

x

19° 51'

86.7 m

10.4 THE TANGENT RATIO: EVALUATING AN ANGLE

On graph paper or ordinary paper, construct a **large** right-angled triangle in order to deduce the size of the angle that has a tangent of 2.30. The lengths of two sides (not including the hypotenuse) will need to be in the ratio 2.3 : 1.

Using a protractor, measure as accurately as possible the size of the angle that has a tangent of 2.3. Record your result. Does your calculator agree that the tangent of this angle is 2.30? What is your percentage error? (With careful drawing and measurement you should not have an error of more than about 2% or 3%, i.e. the tangent of your angle should lie between about 2.25 and 2.36.)

If we are told that $\tan \theta = 0.7269$, then, although we don't know the size of the angle θ, we *do* know that it is the angle that has a tangent of 0.7269. This angle is called $\tan^{-1} 0.7269$ or arc tan 0.7269.

Use your calculator to verify that $\tan^{-1} 0.7269 \approx 36.0134° \approx 36° 00'48''$.

Exercises 10.4

1 Use your calculator to verify the following results:
   a $\tan^{-1} 1.2345 \approx 50.9910°$, which $\approx 50° 59'28''$
   b the angle whose tangent is $0.8692 \approx 40.9972°$, which $\approx 40° 59'50''$
   c the angle whose tangent is exactly 1 is $45°$

2 Find the acute angles that have the following values for their tangents (answer in decimal degrees correct to 4 significant figures):
   a 0.7186   b 0.1246   c 0.0365

3 Use your calculator to find the following angles (in decimal degrees):
   a $\tan^{-1} 3.457$ b arc tan 1.573   c $\tan^{-1} 0.006\,421$

4 Find, in degrees and minutes, the angles that have the following tangents:

   a 0.6315   **b** 17.84   **c** 5.4120

5 Find the size of the angle θ in the adjacent right-angled triangle.

Set your work out thus:

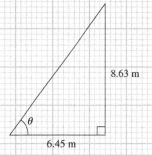

$$\tan \theta = \frac{8.63}{6.45}$$

$$\approx \dots$$

$$\therefore \theta = \dots$$

8.63 m

6.45 m

6 Setting your work out as shown in question 5, find the angles marked by pronumerals in the following triangles. Record answers both in decimal degrees and in degrees and minutes.

  a

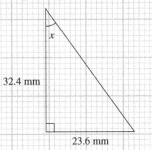

x

32.4 mm

23.6 mm

  b

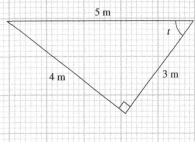

5 m

t

4 m

3 m

7 Evaluate the following in degrees and minutes:

  a $\text{arc tan} = \dfrac{2.346}{1.254}$ **b** $\tan^{-1}(3.25 \times 0.492)$

8 **a** In $\triangle ABC$: $\angle C = 90°$, $\angle B = 61°24'$ and $a = 2.46$ m; evaluate b.

    b In $\triangle KTD$: $\angle K = 90°$, $d = 83.7$ m and $t = 76.2$ m; evaluate $\angle D$.

    c In $\triangle TPW$: $\angle T = 90°$, $\angle P = 63°34'$ and $p = 7.62$ mm; evaluate w.

    d In $\triangle ABC$: $\angle B = 90°$, $a = 6.83$ m and $c = 4.61$ m; evaluate $\angle C$.

10.5 THE SINE AND COSINE RATIOS

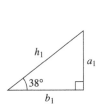

h_1

a_1

38°

b_1

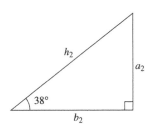

h_2

a_2

38°

b_2

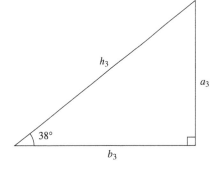

h_3

a_3

38°

b_3

For right-angled triangles containing an angle of 38°, the ratio $\dfrac{\text{length of opposite side}}{\text{length of adjacent side}}$ is a constant,

i.e. $\dfrac{a_1}{b_1} = \dfrac{a_2}{b_2} = \dfrac{a_3}{b_3}$. We called this ratio tan 38° (≈ 0.781).

Similarly, the ratio $\dfrac{\text{length of opposite side}}{\text{length of hypotenuse}}$ is a constant:

$$\frac{a_1}{h_1} = \frac{a_2}{h_2} = \frac{a_3}{h_3} = \cdots$$

We call this ratio the **sine** of 38° or sin 38°.

Similarly, the ratio $\dfrac{\text{length of adjacent side}}{\text{length of hypotenuse}}$ is a constant:

$$\frac{b_1}{h_1} = \frac{b_2}{h_2} = \frac{b_3}{h_3} = \cdots$$

We call this ratio the **cosine** of 38° or cos 38°.

The sine and cosine of any angle are obtained from the calculator by pressing the sin and cos keys. The angles that have a given sine or cosine are obtained by pressing the $\sin^{-1}$ and $\cos^{-1}$ keys (or arc sin, arc cos, INV sin or INV cos keys).

Exercises 10.5

1 Evaluate, correct to 4 significant figures:

   a sin 57°   **b** sin 46° 37′   **c** cos 12°   **d** cos 53.62°   **e** cos 76° 51′

2 Evaluate, correct to 4 significant figures:

   a 6.483 cos 38°   **b** $\dfrac{\sin 15°}{3.245}$ **c** $\dfrac{17.82}{\cos 72°13'}$ **d** $\dfrac{2.385 \sin 41°}{\sin 69°}$

3 Evaluate in decimal degrees (correct to 4 significant figures) and also in degrees and minutes:

   a $\cos^{-1} 0.7631$ **b** arc sin 0.3427   **c** arc cos 0.0436   **d** $\sin^{-1} 0.9138$

Right-angled triangles are solved using the sine and cosine ratios, in the same way as when using the tangent ratio.

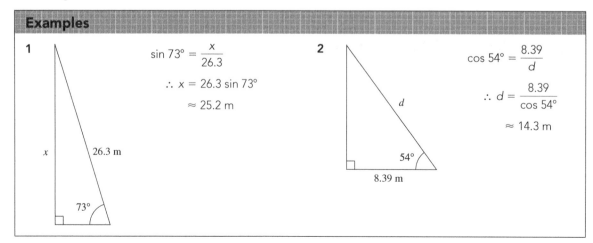

Examples

1
$$\sin 73° = \frac{x}{26.3}$$
$$\therefore x = 26.3 \sin 73°$$
$$\approx 25.2 \text{ m}$$

x   26.3 m   73°

2
$$\cos 54° = \frac{8.39}{d}$$
$$\therefore d = \frac{8.39}{\cos 54°}$$
$$\approx 14.3 \text{ m}$$

d   54°   8.39 m

Rule: $\tan \theta = \dfrac{\text{opposite}}{\text{adjacent}}$

$\sin \theta = \dfrac{\text{opposite}}{\text{hypotenuse}}$

$\cos \theta = \dfrac{\text{adjacent}}{\text{hypotenuse}}$

Exercises 10.5 *continued*

4

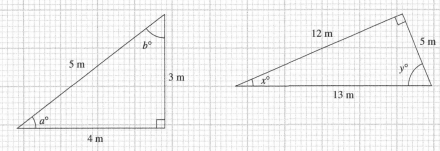

Write down, as fractions, the values of the following ratios, where the angles $a°$, $b°$, $x°$ and $y°$ are as shown in the triangles above:

a $\tan a°$ **b** $\sin y°$ **c** $\cos b°$ **d** $\sin x°$

e $\tan y°$ **f** $\cos x°$ **g** $\sin b°$ **h** $\cos a°$

5 Evaluate the pronumerals in the following figures correct to 3 significant figures:

a   **b**   **c**

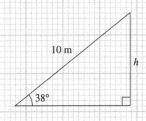 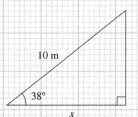 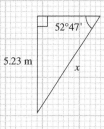

d   **e**   **f**

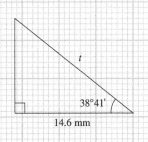 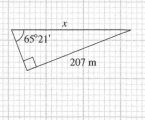 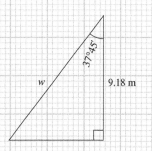

continued

continued

6   In the figures below, evaluate the angles labelled x correct to the nearest minute:

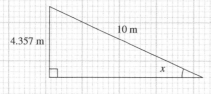

a   10 m
     4.357 m
            x

b   14.8 mm
         x
     20 mm

7   In $\triangle KTW$:
    a $\angle W = 90°$, $\angle K = 34°23'$ and $w = 43.8$ m; evaluate k.
    b $\angle W = 90°$, $\angle T = 68°14'$ and $t = 3.82$ m; evaluate w.

8   In $\triangle ABE$:
    a $\angle B = 90°$, $b = 17.6$ m and $e = 10.7$ m; evaluate $\angle E$ correct to the nearest minute.
    b $\angle A = 90°$, $a = 32.8$ mm and $b = 23.4$ mm; evaluate $\angle E$ correct to the nearest minute.

9   The midpoint M of one side AB of a square $ABCD$ is joined to the opposite corner of the square, D. Find the angle that MD makes with the side AB correct to the nearest minute.

10 A triangle has sides that measure 3 m, 4 m and 5 m. Find the measures of the three angles, correct to the nearest minute.

11 Find the altitude of an isosceles triangle whose base measures 14.8 m and whose base angles are each $63°24'$.

10.6 APPLICATIONS

Exercises 10.6

State results correct to 3 significant figures and angles to the nearest minute.

1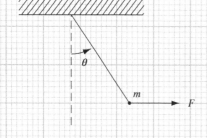
    m
       F
    θ

A mass m kg is suspended from a cable. The horizontal force required to hold this mass so that the cable makes an angle θ with the vertical is given by $F = mg \tan \theta$ N, where $g \approx 9.8$ m/s^2.
Find:

a   the force F required to hold a mass of 1 t so that $\theta = 35°$

b   the angle θ when $m = 570$ kg and the force F is 3 kN

2   The ideal velocity of a vehicle (i.e. the velocity that gives no tendency to sideslip) when travelling in a curve of radius r m, on a track that is banked at an angle θ to the horizontal, is given by $v = \sqrt{gr \tan \theta}$ m/s, where $g \approx 9.8$ m/s^2. Find:

a the ideal velocity for a vehicle travelling on a circular track of radius 200 m, banked at an angle of 20° (answer in km/h)

b the ideal angle of bank for a track of radius 150 m, designed for a vehicle travelling at 80 km/h

3 The cross-section of a wedge is in the form of an isosceles triangle with sides 324 mm and base 53 mm. Find $\angle A$ of the wedge.

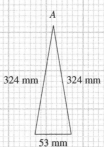

4

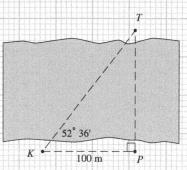

The connecting rod AB of an engine is 0.8 m long and the length of the crank BC is 0.3 m. Calculate the total angle through which AB oscillates.

5 A cylindrical roller has diameter 0.64 m and its handle, attached to its axis, has length 2.3 m. If the roller rests on level ground, find the greatest angle through which the handle can rotate.

6 To find the width of a river, a tree T on the opposite bank is sighted from P. A peg is placed at K, 100 m from P so that $\angle TPK = 90°$.
The angle TKP is then measured as 52° 36′. Find the measure of PT.

7

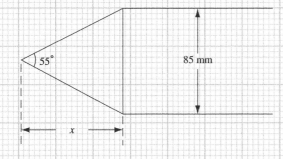

Calculate the dimension x.

+++ 8 Find the diameter of the cylindrical plug (shaded) in the diagram.
(If necessary, revise the results in section 8.5 concerning two tangents drawn to a circle from an external point.)

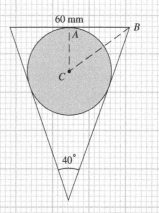

continued

continued

9

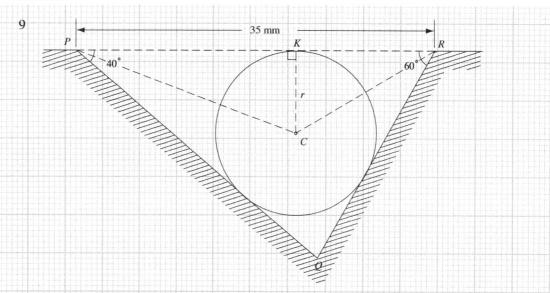

A groove *PQR* is to be cut in a block of metal, where $\angle RPQ = 40°$, $\angle PRQ = 60°$ and the width of the groove, *PR*, is 35 mm, as shown. To test the accuracy of the cut, a cylindrical plug is to be fitted, which will be flush with the top surface of the metal, *PKR*.

   a Find the measure of *PK* in terms of *r*.

   b Find the measure of *RK* in terms of *r*.

   c Using the fact that $PK + RK = 35$ mm, find the radius of the plug to be used.

10

This diagram shows the relationship between the applied AC voltage, the voltages across the resistor and the inductance and the phase angle ϕ by which the current *I* lags the applied voltage.

   a Given that $\phi = 18.6°$ and $v = 243$ v, evaluate v_L.

   b Given that $\phi = 41.3°$ and $v_L = 6.24$ v, evaluate v_R.

   c Given that $\phi = 27.5°$ and $v_R = 23.8$ v, evaluate v.

   d Given that $v = 126$ V and $v_R = 89.3$ v, evaluate ϕ.

11

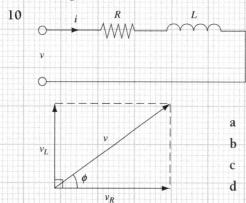

This diagram shows the relationships between the current *i*, the currents through the resistor and the capacitor, and the phase angle ϕ by which the current *i* leads the applied AC voltage.

   a Given that $i_R = 2.15$ A and $\phi = 32.8°$, evaluate i_C.

   b Given that $i_C = 16.3$ mA and $\phi = 71.4°$, evaluate *i*.

   c Given that $i = 5.26$ mA and $i_R = 4.82$ mA, evaluate ϕ.

   d Given that $i = 1.23$ A and $i_C = 687$ mA, evaluate ϕ.

10.7 HOW TO FIND A TRIGONOMETRICAL RATIO FROM ONE ALREADY GIVEN

If we are given the value of one trigonometrical ratio of an unknown angle, we can find the value of any other trigonometrical ratio by means of Pythagoras' theorem. We place the angle in a right-angled triangle such that it has the given trigonometrical ratio and use Pythagoras' theorem to evaluate the third side of the triangle.

Examples

1 Given that $\sin \theta = k$, express $\tan \theta$ in terms of k.

$\sin \theta = k$ gives the angle θ as shown in the adjacent diagram.

The third side of the triangle $= \sqrt{1 - k^2}$.

$$\therefore \tan \theta = \frac{k}{\sqrt{1 - k^2}}$$

2 The same method can be used when the given ratio is a number.

If $\tan \beta = 0.8572$, this gives the angle β as shown in the adjacent diagram.

The hypotenuse is $\sqrt{1^2 + (0.8572)^2} \approx 1.3171$

$$\therefore \sin \beta = \frac{0.8572}{1.3171}$$

$$= 0.6508$$

However, it is much simpler to use a calculator without drawing any diagram.

$$\tan \beta = 0.8572$$

$$\therefore \beta = \tan^{-1} 0.8572$$

$$\sin \beta = \sin(\tan^{-1} 0.8572)$$

$$= 0.6508$$

Exercises 10.7

1 Given that $\cos \theta = t$:

  a Express $\sin \theta$ in terms of t.

  b Express $\tan \theta$ in terms of t.

  c Express $\sin \theta \times \tan \theta$ in terms of t.

2 Given that $\beta = \frac{2}{3}$, find the *exact* value of $\cos \beta$.

3 Given that $\cos \phi = 0.25$ exactly, find the *exact* value of $\tan \phi$.

Example

Given that   cos A = 0.6413, evaluate tan A.

$$\cos A = 0.6413$$

$$\therefore A = 50.1112°$$

$$\therefore \tan A = 1.196$$

Exercises 10.7 *continued*

4 **a** Given that $\tan \theta = 1.8604$, evaluate $\sin \theta$.

   b Given that $\cos R = 0.8405$, evaluate $\tan R$.

5 **a** Given that $\cos A = \dfrac{234.6}{719.8}$, evaluate $\sin A$.

   b Given that $\dfrac{1}{\cos K} = 3.5072$, evaluate $\sin K$

6 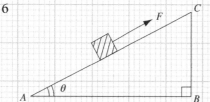 In the adjacent diagram, $AB = 1$ m and $BC = 385$ mm.
   Neglecting friction, the force required to pull a mass of
   8.40 kg up the plane with constant velocity is given by:
   $F = 8.40 \times 9.81 \times \sin \theta$ newtons.
   Find the size of this force.

7 A circuit has an impedance, Z, of 745 Ω.
   Given that the power factor of the circuit, $\cos \phi$, is 0.860, find the reactive component of the
   impedance, $Z \sin \phi$.

10.8 COMPASS DIRECTIONS; ANGLES OF ELEVATION AND DEPRESSION

Compass directions

A compass direction may be specified in either of two ways:

1 by stating its direction in terms of north, south, east and west

2 by stating its **bearing**, the angle of *clockwise* rotation from direction *north* to the direction
   concerned. Three digits are always stated, e.g. 237°, 068°, 007°.

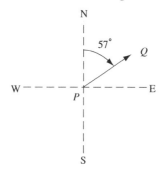

Direction *PQ* may be expressed as '57° E of N', or 'N57°E' or
'having a bearing of 057°'.

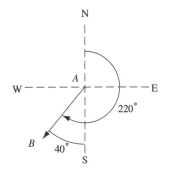

Direction *AB* may be expressed as '40° W of S', or 'S40°W' or 'having a bearing of 220°'.

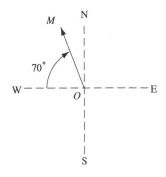

Direction *OM* may be expressed as '20° W of N', or 'N20°W' or 'having a bearing of 340°'.

Note: When using method 1, the rotation stated is always from north or south towards east or west, and not as a rotation from east or west. If the angle is not stated, 45° is implied, e.g. NE means N45°E and SW means S45°W. When using method 2, the angle of rotation is always *clockwise* from *north*.

Exercises 10.8

1 Express each of the following directions as a bearing:
  a S20°E   **b** N10°W   **c** SW
  d 40° W of S   **e** 70° east of north   **f** NW

2 A ship is 23.4 km due west of a certain radio direction finding station. After sailing due north for some time, the bearing of the same station is measured to be 143°. How far did the ship sail?

3 *A* and *B* are two radio receivers situated 253 km apart, *B* being due east of *A*. From *A* the bearing of a certain transmitter is measured as 068° and from *B* the bearing of the same transmitter is measured as 338°. How far is the transmitter from *B*?

4 A ship sails for 42.0 km in a direction bearing 048° and then sails for 57 km in a direction bearing 063°. How far eastwards has the ship sailed?

Angles of elevation and depression

From a given point *P*, the **angle of elevation** or **depression** of another point *Q* is the angle that the line *PQ* makes with the *horizontal*.

Example

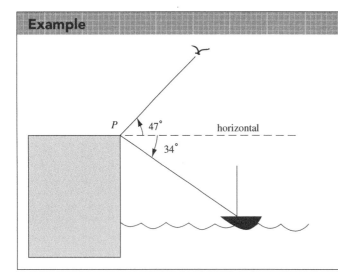

From point P (on top of a cliff), the angle of elevation of the seagull is 47° and the angle of depression of the boat is 34°.

Exercises 10.8 *continued*

5 From a certain point P on the same horizontal level as the base of a tower, and 55.0 m from it, the angle of elevation of the top of the tower is measured as 49° 37′.

   a Find the height of the tower.

   b A wire stay runs from a point 12.23 m below the top of the tower to the point P. Assuming the stay to be straight, find the angle that the stay makes with the ground.

   c Find the length of the stay.

6 From a point at the top of a vertical cliff, 63.45 m above sea level, the angle of depression of a boat is measured as 23°14′. Find the distance of the boat from the base of the cliff.

 10.9 AREA OF A TRIANGLE

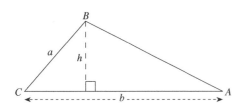

$$\sin C = \frac{h}{a}$$

$$\text{Area of } \Delta = \tfrac{1}{2}bh$$
$$= \tfrac{1}{2}b \times a \sin C$$
$$= \tfrac{1}{2}ab \sin C$$

Rule: The area of a triangle is given by one-half of the product of the lengths of any two sides and the sine of the included angle.

Note that the formula involves the three letters by which the triangle is named. For example, for a triangle *KBT*,

$$\text{the area} = \tfrac{1}{2}kb \sin T$$
$$\text{or} = \tfrac{1}{2}bt \sin K$$
$$\text{or} = \tfrac{1}{2}tk \sin B$$

Exercises 10.9

1 Find the areas of the following triangles:

  a $\triangle ABC$, where $a = 23.6$ m, $b = 17.8$ m, $\angle BCA = 23.7°$

  b $\triangle PTK$, where $t = 3.46$ m, $k = 4.89$ m, $\angle TPK = 43°51'$

  c $\triangle NEW$, where $w = 4.53$ km, $n = 561$ m, $\angle NEW = 62°43'$

SELF-TEST

1 a

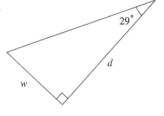

Evaluate $\dfrac{w}{d}$

b

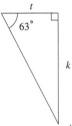

Evaluate $\dfrac{t}{k}$

c

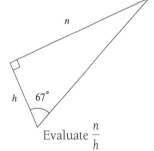

Evaluate $\dfrac{n}{h}$

2 Evaluate the pronumerals:

  a

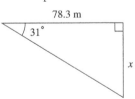

  b

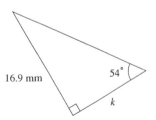

3 Use your calculator to evaluate, correct to 4 significant figures:

  a $\tan 43°16'$ **b** $\tan 18°39'$ **c** $\dfrac{468.4 \sin 56°47'}{723.6}$

4

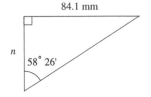

Evaluate *n*.

5 **a** In $\triangle PQR$: $\angle R = 90°$, $\angle Q = 26°52'$ and $p = 1.48$ m; evaluate q.

   b In $\triangle ABC$: $\angle C = 90°$, $a = 5.63$ m and $b = 8.64$ m; evaluate $\angle B$.

6 Evaluate $\dfrac{3.842 + \cos 47°32'}{1.638 - \cos 38°41'}$ correct to 3 significant figures.

7 **a** In $\triangle KTW$: $\angle T = 90°$, $\angle K = 31°18'$ and $t = 765$ m; evaluate w.

   b In $\triangle KTW$: $\angle K = 90°$, $\angle T = 71°13'$ and $t = 3.28$ m; evaluate k.

8 **a** In $\triangle ABE$: $\angle E = 90°$, $a = 345$ m and $e = 864$ m; evaluate $\angle A$.

   b In $\triangle ABE$: $\angle A = 90°$, $b = 34.6$ mm and $e = 51.8$ mm; evaluate $\angle B$.

9 From a ship, a certain lighthouse at a distance of 5.20 km bears 214°. How far does the ship have to sail in direction due south before the lighthouse bears 270°?

10 In $\triangle ABC$: $\angle A = 40°00'$, $\angle B = 50°00'$, and $BC = 3.184$ m.

   Find the length of MB where M is the midpoint of CA.

PART 3

APPLIED MATHEMATICS

CHAPTER 11

INDICES AND RADICALS

11.1 RADICALS

The square root of 25 is 5, because $5 \times 5 = 25$. $\qquad$ $\sqrt{25} = 5$

The cube root of 27 is 3, because $3 \times 3 \times 3 = 27$. $\qquad$ $\sqrt[3]{27} = 3$

The fourth root of 16 is 2, because $2 \times 2 \times 2 \times 2 = 16$. $\qquad$ $\sqrt[4]{16} = 2$

The **inverse** of an operation undoes the operation. If we perform an operation on a number, then perform its inverse operation, we return to the original number. The following are inverse operations:

- addition and subtraction (e.g. $a + 371 - 371 = a$)
- multiplication and division (e.g. $a \times 69 \div 69 = a$)
- squaring and finding the square root (e.g. for $a > 0$, $\sqrt{a^2} = a$ and $(\sqrt{a})^2 = a$).

  Similarly, $(\sqrt[3]{x})^3 = x$, $\sqrt[5]{n^5} = n$, etc.

  An expression of the form $\sqrt[n]{a}$, the nth root of the number a, is called a radical.

Exercises 11.1

1 Evaluate:

   a $(\sqrt{9})^2$ **b** $\sqrt{7^2}$ **c** $(\sqrt{163.72})^2$

   d $(\sqrt{893.4})^2$ **e** $\sqrt{3} \times \sqrt{2} \times \sqrt{3} \times \sqrt{2}$ **f** $(\sqrt{5})^2 \times \sqrt{7^2}$

2 Simplify:

   a $(3\sqrt{x})^2$ **b** $\sqrt{a}\sqrt{b} \times \sqrt{a}\sqrt{b}$ **c** $(a\sqrt{b})^2$

   d $(\sqrt{a}\sqrt{b})^2$ **e** $(\tfrac{1}{2}k\sqrt{n})^2$ **f** $3(2\sqrt{4ab})^2$

Since $(\sqrt{a}\sqrt{b})^2 = (\sqrt{ab})^2$, both being equal to ab, $\therefore \sqrt{a}\sqrt{b} = \sqrt{ab}$.

Example

$\sqrt{5} \times \sqrt{20} = \sqrt{100}$
$\qquad\qquad = 10$

Exercises 11.1 *continued*

3 Evaluate:

  a $\sqrt{2} \times \sqrt{8}$ **b** $\sqrt{12} \times \sqrt{3}$ **c** $\sqrt{3} \times \sqrt{27}$ **d** $\sqrt{3} \times \sqrt{48}$

4 Simplify:

  a $\dfrac{t}{m} \times \dfrac{x}{k}$ **b** $\dfrac{\sqrt{a}}{\sqrt{b}} \times \dfrac{\sqrt{a}}{\sqrt{b}}$ **c** $\left(\dfrac{\sqrt{a}}{\sqrt{b}}\right)^2$ **d** $\left(\sqrt{\dfrac{a}{b}}\right)^2$

Since $\left(\dfrac{\sqrt{a}}{\sqrt{b}}\right)^2 = \left(\sqrt{\dfrac{a}{b}}\right)^2$, both being equal to $\dfrac{a}{b}$, $\therefore \dfrac{\sqrt{a}}{\sqrt{b}} = \sqrt{\dfrac{a}{b}}$

Example

$\dfrac{\sqrt{45}}{\sqrt{5}} = \sqrt{\dfrac{45}{5}} = \sqrt{9} = 3$

Exercises 11.1 *continued*

5 Evaluate mentally:

  a $\dfrac{\sqrt{18}}{\sqrt{2}}$ **b** $\dfrac{\sqrt{3}}{\sqrt{27}}$ **c** $\sqrt{32} \div \sqrt{2}$

  d $\dfrac{\sqrt{10}}{\sqrt{5}} \times \dfrac{\sqrt{6}}{\sqrt{3}}$ **e** $\dfrac{\sqrt{24}}{\sqrt{2}} \div \dfrac{\sqrt{5}}{\sqrt{15}}$ **f** $\sqrt{\dfrac{54}{2}} \times \sqrt{\dfrac{21}{7}}$

  g $\dfrac{\sqrt{8}}{\sqrt{3}} \div \dfrac{\sqrt{2}}{\sqrt{27}}$ **h** $\dfrac{\sqrt{6}}{\sqrt{7}} \times \sqrt{\dfrac{14}{3}}$ **i** $\sqrt{\dfrac{10}{3}} \div \dfrac{\sqrt{5}}{\sqrt{24}}$

6 Using the results $(a + b)^2 = a^2 + 2ab + b^2$ and $(a - b)^2 = a^2 - 2ab + b^2$, expand the following (i.e. express without brackets):

  a $(x + \sqrt{x})^2$ **b** $(3\sqrt{m} + 2)^2$ **c** $(5 - \sqrt{a})^2$

  d $(1 - 2\sqrt{x})^2$ **e** $(2n + 3\sqrt{n})^2$ **f** $(x\sqrt{x} - 1)^2$

continued

continued

7 Using the result $(a + b)(a - b) = a^2 - b^2$, simplify the following:

a $(\sqrt{5} + \sqrt{2})(\sqrt{5} - \sqrt{2})$

b $(x - \sqrt{3})(x + \sqrt{3})$

c $\left(-b + \dfrac{\sqrt{x}}{y}\right)\left(-b - \dfrac{\sqrt{x}}{y}\right)$

d $\left(\dfrac{-b}{2a} + \dfrac{\sqrt{b^2 - 4ac}}{2a}\right)\left(\dfrac{-b}{2a} - \dfrac{\sqrt{b^2 - 4ac}}{2a}\right)$

+++ e $a\left(\dfrac{-b + \sqrt{b^2 - 4ac}}{2a}\right)\left(\dfrac{-b - \sqrt{b^2 - 4ac}}{2a}\right)$

Remember: $(\sqrt[n]{x})^n = \sqrt[n]{x^n} = x; \quad \sqrt{a} \times \sqrt{b} = \sqrt{ab}; \quad \dfrac{\sqrt{a}}{\sqrt{b}} = \sqrt{\dfrac{a}{b}}$

Beware: $\sqrt{a} + \sqrt{b} \neq \sqrt{a + b}$

e.g. $\sqrt{4} + \sqrt{9} = 2 + 3 = 5$, which is *not* $\sqrt{13}$

Similarly, $\sqrt{a} - \sqrt{b} \neq \sqrt{a - b}$

e.g. $\sqrt{16} - \sqrt{9} = 4 - 3 = 1$, which is *not* $\sqrt{16 - 9}$

11.2 POSITIVE INTEGRAL INDICES

You have already learnt that a^n is a 'shorthand' notation for $a \times a \times a \times a \times a \times \ldots$ (n terms).

Examples

1 $2^7 = 2 \times 2 \times 2 \times 2 \times 2 \times 2 \times 2 \ (= 128)$

2 $10^{13} = 10\ 000\ 000\ 000\ 000$, so that, for example, $68\ 300\ 000\ 000\ 000 = 6.83 \times 10^{13}$

When a number is written in the form a^n, it is said to be written in **exponential form:** a is called the base and n is called the **exponent** or the **index.**

Example

The number 64 may be written in exponential form as:

8^2 (using base 8) 4^3 (using base 4) 2^6 (using base 2)

You must note negative signs carefully when evaluating numbers in exponential form.

Examples

1 $3^2 = 3 \times 3$
    $= 9$

2 $(-3)^2 = (-3) \times (-3)$
    $= 9$

3 $(-3)^3 = (-3) \times (-3) \times (-3)$
    $= -27$

4 $\dfrac{3^2}{4} = \dfrac{3 \times 3}{4}$
    $= \dfrac{9}{4}$
    $= 2\frac{1}{4}$

5 $\left(\dfrac{3}{4}\right)^2 = \dfrac{3}{4} \times \dfrac{3}{4}$
    $= \dfrac{9}{16}$

Exercises 11.2

1 Evaluate mentally:

    a 3^4
    b 1.2^2
    c 0.2^3
    d $(\frac{2}{3})^2$
    e $(\frac{1}{2})^4$

    f -3^2
    g $(-3)^2$
    h -2^3
    i $(-2)^3$
    j $(-\frac{3}{4})^2$

2 Evaluate mentally:

    a $(-2)^2$
    b $(-2)^3$
    c $(-2)^4$
    d $(-2)^5$
    e $(-2)^6$

3 *Without* evaluating, state whether the following numbers are positive or negative (write P or N):

    a $(-3)^{10}$
    b $(-7)^9$
    c $(-6.38)^7$
    d $(-81.1)^8$
    e $(-4.38 \times 10^3)^6$

4 Evaluate mentally:

    a 1^8
    b $(-1)^8$
    c 1^9
    d $(-1)^9$

5 Evaluate mentally:

    a $(0.3)^2$
    b $(-0.2)^3$
    c $(-0.1)^2$
    d $-(0.3)^2$
    e $(-0.3)^2$

11.3 SUBSTITUTIONS AND SIMPLIFICATIONS

Remember: An index (exponent) refers only to the pronumeral above which it is placed.

Examples

$ab^2 = a \times b^2 = a \times b \times b$

$a^2b = a^2 \times b = a \times a \times b$

$(ab)^2 = ab \times ab = a \times b \times a \times b$

Note: $(ab)^2 = a^2b^2$
          $3(xy)^2 = 3x^2y^2$
       $(tmk)^3 = t^3m^3k^3$
          $(3xy)^2 = 9x^2y^2$

Exercises 11.3

1 Given that $m = 2$ and $k = -3$, evaluate mentally:

   a mk^2 **b** $(mk)^2$ **c** m^3k^2 **d** $10(m^2k)^2$

   e $2m^3 + m^2k$ **f** $(m - k)^2$ **g** $(k - 2m)^2$ **h** $5(m + k)^3$

2 Given that $a = -1$, $b = -2$ and $c = 3$, evaluate mentally:

   a a^2b^2 **b** $2abc^2$ **c** $5a^4c$ **d** ab^3c^3

   e $2(b - a)^3$ **f** $3(2b + c)^2$ **g** $(b + c - a)^4$ **h** $2(3a^2b)^2$

3 Express without brackets:

   a $(2x)^3$ **b** $(\frac{1}{2}x)^2$ **c** $3x^2 + (3x)^2$

   d $(-3x)^2$ **e** $(-2x)^3$ **f** $3(-2x)^2$

   g $(4x)^2 - (-3x)^2$ **h** $x^2 - (-2x)^2$ **i** $x(2x)^3$

11.4 THE FOUR LAWS FOR INDICES

The first law

$2^3 \times 2^2 = (2 \times 2 \times 2) \times (2 \times 2) = 2^5$

$7^2 \times 7^4 = (7 \times 7) \times (7 \times 7 \times 7 \times 7) = 7^6$

$x^m \times x^n = x \times x \times x \times x \ldots (m \text{ terms}) \times x \times x \times x \times \ldots (n \text{ terms})$

$\qquad\qquad = x \times x \times x \times x \ldots (m + n \text{ terms})$

$\qquad\qquad = x^{m+n}$

> **Rule:** The first law is: $a^m \times a^n = a^{m+n}$

> **Note:** When the bases are the same, we can simplify a product using this rule.

Examples

1 $k^3 \times k^7 = k^{10}$ 2 $x^3 \times y^2 \times x \times x^4 \times y$

3 $2t^5 \times 3n \times 4t$ $= (x^3 \times x^1 \times x^4) \times (y^2 \times y^1)$

   $= (2 \times 3 \times 4) \times (t^5 \times t^1) \times n$ $= x^8y^3$

   $= 24t^6n$

Exercises 11.4

1 Simplify:

   a $x^5 \times y^2 \times x^7 \times x^4 \times y$ **b** $ak^3 \times 3m^5 \times 4k \times m$

   c $(2x)^3 \times x \times 3x^4$ **d** $2x^2 \times (3x)^2 \times x \times 2x$

2 Simplify:

a $a^2b \times a^3b^2$ **b** $mk^2 \times mk^3t$

c $2a^3xy^2 \times 3a^2x^3$ **d** $mk(m + k) - k(m^2 - mk)$

3 Simplify, expressing without brackets:

a $a^3(a^2 + a^5)$ **b** $3x^2(2x + 1)$

c $2k^3(3k - 5) + k^2(2k^2 - k)$ **d** $m^2(l - 2m) - 3m(m^2 - 2m)$

4 Simplify, expressing without brackets:

a $(a^2 + b)(a^2 - b)$ **b** $(a^3 - 1)(3 - a^5)$

c $(2 - k^2)(3 - k^2)$ **d** $(5 + m^3)(5 - m^3)$

5 Expand, expressing without brackets:

a $(a^2 + 1)(a^3 - a)$ **b** $(k - k^4)(2 - k^3)$

c $(2m - 1)(3m^2 - 1)$ **d** $(t^2 - t^3)(t^3 + t^2)$

The second law

$$7^6 \div 7^2 = \frac{7 \times 7 \times 7 \times 7 \times 7 \times 7}{7 \times 7} = 7^4$$

$$3^5 \div 3^3 = \frac{3 \times 3 \times 3 \times 3 \times 3}{3 \times 3 \times 3} = 3^2$$

$$a^m \div a^n = \frac{a \times a \times a \times \ldots (m \text{ terms})}{a \times a \times a \times \ldots (n \text{ terms})} = a^{m-n}$$

> **Rule:** The second law is: $a^m \div a^n = a^{m-n}$

Exercises 11.4 *continued*

6 Simplify:

a $m^8 \div m^4$ **b** $6k^3 \div 6k$ **c** $3a^6 \div \frac{1}{2}a^2$ **d** $\frac{1}{4}x^{12} \div 4x^6$

7 Simplify:

a $12a^2b^3 \div 4ab^2$ **b** $6mt^4 \div 6t^3$ **c** $4k^3h \div \frac{1}{2}kh$

d $m^3w^2 \div \frac{1}{3}mw^2$ **e** $\dfrac{a^2 + 2ab + b^2}{a + b}$ **f** $\dfrac{x^2 - 2kx + k^2}{k - x}$

8 Simplify, using the first and second laws:

a $\dfrac{x^3 \times x^4}{x^5}$ **b** $\dfrac{3m^4 \times 4m^5}{6m^2}$ **c** $\dfrac{2a^2b^4 \times 4a^3c}{4ab^3}$ **d** $\dfrac{m^2kt \times m^3t^2}{m^2 \times mk}$

The third law

$(5^3)^2 = 5^3 \times 5^3 = 5^6$

$(7^5)^3 = 7^5 \times 7^5 \times 7^5 = 7^{15}$

$(x^m)^n = x^m \times x^m \times x^m \ldots (n \text{ terms}) = x^{mn}$

Rule: The third law is: $(a^m)^n = a^{mn}$

Exercises 11.4 *continued*

9 Simplify:

a $(x^3)^5$

b $2(a^2)^3$

c $(2a^2)^3$

d $(ab^3)^2$

e $(m^2b^5)^3$

f $(-x^4)^3$

g $(-m^3)^6$

h $(-2m^2t^3w^4)^3$

i $2(-xy^2)^3$

10 Simplify:

a $(2x^2)^3 - 4(x^3)^2$

b $3x^{12} + (3x^4)^3$

c $(ab^2)^4 - (a^2b^4)^2$

d $(m^3x)^6 - (-m^6x^2)^3$

11 Express without brackets:

a $(p^3 + 3)^2$

b $(2t^3 - 3t^2)^2$

c $(1 - 3a^2b)^2$

The fourth law

$(ab)^m = ab \times ab \times ab \times \ldots (m \text{ terms})$

$\quad = [a \times a \times a \times a \ldots (m \text{ terms})] \times [b \times b \times b \times \ldots (m \text{ terms})]$

$\quad = a^m b^m$

Similarly, $\left(\dfrac{a}{b}\right)^m = \dfrac{a^m}{b^m}$

Rule: The fourth law is: $(ab)^m = a^m b^m$ and $\left(\dfrac{a}{b}\right)^m = \dfrac{a^m}{b^m}$

Exercises 11.4 *continued*

12 Express as fractions without brackets:

a $\left(\tfrac{2}{3}\right)^4$ b $\left(\tfrac{1}{2}\right)^5$ c $\left(\dfrac{3n}{5}\right)^2$ d $\left(\dfrac{2t}{3n^2}\right)^3$ e $\left(-\dfrac{4a^3}{b^2}\right)^2$ f $\left(\dfrac{t^2 - 2t}{3k^3}\right)^2$

13 Simplify:

a $2^n \times 3^n \times 4^n$

b $\dfrac{12^k}{3^k}$

c $2^x \times 3^t \times 4^x \times 5^t \times 2^x$

d $\dfrac{(4x + 6y)^t}{2^t}$

e $\dfrac{(7a + 7b)^x}{(a + b)^x}$

f $\dfrac{(a^2 - 2ab + b^2)^7}{(a - b)^7}$

g $\dfrac{(x^2 - y^2)^5}{(x - y)^5}$

> **Note:** ■ **When the bases are the same,** we can use the first law to simplify a product.
>
> $$a^x \times a^y = a^{x+y}, \quad k^t \times k^{3t} = k^{4t}$$
>
> ■ **When the indices are the same,** we can use the fourth law to simplify a product:
>
> $$2^{13} \times 4^{13} = 8^{13}, \quad a^7 \times b^7 \times c^7 = (abc)^7$$
>
> ■ **However, when both the bases and the indices are different,** *no simplification is possible:*
>
> $2^5 \times 3^7$ There is no simpler way to write this number, except by evaluating it.
>
> $x^5 \times y^3$ No simplification is possible.
>
> $a^k \times b^n$ No simplification is possible.

Exercises 11.4 *continued*

14 Simplify if possible:

a $2^n \times 3^n$ b $n^2 \times n^3$ c $2^n \times 3^t$

d $2^k x \times 4^k y^2$ e $a^n \times a^n$ f $x^y \times y^x$

11.5 ZERO, NEGATIVE AND FRACTIONAL INDICES

We have now established four basic laws for indices that are positive integers (i.e. positive whole numbers). However, we have not given any meaning to zero, fractional or negative indices such as $a^0, a^{-3}, a^{\frac{2}{3}}, a^{-\frac{3}{4}}$. Since it is convenient not to have any exception to our four basic laws, the meanings of these indices are defined so that they obey these same laws.

For example, since the same laws are to apply, statements such as the following must be true:

$a^0 \times a^3 = a^3$	(Law 1)	$x^{-2} \times x^5 = x^3$	(Law 1)
$x^5 \div x^0 = x^5$	(Law 2)	$(a^{-3})^4 = a^{-12}$	(Law 3)
$m^{\frac{2}{3}} \times m^{\frac{1}{3}} = m^{\frac{3}{3}}$	(Law 1)	$(x^{-2})^{-3} = x^6$	(Law 3)

Exercises 11.5

1 Simplify, expressing the result in the form ax^n:

a $k^{-2} \times k^{-3}$ b $2t^3 \times 4t^{-4}$ c $6m^2 \div 2m^{-4}$

d $a^{-\frac{1}{2}} \times a^{\frac{1}{2}}$ e $8m^{-\frac{1}{2}} \div 8m^{\frac{3}{2}}$ f $12b^{-\frac{1}{3}} \div 3b^{-\frac{2}{3}}$

2 Expand (i.e. multiply out):

a $x^3(x^2 - x^{-1})$ b $a^{-3}(a + a^{-2})$ c $2k^{\frac{1}{2}}(3k + k^{-\frac{1}{2}})$

d $e^x(e^{-2x} + e^x)$ e $(e^{2x} - e^x)(e^{-2x} + e^{-x})$ f $(2k^{\frac{1}{2}} - 3)(k - k^{\frac{1}{2}})$

continued

continued

3 Multiply and divide out, expressing in simplest exponential form:

a $\dfrac{a^2(a^3 + 2)}{a^{-1}}$

b $\dfrac{k^{-1} - k}{k^{-1}}$

c $\dfrac{e^x(e^{2x} - 1)}{e^{2x}}$

d $\dfrac{x - x^{\frac{1}{2}}}{x^{-\frac{1}{2}}}$

e $\dfrac{t^{\frac{1}{2}} + t^{-\frac{1}{2}}}{t}$

f $\dfrac{6m^0 + 4m}{2m^{-\frac{1}{2}}}$

Having decided that the zero, negative and fractional indices are to be defined so that they obey the same four basic laws as positive indices, we now examine the meanings of these indices so that we can *evaluate* a^0, a^{-n}, and $a^{\frac{1}{n}}$ when a is a real number.

The zero index

Since $a^n \times a^0 = a^n$ (Law 1), it follows that $a^0 = 1$.

Because a may be replaced by any numeral or pronumeral, we have:

$(a^2b)^0 = 1$, $(32.6)^0 = 1$, $(-5k)^0 = 1$, $(a - b^2 + 3c)^0 = 1$, etc.

Exercises 11.5 *continued*

4 Evaluate mentally:

a $a^0 \times b^0$

b $3x^0$

c $(3x)^0$

d $\dfrac{k^0}{t^0}$

e $(m^2 - 3k^2)^0$

f $\left(\dfrac{a^0}{b^0}\right)^3$

g $12 \div 6^0$

h $\dfrac{4^0 + 6^0}{2^0 + 4^0}$

i $\dfrac{5^0 - 2^0}{5^0 + 2^0}$

j $(2^3 + 2^0)(3^2 - 3^0)$

k $\dfrac{10x^0}{2}$

l $1^0 + 0^1 + 2^0 + 2^1$

Fractional indices

$(a^{\frac{1}{2}})^2 = a$ \qquad $(a^{\frac{1}{3}})^3 = a$ \qquad $(a^{\frac{1}{n}})^n = a$

$\therefore a^{\frac{1}{2}} = \sqrt{a}$ \qquad $\therefore a^{\frac{1}{3}} = \sqrt[3]{a}$ \qquad $\therefore a^{\frac{1}{n}} = \sqrt[n]{a}$

Examples

1 $9^{\frac{3}{2}} = (9^{\frac{1}{2}})^3 = 3^3 = 27$

2 $8^{\frac{5}{3}} = (8^{\frac{1}{3}})^5 = 2^5 = 32$

Exercises 11.5 *continued*

5 Evaluate mentally:

a $16^{\frac{1}{2}}$

b $27^{\frac{1}{3}}$

c $1^{\frac{1}{2}}$

d $(-8)^{\frac{1}{3}}$

e $-8^{\frac{1}{3}}$

f $4^{\frac{1}{2}}$

g $(-1)^{\frac{1}{3}}$

h $16^{\frac{3}{4}}$

i $9^{\frac{5}{2}}$

j $(-8)^{\frac{2}{3}}$

k $-8^{\frac{2}{3}}$

l $100^{\frac{3}{2}}$

6 Evaluate mentally:

a $9^{\frac{3}{8}} \times 9^{\frac{5}{8}}$ **b** $4^{\frac{1}{3}} \times 4^{\frac{1}{6}}$ **c** $25^{\frac{7}{8}} \div 25^{\frac{3}{8}}$ **d** $8^{\frac{1}{2}} \times 8^{\frac{1}{6}}$

e $27^{\frac{7}{12}} \div 27^{\frac{1}{4}}$ **f** $16^{\frac{3}{16}} \times 16^{\frac{1}{16}}$ **g** $9^{\frac{5}{6}} \div 9^{\frac{1}{3}}$ **h** $9^{\frac{1}{4}} \times 9^{\frac{5}{4}}$

7 Evaluate mentally:

a $(8^{\frac{1}{6}})^2$ **b** $(7^{\frac{1}{2}})^2$ **c** $(2^0)^{\frac{1}{2}}$ **d** $(3^{\frac{1}{2}})^4$ **e** $(27^{\frac{1}{6}})^2$

Negative indices

$x^n \times x^{-n} = x^0$ (Law 1)

$\qquad\qquad = 1$

But $x^n \times \dfrac{1}{x^n} = 1$

$\qquad \therefore x^{-n} = \dfrac{1}{x^n}$

Examples

1 $\quad 2^3 = 8$

$\quad \therefore 2^{-3} = \frac{1}{8}$

2 $\quad 9^{\frac{1}{2}} = 3$

$\quad \therefore 9^{-\frac{1}{2}} = \frac{1}{3}$

3 $\quad 8^{\frac{2}{3}} = 4$

$\quad \therefore 8^{-\frac{2}{3}} = \frac{1}{4}$

Note: $\quad x^1 = x$

$\qquad x^{-1} = \dfrac{1}{x}$

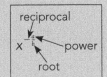

The order in which we perform the three operations does not matter.

Example

To evaluate $4^{-\frac{5}{2}}$, we can first take the square root (obtaining 2). Then we raise this result to the 5th power (obtaining 32). Finally we take the reciprocal (obtaining $\frac{1}{32}$).

$\quad \therefore 4^{-\frac{5}{2}} = \frac{1}{32}$

Exercises 11.5 *continued*

8 Evaluate mentally:

a 2^{-3} **b** 4^{-2} **c** 5^{-1} **d** 3^{-4} **e** 2^{-5}

9 Evaluate mentally:

a $4^{-\frac{1}{2}}$ **b** $8^{-\frac{1}{3}}$ **c** $16^{-\frac{1}{2}}$ **d** $27^{-\frac{1}{3}}$ **e** $1^{-\frac{1}{2}}$

continued

continued

10 Evaluate mentally:

   a $8^{-\frac{2}{3}}$ **b** $9^{-\frac{3}{2}}$ **c** $1^{-\frac{2}{3}}$ **d** $16^{-\frac{3}{2}}$ **e** $16^{-\frac{3}{4}}$

11 Evaluate mentally:

   a $\dfrac{1}{2^{-3}}$ **b** $\dfrac{1}{9^{-\frac{1}{2}}}$ **c** $\dfrac{1}{3^{-2}}$ **d** $\dfrac{2}{4^{-1}}$ **e** $\dfrac{1}{9^{-2}}$

12 Express as a single fraction without negative exponents:

   a $(3x)^{-2} + 2x^{-2}$ **b** $3t^{-1} - (3t)^{-1}$ **c** $(w-1)^{-1} + (1-w)^{-2}$

   d $3x^{-2} - 2x^{-3}$ **e** $(x^2)^{-1} \div 2x^{-2}$ **f** $x^0 + (2x)^{-1}$

11.6 SCIENTIFIC NOTATION

Indices provide a very convenient way of expressing very large and very small numbers because zeros may be replaced by a multiplier that is a power of 10.*

Examples

1 234 000 000 000 is more conveniently expressed as 234×10^9, or 23.4×10^{10} or 2.34×10^{11}.

2 0.000 000 000 613 is more conveniently expressed as 613×10^{-12}, or 61.3×10^{-11} or 6.13×10^{-10}.

Note: When a number is written in the form $C \times 10^n$ where C is a number between 1 and 10, it is said to be expressed in *scientific notation*. In the two examples above, the first number is expressed in scientific notation as 2.34×10^{11} and the second number as 6.13×10^{-10}.

Exercises 11.6

1 Express each of the following numbers in scientific notation:

   a 12 300   **b** 0.591   **c** 83 600 000   **d** 0.054 700

   e 43.8   **f** 0.000 000 063 9   **g** 56.1×10^6 **h** 0.0963×10^5

   i 4000×10^7 **j** 513×10^{-5} **k** 0.007×10^{-3} **l** 69×10^{-2}

Write all answers in the form 10^n or $C \times 10^n$ where C is an integer less than 10, i.e. in scientific notation. Do *not* use a calculator for these exercises.

2 If $D = 10^2$, evaluate:

   a D^3 **b** $1000D$ **c** $D^2 \div 10^{-3}$

3 If $r = 10^{-6}$, evaluate:

   a $100r$ **b** r^2 **c** $r \div 10^2$ **d** $r \div 10^{-4}$

---

* The age of a very old person $\approx 3 \times 10^9$ seconds ($\approx$ 95 years).

The age of the universe $\approx 10^{18}$ seconds ($\approx$ 30 000 million years).

The mass of an electron $\approx 9 \times 10^{-31}$ kg.

4   If $a = 10^{-4}$, evaluate:

   a $(2a)^3$ b $\sqrt{a}$ c $\sqrt{9a}$ d $\sqrt{a^3}$

5   In each case below, given the length of a rectangle l and the breadth b, write down the area:

   a $l = 10^2$ m, $b = 10^3$ m   b $l = 10^{-1}$ m, $b = 10^4$ m
   c $l = 10^{-3}$ m, $b = 10^{-2}$ m   d $l = 10^{-5}$ m, $b = 10^3$ m

6   If a body travels a distance s during time interval t, its average velocity is given by $v = s \div t$. In
   each case below, write down the average velocity:

   a $s = 6 \times 10^3$ m, $t = 2 \times 10^5$ s   b $s = 3 \times 10^2$ m, $t = 10^{-4}$ s
   c $s = 8 \times 10^{-3}$ km, $t = 4 \times 10^2$ h   d $s = 5 \times 10^{-2}$ mm, $t = 10^{-6}$ s

7   The kinetic energy of a moving body is given by $\frac{1}{2}mv^2$ joules, (J), where m is its mass (in
   kilograms) and v is its velocity (in metres per second). In each of the following cases, write
   down the kinetic energy of the body:

   a $m = 6$ kg, $v = 10^4$ m/s   b $m = 2$ kg, $v = 10^{-3}$ m/s
   c $m = 3$ kg, $v = 2 \times 10^{-2}$ m/s   d $m = 96$ g, $v = 10^{-3}$ m/s

8   If a potential difference of V volts drives a current of I amperes through a resistor, the resistance
   is given by $V \div I$ ohms. In each case below, write down the size of the resistance:

   a $V = 10^3$ V, $I = 10^{-2}$ A   b $V = 10^{-2}$ V, $I = 10^{-3}$ A
   c $V = 10$ V, $I = 10^{-2}$ A   d $V = 10^{-1}$ V, $I = 10^{-5}$ A

9   The power dissipated by a resistance of R ohms when it carries a current of I amperes is given
   by I^2R watts. In each case below, write down the power:

   a $I = 10^2$ A, $R = 10^4$ Ω b $I = 10^{-3}$ A, $R = 10^2$ Ω
   c $I = 10^{-2}$ A, $R = 10^6$ Ω d $I = 10^{-3}$ A, $R = 10^{-2}$ Ω

10   If a potential difference of V volts drives a current of I amperes through a resistance of R ohms,
   the power dissipated is $V^2 \div R$ watts. In each case below, evaluate the power:

   a $V = 10^3$ V, $R = 10^2$ Ω b $V = 10^{-2}$ V, $R = 10$ Ω
   c $V = 10^{-3}$ V, $R = 10^{-2}$ Ω d $V = 10^{-1}$ V, $R = 10^{-5}$ Ω

11   The current through a resistor of R ohms, which is dissipating a power of P watts, is given by

   $\sqrt{\dfrac{P}{R}}$ amperes. In each case below, find the current:

   a $P = 10^5$ W, $R = 10$ Ω b $P = 10$ W, $R = 10^3$ Ω
   c $P = 10^{-3}$ W, $R = 10^3$ Ω d $P = 10^{-2}$ W, $R = 10^{-6}$ Ω

Use of the calculator

It is important that you are able to enter a number in scientific notation into your calculator.

   For example, the number 2.34×10^6 is entered by pressing the keys 2.34 $\boxed{\text{EXP}}$ 6 and will
probably be displayed as 2.34^{06} or 2.34E6.

   Using your calculator, verify that $2.34 \times 10^5 + 7.89 \times 10^4 = 3.129 \times 10^5$.

   If the result of the above calculation was displayed in a different form, this shows that your
calculator has not been set to display results in scientific notation.

The result 3.129×10^5 (in 'scientific notation') can be written as:

$$312\,900.000 \text{ or } 312\,900 \text{ ('floating point' notation)}$$

or as 312.9×10^3 ('engineering' notation, which is explained in the next section).

You should learn how to set your calculator so that results will be given in scientific notation.

Exercises 11.6 *continued*

12 Evaluate the following correct to 3 significant figures, expressing the answers in scientific notation:

a $2.67 \times 10^{15} \times 5.11 \times 10^8$

b $8.34 \times 10^{-7} + 9.35 \times 10^{-7}$

c $-7.22 \times 10^{14} - 9.54 \times 10^{14}$

d $\dfrac{3.13 \times 10^{11}}{4.12 \times 10^{-6}}$

e $\dfrac{6.01 \times 10^{-6}}{8.87 \times 10^5} + \dfrac{1}{3.92 \times 10^{11}}$

f $-6.47 \times 10^{-7} + 4.47 \times 10^{-7}$

g $4.23 \times 10^{-5} + 5.93 \times 10^{-3} \times 8.23 \times 10^{-3}$

h $-4.11 \times 10^7 \times 8.38 \times 10^9 + 4.94 \times 10^{17}$

i $\dfrac{8.67 \times 10^5 + 4.21 \times 10^5}{9.34 \times 10^7 - 9.04 \times 10^7}$

j $\dfrac{6.34 \times 10^{11} - 4.51 \times 10^{11}}{4.36 \times 10^7 \times 9.49 \times 10^4}$

11.7 ENGINEERING NOTATION

The values of *physical* quantities (*measured* quantities, quantities with *units*) are most conveniently expressed in the form $C \times 10^n$ where C is a number between 1 and 1000 and n is a multiple of 3 (n has one of the values . . ., $-12, -9, -6, -3, 0, 3, 6, 9, 12, \ldots$). This is called *engineering notation*.

The advantage of expressing values in engineering notation is that a conversion can easily be made to express the quantity in terms of the base unit for that quantity or in terms of a unit with one of the preferred prefixes. These preferred prefixes are:

Prefix	Symbol	Meaning
tera-	T	$\times 10^{12}$
giga-	G	$\times 10^9$
mega-	M	$\times 10^6$
kilo-	k	$\times 10^3$
no prefix-the base unit		$\times 10^0$
milli-	m	$\times 10^{-3}$
micro-	μ	$\times 10^{-6}$
nano-	n	$\times 10^{-9}$
pico-	p	$\times 10^{-12}$

Each step up or down this table decreases or increases the exponent by 3.

Examples

1 2.63 MW = 2.63×10^{12} μW: the exponent changes by 12 because we made 4 'jumps'—from mega- → kilo- → base unit → milli- → micro-. We have changed to a *smaller* unit, so there will be *more* of them. Hence the exponent *increases*.

2 81.7 mg = 81.7×10^{-6} kg. The exponent changes by 6 because we have made 2 'jumps'—from milli- → base unit → kilo-. We have changed to a *bigger* unit so there will be *fewer* of them. Hence the exponent decreases.

A	B	C	D
Given quantity	In same given unit but in engineering notation	In the basic SI unit (eng. notation)	In the most convenient unit for statement
5 360 000 000 mg	5.36×10^9 mg	5.36×10^3 **kg**	5.36 Mg (or t)
5 360 000 000 mN	5.36×10^9 mN	5.36×10^6 **N**	5.36 MN
0.0275 ms	27.5×10^{-3} ms	27.5×10^{-6} **s**	27.5 μs
3 250 000 mm	3.25×10^6 mm	3.25×10^3 **m**	3.25 km
0.000 013 5 μF	13.5×10^{-6} μF	13.5×10^{-12} **F**	13.5 pF

Column A: The value is stated in a clumsy manner. The unit is inappropriate making the size of the quantity difficult to 'visualise' or compare with familiar quantities.

Column C: ▪ The value is stated in the *base* SI unit, in engineering notation. Unless otherwise stated, it is always assumed that a given formula is formulated for the substitution of quantities expressed in *base* SI units.

▪ Note that the only unit in this column that bears a *prefix* is the *kilogram* (kg). The base unit of mass is the *kilogram* and *not* the gram (g).

Column D: The multiplier is abolished by a change in the prefix. This is the most convenient way to express the quantity when stating its value either verbally or in writing. (Note, however, that these values could not be substituted into a formula because none of them is in the *base* unit for the quantity concerned.)

Exercises 11.7

1 For each of the following given quantities, state its value:

   i in the same unit but in engineering notation

   ii in the *base* unit, in engineering notation

   iii with the multiplier abolished by a change to the unit preferred prefix.

   a 26 800 μg   b 26 800 μA   c 0.0825 mm

   d 6840 kW   e 0.637 MHz   f 3800 μH

   g 0.0745 MV   h 62 000 nm   i 13 800 g

   j 4250 kHz   k 0.83 kg   l 68 000 mm

continued

continued

2　Set your calculator to express results in engineering notation and obtain these results directly
from your calculator correct to 4 significant figures:

　　a 5210 × 2173　　　　　　　　　　　　**b** 0.004 328 × 0.021 87

　　c 321.8 × 563.4　　　　　　　　　　　　**d** 0.0837 ÷ 194.9

3　Use your calculator to perform the given operation and express the result in (i) scientific notation
and (ii) engineering notation:

　　a 123.0 × 456.0　　　　　**b** 1234 × 6789　　　　　**c** 3456 × 6789

　　d 462.0 × 1981　　　　　**e** 0.0456 ÷ 0.001 23　　　**f** 0.000 123 4 × 0.004 567

11.8 TRANSPOSITION IN FORMULAE INVOLVING EXPONENTS AND RADICALS

$(N^{\frac{a}{b}})^{\frac{b}{a}} = N^1 = N$. This result is often useful when transposing terms in a formula.

Examples

1　If $N^{\frac{2}{3}} = 4.31$

　　then $N = (4.31)^{\frac{3}{2}}$

　　　　$= (4.31)^{1.5}$

　　　　≈ 8.95

2　If $x^{2.6} = 137$

　　then $x^{\frac{13}{5}} = 137$

　　　$x = (137)^{\frac{5}{13}}$

　　　≈ 6.63

3　If $t^7 = 593$

　　then $t = (593)^{\frac{1}{7}}$

　　　　≈ 2.49

Example

To Make A the subject of the formula $kA^{-\frac{2}{3}} = M$, we proceed thus:

$$A^{-\frac{2}{3}} = \frac{M}{k}$$

$$\therefore A = \left(\frac{M}{k}\right)^{-\frac{3}{2}}, \text{ which is more simply written as } A = \left(\frac{k}{M}\right)^{\frac{3}{2}}, \text{ or } A = \left(\frac{k}{M}\right)^{1.5}.$$

Exercises 11.8

In each of the following exercises a formula is given.

　　i Make the variable in part **a** of the question the subject of the formula.

　　ii Evaluate this new subject using the data provided.

1　$h = \dfrac{l^3}{d^2}$, where h, l and d are lengths.

　　a l　　　　　　　　　　　　　　　　　**b** $d = 47.0$ mm, $h = 2.35$ m

2 $V = \dfrac{S^{1.5}}{2\sqrt{\pi}}$, where S is the surface area of a sphere and V is its volume.

    a S **b** $V = 450 \times 10^{-3}\,\text{m}^3$

3 $A = P\left(\dfrac{R + 100}{100}\right)^n$, where $R\%$ is a rate of interest.

    a R **b** $P = \$255,\ n = 3.5,\ A = \$410$

+++ 4 $T = 2\pi\sqrt{\dfrac{r^3}{GM}}$, where r is the radius of orbit of a planet.

    a r **b** $T = 27.4$ days, $M = 5.98 \times 10^{24}\,\text{kg}$, $G = 66.7 \times 10^{-12}$ SI units

+++ 5 $C = \dfrac{1}{\dfrac{1}{C_1} + \dfrac{1}{C_2}}$, capacitance when C_1 and C_2 are connected in series.

    a C_2 **b** $C = 865\,\text{mF}$, $C_1 = 1.58\,F$

SELF-TEST

1 Expand and express in exponential form (i.e. replacing any roots by indices):

    a $2\sqrt{a}(3a - \sqrt{a})$ **b** $(x - \sqrt{x})^2$

    c $(1 + m\sqrt{m})^2$ **d** $(1 - \sqrt{a})(\sqrt{a} - 1)$

2 Evaluate mentally:

    a 9^{-2} **b** $9^{-\frac{1}{2}}$ **c** 9^0 **d** $27^{\frac{1}{3}}$ **e** $8^{-\frac{2}{3}}$

3 Evaluate mentally:

    a $8^{-\frac{1}{2}} \times 8^{-\frac{1}{2}}$ **b** $7^{\frac{2}{3}} \times 7^{-\frac{2}{3}}$ **c** $5^{-\frac{2}{3}} \div 5^{\frac{1}{3}}$

    d $(3^0)^2$ **e** $(4^{-2})^{-\frac{1}{2}}$

4 Given that $k = 2.36$, $n = -1.74$ and $t = 0.825$, use your calculator to evaluate the following correct to 3 significant figures:

    a $k^2 + 2t^3$ **b** $(k - 2t^2)^3$ **c** k^{nt} **d** $k^n - k^t$ **e** $3t^k - n^3$

5 Solve:

    a $2^{x-2} = \sqrt{2}$ **b** $3^{2n} = 1$ **c** $3^k \times 3^{-7} = \frac{1}{3}$

    d $5^{2n+3} = 1$ **e** $2^t = \dfrac{1}{2}$ **f** $7^{x+1} = \dfrac{1}{\sqrt{7}}$

6 Given that $m = 1$, $k = -2$ and $t = 3$, evaluate:

    a $k^m - k^t$ **b** $(t - k)^2$ **c** $\left(\dfrac{mk}{t}\right)^2$ **d** $m^t - t^m$

7 Evaluate:

    a $(-1\frac{1}{2})^2$ **b** $(-1)^{73}$ **c** $3^{-11} \times 3^{13}$

    d $2^{-37} \div 2^{-40}$ **e** $5^{\frac{3}{8}} \div 5^{-\frac{5}{8}}$ **f** $12^0 \div 12^{-2}$

    g $3^{\frac{1}{2}}(3^{\frac{1}{2}} - 3^{\frac{3}{2}})$ **h** $(3^0 - 3^{-1}) \div 3^{-2}$ **i** $(2^{\frac{1}{2}} + 2^{\frac{5}{2}}) \div 2^{-\frac{1}{2}}$

8 Simplify:

a $6tw \div 2t^{-3}w$

b $4m^{-\frac{1}{2}} \times 3m^0 \div 6m^{\frac{1}{2}}$

c $8a^{-\frac{1}{2}} \div 2a^{-\frac{2}{3}}$

d $\dfrac{3k^2 \times 2k^{-5}}{6k^{-4}}$

e $\dfrac{6t^{-2} + 8t}{2t^{-3}}$

f $\dfrac{n^{-3}(2n^{-2} + n^3) - 1}{n}$

g $\dfrac{(a^2 + 2ab + b^2)^7}{(a + b)^5}$

h $\dfrac{(a^2 - b^2)^5}{(a - b)^4(a + b)^4}$

9 Express without brackets:

a $mt(t^{\frac{1}{2}} + 3m)$

b $(a^{\frac{1}{2}} - a^{-\frac{1}{2}})(a^{\frac{1}{2}} + a^{-\frac{1}{2}})$

c $(n^{\frac{3}{2}} - 2n^{-\frac{1}{2}})^2$

d $y^2(3y^0 - y^{-1})$

e $(y - y^0)(y^3 + y^2)$

f $(b^{\frac{1}{3}} - 1)(b^{\frac{2}{3}} + b^{\frac{1}{3}})$

10 Make V the subject of the formula $E = \dfrac{V^2 t}{R}$ J. Then evaluate V when $E = 295\ \mu$J, $R = 24.8$ kΩ and $t = 17.5$ min. Express the results in millivolts.

+++ 11 Make t_2 the subject of the formula $V_2 = V_1\left(\dfrac{t_1 + 273}{t_2 + 273}\right)^{\frac{1}{\gamma - 1}}$ where t_1 and t_2 are temperatures of a gas in degrees Celsius. Then evaluate t_2 when $t_1 = 20°$C, $V_1 = 650$ mL, $V_2 = 345$ mL and $\gamma = 1.41$. State the result correct to 3 significant figures.

CHAPTER 12

POLYNOMIALS

Learning objectives

- Perform basic operations on polynomials.
- Factorise given polynomials by inspection.
- Solve quadratic equations by factorising.
- Solve quadratic equations by the method of 'completing the square'.
- Derive the quadratic formula by the method of completing the square and use this formula to solve quadratic equations.
- Simplify complex algebraic fractions by factorising.
- Solve quadratic equations graphically.

12.1 DEFINITION

A polynomial is a function in which each term has the form ax^n where n is a non-negative integer. If the polynomial has only one term, it is called a *monomial*. A polynomial with two terms is a *binomial* and with three terms is a *trinomial*.

The term involving the greatest value of n is called the *leading* term and this greatest exponent gives the *degree* of the polynomial.

Examples

1 $2.3x^{15} + 7.6x^8 - 4x^2$ is a trinomial of degree 15.
2 $13x^2 - 6x - 8$ is a trinomial of degree 2. Such a trinomial is called a *quadratic* function.
3 $5x + 3$ is a binomial of degree 1 because $5x = 5x^1$. This is called a *linear* function because its graph is a straight line.
4 $4x^3$ is a monomial of degree 3.
5 7 is a monomial of degree 0 (because $7 = 7x^0$).

12.2 MULTIPLICATION OF POLYNOMIALS

To expand (to multiply out) two polynomials, we use the *distributive law* (see section 4.5). For example, to expand $(2x^3 - 4x + 1)(3x^3 - x^2 - 2)$, the second polynomial is multiplied by $2x^3$, then by $-4x$ and then by 1. This gives:

$$6x^6 - 2x^5 - 4x^3 - 12x^4 + 4x^3 + 8x + 3x^3 - x^2 - 2$$
$$= 6x^6 - 2x^5 - 12x^4 + 3x^3 - x^2 + 8x - 2.$$

This is conveniently set out as a 'long multiplication':

$$2x^3 - 4x + 1$$
$$\underline{3x^3 - x^2 - 2}$$
$6x^6 - 0x^5 - 12x^4 + 3x^3$............................. multiplying by $3x^3$
$\quad\quad\quad -2x^5 \quad\quad\quad + 4x^3 - x^2$..................... multiplying by $-x^2$
$\quad\quad\quad\quad\quad\quad\quad\quad - 4x^3 \quad\quad + 8x - 2$........ multiplying by -2
$$\underline{6x^6 - 2x^5 - 12x^4 + 3x^3 - x^2 + 8x - 2}$$

> **Note:** In the first product (the multiplication by $3x^3$), the term $0x^5$ is included. It is best not to omit any powers of x, holding places for them all in case any appear in later lines of the multiplication.

If we require only the coefficient of a particular term, then it is a waste of time to perform the whole expansion. For example, if in the above expansion we were only interested in the coefficient of x^3, we would observe that an x^3 term can only arise in three ways:

$$(2x^3 - 4x + 1)(3x^3 - x^2 - 2)$$

Hence, the x^3 term will be $-4x^3 + 4x^3 + 3x^3 = 3x^3$.

Exercises 12.1

1 Expand:

   a $(2x - 3)(x^2 - 4x + 2)$ **b** $(x^3 - x + 1)(x^3 - 2x^2 - 1)$

   c $(x^2 - 1)(3x^2 - 2x - 1)$ **d** $(x - 1)(x^2 + x + 1)$

2 Find the coefficient of x^3 in the expansion of:

   a $(x^2 + 3)(3x^3 - 5x + 2)$ **b** $(2x^3 + x^2 + 3x)(x^2 - x + 3)$

We can evaluate a polynomial for a particular value of x. For example, to determine the value of the polynomial $x^2 - 3x + 6$ when $x = 4$, we replace x by 4 in the polynomial and evaluate the resulting expression. That is, when $x = 4$ the value of the polynomial is $4^2 - 3 \times 4 + 6$ which equals 10. For more complicated polynomials, or for values of x that are not integers, evaluating polynomials manually can become quite difficult. This process can be simplified by using the memory facility on your calculator. Check your calculator's manual on the method of storing a number in your calculator's memory. Numbers are recalled from memory by pressing $\boxed{\text{MR}}$ (memory recall) or $\boxed{\text{RCL}}$.

Example

Evaluate the polynomial $4x^3 + 5x^2 - 7x + 3$ when $x = 2.837$ correct to 4 significant figures. After you have stored 2.837 in your calculator's memory the following key strokes will evaluate the polynomial:

$4 \boxed{\times} \boxed{\text{MR}} \boxed{x^y} 3 \boxed{+} 5 \boxed{\times} \boxed{\text{MR}} \boxed{x^2} \boxed{-} 7 \boxed{\times} \boxed{\text{MR}} \boxed{+} 3 \boxed{=}$

giving the answer of 114.7.

Exercises 12.1 *continued*

3 Evaluate $2x^2 + 7x - 9$ when $x = 4$.
4 Evaluate $x^3 - 5x^2 + 9x - 17$ when $x = 3.15$, correct to 3 significant figures.
5 Evaluate $2.1x^3 + 1.6x^2 - 6.3x - 1.8$ when $x = -1.523$, correct to 4 significant figures.

12.3 FACTORISING

When a number or algebraic expression is written as the product of two or more factors, it is said to be written in **factorised form.**
The number 6 written in factorised form is 2×3.
The number 12 written in factorised form is 2×6, or 3×4, or $2 \times 2 \times 3$.
The expression $3k + 2k$ written in factorised form is $5 \times k$, or $5k$.

Note: A factor of an algebraic expression is not necessarily a single pronumeral.

Examples

1 $2(x + y)$ is written in factorised form, the factors being 2 and $x + y$.
2 $k(a + b)(m^2 - t)$ is written in factorised form, the factors being k, $a + b$, and $m^2 - t$.

12.4 COMMON FACTORS

When the terms of an algebraic expression have a common factor, the expression may be factorised (i.e. written in factorised form) by extracting the common factor.

> **Example**
>
> We can express $8ab + 12ac$ in factorised form   as $2(4ab + 6ac)$
>
> or   as $4(2ab + 3ac)$
>
> or   as $2a(4b + 6c)$
>
> or   as $4a(2b + 3c)$

It is accepted practice always to extract the *highest* common factor (the HCF) of the terms. So if you are required to factorise $8ab + 12ac$, the result you should state is the last one above, $4a(2b + 3c)$.

> **Examples**
>
> **1** $9km - 6k = 3k(3m - 2)$ **2** $2a^2b + 4abc = 2ab(a + 2c)$

Exercises 12.2

Factorise the following expressions:

1 **a** $4x + 6$ **b** $9 - 12a$ **c** $ab + a$

   d $6xy - 9x^2$ **e** $10a - 5$ **f** $16a - 24a^2x$

   g $3a^3 - 6a^3b^2$ **h** $5x^2 y - 10ax^2 + 5x^2$ **i** $3a^2b^2 + 6ab^2$

> **Examples**
>
> **1** $km + tm$ **2** $k(x + y) + t(x + y)$
>
>   common factor is m common factor is $x + y$
>
>   factorised form is $m(k + t)$ factorised form is $(x + y)(k + t)$

Exercises 12.2 *continued*

2 **a** $a(p + q) + b(p + q)$ **b** $a(b - c) + 2(b - c)$

   c $x(2a - y) + 3b(2a - y)$ **d** $2l(3 - x) - m(3 - x)$

> **Example**
>
> $n(x - t) + x - t = n(x - t) + 1(x - t)$
>
> $= (x - t)(n + 1)$

Exercises 12.2 *continued*

3 **a** $a(b + c) + (b + c)$

b $l(m + k) + m + k$

c $x + y + m(x + y)$

d $x(2 - y) + 2 - y$

Examples

1 $x(a + b) - a - b = x(a + b) - 1(a + b)$
$$= (a + b)(x - 1)$$

2 $a(k - t) + b(t - k) = a(k - t) - b(k - t)$
$$= (k - t)(a - b)$$

Exercises 12.2 *continued*

4 **a** $p(q + k) + q + k$

b $p(q + k) - q - k$

c $a(b - 3) - b + 3$

d $y(x - y) + y - x$

e $(t - m) - k(m - t)$

f $k(2x - 1) - m(1 - 2x)$

Examples

1 $a^n + a^{n + 3} = a^n(1 + a^3)$

2 $6x^{3n} - 4x^{5n - 7} = 2x^{3n}(3 - 2x^{2n - 7})$

Exercises 12.2 *continued*

5 Copy and complete the following factorised forms:

a $a^k - a^{k + 2} = a^k(\ldots)$

b $e^x + e^{-x} = e^x(\ldots)$

c $e^x + e^{-x} = e^{-x}(\ldots)$

d $6e^{2x} - 4e^x = 2e^x(\ldots)$

6 Simplify the following fractions by factorising as indicated:

a $\dfrac{e^x + e^{-x}}{e^{2x} + 1} = \dfrac{e^{-x}(\ldots)}{e^{2x} + 1} = \ldots$

b $\dfrac{e^x + 3e^{3x}}{e^{-x} + 3e^x} = \dfrac{e^{2x}(\ldots)}{e^{-x} + 3e^x} = \ldots$

12.5 THE DIFFERENCE OF TWO SQUARES

Rule: $(a + b)(a - b) = a^2 - ab + ab - b^2$
$$= a^2 - b^2$$

Hence, $a^2 - b^2$ can be factorised to $(a + b)(a - b)$.

Note: a and b can be replaced by any numerals or pronumerals.

> **Examples**
>
> **1** $x^2 - k^2 = (x + k)(x - k)$ **2** $7^2 - (5t)^2 = (7 + 5t)(7 - 5t)$

Exercises 12.3

1 Factorise the following:

a $E^2 - V^2$
b $L^2 - 3^2$
c $(3Q)^2 - 5^2$

d $(2a)^2 - (3b)^2$
e $(x + y)^2 - y^2$
f $(m - 5)^2 - 4^2$

g $G^2 - (3 - G)^2$
h $(m + 2t)^2 - (m - t)^2$
i $(V + e)^2 - (V - 2e)^2$

2 Evaluate mentally (write nothing down except the answer):

a $45^2 - 44^2$
b $79^2 - 78^2$
c $36^2 - 34^2$

d $51^2 - 48^2$
e $101^2 - 91^2$
f $(8.76)^2 - (8.75)^2$

> **Examples**
>
> **1** $9m^2 - 16 = (3m)^2 - (4)^2$
> $= (3m + 4)(3m - 4)$
> **2** $1 - 4a^2b^2 = (1)^2 - (2ab)^2$
> $= (1 + 2ab)(1 - 2ab)$
>
> It is important that you always decide correctly on just what the quantities are that are squared, and it is always advisable to write the given expression to be factorised in the form $(\ldots)^2 - (\ldots)^2$ before factorising.

Exercises 12.3 *continued*

3 Factorise:

a $V^2 - 9$
b $4C^2 - 25$
c $9L^2 - M^2$

d $16 - 9Q^2$
e $49R^2 - 1$
f $a^2b^2 - 4$

g $1 - r^2$
h $25E^2 - 16$
i $16a^2 - 25b^2c^2$

12.6 TRINOMIALS

First you should practise some more binomial multiplications.

$$(x + 2)(x + 5) = x^2 + 5x + 2x + 10$$
$$= x^2 + 7x + 10$$

Notice how the four products arise:

$(x + \ldots)(x + \ldots): x^2$

$(\ldots + 2)(\ldots + 5): 10$

$(x + 2)(x + 5): 5x + 2x$

You should be able to obtain the result $(x + 2)(x + 5) = x^2 + 7x + 10$ mentally, without having to write down $5x + 2x$.

Exercises 12.4

Write down the following products, obtaining the results mentally:

1 a $(x + 2)(x + 3)$ b $(x + 3)(x + 4)$ c $(x + 3)(x + 5)$
   d $(x + 4)(x + 5)$ e $(x + 1)(x + 8)$ f $(x + 5)(x + 7)$
   g $(x + 7)(x + 3)$ h $(x + 6)(x + 10)$ i $(x + 5)(x + 8)$

2 a $(x + 3)(x + 8)$ b $(x + 2)(x + 12)$ c $(x + 6)(x + 4)$

For question 2, above, notice that in each case the result is $x^2 + \ldots + 24$, since the product of the numerals is 24 in each case.

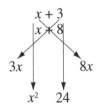

An alternative method for a binomial multiplication is to place one of the factors *under* the other, and then to multiply vertically and diagonally, as shown.

Notice that the middle term of the product, $11x$, is obtained from the two *diagonal* products ($3x$ and $8x$).

To factorise the expression $x^2 + 7x + 12$, we know that the only ways to obtain the 12 are from 1×12, 2×6 or 3×4. Hence, the factorised form must be:

$$(x + 1)(x + 12) \quad or \quad (x + 2)(x + 6) \quad or \quad (x + 3)(x + 4)$$

We now choose the *correct* one by testing each of these to discover which one gives the required middle term.

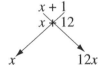

$(x + 1)(x + 12) = x^2 + 13x + 12$ ✗
$(x + 2)(x + 6) = x^2 + 8x + 12$ ✗
$(x + 3)(x + 4) = x^2 + 7x + 12$ ✓

Hence, the correct factorised form is $(x + 3)(x + 4)$.

Exercises 12.4 *continued*

3 Factorise the following:
   a $x^2 + 4x + 3$ b $x^2 + 8x + 15$ c $R^2 + 8R + 7$
   d $Q^2 + 8Q + 12$ e $L^2 + 5L + 6$ f $E^2 + 5E + 4$
   g $Z^2 + 7Z + 10$ h $F^2 + 17F + 16$ i $Q^2 + 12Q + 32$
   j $V^2 + 7V + 12$ k $E^2 + 11E + 24$ l $R^2 + 10R + 16$

continued

continued

4 Write down the following products, obtaining the result mentally:

   a $(x - 2)(x - 3)$ **b** $(x - 4)(x - 5)$

   c $(x - 3)(x - 4)$ **d** $(x - 1)(x - 7)$

In question 4 above, the last term of the product (the numeral) is positive, but the middle term is negative.

Consider $x^2 - 7x + 12$. The two factors must each have a *negative* numeral in order to obtain the negative middle term $(-7x)$.

The possibilities to test are: $(x - 1)(x - 12)$ ✗

$(x - 2)(x - 6)$ ✗

and $(x - 3)(x - 4)$ ✓

Exercises 12.4 *continued*

5 Factorise the following:

   a $x^2 - 11x + 24$ **b** $R^2 - 6R + 8$

   c $C^2 - 10C + 9$ **d** $F^2 - 37F + 36$

   e $r^2 - 6r + 9$ **f** $Q^2 - 52Q + 100$

6 Write down the following products, obtaining the results mentally:

   a $(x + 5)(x - 2)$ **b** $(x - 3)(x + 7)$

   c $(x - 4)(x + 1)$ **d** $(x + 8)(x - 2)$

In question 6 above, the last term is negative (since one of the numerals is positive and the other is negative).

Consider $x^2 - 5x - 24$. The possibilities to obtain the -24 are:

$1 \times -24, -1 \times 24, 3 \times -8, -3 \times 8, 2 \times -12, -2 \times 12, 4 \times -6, -4 \times 6$

On testing, we obtain $(x - 8)(x + 3)$.

Exercises 12.4 *continued*

7 Factorise the following:

   a $x^2 - 3x - 10$ **b** $y^2 + 4y - 12$ **c** $R^2 + R - 12$

   d $C^2 - C - 12$ **e** $E^2 - 2E - 24$ **f** $Q^2 + Q - 6$

   g $t^2 - t - 6$ **h** $L^2 + 2L - 24$ **i** $V^2 - 2V - 15$

   j $r^2 + 3r - 10$ **k** $Z^2 + 5Z - 36$ **l** $F^2 - F - 30$

When factorising $ax^2 + bx + c$ in cases where $a \neq 1$, we must also consider all the possible factors that give the x^2 term.

Example

If factorising $12x^2 + 20x + 3$, the result must be $(\ldots x + 3)(\ldots x + 1)$, since 3 and 1 are the only integers that will give us the constant term of 3. However, to obtain the $12x^2$ term the possibilities are $12x \times x$, $x \times 12x$, $6x \times 2x$, $2x \times 6x$, $3x \times 4x$, and $4x \times 3x$. On testing we obtain $(2x + 3)(6x + 1)$.

Exercises 12.4 *continued*

8 Factorise the following
 a $2x^2 + 3x + 1$ **b** $2x^2 + 5x + 3$ **c** $2x^2 + 7x + 3$
 d $6a^2 + 5a + 1$ **e** $6a^2 + 7a + 1$ **f** $4m^2 + 9m + 5$

9 Factorise the following:
 a $2x^2 - 5x + 2$ **b** $2x^2 - 5x + 3$ **c** $2a^2 - 11a + 5$
 d $6k^2 - 8k + 2$ **e** $6k^2 - 7k + 2$ **f** $6k^2 - 13k + 2$

10 Factorise the following:
 a $6x^2 + 7x - 3$ **b** $6x^2 + 11x - 2$ **c** $6x^2 - x - 2$
 d $36k^2 - 8k - 5$ **e** $24m^2 + 14m - 5$ **f** $24m^2 + 2m - 5$

In factorising $ax^2 + bx + c$, it may even be necessary to consider the different possibilities to obtain the ax^2 term and also the different possibilities to obtain the constant term, c.

Example

In factorising $15x^2 + 61x + 4$, the possibilities to obtain $15x^2$ are $5x \times 3x$, $3x \times 5x$, $15x \times x$, $x \times 15x$, and the possibilities to obtain the 4 are 4×1, 1×4, and 2×2. On testing, we find that the correct combination is $(15x + 1)(x + 4)$.

Experience gained from practice will enable you to choose the correct factors quickly without the need for much laborious testing.

Exercises 12.4 *continued*

11 Factorise:
 a $6b^2 + 11b + 4$ **b** $6b^2 + 25b + 4$ **c** $4a^2 + 15a + 9$
 d $6x^2 + 37x + 6$ **e** $8x^2 + 22x + 9$ **f** $8x^2 + 73x + 9$

12 Factorise:
 a $6x^2 - 13x + 6$ **b** $6x^2 - 19x + 8$ **c** $9x^2 - 9x + 2$
 d $9x^2 - 11x + 2$ **e** $10k^2 - 13k + 3$ **f** $10k^2 - 17k + 3$

13 Factorise:
 a $4a^2 + 4a - 15$ **b** $4a^2 - 4a - 15$ **c** $4a^2 + 7a - 15$
 d $4a^2 - 7a - 15$ **e** $6n^2 - n - 2$ **f** $6n^2 - 11n - 2$

continued

continued

14 Factorise:

a $6n^2 - 7n + 2$ b $6a^2 + 7a - 10$ c $4x^2 + x - 5$
d $4n^2 + 21n + 5$ e $8x^2 - 26x + 6$ f $10t^2 + 21t - 10$

12.7 COMPOSITE TYPES

When factorising, always first look for a common term, and if there is one, extract it before proceeding further. Having extracted any common terms, further factorising may be possible.

> **Examples**
>
> 1 $2x^2 - 18 = 2(x^2 - 9)$ (extracting the common term, 2)
>
> $= 2(x + 3)(x - 3)$ (factorising $x^2 - 9$ to $(x + 3)(x - 3)$)
>
> 2 $ax^2 + 5ax + 6a = a(x^2 + 5x + 6)$
>
> $= a(x + 2)(x + 3)$ 3 $45C^2 - 5 = 5(9C^2 - 1)$
>
> $= 5(3C + 1)(3C - 1)$
>
> 4 $20kx^2 - 60kx - 200k = 20k(x^2 - 3x - 10)$
>
> $= 20k(x + 2)(x - 5)$

Exercises 12.5

Factorise the following:

1    $3 - 3x$ 2 $5x^2 + 30x + 40$ 3 $wt^2 - 4w$
4    $4E^2 - 36$ 5 $6Q^2 + 6Q - 36$ 6 $5L^2 - 15L - 50$
7    $7E^2 + 42E + 56$ 8 $3Z^2 - 15Z + 12$ 9 $2v^2 + 6v - 36$
10   $12x^3 - 3x$ 11 $10t^3 - 50t^2 + 60t$ 12 $30k^4 + 25k^3 + 20k^2$

12.8 ALGEBRAIC FRACTIONS (SIMPLIFICATION BY FACTORISING)

Whenever the numerator and/or denominator of an algebraic fraction can be factorised, it is advisable to express these in factorised form. This often leads to a simplification of the fraction.

> **Examples**
>
> 1 $\dfrac{6x^2 - 24}{4x + 8} = \dfrac{6(x^2 - 4)}{4(x + 2)}$ 2 $\dfrac{8x^2 + 40x + 48}{4x^2 - 36} = \dfrac{8(x^2 + 5x + 6)}{4(x^2 - 9)}$
>
> $= \dfrac{{}^{3}\cancel{6}\,\cancel{(x + 2)}(x - 2)}{{}_{2}\cancel{4}\,\cancel{(x + 2)}}$ $= \dfrac{{}^{2}\cancel{8}(x + 2)\cancel{(x + 3)}}{\cancel{4}\,\cancel{(x + 3)}(x - 3)}$
>
> $= \dfrac{3(x - 2)}{2}$ $= \dfrac{2(x + 2)}{x - 3}$

Exercises 12.6

Factorise where possible and hence simplify:

1 $\dfrac{(x + 1)(x + 2)}{3x + 3}$

2 $\dfrac{2x + 12}{(x + 8)(x + 6)}$

3 $\dfrac{E^2 + 5E + 4}{E + 4}$

4 $\dfrac{C^2 - 4C + 4}{2 - C}$

5 $\dfrac{R^2 - 4}{R - 2}$

6 $\dfrac{R^2 - 4}{2 + R}$

7 $\dfrac{2Q + 1}{4Q^2 - 1}$

+++ 8 $\dfrac{5k^2 - 50k + 120}{80 - 5k^2}$

+++ 9 $\dfrac{60t^2 - 110t + 30}{40t^2 - 80t + 30}$

12.9 QUADRATIC EQUATIONS: INTRODUCTION

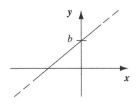

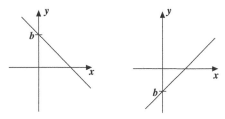

We have learnt that a first-degree expression $(ax + b)$ is called a **linear function** and that its graph is a straight line (with gradient a and y-intercept b).

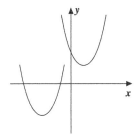

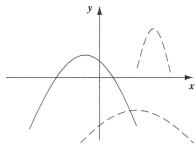

A second-degree expression $(ax^2 + bx + c)$ is called a **quadratic function.** Its graph is a parabola with a vertical axis. It is a very important curve with many applications in technology.

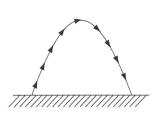

The path of a projectile (in a vacuum) is parabolic.

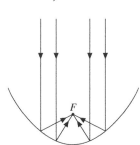

Parabolic shapes are used for receivers of radiation where a parallel beam is required to be focused at a 'point' F, called the focus (e.g. radar receiving antennae, solar furnaces, light telescopes, radiotelescope dishes, dishes for receiving satellite signals).

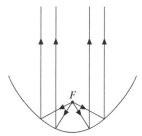

Reflectors of radiation when a parallel beam is required from a 'point' source are parabolic (e.g. radar transmitting antennae, reflectors in searchlights, electric torches, radiators, dishes to send signals into space).

Since the quadratic function $ax^2 + bx + c$ is of such importance, we now investigate the values of the variable x for which this function has a particular value. If, for example, we wish to know the value of x for which $3x^2 + 11x + 20$ has the value 13, we need to solve the equation $3x^2 + 11x + 20 = 13$, i.e. $3x^2 + 11x + 7 = 0$. So we now investigate the solution of the quadratic equation $ax^2 + bx + c = 0$.

12.10 SOLUTION OF THE EQUATION $P \times Q = 0$

If I tell you that I am thinking of two numbers whose product is 12, you cannot deduce what the numbers *must* be because there is no limit to the number of possibilities $(12 \times 1, 6 \times 2, 4 \times 3, 1.5 \times 8, 0.24 \times 50, \ldots)$. However, if I tell you that the product is *zero*, you can then deduce that one or both of the numbers *must* be zero.

If $P \times Q = 0$,	then	$P = 0$ or $Q = 0$ or they are both zero (i.e. $P = Q = 0$)
If $5x = 0$,	then	$x = 0$
If $8(x - 2) = 0$,	then	$x - 2 = 0$
If $x(x - 5) = 0$,	then	$x = 0$ or $x - 5 = 0$
If $(x - 3)(x + 4) = 0$,	then	$x - 3 = 0$ or $x + 4 = 0$

Exercises 12.7

Solve for the pronumeral by inspection:

1 $x(x + 2) = 0$ 2 $7x(x - 6) = 0$ 3 $(k + 2)(k - 1) = 0$
4 $6(t + 3)(t + 4) = 0$ 5 $9E(2E - 5) = 0$ 6 $13C(7 + 2C) = 0$
7 $(1 + Q)(3 + 2Q) = 0$ 8 $(5 - 4x)(7 - 2x) = 0$ 9 $4(5 + 2V)(6 + 5V) = 0$

12.11 SOLUTION OF THE EQUATION $X^2 = C$

I am thinking of a number. When I square this number, the result is 9. Can you deduce the number from this information?

No, you can't. The number may be 3 or it may be -3 (because $3^2 = 9$ and $(-3)^2 = 9$).

Examples

1   Solve: $x^2 = 16$
$$x = -4, 4 \text{ (i.e. } x = \pm 4)$$

2   Solve: $2(x^2 + 3x) + 3 = 3(7 + 2x)$
$$2x^2 + 6x + 3 = 21 + 6x$$
$$2x^2 + 3 = 21$$

$$2x^2 = 18$$
$$x^2 = 9$$
$$\therefore x = \pm 3$$

Exercises 12.8

1 Solve the following (a calculator is not necessary):

  a $V^2 = 25$ **b** $R^2 - 49 = 0$ **c** $R^2 + 12^2 = 13^2$

  d $2L^2 = 32$ **e** $12 - x^2 = 8$ **f** $5 - 2Q^2 = -67$

2 Using a calculator, solve, correct to 3 significant figures:

  a $L^2 = 27.3$ **b** $3R^2 = 41.7$ **c** $5.13x^2 = 8.07$

  d $13.4R^2 = 11.8$ **e** $7(3 - x^2) = 13$ **f** $(x + 7)(x - 7) = 83$

3 Solve the following (a calculator is not necessary):

  a $\dfrac{x}{4} = \dfrac{9}{x}$ **b** $Q - \dfrac{81}{Q} = 0$ **c** $x(x + 3) = 3(x + 27)$

  d $\dfrac{2}{L} = \dfrac{L}{8}$ **e** $\dfrac{x - 6}{2} = \dfrac{8 - 3x}{x}$ **f** $\dfrac{64}{t} - t = 0$

4 Using a calculator, solve, correct to 3 significant figures:

  a $\dfrac{6}{m} = \dfrac{2m}{5}$ **b** $3x(x - 1) - x(x - 3) = 6$ **c** $\dfrac{2t - 1}{3} = \dfrac{9 - 2t}{6t}$

  d $\dfrac{6 - 2k}{3} = \dfrac{2(k - 7)}{k}$ **e** $\dfrac{x - 4}{3} = \dfrac{5 - 2x}{x - 2}$ **f** $\dfrac{2}{5 - n} = \dfrac{5 + n}{6}$

If we know that $x^2 = 9$
    then $x = -3$ or 3
If we know that $(t - 7)^2 = 9$
    then $t - 7 = -3$ or 3
                   $t = 4$ or 10

Examples

1 Solve: $3(m + 5)^2 = 12$
            $(m + 5)^2 = 4$
             $m + 5 = -2$ or 2
                $m = -7, -3$

2 Solve: $\dfrac{2t - 3}{25} = \dfrac{4}{2t - 3}$
            $(2t - 3)^2 = 100$
            $2t - 3 = -10, 10$
                $t = -3.5, 6.5$

Exercises 12.8 continued

5 Solve the following (a calculator is not necessary):

a $(x - 2)^2 = 9$　　　　　　　　　　　b $(R + 3)^2 = 1$

c $50 - 2(C - 1)^2 = 0$　　　　　　　d $\dfrac{5}{R + 3} = \dfrac{3 + R}{5}$

e $\dfrac{73.9}{k - 1} = \dfrac{k - 1}{73.9}$　　　　　　　f $\dfrac{2 + t}{3.41} = \dfrac{3.41}{2 + t}$

6 Using a calculator, solve, correct to 3 significant figures:

a $(k + 3)^2 = 17$　　　b $(2 - C)^2 = 134$　　　c $\dfrac{1}{x - 3} = \dfrac{x - 3}{15.7}$

d $\dfrac{2 - 3n}{6.84} = \dfrac{3.13}{2 - 3n}$　+++　e $\dfrac{k + 3}{2.83} = \dfrac{1.48}{2k + 6}$　+++　f $\dfrac{3x - 6}{16.7} = \dfrac{24.3}{9x - 18}$

12.12 SOLUTION OF THE EQUATION $ax^2 + bx = 0$

Quadratic equations of the form $ax^2 + bx = 0$ are solved by factorising the left-hand side.

Example

If　$x^2 - 3x = 0$

　$x(x - 3) = 0$

　　　$\therefore x = 0$ or $x = 3$

Note: A common error is:　$x^2 - 3x = 0$

　　　　　　　　　　$x - 3 = 0$　　　　　　　(dividing both sides by x)

　　　　　　　　　　$\therefore x = 3$

This loses the solution $x = 0$. Dividing both sides by x is valid *only* if $x \neq 0$. We must examine the possibility that x *does* equal zero, and in this case this is another solution to the equation. This complication can be avoided by always factorising first and not dividing both sides by the pronumeral.

Examples

1 Solve:

　　　　$1.2R + 0.8R^2 = 0$

　　　　$12R + 8R^2 = 0$

　　　　$3R + 2R^2 = 0$

　　　　$R(3 + 2R) = 0$

　　$R = 0$ or $3 + 2R = 0$

　　　　　$\therefore R = 0$ or $2R = -3$

　　　　　$\therefore R = 0, -1.5$

2 Solve:

　　　　$\frac{1}{3}k^2 - \frac{5}{6}k = 0$

　　　　$2k^2 - 5k = 0$

　　　　$k(2k - 5) = 0$

　　　　　$k = 0$ or $2k - 5 = 0$

　　　　　$\therefore k = 0$ or $2k = 5$

　　　　　$\therefore k = 0, 2.5$

Exercises 12.9

1 Solve:

a $V^2 - 6V = 0$ **b** $W^2 + 2W = 0$ **c** $5M - 2M^2 = 0$

d $2L^2 + 3L = 0$ **e** $\dfrac{Q^2}{3} + \dfrac{Q}{6} = 0$ **f** $2.5R - 0.5R^2 = 0$

Always express a quadratic equation with *zero* on the right-hand side before attempting to solve it.

Examples

1 Solve: $E^2 = 2E$

$E^2 - 2E = 0$

$E(E - 2) = 0$

$\therefore E = 0, 2$

2 Solve: $\dfrac{M^2}{2} = \dfrac{2M}{3}$

$3M^2 = 4M$

$3M^2 - 4M = 0$

$M(3M - 4) = 0$

$\therefore M = 0 \text{ or } 3M - 4 = 0$

$\therefore M = 0 \text{ or } 1\tfrac{1}{3}$

Exercises 12.9 *continued*

2 Solve:

a $R^2 = 5R$ **b** $3V = 5V^2$

c $\dfrac{x}{2} = \dfrac{5x^2}{3}$ **d** $0.3K + 1.5K^2 = 0$

+++ **e** $R(R - 2) = 3(R^2 - R)$ +++ **f** $\dfrac{t - 3t^2}{3 - t} = \dfrac{t}{2}$

12.13 SOLUTION OF THE EQUATION $ax^2 + bx + c = 0$

Equations of the form $ax^2 + bx + c = 0$ can sometimes be solved by factorising the left-hand side.

Examples

1 Solve: $x^2 - 4x + 3 = 0$

$(x - 1)(x - 3) = 0$

$\therefore x = 1, 3$

2 Solve: $6R^2 - 7R - 3 = 0$

$(3R + 1)(2R - 3) = 0$

$R = -\dfrac{1}{3}, \dfrac{3}{2}$

Exercises 12.10

1 Solve the following quadratic equations by factorising the left-hand side:

    a $R^2 - R - 6 = 0$ **b** $t^2 - 5t + 6 = 0$

    c $W^2 + 10W - 24 = 0$ **d** $d^2 - 8d + 16 = 0$

2 Solve, by factorising:

    a $C(C + 8) + 12 = 0$ **b** $T(T - 5) - 36 = 0$

    c $x^2 - 12(x - 3) = 0$ **d** $24 - R(5 + R) = 0$

3 Solve by factorising:

    a $4k^2 - 16k + 15 = 0$ **b** $8V^2 + 10V + 3 = 0$

    c $6C^2 - 7C + 2 = 0$ **d** $12x^2 + 16x - 3 = 0$

As mentioned earlier, always express a quadratic equation with zero on the *right-hand side* before attempting to solve it.

Examples

1 Solve: $x^2 = 5x - 6$

               $x^2 - 5x + 6 = 0$

               $(x - 2)(x - 3) = 0$

                  $\therefore x = 2, 3$

2 Solve: $3(1 + 2x) = x(x + 4)$

             $3 + 6x = x^2 + 4x$

            $x^2 - 2x - 3 = 0$

          $(x - 3)(x + 1) = 0$

               $\therefore x = -1, 3$

Exercises 12.10 *continued*

4 Solve by factorising:

    a $x^2 = 4(2x - 3)$ **b** $W^2 + 6^2 = 20W$ **c** $\dfrac{2x + 1}{x - 1} = \dfrac{x + 2}{x + 3}$

    d $(x - 2)(x - 3) = (2x + 1)(x + 6)$ **e** $\dfrac{R - 2}{R - 1} = \dfrac{2R - 4}{R + 1}$

Remember: The first step in factorising is always to extract any common factors.

+++ 5 Solve by factorising:

    a $70x^2 + 140x - 560 = 0$ **b** $15R = 180R - 300$

    c $(100x + 600)(4x + 16) = 800x^2$ **d** $(x - 1)(x - 3) = x - 3$

Examples

1 Solve: $5x^3 - 5x = 0$

$5x(x^2 - 1) = 0$

$5x(x + 1)(x - 1) = 0$

$\therefore x = -1, 0, 1$

(**Note:** *Three* solutions.)

2 Solve: $3x^3 - 21x^2 + 36x = 0$

$3x(x^2 - 7x + 12) = 0$

$3x(x - 3)(x - 4) = 0$

$\therefore x = 0, 3, 4$

(**Note:** *Three* solutions.)

Exercises 12.10 *continued*

+++ **6** Solve by factorising:

a $600V^3 + 100V^2 - 1200V = 0$

b $40C^3 + 70C^2 - 75C = 0$

c $40a^3 - 70a^2 - 45a = 0$

d $30x^3 - 35x^2 - 15x = 0$

12.14 'ROOTS' AND 'ZEROS'

If $x^2 - 5x + 6 = 0$

then $(x - 2)(x - 3) = 0$

$\therefore x = 2, 3$

2 and 3, the values that satisfy the quadratic equation, are called the **roots** of the equation. They are also called the **zeros** of the quadratic expression $x^2 - 5x + 6$, since if either of these values is substituted for x in the expression $x^2 - 5x + 6$, the expression becomes equal to zero.

Exercises 12.11

1 What values of m make the expression $m^2 - m - 2$ equal to zero?

2 What are the zeros of the expression $3x^2 - 12x - 36$?

3 What are the roots of the quadratic equation $x^2 + 5x + 6 = 0$?

4 Write down a quadratic equation that has roots 2 and 5 (write the equation in the form $x^2 + bx + c = 0$).

12.15 THE QUADRATIC FORMULA

Most quadratic expressions do not factorise easily and so to solve them we use a formula derived by the method shown below, called 'completing the square'.

	A particular case $3x^2 - 5x - 7 = 0$	The general case $ax^2 + bx + c = 0$
1 Divide through by a	$x^2 - \dfrac{5}{3}x - \dfrac{7}{3} = 0$	$x^2 + \dfrac{b}{a}x + \dfrac{c}{a} = 0$
2 Move the constant term to the right side	$x^2 - \dfrac{5}{3}x = \dfrac{7}{3}$	$x^2 + \dfrac{b}{a}x = -\dfrac{c}{a}$
3 Add to both sides the square of half the coefficient of x	$x^2 - \dfrac{5}{3}x + \left(-\dfrac{5}{6}\right)^2 = \dfrac{7}{3}x + \left(-\dfrac{5}{6}\right)^2$ $x^2 - \dfrac{5}{3}x + \dfrac{25}{36} = \dfrac{109}{36}$	$x^2 + \dfrac{b}{a}x + \left(\dfrac{b}{2a}\right)^2 = -\dfrac{c}{a} + \left(\dfrac{b}{2a}\right)^2$ $x^2 + \dfrac{b}{a}x + \dfrac{b^2}{4a^2} = \dfrac{b^2 - 4ac}{4a^2}$
4 Express the left side in the form $(x + k)^2$	$\left(x - \dfrac{5}{6}\right)^2 = \dfrac{109}{36}$	$\left(x^2 + \dfrac{b}{2a}\right)^2 = \dfrac{b^2 - 4ac}{4a^2}$
5 Hence find $x + k$	$x - \dfrac{5}{6} = \pm\sqrt{\dfrac{109}{36}}$	$x + \dfrac{b}{2a} = \pm\dfrac{\sqrt{b^2 - 4ac}}{2a}$
6 Hence find x	$x = \dfrac{5}{6} \pm \sqrt{\dfrac{109}{36}}$ $= -0.907, 2.57$	$x = -\dfrac{b}{2a} \pm \dfrac{\sqrt{b^2 - 4ac}}{2a}$ $x = \dfrac{-b \pm \sqrt{b^2 - 4ac}}{2a}$

A particular equation has been solved above by completing the square, but since it would be tedious to solve every quadratic equation by working through the six steps involved, we use the formula derived by solving the general equation $ax^2 + bx + c = 0$.

This important and famous result should be memorised:

Rule: The solution of the quadratic equation $ax^2 + bx + c = 0$ is

$$x = \frac{-b \pm \sqrt{b^2 - 4ac}}{2a}$$

To use this formula, it must be remembered that:

1 The equation must first be written in the form $ax^2 + bx + c = 0$.
2 a is the coefficient of x^2 (i.e. the number in front of x^2); b is the coefficient of x (i.e. the number in front of x); and c is the constant term.

Write down: $x = \dfrac{-b \pm \sqrt{b^2 - 4ac}}{2a} = \dfrac{-() \pm \sqrt{()^2 - 4()()}}{2()}$

and then place the values of a, b and c in the brackets. This will help to avoid errors that are commonly made when negatives and double negatives occur.

Examples

1 Solve: $x^2 - 4x + 3 = 0$

$$x = \frac{-b \pm \sqrt{b^2 - 4ac}}{2a}$$

$$\begin{cases} a = 1 \\ b = -4 \\ c = 3 \end{cases}$$

$$\therefore x = \frac{-(-4) \pm \sqrt{(-4)^2 - (4)(1)(3)}}{2(1)}$$

$$= \frac{4 \pm \sqrt{16 - 12}}{2}$$

$$= \frac{4 \pm \sqrt{4}}{2}$$

$$= \frac{4 \pm 2}{2} = \frac{4 + 2}{2} \text{ or } \frac{4 - 2}{2} = 1 \text{ or } 3$$

In this simple case, we can see by inspection that:

If $x^2 - 4x + 3 = 0$

then $(x - 3)(x - 1) = 0$

$$\therefore x = 1, 3$$

2 Solve: $(2x - 3)(2x - 1) = 8$

$$4x^2 - 8x + 3 = 8$$

$$\therefore 4x^2 - 8x - 5 = 0$$

$$x = \frac{-b \pm \sqrt{b^2 - 4ac}}{2a}$$

$$\begin{cases} a = 4 \\ b = -8 \\ c = -5 \end{cases}$$

$$\therefore x = \frac{-(-8) \pm \sqrt{(-8)^2 - 4(4)(-5)}}{2(4)}$$

$$= \frac{8 \pm \sqrt{64 + 80}}{8}$$

$$= \frac{8 \pm 12}{8}$$

$$= \frac{-4}{8} \text{ or } \frac{20}{8} = -\frac{1}{2} \text{ or } 2\frac{1}{2}$$

In this case it would have been very difficult to find the solutions by inspection.

Exercises 12.12

Given that the roots of the quadratic equation $ax^2 + bx + c = 0$ are $x = \dfrac{-b \pm \sqrt{b^2 - 4ac}}{2a}$, solve the following equations giving the roots correct to 3 significant figures:

1 a $x^2 + 2x - 5 = 0$ **b** $t^2 - 3t - 1 = 0$

c $2x^2 + 3x - 4 = 0$ **d** $3Q^2 - Q - 1 = 0$

continued

continued

2 a $1 - 5x - 2x^2 = 0$ b $2 + 3R - 4R^2 = 0$

   c $100C^2 + 300C + 100 = 0$ d $50E^2 + 150E - 100 = 0$

3 a $\frac{1}{4}x^2 - \frac{1}{2}x - \frac{1}{2} = 0$ b $C(0.2C + 0.7) + 0.3 = 0$

4 a $\frac{x - 1}{x - 2} = \frac{x - 3}{4}$ b $(d - 1)(d - 2) = 3$

   c $2V + 1 = \frac{2V + 5}{2V + 3}$ d $(E - 2)(E + 3) = 1$

Exercises 12.13

Mechanical and general applications

1 For a certain uniform beam, the bending moment at a distance x m from one end is given by
   $M = 25x - 2.5x^2$ Nm. Find the distances from the end of the points where the bending moment is:
   a 60 Nm (factorise)   b zero (factorise)   c 50 Nm (use the formula)

2 A metal plate is in the shape of an isosceles triangle whose height is 3 m greater than the
   base length.
   a Taking the length of the base as b m, write down an expression for the area of the plate, in
      terms of b.
   b By putting the above expression for the area equal to 14 and solving the equation obtained,
      find the base length of the plate when the area is 14 m^2 (factorise).
   c Find the length of the base when the area is 8 m^2.

3 A rectangular piece of sheet metal is to be cut so that its breadth is 2 mm less than its length.
   a Taking the length of the rectangle as l mm, what is to be the breadth?
   b Write down an expression for the area, in terms of l.
   c By putting the expression for the area equal to 24 mm^2, find the length when the area is
      24 mm^2 (factorise).
   d Find the length when the area is 13 mm^2 (use the formula).
   e Find the breadth when the area is 7 mm^2.

4 A boat, with its engine running at constant speed, travels 12 km upstream in still water and later
   travels downstream 12 km when assisted by a current moving at 2 km/h. If the speed of the boat
   in still water is v km/h:
   a how long did the boat take for its journey upstream?
   b how long did it take for its journey downstream?
   c Given that the difference between the two times was 0.2 h, evaluate v (factorise).
   d If the difference between the two times was 0.4 h, evaluate v.

+++ **5** **a** What time is taken to travel a distance of 200 km at an average speed of **i** v km/h?

ii $v + 10$ km/h?

b A man who usually travels a certain journey of 200 km at an average speed of v km/h finds that if he increases his average speed by 10 km/h, he saves 1 hour. Find his usual average speed.

Exercises 12.14

Electrical applications

In this set of exercises for electrical students, all quantities and answers are in SI units, hence no units are actually stated.

1 If the power dissipated by the resistor R is 1 watt, it can be shown that:

$$R^2 + (2r - V^2)R + r^2 = 0$$

In each case below, given the values of r and V:

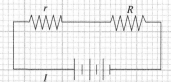

i write down the quadratic equation in the form $R^2 + bR + c = 0$, where b and c are numbers

ii solve the equation to find possible values of R.

Factorise for parts **a, b** and use the formula for part **c**.

a $r = 1, V = 2$ **b** $r = 6, V = 5$ **c** $r = 5, V = 5$

2 For the same circuit as question 1, if the power dissipated in resistor R is P watts, it can be shown that:

$$PR^2 + (2Pr - V^2)R + Pr^2 = 0$$

In each case below, form the quadratic equation (given the values of P, r and V) and then solve for R. Factorise for parts **a, b, c** and **d**. Remember to extract any common factor as the first step when factorising.

a $P = 2, r = 2, V = 4$ **b** $P = 4, r = 2, V = 6$

c $P = 2, r = 8, V = 8$ **d** $P = 100, r = 6, V = 50$

e $P = 1, r = 8, V = 8$ **f** $P = 100, r = 6, V = 60$

3 The total power dissipated by this circuit is given by:

$$P = \frac{V^2}{R} + Vi$$

In each of the following cases, given the values of P, R and i:

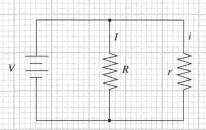

continued

continued

    i solve the quadratic equation to evaluate V (the positive root of the equation) (factorise for parts **a**, **b** and **c**)

    ii hence find the value of the other resistor (use $r = \dfrac{V}{i}$)

    iii find the value of the current through resistor R (use $I = \dfrac{V}{R}$).

    a $P = 4, r = 1, i = 3$ **b** $P = 4, r = 9, i = 1$

    c $P = 30, r = 2, i = 2$ **d** $P = 7, r = 2, i = 3$

+++ 4 When two particular resistors R_1 and R_2 are in series, the total resistance $(R_1 + R_2)$ is known to be 90 Ω. When these same resistors are in parallel, the total resistance

$$\left(\frac{R_1 R_2}{R_1 + R_2} \right)$$

is known to be 10 Ω. Find the values of the two resistors.

Note: Before substituting any value into a formula, make sure that the quantity is expressed in SI units.

Examples

1 If using the formula $v = \dfrac{Ft}{m}$ and the data state that $F = 2.34$ kN, $m = 750$ g and $t = 3.5$ min, the values that must be substituted into the formula are: $F = 2340$ (newtons), $m = 0.75$ (kilograms) and $t = 210$ (seconds), and v will be in metres per second (not, for example, kilometres per hour).

2 If using the formula $W = \frac{1}{2}CV^2$ and the data state that $C = 25\ \mu$F and $V = 850$ mV, the values substituted must be $C = 25 \times 10^{-6}$ (farads) and $V = 0.850$ (volts). W will be in joules.

Exercises 12.15

1 The height above the ground (h m) of a projectile t s after its release is given by $h = ut - \frac{1}{2}gt^2$, where u m/s is the vertical component of its release velocity and g m/s² is the acceleration due to gravity. Write this equation in the form $at^2 + bt + c = 0$, and taking $g = 10$ m/s², find the approximate times after its release when the projectile is 20 m above the ground, given that:

    a $u = 90$ km/h.   **b** $u = 126$ km/h.

2 A particular process requires the expenditure of 18 kJ of energy.

    a How long does it take to complete this process using a power input of p W (i.e. p J/s)?

    b How long does it take to complete the process if the power input is increased by 500 mW?

    c Given that the time saved by the above increase in power is 50 min, write this information in the form of an equation in p and solve the equation to find the power p and the time taken using this power. Solve the quadratic equation by factorising.

3

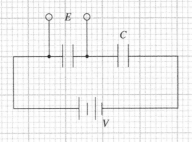

The total energy stored in these charged capacitors is given by $W = \frac{1}{2}QE + \frac{1}{2}\frac{Q^2}{C}$, where Q is the charge stored in each capacitor.

a Write this quadratic equation in the form $aQ^2 + bQ + c = 0$.

b Find the charge stored when $C = 4.00$ mF, $E = 575$ V and $W = 1.50$ kJ.

c Hence find the emf of the battery $\left(V = E + \dfrac{Q}{C} \right)$.

12.16 GRAPHICAL SOLUTION OF A QUADRATIC EQUATION

The graph of a quadratic function $f(x) = ax^2 + bx + c$ gives the values of the function for all the values of the variable x. The values of the variable for which $f(x) = 0$ are the values of x at the points where the curve crosses the x-axis. These values of x are the zeros of the function $ax^2 + bx + c$ and the roots of the equation $ax^2 + bx + c = 0$.

Exercises 12.16

1 For the function $y = x^2 + 2x - 5$, complete the table of values shown below. Hence plot an accurate graph of the function and use your graph to solve the quadratic equation $x^2 + 2x - 5 = 0$, stating the roots correct to the number of significant figures you consider justified. (Check your solution by solving the equation using the quadratic formula.)

x	−5	−4	−3	−2	−1	0	1	2	3
y									

2 Plot the graph of each of the following functions $f(x)$ for integral values of x over the domain specified. In each case solve the equation $f(x) = 0$. Then compare your solution with that obtained using the quadratic formula.

a $x^2 + 4x + 2 = 0$, for $-5 \le x \le 2$ **b** $21 - 2.3x - x^2$, for $-8 \le x \le 6$

SELF-TEST

1　Expand:

   a　$(x^2 - 3)(x^3 + 3x - 1)$　　　　　　　　**b**　$(x^2 + 3)(2x^3 - x^2 + 3)$

2　Find the coefficient of x^3 in the expansion of $(5x^3 + 2x^2 + 1)(2x^3 - 3x - 1)$.

   Factorise the following expressions:

3　**a**　$3x + 12ax - 9ax^2$　　　　　　**b**　$12pq^2 - 18q$　　　　　　**c**　$a(k + t) + 3(k + t)$

   d　$t - 3 - b(t - 3)$　　　　　　　**e**　$12e^{-x} + 8e^{-3x}$　　　　　**f**　$2a(x - 3y) + b(3y - x)$

   g　$p(2a - b) + b - 2a$　　　　　　**h**　$(a - b)(x - y) + (b - a)(y - x)$

4　**a**　$(4a + 3b)^2 - (a - 4b)^2$　　　　　　　　**b**　$1 - 36R^2$

5　**a**　$R^2 + 33R + 32$　　　　　　　**b**　$t^2 + 10t + 24$　　　　　**c**　$Z^2 + 37Z + 36$

   d　$Q^2 + 9Q + 18$　　　　　　　**e**　$4k^2 - 5k - 6$　　　　　　**f**　$9n^2 - 21n + 10$

   g　$4x^2 - 4x - 3$　　　　　　　　**h**　$10t^2 + 59t - 6$　　　　　**i**　$6a^2 - 17a + 12$

6　**a**　$2V^2 + 34V - 36$　　　　　　**b**　$7 - 28R^2$　　　　　　　**c**　$aK^2 - 7aK + 6a$

   d　$2kQ^2 - 14kQ - 36k$　　　　**e**　$3ab^2 - 3a^3c^2$　　　　　**f**　$8x^3 - 18xy^2$

7　Factorise where possible and hence simplify:

   a　$\dfrac{C^2 - C - 6}{C^2 - 2C - 3}$　　　　　　**b**　$\dfrac{3y - 12y^3}{2y^2 - y}$　　　　　　**c**　$\dfrac{K^2 - 9}{9 - K^2}$

8　Solve for the pronumeral:

   a　$0.3R - 0.06R^2 = 0$　　　　　　　　**b**　$\dfrac{2d^2 + d}{3 + d} = \dfrac{d}{3}$

   c　$x(5 - 0.4x) = 0.2(x^2 - 2x)$　　　　　**d**　$\dfrac{R^2 + 5R - 4}{R + 2} = 4$

   e　$20V^2 - 5V^4 = 0$　　　　　　　　　**f**　$6I^2 + 7I - 10 = 0$

   g　$24E^2 + 99E + 12 = 0$　　　　　　　**h**　$2.4x^2 - 3.7x + 1.4 = 0$

9　Give the roots correct to 3 significant figures:

   a　$3F^2 + 4F - 2 = 0$　　　　　　　　**b**　$2 - d - 5d^2 = 0$

   c　$60 = 40(5V - 2V^2)$　　　　　　　　**d**　$\frac{3}{4}F^2 = \frac{5}{6}F + \frac{2}{3}$

   e　$0.7k^2 = 0.3(1 + 4k)$　　　　　　　**f**　$(C - 2)(C - 3) = (3C - 2)(C + 4)$

10　Find the length and the breadth of a rectangle given that it has a perimeter of 6.82 m and an area of 2.70 m^2.

CHAPTER 13

FUNCTIONS AND THEIR GRAPHS

Learning objectives

■ Sketch the graph of a parabola, circle or rectangular hyperbola, if given its equation, and find the equation if given the sketch-graph.

■ Solve analytically and graphically a system consisting of a quadratic equation and a linear equation or a system consisting of two quadratic equations.

■ Use the sketch-graph of a parabola to find the maximum or minimum value of a quadratic function and apply this skill to solving practical problems.

13.1 FUNCTION NOTATION

Consider the equation $y = 2x^2 - 3$.

When $x = 0, y = -3$

When $x = 2, y = 5$

When $x = -1.3, y = 0.38$, etc.

For *any* value of x, the corresponding value of y can be calculated, and since there is only *one* value of y for each value of x, y is said to be a *function* of x. In mathematical language we write this as $y = f(x)$. For example, $W = f(t)$ means that for every value of t, a single corresponding value for W exists. If we are dealing with several different functions of a variable, we give them different names for identification. For example, $x + 3, 5x - 2, x^2 + 1$ are different functions of the variable x. We could identify them thus:

$$f(x) = x + 3 \qquad g(x) = 5x - 2 \qquad K(x) = x^2 + 1$$

(You could think of these as 'names' in the same way that we name people for identification purposes, e.g. F for Fred, G for Gabrielle and K for Keith.)

If we identify these three functions as above, then:

$$f(x) + K(x) = (x + 3) + (x^2 + 1) = x^2 + x + 4$$
$$K(x) - g(x) = (x^2 + 1) - (5x - 2) = x^2 - 5x + 3$$

Exercises 13.1

1 Given that $A(x) = x - 1$
$$h(x) = 3x - 2$$
$$t(x) = 1 - x^2$$

state the function of x in each case below, in simplest form.

a $A(x) + h(x)$ **b** $h(x) - A(x)$

c $t(x) - h(x)$ **d** $t(x) + h(x) - A(x)$

If $f(x)$ is a given function of x, $f(a)$ is the function obtained when a is substituted for x.

Examples

1 If $f(x) = x - 3$

then $f(17) = 17 - 3 = 14$ (17 is substituted for x)

2 If $g(x) = 4 + x^2$

then $g(3) = 4 + 3^2 = 13$ (3 is substituted for x)

3 If $F(t) = 4 - 2t$

then $F(t^2 - 4) = 4 - 2(t^2 - 4) = 12 - 2t^2$ ($t^2 - 4$ is substituted for t)

Exercises 13.1 *continued*

2 If $g(x) = 2 - 3x$ evaluate:

a $g(1)$ **b** $g(5)$ **c** $g(0)$ **d** $g(-2)$

3 If $f(x) = \dfrac{1}{x + 1}$:

a express $f(a)$ in simplest form   **b** evaluate $f(1)$

c express $f\left(\dfrac{1}{x}\right)$ in simplest form   **d** evaluate $f\left(-\dfrac{1}{2}\right)$

4 If $f(x) = \dfrac{x + 1}{x - 1}$:

a evaluate $f(-1)$ **b** express $f\left(\dfrac{1}{x}\right)$ in simplest form

c evaluate $f(0)$ **d** express $f(2x)$ in simplest form

5 If $g(x) = 3x - 7$:

a find $g(y)$ **b** evaluate $g(-2)$ **c** find $g(3 - x)$

d evaluate $g(0)$ **e** find $g(x^2 - 1)$ **f** find $g(2x + 3)$

6 If $f(x) = g(x) = \dfrac{1}{x}$, evaluate:

a $f(2) + g(2)$ **b** $f(1) + g(3)$ **c** $f(\tfrac{1}{3})$ **d** $f(\tfrac{1}{4}) - g(\tfrac{1}{3})$

7 If $f(x) = 2x - 3$, solve for x:

  a $f(x) = -5$ **b** $f(x) = 0$ **c** $f(x) = x + 1$

8 If $g(x) = 3x - 1$, solve for x:

  a $g(x) = x$ **b** $g(x) = -1$ **c** $\dfrac{1}{g(x)} = 2$

+++   **9** If $g(y) = 3y - 2$, solve $g(y - 1) = 7$

+++   **10** If $g(x) = 3x - 2$, solve: $\dfrac{g(x - 1) - 6}{g(2 - x)} = g(x + 1)$

13.2 THE PARABOLA: GRAPHING THE CURVE FROM ITS EQUATION

The graph of the quadratic function $ax^2 + bx + c$ is a parabola with a vertical axis.

  It is important to be able to sketch a parabola quickly from its equation $y = ax^2 + bx + c$.

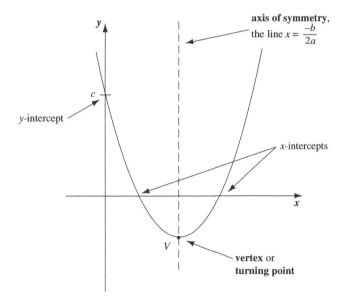

1 Find its **axis of symmetry,** $x = \dfrac{-b}{2a}$. This is also the x-value (abscissa) of the **vertex.**

2 Compute the value of y when $x = \dfrac{-b}{2a}$, this being the y-coordinate (ordinate) of the vertex.

3 Note the y-intercept, the value of c.

4 Draw axes. Mark the positions of the axis of symmetry, the vertex (turning point) and the y-intercept.

5 Sketch the parabola, making sure that the curve is symmetrical about the axis of symmetry. (You will find that the curve is ∪-shaped when $a > 0$, but ∩-shaped when $a < 0$.)

Examples

1 $y = 2x^2 - 12x + 13$

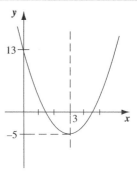

- Axis of symmetry is the line $x = \dfrac{-(-12)}{2(2)}$, i.e. the line $x = 3$.
- When $x = 3$, $y = 2(3^2) - 12(3) + 13$, i.e. $y = -5$. The vertex is the point $(3, -5)$.
- The y-intercept $= 13$.

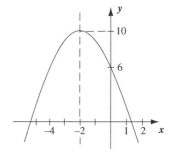

2 $y = -x^2 - 4x + 6$

- Axis of symmetry is the line $x = \dfrac{-(-4)}{2(-1)}$, i.e. the line $x = -2$.
- When $x = -2$, $y = 10$. The vertex is the point $(-2, 10)$.
- The y-intercept $= 6$.

The x-intercepts could be found by computing the values of x when $y = 0$.

In example 1 above, when $y = 0$, we have:

$$2x^2 - 12x + 13 = 0$$
$$\therefore x = \frac{12 \pm \sqrt{144 - 104}}{4}$$
$$= 1.42, 4.58$$

In example 2 above, when $y = 0$, we have:

$$-x^2 - 4x + 6 = 0$$
$$\therefore x = \frac{4 \pm \sqrt{16 + 24}}{-2}$$
$$= -5.16, 1.16$$

It is not necessary to find these points in order to draw the graph; however, if the quadratic factorises easily, it is useful to find them.

For example, for the graph of $y = 2x^2 - 12x + 10$, the x-intercepts occur where:

$$2x^2 - 12x + 10 = 0$$
i.e. where $x^2 - 6x + 5 = 0$
$$(x - 1)(x - 5) = 0$$
$$x = 1, 5$$

In order to draw an *accurate* graph of a curve, rather than just a 'sketch-graph', it is necessary to compute the coordinates of enough points to give the shape accurately. It is suggested that a sketch-graph be drawn first, then the values computed so that the final graph will include the vertex (turning point) and any intercepts on the axes.

Example

To plot accurately the graph in example 1 above, the sketch-graph shows that we should plot points between x-values of, say, -1 and 7. It is suggested that the calculations be made in tabular form:

$y = 2x^2 - 12x + 13$

x	-1	0	1	2	3	4	5	6	7	
x^2	1	0	1	4	9	16	25	36	49	
$2x^2$	2	0	2	8	18	32	50	72	98	} add
$-12x$	12	0	-12	-24	-36	-48	-60	-72	-84	
13	13	13	13	13	13	13	13	13	13	
y	27	13	3	-3	-5	-3	3	13	27	

The points should then be plotted according to the instructions in section 9.7.

Exercises 13.2

1 For the graphs of each of the following parabolas:

   i state the equation for the axis of symmetry

  ii state the coordinates of the turning point (vertex)

 iii state the value of the y-intercept

  iv sketch the curve, showing the turning point and any intercepts

   v draw an accurate graph of the curve (plot about 7-9 points).

   a $y = x^2 + 6x + 4$ **b** $y = -2x^2 + 4x + 7$ **c** $y = x^2 - 20x - 50$

2 For the graphs of each of the following parabolas:

   i state the equation for the axis of symmetry

  ii state the coordinates of the turning point (vertex)

 iii state the value of the y-intercept

  iv state the values of the x-intercepts

   v sketch the curve, showing the turning point and any intercepts

  vi draw an accurate graph of the curve (plot 7–9 points).

Note: The x-intercepts are computed in these cases because the functions factorise easily.

   a $y = x^2 + 4x + 3$ **b** $y = -2x^2 - 16x - 24$

   c $y = -3x^2 + 18x - 15$ **d** $y = 2x^2 - 16x + 24$

A simpler case

$y = ax^2 + bx$ is a simpler graph because the y-intercept $= 0$ and hence the graph passes through the origin. Also, for these functions, factorising is always possible (common factor x), so the x-intercepts can always be found easily.

Example

$y = 5x^2 - 37x \, [= x(5x - 37)]$

The x-intercepts are $0, \frac{37}{5}$, i.e. $0, 7.4$.

The axis of symmetry is $x = 3.7$ (halfway

between the x-intercepts).

The turning point is $(3.7, -68.45)$.

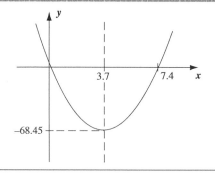

Exercises 13.2 *continued*

3   For each of the following graphs:

    i state the x-intercepts

    ii state the equation for the axis of symmetry

    iii state the coordinates of the turning point

    iv sketch the curve.

    a $y = 3x^2 - 6x$ **b** $y = -10x^2 + 50x$ **c** $y = -x^2 - 13x$

The simplest case

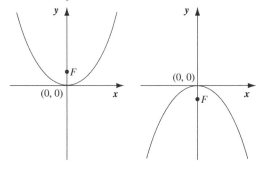

$y = ax^2$ is an even simpler graph because the turning point is the origin, $(0, 0)$.

The curve is ∪-shaped when $a > 0$ and is ∩-shaped when $a < 0$.

F, the 'focus', is described in section 12.9. The distance OF is the 'focal length' of the parabola.

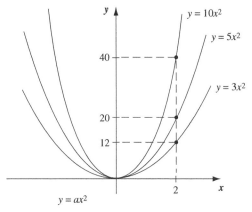

As the constant a increases, the curve 'closes up' because the y-value increases for any particular value of x.

Exercises 13.2 *continued*

4 On the same axes draw accurate graphs of the curves $y = x^2$ and $y = 3x^2$, for $-3 \leq x \leq 3$.

5 On the same axes draw accurate graphs of the curves $y = -x^2$ and $y = -2x^2$, for $-4 < x < 4$.

13.3 THE PARABOLA: FINDING THE EQUATION KNOWING THE COORDINATES OF THREE POINTS

If we know that a certain curve is a parabola, we know that its equation is $y = ax^2 + bx + c$, where a, b and c are constants. If we are given three points on the curve (either as ordered pairs or by reading from the given graph), we can evaluate the constants a, b and c and hence write down the equation for the curve. If we are given the y-intercept, we immediately have the value of c and we can then substitute the coordinates of the other two points to give two simultaneous equations that we can solve for a and b.

Example

Find the equation for the parabola whose sketch-graph is shown adjacent.

$c = -30$, the y-intercept (the value of y when $x = 0$).

$\therefore y = ax^2 + bx - 30$

Substituting $(-5, 0)$: $\quad 0 = 25a - 5b - 30$
Substituting $(2, 0)$: $\quad 0 = 4a + 2b - 30$

Solving these simultaneous equations gives $a = 3$, $b = 9$, so the equation is:

$\quad y = 3x^2 + 9x - 30$

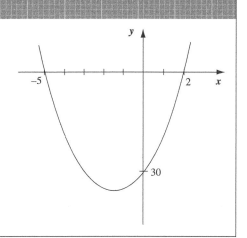

Note: This question could have been worded:

'Find the equation for the parabola that contains the points $(0, -30)$, $(-5, 0)$ and $(2, 0)$' or 'Find the equation for the parabola whose x-intercepts are -5 and 2 and whose y-intercept is -30'.

Exercises 13.3

1 Find the equation for each of the parabolas whose sketch-graph is shown below:

a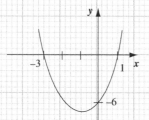

b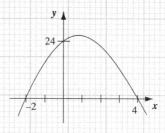

c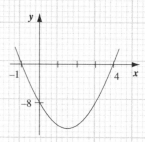

2 Find the equation for the parabola that has:

a y-intercept of 6 and x-intercepts of -2 and 3

b y-intercept of -9 and x-intercepts of 1 and 3

c y-intercept of 12 and x-intercepts of -3 and -1

3 Find the equation for the parabola that passes through the three given points:

a $(0, 4)$, $(1, 0)$ and $(2, 0)$ b $(0, 9)$, $(-3, 0)$ and $(1, 0)$

c $(0, -12)$, $(-6, 0)$ and $(-1, 0)$ d $(0, -45)$, $(-3, 0)$ and $(3, 0)$

If there are no x-intercepts but the turning point is given, use a third point found by symmetry.

Example

Given the y-intercept is 5 and the turning point
is $(2, 3)$, the third known point is $P(4, 5)$,
by symmetry.

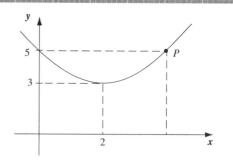

Exercises 13.3 *continued*

4 Find the equation for the parabola that has:

a y-intercept of -5 and turning point at $(2, -1)$

b y-intercept of 3 and turning point at $(-1, 2)$

13.4 PRACTICAL APPLICATIONS OF THE PARABOLA

In section 13.2 we established that the equation for a parabola with a vertical axis with its vertex
(turning point) at the origin is $y = ax^2$. F is the 'focus' and the distance from the origin to the focus
is called the 'focal length'.

It can be shown that the local length of a parabola $y = ax^2$ is given by $f = \dfrac{1}{4a}$

The coordinates of every point on the parabola satisfy the equation $y = ax^2$, hence since $a = \dfrac{y}{x^2}$, the value of a may be calculated if the coordinates of any point on the curve are known.

Example

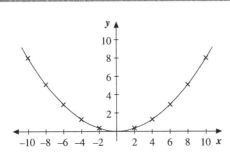

Given that the point (5, 2) lies on the parabola $y = ax^2$, we can deduce that:

1 $a = \dfrac{y}{x^2} = \dfrac{2}{25} = 0.08$

2 the focal length $f = \dfrac{1}{4a} = \dfrac{1}{0.32}$

3 the equation is $y = 0.08\,x^2$

x	2	4	6	8	10
y	0.32	1.3	2.9	5.1	8

Exercises 13.4

+++ **1** A truck 3.00 m tall drives under a parabolic arch which is 5.00 m high and 8.00 m wide at road-level.

    a What is the minimum horizontal distance at the base of the tunnel between the side of the truck and the wall?

    b What is the width of the arch at the level of the top of the truck?

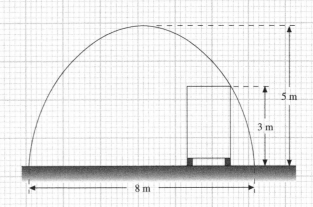

2 PQ is a parabolic satellite dish having a diameter of 3.00 m and a focal length of 80.0 cm. Find d, the depth of the dish.

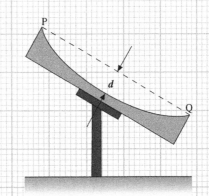

continued

continued

3 A rectangular sheet of polished stainless steel is bent into a parabolic shape to reflect the sun's heat rays to a pipe placed at the focus. At what distance from the vertex must the pipe be placed?

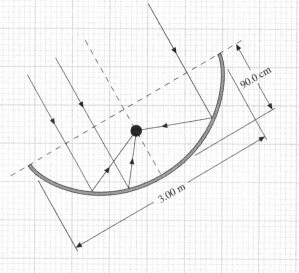

+++ 4 If a beam supported at each end is deflected by a concentrated load at its midpoint it will take the shape of a parabola. If the length of the beam is 10 m and the deflection at its midpoint is 4.0 cm, find the distance from each end where the deflection is 1.0 cm.

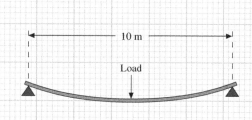

 13.5 THE CIRCLE

The coordinates of every point on the circle satisfy the equation $x^2 + y^2 = r^2$

Note: ∴ The equation for a circle with centre at the origin and radius r units is: $x^2 + y^2 = r^2$

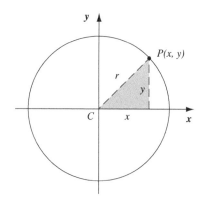

Example

$x^2 + y^2 = 16$ is the equation for a circle with centre at the origin and radius 4 units.

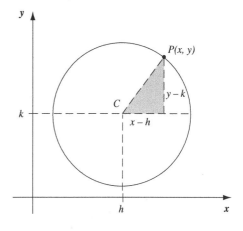

Note: The equation for a circle with centre at the point (h, k) and radius r units is:

$$(x - h)^2 + (y - k)^2 = r^2$$

Example

$(x - 5)^2 + (y + 3)^2 = 49$ is the equation for a circle with centre at the point $(5, -3)$ and radius 7 units.

Once an equation is recognised as representing a circle, the quickest way to draw either a sketch-graph or an accurate graph is by using a compass. There is no need to plot any points except for the centre. The scales on both axes must be the same.

Exercises 13.5

1 Describe the graph that represents each of the following equations and draw a sketch-graph.

  a $x^2 + y^2 = 9$

  b $(x + 2)^2 + (y - 3)^2 = 1$

  c $(x + 4)^2 = 1 - (y + 1)^2$

  d $y^2 = (1 + x)(1 - x)$

2 Write down the equation for the circle that has:

  a centre at the origin and radius 8 units

  b centre at the point $(7, 3)$ and radius 1 unit

  c centre at the point $(-2, 6)$ and radius 3 units

3 Write down the equation for the circle shown in each of the following sketch-graphs:

  a

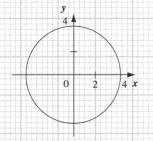

  b

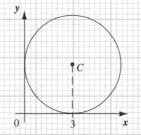

  c

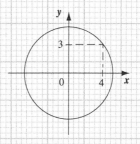

continued

continued

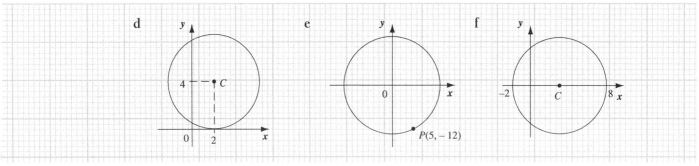

 13.6 THE RECTANGULAR HYPERBOLA

The equation $xy = C$ (or $y = C \times \dfrac{1}{x}$), where C is a constant, represents a **rectangular hyperbola**. If $xy = C$, x and y vary inversely with one another. When one increases the other decreases, in the same ratio, the product remaining constant.

Example

$xy = 16$

x	16	8	4	2	1
y	1	2	4	8	16

As $x \to \infty$, $y \to 0$.

As $y \to \infty$, $x \to 0$.

As x or y increases, the curve approaches the axis more and more closely, but never touches it. The axes are called **asymptotes** and the curve is said to be **asymptotic** to the axes.

If $xy = 16$, x and y can both have *negative* values, so a second branch of the curve exists in the third quadrant (the broken curve in the figure).

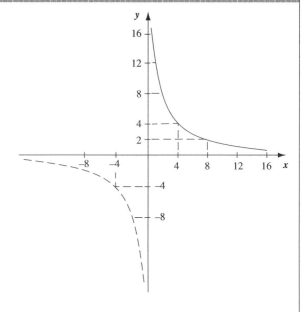

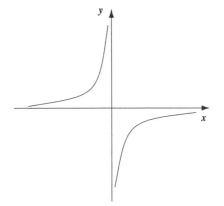

If C has a negative value, the two branches of the curve are in the second and fourth quadrants because when one of the variables is positive the other must be negative.

Exercises 13.6

Draw an accurate graph of each of the following curves:

1 $xy = 36$ **2** $xy = -100$ **3** $y = \dfrac{24}{x}$ **4** $y = -48 \times \dfrac{1}{x}$

13.7 ALGEBRAIC SOLUTION OF SIMULTANEOUS EQUATIONS THAT INVOLVE QUADRATIC FUNCTIONS

One quadratic equation and one linear equation

Example

Solve for x and y: $\begin{cases} y = x^2 + x - 2 \\ y = 2x + 10 \end{cases}$

$x^2 + x - 2 = 2x + 10$

$\therefore x^2 - x - 12 = 0$

$\therefore (x + 3)(x - 4) = 0$

$\therefore x = -3, 4$

When $x = -3, y = 4$

When $x = 4, y = 18$

(It is advisable to check both of these solutions in both equations.)

$\begin{cases} x = -3 \\ y = 4 \end{cases}$ $\begin{cases} x = 4 \\ y = 18 \end{cases}$

Exercises 13.7

Solve for the pronumerals:

1 $\begin{cases} y = x^2 - 12x + 27 \\ y = x + 5 \end{cases}$ **2** $\begin{cases} y = \dfrac{1}{x} + 3x \\ xy = 28 \end{cases}$ **3** $\begin{cases} y = x - \dfrac{5}{x} + 4 \\ xy = 8x + 7 \end{cases}$

Two quadratic equations

Where there are two quadratic equations to be solved simultaneously, they can be solved by the same method as that used above.

Example

$$\begin{cases} y = 5x^2 - 2x + 3 \\ y = 4x^2 - x + 5 \end{cases}$$

$$5x^2 - 2x + 3 = 4x^2 - x + 5$$

$$\therefore x^2 - x - 2 = 0$$

$$\therefore (x - 2)(x + 1) = 0$$

$$\therefore x = -1, 2$$

$$\begin{cases} x = -1 \\ y = 10 \end{cases} \qquad \begin{cases} x = 2 \\ y = 9 \end{cases}$$

Exercises 13.7 *continued*

Solve for the pronumerals:

4 $\begin{cases} m - 3 = k(4k - 5) \\ m + 9 = k(5k - 1) \end{cases}$

5 $\begin{cases} g + 1 = 7n + \dfrac{3}{n} \\ \\ g - 2 = 5n + \dfrac{8}{n} \end{cases}$

13.8 GRAPHICAL SOLUTION OF SIMULTANEOUS EQUATIONS THAT INVOLVE QUADRATIC FUNCTIONS

We have learnt how to plot an accurate graph of a quadratic function. (A *sketch-graph* should be drawn first and the ordered pairs to be plotted should be computed in a systematic tabulation.)

Also, it has been shown that the solutions to any pair of simultaneous equations are given by any points of intersection when the two curves are plotted.

We can now apply this knowledge to the graphical solution of simultaneous equations that involve quadratics.

Note: Values read from graphs can be only approximate, the accuracy depending upon how carefully the graph is plotted and its size. In exercises in which results are obtained from a graph, do not expect your result to agree exactly with the given answers.

Note: All graphing exercises could be investigated using a graphics calculator, in which case coordinates of points of intersection can be evaluated *accurately*.

One quadratic equation and one linear equation

Example

Find, by graphical means, the approximate solutions of the simultaneous equations:

$$\begin{cases} y = x^2 - 5x + 4 \\ 3y + 2x - 6 = 0 \end{cases}$$

$y = x^2 - 5x + 4 = (x - 4)(x - 1)$

x-intercepts are 4, 1

Axis of symmetry is $x = 2\frac{1}{2}$.

Minimum turning point is at $(2\frac{1}{2}, -2\frac{1}{4})$.

y-intercept is 4

$3y + 2x - 6 = 0$

$y = -\frac{2}{3}x + 2$

$\therefore$ From the graph the solutions are:

$x = 0.5, y = 1.7$ and $x = 3.8, y = -0.6$.

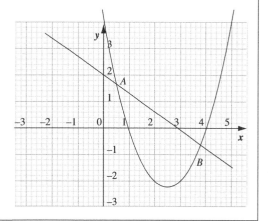

Exercises 13.8

1   Using the graph of the parabola $y = x^2 - 5x + 4$ in the above example (and ignoring the straight line $3y + 2x - 6 = 0$), find approximate solutions to the equations:

   a $x^2 - 5x + 4 = 0$ b $x^2 - 5x = -4$ c $10x^2 - 50x + 40 = 0$

   d $-x^2 + 5x - 4 = 0$ e $x(5 - x) = 4$ f $x(x - 5) = -4$

   g $x^2 - 5x + 3 = 0$ (i.e. $x^2 - 5x + 4 = 1$)

   h $x^2 - 5x + 2 = 0$ (i.e. $x^2 - 5x + 4 = 2$)

   i $x^2 - 5x + 5 = 0$ (i.e. $x^2 - 5x + 4 = -1$)

2   Draw an accurate graph of the parabola $y = -x^2 - 2x + 3$ and use it to find the approximate values of the solutions to each pair of the following simultaneous equations. (Note that the same quadratic function occurs in each part of this question.)

   a $\begin{cases} y = -x^2 - 2x + 3 \\ y = 0 \end{cases}$ b $\begin{cases} y = -x^2 - 2x + 3 \\ y = 2 \end{cases}$

   c $\begin{cases} y = -x^2 - 2x + 3 \\ y = 4 \end{cases}$ d $\begin{cases} y = -x^2 - 2x + 3 \\ y = 5 \end{cases}$

3   For each pair of simultaneous equations below, plot on the same axes an accurate graph of each of the functions and hence find approximate values of the solutions:

   a $\begin{cases} y = x^2 - 4x - 12 \\ y = -3x + 5 \end{cases}$ b $\begin{cases} y = x^2 - 2x - 3 \\ 2y - x + 2 = 0 \end{cases}$

Two quadratic equations

Example

Find, by graphical means, the approximate solutions of the simultaneous equations:

$y = x^2 - 3x$ and $y = 2 + x - x^2$.

$y = x^2 - 3x$ $= x(x - 3)$

x-intercepts are 0, 3; y-intercept is 0.

Axis of symmetry is $x = 1.5$

Minimum turning point is at $(1\frac{1}{2}, -2\frac{1}{4})$.

$y = 2 + x - x^2$ $= (2 - x)(1 + x)$

x-intercepts are 2, -1; y-intercept is 2.

Axis of symmetry is $x = \frac{1}{2}$

Maximum turning point is at $(\frac{1}{2}, 2\frac{1}{4})$.

From the graph the solutions are the coordinates of A and B.

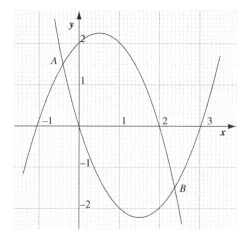

$$\therefore \left. \begin{array}{l} x = -0.4 \\ y = 1.4 \end{array} \right\} \quad \text{and} \quad \left. \begin{array}{l} x = 2.4 \\ y = -1.4 \end{array} \right\}$$

Exercises 13.8 *continued*

For each pair of simultaneous equations below, first draw on the same axes a sketch graph of each function and then plot accurate graphs which include any points of intersection. Hence, find approximate solutions to the simultaneous equations:

4 $\begin{cases} y = x^2 - 6x \\ y = -x^2 + 4x \end{cases}$

5 $\begin{cases} y = x^2 + 2x - 3 \\ y = 4 - x^2 \end{cases}$

6 $\begin{cases} y = x^2 + 2x - 3 \\ y = 8 - 2x - x^2 \end{cases}$

13.9 VERBALLY FORMULATED PROBLEMS INVOLVING FINDING THE MAXIMUM OR MINIMUM VALUE OF A QUADRATIC FUNCTION BY GRAPHICAL MEANS

These exercises are all based on the fact that the maximum or minimum value of a quadratic function occurs at the 'turning point' of the graph of the function and the turning point lies on the 'axis of symmetry'. If there are x-intercepts, the axis of symmetry lies halfway between them.

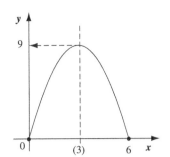

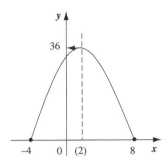

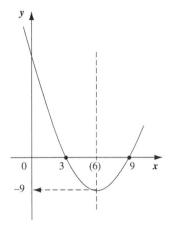

$$y = -x(x - 6)$$
$$y_{max} = -3(3 - 6)$$
$$= 9$$

$$y = -(x + 4)(x - 8)$$
$$y_{max} = -(2 + 4)(2 - 8)$$
$$= 36$$

$$y = (x - 3)(x - 9)$$
$$y_{min} = (6 - 3)(6 - 9)$$
$$= -9$$

Examples

1   Find the two numbers A and B such that when A is added to twice B the sum is 24 and the product of the numbers A and B is to be as large as possible.

We express the product in terms of one of the variables and then draw a sketch-graph of the product, P, against values of that variable.

$$A + 2B = 24$$
$$A = 24 - 2B$$
$$P = A \times B$$
$$= (24 - 2B) \times B$$
$$= 2B(12 - B)$$

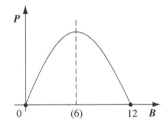

P is a maximum when $B = 6$

$\therefore A = 12$

Hence, the two numbers are $A = 12$, $B = 6$.

2   A farmer wishes to enclose a rectangular yard *EFGH* against an existing wall *PQ*. He has 20 m of fencing available. Find the dimensions of the rectangle he would need to fence in order to maximise the area enclosed.

$$l + 2b = 20$$
$$\therefore l = 20 - 2b$$

Area, $A = l \times b$
$$= (20 - 2b) \times b$$
$$= 2b(10 - b)$$

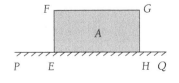

continued

continued

From the sketch-graph it is seen that A_{max} occurs when $b = 5$.

When $b = 5$, $l = 20 - 2b$

                          $= 10$

Hence, the area to be fenced has a length of 10 metres and a breadth of 5 metres (therefore, area enclosed: $A = 50 \text{ m}^2$).

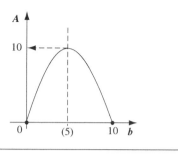

Exercises 13.9

1   A rectangular sheet of metal having dimensions 10 m × 6 m is to be bent to form an open channel of rectangular cross-section, as shown in the diagram.

   a   Express W in terms of h.

   b   Express the cross-sectional area of the channel, A, in terms of h.

   c   Draw a sketch-graph of A against h (showing how the cross-sectional area depends upon the choice of h).

   d   If the cross-sectional area is to have the greatest possible value, find the dimensions of the channel (i.e. the values of W, h and A).

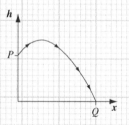

+++ 2   A ball thrown from point P follows a parabolic path striking the ground at point Q. The equation for its trajectory is $h = -0.05(x^2 - 16x - 36)$ m.

   a   Draw a sketch graph of the trajectory.

   b   From your sketch-graph, find approximate values for:

      i   the height of P above the ground

      ii   the maximum height above the ground attained by the ball

      iii   the range of the ball (i.e. the horizontal distance travelled)

      iv   the significance of the negative value of x which makes h equal to zero

      v   the significance of the value $h = -2.2$ when $x = 20$.

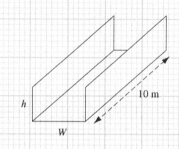

SELF-TEST

1   If $f(x) = x^3 - 2x$, evaluate:

   a $f(1)$ b $f(-2)$ c $f(0)$ d $f(-1)$ e $f(2t)$

2   It $f(x) = 5 - x^2$ and $g(x) = x^2 - 3$, evaluate:

   a $\dfrac{g(-1)}{f(-2)}$ b $\dfrac{f(1)}{g(2)}$ c $f(x) + g(x)$ d $f(-1) - g(-1)$

3 If $f(y) = 2y - 3$, solve the equations:

    a $f(t^2) + f(t + 1) = 0$
                              b $f(k^2 - 1) - f(k + 2) = 6$

    c $\dfrac{f(m + 3) + 2}{f(m + 2)} = f(m + 1)$

4 Draw a sketch-graph of each of the following parabolas, including the turning point:

    a $y = 2x^2 + 5x$
          b $y = x^2 - 2x - 8$
          c $y = -2x^2 - 12x - 10$

5 Find the equation for the parabola that:

    a has y-intercept -8 and x-intercepts -2 and 4

    b contains the three points $(0, -10)$, $(2, 0)$ and $(-5, 0)$

    c has y-intercept -3 and turning-point $(-2, -1)$

6 Describe the graph that represents the equation:

    a $x^2 + (y - 3)^2 = 4$
                             b $xy = -7$

CHAPTER 14

LOGARITHMS AND EXPONENTIAL EQUATIONS

Learning objectives

- Interpret a logarithm as an exponent and vice versa.
- Simplify logarithmic expressions using the three *laws of logarithms*.
- Solve equations involving logarithms.
- Solve exponential equations.
- Evaluate the logarithm of a number to any base.
- Transpose terms in an equation involving logarithms.
- Sketch the graph of a logarithmic function.
- Apply your knowledge of logarithms to the solution of practical problems.

14.1 DEFINITION OF A LOGARITHM: TRANSLATION BETWEEN EXPONENTIAL AND LOGARITHMIC LANGUAGES

We shall learn shortly that a logarithm is an exponent (i.e. an index), so to understand logarithms you must be sure of the facts you have learnt about indices (exponents) in Chapter 11. If you have trouble with the following exercises, go back to Chapter 11 and revise the work on indices.

The logarithm of a number to any particular base is the index ('exponent') to which that base must be raised in order to obtain the number.

We can write the number 8 in exponential form as 2^3, i.e. when the base is 2, the exponent (index) is 3.

The index, 3, is also called the **logarithm** of 8 to the base 2, i.e. the logarithm of 8 to the base 2 is 3 or $\log_2 8 = 3$.

Note: The base, 2, is placed as a subscript to the word 'log'.

Remember: A *logarithm* is an *index*. To understand logarithms, it is *essential* that you keep this fact constantly in mind.

SUMMARY

Stating this same fact in four different ways:

- When we are looking for the value of a *log*, we are looking for the value of an *index*.
- When we are looking for the value of an *index*, we are looking for the value of a *log*.
- When we are told the value of a *log*, we are being told the value of an *index*.
- When we are told the value of an *index*, we are being told the value of a *log*.

$8 = 2^3$ states (in exponential language) the fact that if 8 is expressed with base 2, the index is 3. $\log_2 8 = 3$ states the *same* fact in *logarithmic* language: when the base is 2, the index (i.e. the logarithm) is 3.

Note: These two statements are equivalent: $\log_b N = L$
$$b^L = N$$

Note the position of the base *b* in each statement.

Exercises 14.1

1   Write in exponential language ('exponential form'):

a $\log_3 9 = 2$ b $y = \log_2 7$ c $\log_7 1 = 0$

d $\log_2 \tfrac{1}{8} = -3$ e $\log_9 \tfrac{1}{27} = -\tfrac{3}{2}$ f $\log_4 4 = 1$

2   Write in logarithmic language ('logarithmic form'):

a $5^2 = 25$ b $4^{-2} = \tfrac{1}{16}$ c $3^0 = 1$

d $16^{\frac{1}{2}} = 4$ e $9^1 = 9$ f $4^{-\frac{5}{2}} = \tfrac{1}{32}$

3   Write each of the following equations in exponential form (i.e. in the form $b^n = L$):

a $y = \log_2 7$ b $\log_{10} y = k$ c $t = \log_e M$

d $\log_a Q = 2$ e $y = \log_{10} x$ f $\log_e x = 10$

4   Write each of the following equations in logarithmic form (i.e. in the form $\log_b N = L$):

a $y = t^2$ b $e^m = k$ c $10^p = Q$ d $2^{-3} = \tfrac{1}{8}$

14.2 EVALUATIONS USING THE DEFINITION (LOGS AND ANTILOGS)

Using the definition of a logarithm, in simple cases a logarithm may be evaluated thus:

Step 1: Let *x* be the value of the logarithm.

Step 2: Translate into exponential form.

Step 3: Solve the exponential equation by making the bases the same.

Examples

1 Evaluate $\log_3 81$

Let $\log_3 81 = x$

$\therefore 3^x = 81$

$\therefore 3^x = 3^4$

$\therefore x = 4$

$\therefore \log_3 81 = 4$

2 Evaluate $\log_9 \left(\frac{1}{3}\right)$

Let $\log_9 \left(\frac{1}{3}\right) = x$

$\therefore 9^x = \frac{1}{3}$

$\therefore \left(3^2\right)^x = 3^{-1}$

$\therefore 3^{2x} = 3^{-1}$

$\therefore 2x = -1$

$\therefore x = -\frac{1}{2}$

3 Evaluate $\log_3 1$

Let $\log_3 1 = x$

$\therefore 3^x = 1$

$\therefore x = 0$

$\therefore \log_3 1 = 0$

If you can manage it, it is quicker to carry out the above evaluations mentally, immediately from the definition, avoiding all the steps shown in the above examples.

Examples

1 $\log_3 81$ is the index to which the base 3 must be raised in order to obtain the number 81. Now $3^4 = 81$,
$\therefore \log_3 81 = 4$

2 $\log_2 \frac{1}{16}$ is the index to which 2 must be raised in order to obtain $\frac{1}{16}$ $\left(2^? = \frac{1}{16}\right)$.
Now $2^4 = 16$ and $2^{-4} = \frac{1}{16}$, $\therefore \log_2 \frac{1}{16} = -4$

3 $\log_9 \frac{1}{3}$ is the index to which 9 must be raised in order to obtain $\frac{1}{3}$.
Now $9^{\frac{1}{2}} = 3$ and $9^{-\frac{1}{2}} = \frac{1}{3}$, $\therefore \log_9 \frac{1}{3} = -\frac{1}{2}$

Note: The logarithm of 1 to any base is 0. $b^0 = 1, \therefore \log_b 1 = 0$.

Exercises 14.2

1 Evaluate mentally:

a $\log_2 16$

b $\log_4 16$

c $\log_{81} 81$

d $\log_3 \frac{1}{9}$

e $\log_{13} 13$

f $\log_2 0.5$

g $\log_{10} 10\ 000$

h $\log_3 \frac{1}{3^{15}}$

i $\log_2 2^{-8}$

2 Evaluate mentally:

a $\log_2 \sqrt{2}$

b $\log_3 1$

c $\log_5 \frac{1}{\sqrt{5}}$

d $\log_{10} 0.1$

e $\log_5 \sqrt[3]{5}$

f $\log_b 1$

g $\log_N N$

h $\log_x x^7$

I $\log_3 \frac{1}{3^{15}}$

14.3 POWERS AND LOGARITHMS WITH BASE 10 or e

In Chapter 16 we will meet a very important number called e (≈ 2.718). We will not now examine the *significance* of this number, but at this stage you should practise obtaining powers of 10 and powers of e using the $y = k\log_b Cx$ and the e^x keys on your calculator.

The commonly used bases when working with logarithms are 10 and e. Note that $\log_{10}$ is usually written as log and $\log_e$ as ln ('ln' stands for 'natural logarithm').

To display the value of e on your calculator, use the e^x key to find e^1. Check with your calculator that:

a $\log 5.2 \approx 0.716$ (i.e. $\log_{10} 5.2 \approx 0.716$)

b $\ln 8.5 \approx 2.14$ (i.e. $\log_e 8.5 \approx 2.14$)

It follows from **a** that $10^{0.716} \approx 5.2$ and from **b** that $e^{2.14} \approx 8.5$.

Check these with your calculator.

Exercises 14.3

1   Evaluate correct to 3 significant figures, using the method shown in parentheses in each case:

  a 10^2 (mentally), $10^{2.06}$ (calculator)

  b 10^{-1} (mentally), $10^{-0.928}$ (calculator)

  c 10^{-3} (mentally), $10^{-2.87}$ (calculator)

  d $\log_{10} 100$ (mentally), $\log_{10} 106$ (calculator)

  e $\log_{10} 1000$ (mentally), $\log_{10} 987$ (calculator)

  f $\log_{10} 0.01$ (mentally), $\log_{10} 0.013$ (calculator)

2   Use your calculator to evaluate correct to 3 significant figures:

  a $e^{2.34}$ b $\log_e 8.62$ c $e^{-4.26}$ d $\ln 0.0641$

  e $\log_e 10^{13}$ f $e^{-0.123} + 2e$ g $\ln(6.72 \times 10^{11})$ h $\ln(2.83 \times 10^{-7})$

3   Use your calculator to evaluate correct to 3 significant figures:

  a $6.24e^{2.81} - 3.47e$ b $17.6 \log_{10} 8.15$ c $\dfrac{e^{1.46} + e^{-1.46}}{1.85}$

  d $864 \times 10^{-1.61}$ e $e^{1.48} \times \log_e 8.47$ f $\dfrac{\log_{10} 407}{\log_{10} 89.2}$

  g $26.3e^{-2.44} - e^{-1.82}$ h $\ln 43.7 \times \ln 17.8$ i $e^{2.61} + \sqrt{76.4}$

  j $10^{1.28} + 10^{1.43}$ k $\dfrac{12.6}{\ln 8.73}$ l $10^{3.5}{:}1$

4   Use your calculator to evaluate, correct to 3 significant figures:

  a $\log_{10} e$ b $\log_{10}(2e + 1)$ c $\log_{10}(356 - 8e^3)$

14.4 DECIBELS

The output voltage or power of an amplifier may be many thousands times as great as the input, i.e. the ratio (called the 'amplification'), may be a very large number. It is often convenient to express the value of a large ratio in a unit called the 'bel', (B), and this value is obtained by taking the logarithm of the ratio (to the base 10).

Ratio		Ratio expressed in bels	
10:1	10^1:1	1 bel	(1 B)
100:1	10^2:1	2 bels	(2 B)
1000:1	10^3:1	3 bels	(3 B)
1 000 000:1	10^6:1	6 bels	(6 B)

$$\text{Ratio } \frac{Q_1}{Q_2} = \log_{10}\frac{Q_1}{Q_2} \text{ bels}$$

For example, a ratio of 10^4:1 $= \log_{10} 10^4 = 4$ bels

A ratio of 8570:2.73 $(=\dfrac{8570}{2.73} \approx 3140)$; $\log_{10} 3140$ bels ≈ 3.5 bels

[3.5 B = 35 decibels (35 dB).]

Note: A ratio of 3.5 bels is a ratio of $10^{3.5}$:1.

Ratios are usually expressed in decibels (dB). Remember, 1 B = 10 dB.

Example

If an amplifier has an input power of 1.3 mW and an output power of 6.5 W, the gain ratio
(amplification) $= 10^{-6} = 5000$:1 $= \log 5000$ B ≈ 3.7 B = 37 dB

Exercises 14.4

1   An amplifier has an input voltage of 1.7 V and an output voltage of 4.5 kV. What is the
amplification (expressed in decibels)?

2   An amplifier has a power gain of 62 dB. What is the (arithmetic) ratio of output power to
input power?

+++ 3   An input of 0.2 μW is fed to a 42 dB amplifier. What is the output power?

+++ 4   What is the arithmetic ratio corresponding to 21 db?

The *intensity* of sound, I, at a particular location is defined as the power received from the sound
wave per unit area. Hence the basic unit for sound intensity is W/m^2.

The intensity of the softest sound that can just be heard by a healthy ear is approximately
10^{-12} W/m^2. The sound level (β), in bels, is defined as the logarithm of a *special* ratio which is the
ratio of I, the intensity of the sound being received (W/m^2), to I_0, the intensity at the threshold
level (10^{-12} W/m^2).

A sound with an intensity of I W/m^2 is said to have a sound level of $\beta = \log_{10}\dfrac{I}{I_0}$ bels, where
$I_0 = 10^{-12}$ W/m^2.

	Approximate power received I (intensity)	Sound level $\beta = \log_{10}\dfrac{I}{10^{-12}}$ B	dB
Threshold of audibility	10^{-12} W/m^2	$\log_{10}\dfrac{10^{-12}}{10^{-12}} = \log_{10} 1 = 0$ bels	0
Conversational speech Typical air conditioner	10^{-6} W/m^2	$\log_{10}\dfrac{10^{-6}}{10^{-12}} = \log_{10} 10^{6} = 6$ bels	60
Near an unmuffled jackhammer	1 W/m^2	$\log_{10}\dfrac{1}{10^{-12}} = \log_{10} 10^{12} = 12$ bels	120

The power received by a person engaged in normal conversation is approximately 1 million times as great as the minimum power that can be heard by the human ear (the 'threshold of audibility') and the power received by a person standing close to an unmuffled jackhammer is approximately 1 million times as great again.

The intensity of sound

Sound intensity, $I = \log_{10}\dfrac{P}{10^{-12}}$ bels, where the power being received is P W/m^2.

Exercises 14.4 continued

+++   5   At a certain location the noise from a passing train is measured as 73 dB. What is the sound intensity being produced at this location?

+++   6   If at a certain locality the sound level produced by the playing of one violin is 64 dB:

    a   What is the intensity of the sound at that location?

    b   What would be the intensity of the sound produced by 15 violins playing simultaneously, all at the same volume?

    c   What would be the sound level produced by the 15 violins playing simultaneously?

Loudness

The 'loudness' of a sound is subjective and has no scientific definition. Most people with young, healthy ears perceive a difference in sound level of about 2 dB as barely noticeable and judge a sound to double in 'loudness' if the energy received becomes about ten times as great; that is if the sound level difference is about 1 B (10 dB). It requires about ten violins playing at the same volume to produce a sound judged to be about twice as loud as a single violin, and ten times as many again to double the 'loudness' again.

14.5 THE THREE LAWS OF LOGARITHMS

The first law

> **Rule:** $\log (A \times B) = \log A + \log B$

The first law of indices (see section 11.4) is $b^x \times b^y = b^{x + y}$

This states that if two numbers b^x and b^y are multiplied, the product is $b^{x + y}$.

Now $x =$ the log of the first number (using base b)

$y =$ the log of the second number

$x + y =$ the log of the product

Hence, the log of the product equals the sum of the logs of the two numbers, i.e.

$\log (A \times B) = \log A + \log B$

This result should not be surprising, since we know that logs are *indices,* and when we multiply two numbers (expressed with the same base), we *add* the indices.

Using the result $\log (A \times B) = \log A + \log B$ (the logs both being to any base, b), we can express the log of a number as the sum of two logs.

Example

$\log 24 = \log (2 \times 12) = \log 2 + \log 12$

$\log 24 = \log (4 \times 6) = \log 4 + \log 6$

$\log 24 = \log (3 \times 8) = \log 3 + \log 8$

$\log 24 = \log (9.6 \times 2.5) = \log 9.6 + \log 2.5$

These results can be checked on your calculator, using base 10. They are, however, true, no matter *what* base is used. Check for example that $\ln 24 = \ln 9.6 + \ln 2.5$

Exercises 14.5

1   a   Express log 30 as the sum of two logs in three different ways.

     b   Hence, simplify $\log 30 - \log 5$.

Remember: Although no bases are indicated here, they must all be the same.

Exercises 14.5 *continued*

2   a   Express log 20 as the sum of two logs in two different ways.

     b   Hence, simplify $\log 20 - \log 4$.

Note: $\log(A \times B \times C) = \log[(A \times B) \times (C)]$

$$= \log(A \times B) + \log C$$

$$= (\log A + \log B) + \log C$$

$$= \log A + \log B + \log C$$

Result: $\log(A \times B \times C \times \ldots) = \log A + \log B + \log C + \ldots$

Or, writing this result in reverse:

$\log A + \log B + \log C + \ldots = \log(A \times B \times C \times \ldots)$

Examples

1 $\log_2 3 + \log_2 5 + \log_2 6$

$= \log_2 (3 \times 5 \times 6)$

$= \log_2 90$

2 $\log_2 5 + \log_2 8 + \log_2 0.2$

$= \log_2 (5 \times 8 \times 0.2)$

$= \log_2 8$

$= 3$

Exercises 14.5 *continued*

3 Simplify:

a $\log_3 5 + \log_3 1.8$

b $\log_6 2 + \log_6 3$

c $\log_2 3 + \log_2 \frac{1}{24}$

d $\log_3 6 + \log_3 2 + \log_3 2.25$

e $\log_2 \frac{1}{3} + \log_2 6$

f $\log_5 4 + \log_5 0.25$

The second law

Rule: $\log \dfrac{A}{B} = \log A - \log B$

This law can be proved in a similar way to the proof of the first law.

Examples

1 $\log 2 = \log \frac{6}{3} = \log 6 - \log 3$

$\log 2 = \log \frac{12}{6} = \log 12 - \log 6$

$\log 2 = \log \frac{7}{3.5} = \log 7 - \log 3.5$

$\log 2 = \log \frac{4.68}{2.34} = \log 4.68 - \log 2.34$

These results can be checked on your calculator using base 10. However, they are true no matter *what* base is used.

2 Evaluate: $\log_2 24 - \log_2 3$

$\log_2 \left(\frac{24}{3}\right) = \log_2 8 = 3$

3 Simplify: $2 + \log_3 \left(\frac{4}{9}\right)$

$2 + (\log_3 4 - \log_3 9) = 2 + (\log_3 4 - 2) = \log_3 4$

Combining the first two laws, we obtain:

$$\log \frac{A \times B \times C \times \ldots}{P \times Q \times R \times \ldots} = \log (A \times B \times C \times \ldots) - \log (P \times Q \times R \times \ldots)$$

$$= \log A + \log B + \log C + \ldots - \log P - \log Q - \log R \ldots$$

Exercises 14.5 continued

4 Evaluate:

a $\log_2 3 - \log_2 6$ **b** $\log_3 24 - \log_3 2 - \log_3 4$

c $\log_4 3 - \log_4 2 - \log_4 6$ **d** $\log_2 3 - \log_2 30 + \log_2 5$

e $\log_2 \sqrt{6} - \log_2 \sqrt{3}$ **f** $\log_2 \sqrt{5} - \log_2 \sqrt{10}$

The third law

Rule: $\log A^n = n \times \log A$

You should be able to deduce this from the first law.

Examples

1 $\log_5 8 = \log_5 2^3 = 3 \log_5 2$

2 Evaluate: $\log_3 36 - 2\log_3 2$

    $\log_3 36 - \log_3 2^2 = \log_3 36 - \log_3 4$

3 $2 \log_3 5 = \log_3 5^2$

    $= \log_3 25$

    $= \log_3 \frac{36}{4}$

    $= \log_3 9$

    $= 2$

Exercises 14.5 continued

5 Evaluate:

a $2 \log_2 3 - \log_2 36$ **b** $3 \log_3 2 - \log_3 24$

c $2 \log_2 6 - \log_2 9$ **d** $\log_3 12 - 3 \log_3 2 + \log_3 6$

6 Write in the form Cx, where C is a decimal correct to 3 significant figures:

a $\log_e 10^x$ **b** $\log_{10} e^x$ **c** $\log_{10} 10^x$

d $\log_e e^x$ **e** $\log_e 10^{3x}$ **f** $\log_{10} e^{7x}$

SUMMARY

- $\log (A \times B) = \log A + \log B$

- $\log \dfrac{A}{B} = \log A - \log B$

- $\log A^n = n \log A$

It is important that you are able to change the form of a logarithmic expression.

Examples

Write in the form $a \log x + b \log y + \dots$

1 $\log \dfrac{x}{y^3}$

$\log \dfrac{x}{y^3} = \log x - \log y^3$ (Law 2)

$\quad\quad\quad = \log x - 3 \log y$ (Law 3)

2 $\log (P^3 \times \sqrt{Q})$

$\log (P^3 \times \sqrt{Q}) = \log P^3 + \log \sqrt{Q}$ (Law 1)

$\quad\quad\quad\quad\quad = \log P^3 + \log Q^{\frac{1}{2}}$

$\quad\quad\quad\quad\quad = 3 \log P + \tfrac{1}{2} \log Q$ (Law 3)

Exercises 14.5 *continued*

7 Write each of the following in the form $a \log x + b \log y + \dots$

a $\log N^k$

b $\log (M \times N)$

c $\log \dfrac{A}{B}$

d $\log (k \times x^n)$

e $\log(m \times \sqrt[3]{w})$

f $\log 5x^3$

g $\log \left(\dfrac{\sqrt{x}}{3} \right)$

h $\log \dfrac{k\sqrt{t}}{m}$

i $\log \dfrac{5\sqrt{x}}{y}$

j $\log e^x$

k $\log cx^n$

l $\log Ke^x$

14.6 CHANGE OF BASE

To evaluate the logarithm of a number to a base other than 10 or e, we express it *in terms of* a log to base 10 or e, then use the calculator.

To evaluate $\log_a N$:

Let $\log_a N = x$

$\therefore a^x = N$

$\therefore \log_b a^x = \log_b N$

$\therefore x \log_b a = \log_b N$

$\therefore x = \dfrac{\log_b N}{\log_b a}$

Rule: $\log_a N = \dfrac{\log_b N}{\log_b a}$

The logarithm of a number to any base a is equal to the logarithm of that number to any other base, b, divided by the logarithm of a to base b.

> ### Example
>
> $$\log_7 13 = \frac{\log_{10} 13}{\log_{10} 7} \text{ or } = \frac{\log_e 13}{\log_e 7}$$
>
> (Note the same base in the numerator and denominator of each fraction.)

Exercises 14.6

1 Evaluate, correct to 3 significant figures:

    a $\log_3 7.61$ **b** $\log_2 100$ **c** $\log_5 10^6$

    d $\log_3 0.0693$ **e** $\log_2 (3.68 \times 10^{11})$

2 Change $\log_b a$ to a logarithm to base a.

14.7 EVALUATIONS USING THE LAWS OF LOGARITHMS

> ### Examples
>
> If $\log_{10} x = 2.3$ and $\log_{10} y = 3.4$,
>
> $$\begin{aligned} \log_{10} x^3 &= 3 \log_{10} x \\ &= 3 \times 2.3 \\ &= 6.9 \end{aligned} \qquad \text{and}$$
>
> $$\begin{aligned} \log_{10} \frac{x}{\sqrt{y}} &= \log_{10} x - \log_{10} y^{\frac{1}{2}} \\ &= \log_{10} x - \tfrac{1}{2} \log_{10} y \\ &= 2.3 - \tfrac{1}{2}(3.4) \\ &= 0.6 \end{aligned}$$

Exercises 14.7

1 Given that $\log_{10} a = 2.6$ and $\log_{10} b = 1.2$, evaluate:

    a $\log_{10} \sqrt{a}$ **b** $\log_{10} ab$ **c** $\log_{10} \dfrac{\sqrt{b}}{a^2}$

    d $\log_{10} a^2 b^3$ **e** $\dfrac{\log_{10} a}{\log_{10} b}$ **f** $\log_{10} \sqrt{\dfrac{a}{b}}$

2 Given that $\log_e x = 4.6$ and $\log_e y = 2.3$, evaluate:

    a $\dfrac{\log_e x}{\log_e y}$ **b** $\log_e \dfrac{x}{y}$ **c** $\log_e xy$ **d** $\log_e x \times \log_e y$

14.8 SOLUTION OF LOGARITHMIC EQUATIONS

Note: To solve such equations first translate the given equality into exponential language. Remember that $\log_b N = L$ and $b^L = N$ are equivalent statements.

Examples

1 If $\log_{10} x = -2$

   $\therefore x = 10^{-2}$

   $\therefore x = 0.01$

2 If $\log_e (x - 3) = 1.84$

   $\therefore x - 3 = e^{1.84}$

   $\therefore x - 3 \approx 6.2965$

   $\therefore x \approx 9.30$

3 If $\log_2 (2x + 3) = 1.86$

   $\therefore 2x + 3 = 2^{1.86}$

   $\therefore 2x + 3 \approx 3.6301$

   $\therefore x \approx 0.315$

Exercises 14.8

1 Solve for the pronumeral without using a calculator:

   a $\log_2 \dfrac{V}{3} = 4$

   b $\log_{10} \dfrac{x}{36.4} = -2$

   c $\log_5 \dfrac{E}{0.004} = 2$

   d $\log_2 (3 - x) = 4$

   e $\log_2 (1 - \tfrac{1}{2}x) = -3$

   f $\log_2 (0.5V + 0.2) = -2$

2 Solve for the pronumeral, using a calculator:

   a $\log_2 x = 3.61$

   b $\log_5 P = -0.0635$

   c $\log_{10} C = -4.61$

   d $\log_2 \dfrac{x}{5.64} = 9.35$

   e $\log_{10} \dfrac{4.31}{P} = 3.33$

   f $\log_{10} (k - 2.45) = 0.762$

14.9 EXPONENTIAL EQUATIONS

An equation in which the variable to be evaluated is contained in an exponent (index) is called an **exponential equation**.

Example

If $2^x + 5 = 13$

   $2^x = 8$

   $\therefore x = 3$

Note: If $a^x = a^n$, x must equal n (unless $a = 0$ or 1).

Examples

1 If $3^x = 3^{17}$, then $x = 17$

3 If $9^k = 9^{13}$, then $k = 13$

2 If $(17.6)^{t+1} = (17.6)^{25}$

then $t + 1 = 25$

$\therefore t = 24$

SUMMARY

■ It is clear that when the *bases* are the *same*, such exponential equations are simple to solve.

■ Some exponential equations where the bases are *not* the same can be rewritten so that the bases *are* the same, then solved as above.

Examples

1 If $\quad 4^x = 8$

$\left(2^2\right)^x = 2^3$

$2^{2x} = 2^3$

$2x = 3$

$\therefore x = 1.5$

3 If $\quad \dfrac{1}{9^n} = 81$

$\therefore 9^{-n} = 9^2$

$-n = 2$

$\therefore n = -2$

2 If $\quad 3^k = \dfrac{1}{9}$

$\therefore 3^k = \dfrac{1}{3^2}$

$3^k = 3^{-2}$

$\therefore k = -2$

4 If $\quad 8^{2t+1} = 16$

$\therefore \left(2^3\right)^{2t+1} = 2^4$

$2^{6t+3} = 2^4$

$6t + 3 = 4$

$6t = 1$

$\therefore t = \tfrac{1}{6}(\approx 0.167)$

Exercises 14.9

1 Solve:

a $3^{k+1} = 3^5$ **b** $5^{n-2} = 5^{-7}$ **c** $4^{2x} = 4^3$ **d** $2^{t-3} = 2$

2 Solve:

a $2^{m+3} = 8$ **b** $7^{3-x} = 49$ **c** $5^t = 1$ **d** $4^{1-2x} = 16$

3 Solve:

a $9^n = 27$ **b** $8^x = 4$ **c** $4^{3x+2} = 8$ **d** $81^{1-2t} = 27$

4 Solve:

a $2^m = \sqrt{2}$ **b** $4^k = 4\sqrt{2}$ **c** $9^x = \dfrac{\sqrt{3}}{3}$ **d** $25^{t+1} = 5\sqrt{5}$

5 Solve:

a $27^{-x} = \tfrac{1}{3}$ **b** $9^t = \left(\tfrac{1}{3}\right)^{-2}$ **c** $\left(\tfrac{1}{3}\right)^{m-3} = \tfrac{1}{9}$

6 Solve:

a $\dfrac{1}{9^x} = \sqrt{3}$ **b** $5^t \div 5^{2-3t} = \tfrac{1}{25}$ **c** $\dfrac{1}{3^x} = 1$

However, usually it is impossible to make the bases the same by inspection, e.g. if $4^t = 7$. In such cases, we solve the exponential equation by taking the logarithm of both sides (which in effect means that we make the base 10 or e on each side).

Examples

1 If $\qquad 4^{1-2t} = 7$

$\therefore \log 4^{1-2t} = \log 7$

$(1 - 2t)\log 4 = \log 7$

$1 - 2t = \dfrac{\log 7}{\log 4}$

$2t = 1 - \dfrac{\log 7}{\log 4}$

$\therefore t \approx -0.202$

2 If $\qquad 2 \times 3^k = 7$

$\therefore \log(2 \times 3^k) = \log 7$

$\log 2 + k \log 3 = \log 7$

$k = \dfrac{\log 7 - \log 2}{\log 3}$

$\therefore k \approx 1.14$

The logs may be taken to *any* base, but in practice we always use base 10 or e. When powers of 10 are involved it is usually simplest to take logs to base 10, since $\log_{10} 10^x = x$, and when powers of e are involved, to take logs to base e, since $\log_e e^x = x$.

Examples

1 If $\qquad 10^{2x} = 23 \times 10^{x-1}$

$\log_{10} 10^{2x} = \log_{10}(23 \times 10^{x-1})$

$2x = \log_{10} 23 + (x - 1)$

$x = \log_{10} 23 - 1$

$\therefore x \approx 0.362$

2 If $\qquad 28e^{1-x} = e^{x+2}$

$\log_e 28^{1-x} = \log_e e^{x+2}$

$\log_e 28 + (1 - x) = x + 2$

$2x = \log_e 28 - 1$

$\therefore x \approx 1.17$

Exercises 14.9 *continued*

Solve the following exponential equations for the pronumerals, giving answers correct to 3 significant figures:

7 Solve:

   a $3^x = 7$

   c $97.24^{2n-3} = 12.86$

   b $5^{x-1} = 13$

   d $1.086^{1-x} = 2.748$

8 Solve:

   a $3^k = 2^{k+1}$

   c $2.73^{1-3k} = 10^{k+1}$

   e $3 \times 4^x = 17$

   b $5^{2-n} = 3^{n+1}$

   d $e^{t-1} = (e+1)^t$

   f $5.61 \times 10^{2n-3} = 674$

14.10 CHANGE OF SUBJECT INVOLVING LOGARITHMS

When the pronumeral required as the subject of a formula appears as a logarithm, translate the formula into exponential form. When it appears as an index, translate it into logarithmic form (i.e. take logs of both sides). When the base is 10, take logs to base 10 (since $\log_{10} 10 = 1$), and when the base is e, take logs to base e (since $\log_e e = 1$).

Examples

Make x the subject of the formula:

1 $\log_{10} \dfrac{k}{k - x} = t$

$$\dfrac{k}{k - x} = 10^t$$

$$\dfrac{k - x}{k} = 10^{-t}$$

$$k - x = k \times 10^{-t}$$

$$x = k - (k \times 10^{-t})$$

$$\therefore x = k (1 - 10^{-t})$$

2 $w = k + 10^x$

$$\therefore 10^x = w - k$$

$$\therefore x = \log_{10} (w - k)$$

3 $w = ke^x$

$$\therefore \log_e w = \log_e k + x$$

$$\therefore x = \log_e w - \log_e k$$

$$\therefore x = \log_e \dfrac{w}{k}$$

Exercises 14.10

1 For each of the following formulae, make the variable in brackets the subject of the formula:

   a $F = k^t$ (t) **b** $y = \log_{10} x$ (x)

   c $y = 3^{-x}$ (x) **d** $x^{\frac{a}{b}} = C$ (a)

   e $m = \log_{10} \dfrac{a}{b - c}$ (b) **f** $\log_{10} \left(\dfrac{x}{y} \right) = n$ (x)

   g $\log x = \log y$ (x) **h** $\log_{10} \dfrac{1}{x} = k$ (x)

+++ 2 Make t the subject of the formula:

   a $y = y_0 \times 10^{-kt}$ **b** $Q = Q_0 e^{-kt}$

   c $y = y_0 \times (1 - A^{kt})$ **d** $Q = Q_0 (1 - e^{kt})$

14.11 APPLICATIONS

Exercises 14.11

1 The number of states, S, that can exist in a binary number of b bits is given by $S = 2^b$.

   a Make b the subject of this formula.

b Hence find the number of bits required in a binary number where at least 500 states are required. (*Note:* The answer must be an integer.)

2 The effort force F N required to support a load of W N suspended from a rope wrapped around a beam of circular cross-section is given by $F = W \times k^t$, where k is a constant and t is the number of turns of the rope around the beam. Show that:

$$t = \frac{\log \dfrac{F}{W}}{\log k}$$

3 The ratio of the atmospheric pressure (p) at a height h km above sea level, to the pressure at sea level p_s, is given by $\dfrac{p}{p_s} = (0.883)^h$. Make h the subject of this formula.

4 When a gas expands adiabatically, $pV^\gamma = k$, where p Pa is the pressure of the gas, V m^3 is the volume of the gas, γ is the adiabatic index of the gas and k is a constant. Show that:

$$\gamma = \frac{\log \dfrac{k}{p}}{\log V}$$

5 If the rate of increase of a quantity Q at any instant is proportional to the amount present at that instant, it may be proved that $Q = Q_0\, e^{kt}$, where Q_0 is the amount present originally, Q is the amount present after t seconds, e and k are constants. Make t the subject of this formula.

6 If P is invested at compound interest, the interest rate being $R\%$ per year, paid yearly, the total amount, A (i.e. principal + interest) after n years is given by $A = PR^n$. Make n the subject of this formula.

7 The characteristic impedance of a parallel conductor (twin-lead), where dry air separates the conductors, is given by $Z = 276 \log_{10} \dfrac{2D}{d}$, where D is the separation of the conductors and d is the diameter of the conductors.

a Make D the subject of this formula.

b Evaluate D when $d = 3.00$ mm and $Z = 350\ \Omega$.

14.12 THE CURVE $y = k \log_b Cx$

First, two points:

1 '$x \to \infty$' is mathematical language for 'x becomes a larger and larger positive number, without limit'. '$x \to -\infty$' means 'x becomes a 'larger and larger' negative number, without limit'.

For example, it is true to say that 'as $x \to \infty$, $\dfrac{1}{x} \to 0$'.

2   If a curve approaches closer and closer to a
    straight line, but never actually contacts it,
    the straight line is said to be an *asymptote* to
    the curve. The curve is said to be *asymptotic*
    to the straight line.

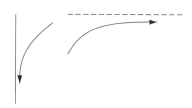

The curve $y = \log_{10} x$

x	0.2	0.4	0.6	0.8	1	2	3	4	5
$y(= \log_{10} x)$	−0.70	−0.40	−0.22	−0.97	0	0.30	0.48	0.60	0.70

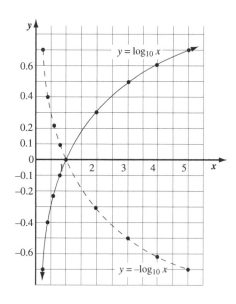

$y = \log_{10} x$ is equivalent to $10^y = x$ and from this it can
be seen that:

when $y = 0$, $x = 1$
when $y \to -\infty$, $x \to 0$
and when $y \to \infty$, $x \to \infty$.

x can never have a negative value because there is no
value of y that gives $10^y < 0$.

As $x \to 0$, the curve approaches closer and closer to
the y-axis but never actually touches it. The curve is
asymptotic to the y-axis; the y-axis is an *asymptote* of the
curve.

To *sketch* the graph of this curve, $y = \log_{10} x$, it is convenient to express the equation in
exponential language, $10^y = x$.

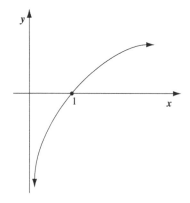

By considering this equation it can be seen that:
■   when $y = 0$, $x = 1$
■   as $y \to \infty$, $x \to \infty$
■   as $y \to -\infty$, $x \to 0$
■ x can never have a negative value. (There is no index y such
    that $10^y < 0$.)

Hence, we obtain the curve shown. As $x \to 0$, the curve
approaches closer and closer to the negative y-axis but never
actually touches it. The curve is asymptotic to the negative
y-axis. The negative y-axis is an asymptote to the curve.

Remember: **i** $\log_b 1 = 0$, because $b^0 = 1$ (the logarithm of 1 to *any* base = 0)

  ii as $x \to 0$, $\log_b x \to -\infty$.

The graphs of *all* functions of the form $k \log_b Cx$, where k is positive, have this shape. The scale on the y-axis depends upon the values of the constants k, b and C.

To sketch the curve $y = k \times \log_b (cx + d)$

1 When $cx + d = b$, $y = k$ (because $\log_b b = 1$, i.e. $b^1 = b$).
2 When $cx + d = 1$, $y = 0$ (because $\log_b 1 = 0$, i.e. $b^0 = 1$).
3 As $cx + d \to 0$, $y \to \pm\infty$ ($-\infty$ when k is positive, $+\infty$ when k is negative).
4 It is useful, as a check, to calculate one value for y corresponding to some convenient value for x.

Examples

$y = 4\log_{10} 5x$

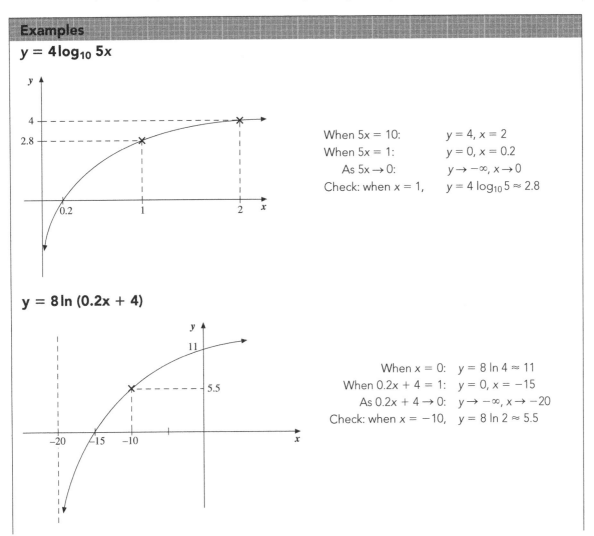

When $5x = 10$: $y = 4$, $x = 2$
When $5x = 1$: $y = 0$, $x = 0.2$
  As $5x \to 0$: $y \to -\infty$, $x \to 0$
Check: when $x = 1$, $y = 4 \log_{10} 5 \approx 2.8$

$y = 8\ln (0.2x + 4)$

When $x = 0$: $y = 8 \ln 4 \approx 11$
When $0.2x + 4 = 1$: $y = 0$, $x = -15$
  As $0.2x + 4 \to 0$: $y \to -\infty$, $x \to -20$
Check: when $x = -10$, $y = 8 \ln 2 \approx 5.5$

continued

continued

$y = -5\log_{10}(2x + 4)$

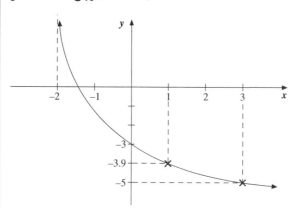

When $2x + 4 = 10$: $y = -5, x = 3$
When $2x + 4 = 1$: $y = 0, x = -1.5$
As $2x + 4 \to 0$: $y \to \infty, x \to -2$
When $x = 0$: $y = -5\log 4 \approx -3$
Check: when $x = 1$, $y = -\log 6 \approx -3.9$

Examples

1 Sketch the curve $y = 12\log_2 5x$.

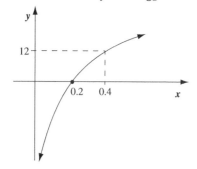

$y = 0$ when $5x = 1$, that is, when $x = 0.2$
$y = 12$ when $5x = 2$, that is, when $x = 0.4$

2 Sketch the curve $y = 8\log_6 2x + 4$.

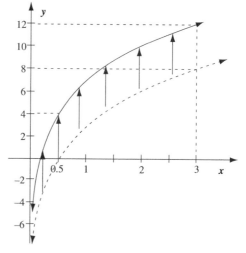

First, sketch $y = 8\log_6 2x$ (the broken curve).
$y = 0$ when $2x = 1$, that is, when $x = 0.5$
$y = 8$ when $x = 3$ ($\therefore 2x = 6$)
Then increase all the ordinates by 4 units.

Examples

3   Sketch the curve $y = -12 \log_2 5x$.

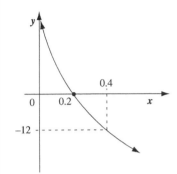

$y = 0$ when $x = 0.2$
$y = -12$ when $5x = 2$, that is, when $x = 0.4$

4   Sketch the curve $y = \log(\tfrac{1}{2}x + 3)$.

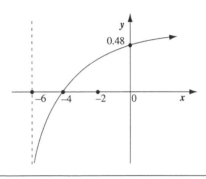

Note: ■ This is different from $y = \log\tfrac{1}{2}x + 3$.
  ■ When no base is specified, base 10 is implied.
$y = 0$ when $\tfrac{1}{2}x + 3 = 1$, that is, when $x = -4$
$y \to -\infty$ as $\tfrac{1}{2}x + 3 \to 0$, that is, as $\tfrac{1}{2}x \to -3$, that is, as
$x \to -6$
When $x = 0$, $y = \log 3 \approx 0.48$

Exercises 14.12

Sketch the following curves, showing the value of the x-intercept and the scales on both axes:
Remember: If no base is specified, then base 10 is implied.

1   $y = 3 \log 2x$ 2 $y = \log(\tfrac{1}{4}x + 2)$
3   $y = -\log 0.2x$ 4 $y = -\log 0.2x + 0.3$

14.13 A QUICK BUT IMPORTANT REVISION OF EXPONENTS AND LOGARITHMS

1 $(x^a)^b = (x^b)^a = x^{ab}$. Examples $\begin{cases} (2^2)^3 = 4^3 = 64 \\ (2^3)^2 = 8^2 = 64 \\ (2)^{2\times3} = 2^6 = 64 \end{cases}$

Hence if $4^{x+2} = 2^{3x}$
$\therefore (2^2)^{x+2} = 2^{3x}$
$2^{2x+4} = 2^{3x}$
$\therefore 2x + 4 = 3x$
$x = 4$

2 $\log x$ implies $\log_{10} x$ (i.e. if no base is specified, base 10 is implied).

3 Because $\log_b x$ is the exponent required on b to obtain x (e.g. $\log_2 8 = 3$), it follows that if we *do* place this exponent on b, we *will* obtain x, i.e. $b^{\log x} = x$.

Example

$3^{\log_3 7} = 7$, $10^{\log 4.61} = 4.61$

Of special interest are logarithms to base 10: $(10)^{\log 17} = 17$, $(10)^{\log(x-3)} = x - 3$.

Note the following examples in which all three of the above revision facts are used:

$(10^2)^{\log 3} = (10^{\log 3})^2 = 3^2 = 9$

$(10^{-3})^{\log 2} = (10^{\log 2})^{-3} = 2^{-3} = \dfrac{1}{8}$

Exercises 14.13

1 Evaluate mentally and then check your answer using your calculator:

a $(2^{1.2})^5$	**b** $(2^{0.1})^{10}$	**c** $(4^5)^{-0.2}$	**d** $(10^{-1})^{-1}$
e $(10^{-1})^2$	**f** $(2^{-3})^{-2}$	**g** $(8^{\frac{5}{2}})^{\frac{2}{3}}$	**h** $(4 \times 1.5^2)^2$

2 Solve for x then check your answer by substitution:

a $2^{-x} = 4^{x-3}$ **b** $3^{x+2} = 27^{x-2}$

c $8^{x^2+3} = 16^{x^2-4}$ **d** $27^{3x^2+x} = 81^{2x^2+1}$

3 Evaluate mentally and then check your answer using your calculator:

a $\log 10$	**b** $\log 10^2$	**c** $(\log 10)^2$	**d** $\log\left(\dfrac{1}{100}\right)$
e $\log 0.001$	**f** $\log \sqrt{10}$	**g** $\log \dfrac{1}{\sqrt{10}}$	**h** $\log \dfrac{100}{\sqrt{10}}$

+++ 4 Evaluate mentally and then check your answer using your calculator:

a $(10^2)^{\log 3}$	**b** $\left(\dfrac{1}{\sqrt{10}}\right)^{\log 16}$	**c** $(10^{-3})^{\log 2}$	**d** $(0.1)^{\log 20}$
e $(\sqrt{10})^{\log 25}$	**f** $\left(\dfrac{1}{100}\right)^{\log 0.5}$	**g** $(10^{\frac{3}{2}})^{\log 4}$	**h** $(10^{0.25})^{\log 81}$

14.14 LIMITS

If the value of x increases and *keeps* increasing we say that '$x \to \infty$'. Infinity (∞) is not a number but a *concept* (an *idea*), so x can never actually *reach* infinity because infinity does not actually *exist* and '$x = \infty$' is a meaningless statement.

   If the value of x becomes closer and closer to 3, the value of $(2x + 5)$ becomes closer and closer to 11. We say that 'as x approaches 3, $(2x + 5)$ approaches 11', which is abbreviated to the statement 'as $x \to 3$, $(2x + 5) \to 11$'. This statement does *not* imply that when $x = 3$, then $(2x + 3) = 11$, although of course this is actually true. It is true that as x increases, $\dfrac{7}{x}$ decreases, i.e. 'as $x \to \infty$, $\dfrac{7}{x} \to 0$', but x can never *equal* infinity and $\dfrac{7}{x}$ can never *equal* 0.

Exercises 14.14

1 Write down the values of the following:

    a $\lim\limits_{x \to 2}(3x + 1)$ **b** $\lim\limits_{x \to \infty}(3x + 1)$ **c** $\lim\limits_{x \to -\infty}(3x + 1)$ **d** $\lim\limits_{x \to \infty}\dfrac{17.6}{x}$

    e $\lim\limits_{x \to 5}\dfrac{3.84}{x - 5}$ **f** $\lim\limits_{x \to \infty}\dfrac{4.1}{2 - 3x}$ **g** $\lim\limits_{x \to \infty} 3^{-x}$ **h** $\lim\limits_{x \to -\infty} 6^{x}$

    i $\lim\limits_{x \to 0} 8^{x}$ **j** $\lim\limits_{x \to \infty}(3 - 2^{x})$

2 **a** $\lim\limits_{x \to \infty} 0.2^{x}$ **b** $\lim\limits_{x \to -\infty} 0.2^{x}$ **c** $\lim\limits_{x \to 0}\left(x - \dfrac{1}{x}\right)$ **d** $\lim\limits_{x \to \infty} 0.2^{-x}$ **e** $\lim\limits_{x \to -\infty} 0.2^{-x}$

SELF-TEST

1 Evaluate:

    a $\log_2 \tfrac{1}{8}$ **b** $\log_2 1$ **c** $\log_8 \tfrac{1}{4}$

    d $\log_3 \dfrac{1}{\sqrt{3}}$ **e** $\log_9 3$ **f** $\log_{10} 10^7$

2 Use your calculator to evaluate, correct to 3 significant figures:

    a $\log_{10} 200 \times \log_{10} 300$ **b** $e^{1.23} - e^{-1.23}$

    c $3.16e^{2.71}$ **d** $187 \div \log_e 9.46$

    e $\log_e 1.98 - e^{-0.657}$ **f** $1.69e^{1.83} - 3.25e^{1.07}$

3 Express each of the following as a single logarithm and hence evaluate mentally:

    a $\log_2 3 + \log_2 5 - \log_2 30$ **b** $\log_3 \sqrt{2} - \log_3 \sqrt{6}$

    c $\log_2 10 - \log_2 45 + 2\log_2 3$ **d** $2\log_5 10 - \log_4 6 - 2\log_5 2 + \log_4 3$

4 Use your calculator to evaluate correct to 3 significant figures:

    a $\log_2 5$ **b** $\log_3 0.5$ **c** $\log_5 \sqrt{7}$

5 Solve without using a calculator:

    a $2^x = 2\sqrt{8}$ **b** $3^{2n} = 1$ **c** $8^k = 4^{k + 3}$

    d $(\tfrac{1}{9})^m = 27$ **e** $\dfrac{1}{2^t} = 4^{1-t} \div 8^{-t}$

6 Solve for the pronumeral without using a calculator:

    a $\log_2 (2 - k) = 3$ **b** $\log_5 \dfrac{100}{t - 3} = 2$

    c $\log_x \dfrac{27}{x} = 2$ **d** $10^{\log 2x} \times e^{\ln 3x}$

7 Solve for the pronumeral:

a $4^{t-1} = 3$

b $2.68^{2-n} = 1.73^n$

+++ c $3e^k = 7e^{1-3k}$

+++ d $10^x = 5.37e^x$

8 Make the variable in brackets the subject of the formula:

a $A^{\frac{k}{t}} = b$ (t)

b $\log_{10} \dfrac{a}{a-x} = k$ (x)

c $ke^{at+b} = e^t$ (t)

+++ d $t = \dfrac{1}{k} \log_e \left(1 - \dfrac{x}{y}\right)$ (y)

9 In a particular circuit a voltage doubles in size each millisecond. Hence in t ms a voltage V_0 increases to $V_0 \times 2^t$, i.e. $V = V_0 \times 2^t$, where V_0 is the original voltage and V is the voltage after t ms.

a Make t the subject of this formula.

b How long will it take for a voltage of 0.075 V to increase to 250 V?

+++ 10 The intensity of a sound is given by $I = 10 \log_{10} \dfrac{P}{10^{-12}}$ dB, where P W/m^2 is the sound power being received per square metre. Find the sound power received per square metre (in μW/m^2) at a place where the sound intensity is 75 dB.

CHAPTER 15

NON-LINEAR EMPIRICAL EQUATIONS

Learning objectives

- Transform non-linear data to a linear form, draw the line of best fit and then determine the non-linear formula for the data.

15.1 INTRODUCTION

Empirical equations are those based on experimental results. If experimental results are plotted on ordinary graph paper and form a straight line, then the equation connecting the two variables can be easily found. The equation is not so easy to find if the relationship between the variables is not linear.

However, various methods can be used to convert non-linear results into a linear form. When this is done, we must only use the equation for values within the range given by the experiment.

Revision of linear equations

The gradient form of the equation of a straight line $(y = mx + b)$ is useful in that it enables us to read off directly the gradient of the line, m, and the y-intercept, b.

Conversely, if we have some points plotted which fall on a straight line, then by measuring the intercept on the vertical axis and the gradient of the line or by substituting the coordinates of two points on the line into the equation, we can determine the equation of the line.

15.2 CONVERSION TO LINEAR FORM BY ALGEBRAIC METHODS

Example

The following values of x and y are believed to satisfy an equation of the type $y = ax^2 + bx$.
Find a linear equation that suits this information and so evaluate a and b:

x	1	2	3	4	5
y	5	16	34	57	84

continued

continued

$$y = ax^2 + bx$$

Divide by x:

$$\frac{y}{x} = ax + b$$

If we now put $Y = \frac{y}{x}$, the relation becomes:

$$Y = ax + b$$

To plot this, we use the following data:

x	1	2	3	4	5
$Y \left(= \dfrac{y}{x} \right)$	5	8	11.3	14.3	16.8

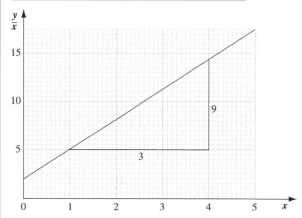

From the graph, the vertical intercept $b = 2$ and the gradient $a = 3$.

That is, $\dfrac{y}{x} = 3x + 2$.

$\therefore$ Required equation is $y = 3x^2 + 2x$.

The line is drawn for two purposes:

1   to check that a straight line *is* obtained so that the form of the original equation is verified (in this case, $y = ax^2 + bx$)

2   to obtain the 'line of best fit'. The values plotted are from instrument readings and therefore must be only approximate. It could happen that some or even *all* of the points plotted will not fall on the line of best fit. To evaluate the constants we use two points on the **line of best fit**.

Exercises 15.1

1   V and P are related by an equation of the form $P = \dfrac{K}{V} + b$. Obtain a linear graph from the following values and so find the values of the constants K and b.

V	1	2	3	4	5
P	6.5	4	3.2	3	2.3

2 The resistance force R on a body moving through water is measured for various velocities v.

R (N)	0	420	1850	3600	6720
v (m/s)	0	5	10	15	20

By plotting R against v^2, verify that the relationship is of the form $R = C \times v^2$. Draw the line of best fit and hence:

a evaluate the constant C

b estimate the resistance force when $v = 7.5$ m/s

c estimate the value of v for which R has the value 2500 N

15.3 THE USE OF LOGARITHMS TO PRODUCE A LINEAR FORM

We consider two types of functions in which logarithms are used to produce a linear form–power functions and exponential functions.

Power functions

If y and x are connected by the equation $y = Kx^n$, where K and n are constants, then taking the logarithms of both sides of the equation gives:

$$\log y = \log K + \log x^n$$
$$= \log K + n \log x$$
$$\therefore \log y = n \log x + \log K$$

That is, $Y = nX + b$

where $\begin{cases} Y = \log y \\ X = \log x \\ b = \log K \end{cases}$

If we plot Y against X we obtain a straight line having gradient n and a vertical intercept of $\log K$.

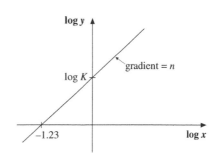

Example

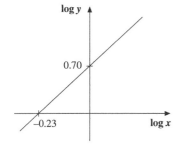

If from measured values of x and y we plot $\log y$ against $\log x$ and obtain the line of best fit as shown, we deduce that:

gradient: $\qquad m = \dfrac{0.70}{0.23} \approx 3.0$

vertical intercept: $\quad \log K = 0.7$

$$\therefore K = 10^{0.7} \approx 5.0$$

Hence, the relationship is $y = 5.0 \times x^{3.0}$.

Exponential functions

Exponential functions are functions in which a variable appears as an exponent.

If the relationship is of the form:

$$y = KC^x \text{ where } K \text{ and } C \text{ are constants}$$

$$\therefore \log y = \log K + \log C^x$$

$$= \log K + x \log C$$

$$\therefore \log y = (\log C)x + \log K$$

That is, $Y = mX + b$

$$\text{where } \begin{cases} Y = \log y \\ m = \log C \\ b = \log K \end{cases}$$

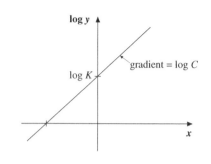

Example

Gradient: $\qquad m = \log C = \dfrac{0.52 - 0.20}{4} = 0.08$

$$\therefore C = 10^{0.08} \approx 1.20$$

vertical intercept: $\log K = 0.20$

$$\therefore K = 10^{0.20} \approx 1.58$$

Hence, the relationship is $y = 1.58 \times (1.20)^x$.

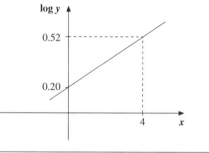

Example (power function)

In an experiment, the following results for P and V were obtained. Show that the relationship is of the form $P = KV^n$ and find the values of K and n.

P	135	110	85	63	47
V	3.69	4.25	5.24	6.5	7.8

If $P = KV^n$, then $\log P = \log K + n \log V$

that is, $\log P = n \log V + \log K$

Hence, if log P was plotted against log V, a straight line would be obtained.

From the above readings, we have

log V	0.567	0.628	0.719	0.813	0.892
log P	2.13	2.04	1.93	1.80	1.67

The fact that the points are clearly collinear (except for minor irregularities which we would expect from instrument readings) verifies that the relationship is of the form $P = KV^n$.

To evaluate the constants, we choose two points *on the line*, preferably well separated, and substitute their coordinates into the equation.

Using the points:

$\begin{cases} \log V = 0.6 \\ \log P = 2.10 \end{cases}$ and $\begin{cases} \log V = 0.9 \\ \log P = 1.67 \end{cases}$

and substituting them into the equation $\log P = n \log V + \log K$, we have:

$\begin{cases} 2.10 = n \times 0.6 + \log K \\ 1.67 = n \times 0.9 + \log K \end{cases}$

Solving these simultaneous equations gives $K = 912$ and $n = -1.43$.

Hence, the relationship is $P = 912 \times V^{-1.43}$.

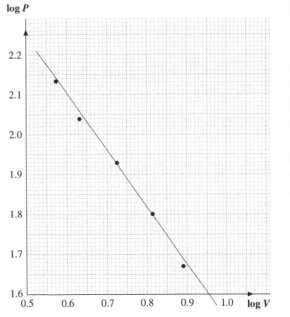

SUMMARY

If $\quad y = K \times x^n$

$\therefore \quad \log y = \log K + n \log x$

i.e. $(\log y) = n (\log x) + \log K$

$\therefore$ if $\log y$ is plotted against $\log x$,

a straight line is obtained

whose $\begin{cases} \text{gradient} = n \\ y\text{-intercept} = \log K \end{cases}$

If $\quad y = K \times C^x$

$\therefore \log y = \log K + x \log C$

$\therefore (\log y) = \log C \times (x) + \log K$

$\therefore$ if $\log y$ is plotted against x,

a straight line is obtained

whose $\begin{cases} \text{gradient} = \log C \\ y\text{-intercept} = \log K \end{cases}$

Exercises 15.2

1 In cooling a body, the following data was obtained. θ represents the temperature of the body above its surroundings and t the time from the point at which cooling begins.

θ	49.4	44.7	33.7	35.5	29.7	27.1	24.5
t	0	1	2	3	4	5	6

Given that the relationship between θ and t is of the form $\theta = Ke^{-at}$, evaluate the constants K and a.

continued

continued

2 The following table gives values for E and V. Show that they satisfy the relationship $E = KV^N$ and find the values of the constants K and N.

V	1.20	1.70	2.51	3.39	5.25	6.31
E	3.80	6.61	11.5	18.2	33.1	52.5

3 If L and i satisfy the equation $i = CL^n$, find the values of the constants C and n from the following values of L and i.

L	1.00	2.50	4.00	10.0	15.0
i	2.50	1.60	1.30	0.80	0.640

4 a If i and t satisfy the equation $i = Kb^t$, use the following sets of values to find K and b.

t	0	1	2	3	4
i	70	33	16	8.0	3.5

   b By expressing b as a power of e, find the equation $i = Ke^{ct}$.

SELF-TEST

1
x	0.95	1.0	1.1	1.2	1.4	1.6	2.0
y	7.4	6.8	6.0	4.9	4.2	3.4	2.6

By plotting y against $\dfrac{1}{x^2}$, show that the ordered pairs in the above table of values verify that the relationship is of the form $y = \dfrac{C}{x^2} + K$.

From the line of best fit evaluate the constants C and K.

2 The following table gives corresponding values for y and x. By plotting a straight line graph and drawing the line of best fit, show that the relationship is of the form $y = C \times x^n$ where C and n are constants and evaluate C and n.

x	1	2	3	4	5
y	7	22	38	55	82

+++ 3 A solution was allowed to cool and readings of its temperature above its surroundings were taken every 2 minutes. From the results, show that the cooling curve is of the form $\theta = Ke^{-ct}$ and find the values of the constants K and c.

θ	33.4	32.5	31.8	31.1	30.4	29.7
t	0	2	4	6	8	10

CHAPTER 16

COMPOUND INTEREST: EXPONENTIAL GROWTH AND DECAY

Learning objectives

- Solve compound interest problems.
- Draw graphs relating to exponential growth and decay.
- Solve exponential growth and decay problems.

16.1 COMPOUND INTEREST

To understand the material in this chapter, you need to apply the knowledge from topics covered in Chapter 2 (section 2.6 Percentages) and Chapter 14 (section 14.9 Exponential equations).

If a number N is increased by $r\%$, it becomes $N + \left(\dfrac{r}{100} \times N \right) = N \times \left(1 + \dfrac{r}{100} \right)$.

If this *result* is increased by $r\%$, it becomes $N \times \left(1 + \dfrac{r}{100} \right) \times \left(1 + \dfrac{r}{100} \right)$.

If this 'compounding' process is performed n times, N increases to:

$$N \times \left[\left(1 + \frac{r}{100} \right) \times \left(1 + \frac{r}{100} \right) \times \left(1 + \frac{r}{100} \right) \times \ldots n \text{ terms} \right] = N \times \left(1 + \frac{r}{100} \right)^n.$$

If money is invested in such a way that after certain regular time periods the interest is calculated and added to the account so that *it* attracts interest during the next period, the interest is said to be 'compounded'.

If $P (the 'principal') is invested for n periods at a compound interest rate of $r\%$ per period, there will be n interest payments into the account, each of $r\%$ of the amount in the account at that time. The amount to which the money grows in these n periods is given by:

$$\textbf{Rule: } A = P \times \left(1 + \frac{r}{100} \right)^n$$

This is the 'compound interest formula' where:

$\begin{cases} P \text{ is the principal sum invested} \\ r\% \text{ is the interest rate over the period of time between interest payments} \\ A \text{ is the amount to which the principal grows after a time interval of } n \text{ of the above periods.} \end{cases}$

Example

Calculate the amount that $5000 grows to when it is invested for three weeks at an interest rate of 13% per year, the interest being compounded daily.

Solution

Interest rate = 13% per year = $\frac{13}{365}$% per day, and there will be 21 interest payments into the account.

$$\therefore A = \$5000 \times \left(1 + \frac{\frac{13}{365}}{100}\right)^{21} \approx \$5037.53$$

Exercises 16.1

1   Money is invested at an interest rate of 6% p.a. (per year), the interest being paid and compounded at the end of each year.
   a   By what percentage does the investment increase in:
       i   3 years?   ii   8 years?
   b   For how many complete years must the money remain invested before it has more than doubled?

2   The sum of $2550 is invested at an interest rate of 0.7% per month, the interest being paid and compounded at the end of each month.
   a   State the *exact* amount on deposit at the end of 5 months and also its value correct to the nearest cent.
   b   By what percentage will the investment increase in 1 year? State the *exact* percentage and also its value correct to 3 significant figures.

+++ 3 $3000 is invested for 26 weeks at an interest rate of 7% p.a., the interest being compounded daily. How much total interest is paid on this investment?

16.2 EXPONENTIAL GROWTH

The more frequently the interest is calculated and compounded, the more rapidly the amount in the account grows.

Example

Calculate the amount that $10 000 grows to when it is deposited for 3 years at an interest rate of 5% per year.

Solution

The amount A to which the principal grows in the three years is given by:

- if interest is compounded yearly: $A = \$10\,000 \times \left(1 + \frac{5}{100}\right)^3$ $= \$11\,576.25$

- if interest is compounded monthly: $A = \$10\,000 \times \left(1 + \frac{\frac{5}{12}}{100}\right)^{36}$ $= \$11\,614.72$

- if interest is compounded weekly: $A = \$10\,000 \times \left(1 + \frac{\frac{5}{52}}{100}\right)^{3 \times 52}$ $= \$11\,617.51$

- if interest is compounded daily: $A = \$10\,000 \times \left(1 + \frac{\frac{5}{365}}{100}\right)^{3 \times 365}$ $= \$11\,618.22.$

It can be shown that as n becomes larger and larger (interest compounded each hour, minute, second, . . .), the amount A does not increase *indefinitely* (i.e. 'without bound') but approaches a limit, which is given by

$$\textbf{Rule: } A = P \times e^{kt}$$

where
$\begin{cases} k \text{ is the rate of interest per year expressed as a decimal} \\ t \text{ is the investment period in years} \\ e \text{ is a constant named after the Swiss mathematician Leonard Euler.} \end{cases}$

Note: The constant e is an irrational number—that is, a non-recurring, non-repeating decimal (like π). Its approximate value is 2.718. This number is so important in technology that, like π, all scientific calculators provide an e^x key. You should check on your calculator that $e^1 (= e) \approx 2.718$. You should be able to obtain, for example, $e^3 \approx 20.09$ and $e^{-5} \approx 6.738 \times 10^{-3}$.

Note: Students who are interested should study Appendix A, which contains more information concerning the constant e and the derivation of the above formula for exponential growth.

You can now find the limit to which the amount in the previous example approaches—the amount if the interest is paid *continuously*:

$$A = P \times e^{kt} = \$10\,000 \times e^{0.05 \times 3} = \$10\,000 \times e^{0.15} = \$11\,618.34$$

Regardless of how often the interest is compounded, the total amount can never exceed this amount. This is the amount if the interest is paid *continuously*. Banks pay interest on some accounts daily or continuously and in both cases the continuous formula would be used since there is very little difference between the amounts in each case.

Exercises 16.2

1   If $P is invested at a compound interest rate of 8% p.a., state the *exact* amount of money on deposit at the end of 1 year:
    a   if the interest is paid at the end of each year
    b   if the interest is paid quarterly
    c   if the interest is paid continuously
2   If $10 000 is invested at an interest rate of 4% p.a., what will this amount become at the end of 1 year, to the nearest cent:
    a   if interest is paid each year?   b   if interest is paid each quarter?
    c   if interest is paid each month?   d   if interest is paid each week?
    e   if interest is paid each day?   f   if interest is paid continuously?

Actually *all* growth must be in 'spurts' but these are often so small and so frequent (e.g. molecule by molecule, quantum by quantum, cent by cent) that the growth may be *considered* to be continuous.

Note: The above formula for exponential growth applies, of course, not only to a quantity of money but to any quantity that grows in this continuous manner where the percentage rate of growth is constant and hence the actual rate of growth is proportional to the amount present at any instant. With a bank deposit invested at compound interest paid continuously, the percentage rate of growth remains constant and so the actual rate of growth (e.g. in dollars per day) increases in proportion to the amount of money in the account.

This type of growth is quite common, an example being population growth (e.g. bacterial, animal or human). When the population of rabbits doubles, the rate of breeding (e.g. in rabbits per second) will double; the rate of growth of the population is proportional to the size of the population at every particular instant. The growth of the population can be regarded as continuous only when the population is large. When the population is small (as was the original population of rabbits in Australia), the growth could not be regarded as continuous. All runaway chain reactions are examples of this type of growth, such as occurs in uncontrolled nuclear fission. This type of growth is sometimes referred to as 'snowballing' because when a snowball rolls down a snow-covered slope, the rate at which it picks up snow at any instant (e.g. in grams per second) is proportional to the weight of the snowball at that instant.

SUMMARY

- Exponential growth is defined as continuous growth in which the actual rate of growth at any instant is proportional to the quantity present at that instant.
- The formula for exponential growth is $Q = Q_0 e^{kt}$,

    where: $\begin{cases} Q_0 \text{ is the quantity originally present (i.e. the value of } Q \text{ when } t = 0) \\ k \text{ is the percentage continuous rate of growth during a specified period of time,} \\ \text{expressed as a decimal} \\ Q \text{ is the quantity present at the end of } t \text{ of the above time periods.} \end{cases}$

Example

If the mass of a culture increases exponentially from an original mass of 3.45 g at a continuous rate of growth of 13% per hour, find the mass present after two days of growth.

Solution

$$k = 13\% = 0.13 \text{ per hour}$$
$$t = 2 \times 24 = 48 \text{ hours}$$
$$Q = Q_0 e^{kt}$$
$$= 3.45 \times e^{0.13 \times 48}$$
$$= 3.45 \times e^{6.24}$$
$$\approx 1770 \text{ g}$$

We could equally well have used 1 *day* as the time period, in which case $k = 0.13 \times 24 = 3.12$ per day, and $t = 2$ days and $e^{kt} = e^{6.24}$ as before.

Note: When there is an exponential decrease ('decay') the value of k is negative.

Exercises 16.3

1 The radioactive element radon has a half-life of 3.82 days. (The *half-life* is the time taken for half of any mass to disintegrate.)

Find:

a   the value of k, the continuous fractional rate of growth

b   how long it will take for 100 g of radon to decay to 45.0 g.

2 The population of a particular town, which is currently 10 400, is estimated to be increasing at a rate of 7% per year. Assuming that this rate of growth continues:

a   what is the estimate of the population in 5 years' time?

b   how long will it take for the population to double?

c   how long will it take for the population to increase by 35%?

3 When the key in this circuit is closed, the current decreases exponentially according to the formula $i = \dfrac{E}{R} \times e^{-\frac{1}{RC}t}$ A/s.

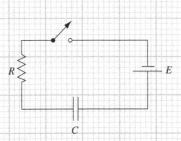

If $E = 180$ V, $C = 120\ \mu$F and $R = 3.8$ kΩ, find the current:

a   immediately after the key is closed

b   730 ms after the key is closed.

4 A manufacturer of yeast finds that their culture grows exponentially at the rate of 13% per hour.

a   What mass will 3.7 g of yeast grow to:

   i   in 7.0 hours?   ii   in two days?

continued

continued

    b What mass must be present initially in order to acquire a mass of 15 kg at the end of two days?

    c How long will it take for any given mass of this yeast to double?

+++ 5 In the inversion of raw sugar, at any particular moment the rate of decrease of the mass of sugar present is proportional to the mass of sugar remaining. (Remember that this means that the rate of decrease is *exponential*.) If after 8 hours 53 kg has reduced to 37 kg, how much sugar will remain after a *further* 24 hours?

16.3 GRAPHS OF EXPONENTIAL FUNCTIONS

Functions in which the variable appears as an exponent (index) are called 'exponential' functions. Examples are 3.2^x, 7.15^{-3x}, $2.86^{4.31x - 2.45}$. The general expression for an exponential function is $K \times a^{bx + c}$, where each of the numbers K, a, b and c may be positive or negative and c may be zero. **A very important property of these functions is that equal increases in the value of x result in equal *percentage* increases in the value of the function.**

$$N^{x+k} = N^x + N^k$$
$$N^{x+2k} = N^{x+k} + N^k$$
$$N^{x+3k} = N^{x+2k} + N^k$$
$$N^{x+4k} = N^{x+3k} + N^k$$

Successive additions of k to the exponent result in a succession of multiplications by the same number, N^k. Hence, as shown in Chapter 2 (section 2.6, Percentages) successive additions of k to the exponent result in a succession of increases by the same percentage.

Consider the function $y = K \times a^{bx + c}$. If x is increased by t, the value of y becomes $K \times a^{b(x + t) + c} = K \times a^{bx + c + bt} = K \times a^{bx + c} \times a^{bt}$. Hence the value of y is doubled when $a^{bt} = 2$, i.e. when $bt \log a = \log 2$, or $t = \dfrac{\log 2}{b \log a}$. There is no need to memorize this result but it will assist you to understand the following work if you can follow the reasoning.

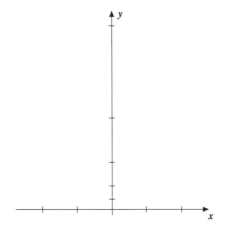

To sketch any curve of the form $y = K \times a^{bx + c}$ when no range of required values is specified for x or y, you can either set up a table of corresponding values of x and y as described in Chapter 9 (section 9.6 Dependent and independent variables) or use the following general method, which is much quicker and also provides an understanding of the properties of exponential functions.

1   Draw axes as shown, placing 2 equally spaced marks on the x-axis on each side of the origin and place 5 marks on the y-axis starting with the top one whose position is the highest you wish your graph to be and each of the succeeding marks below being at half the remaining distance down to the origin. This results in 5 marks, each one being twice the distance from the origin as the mark below it.

2   Calculate the value of y when x = 0 and label this value on the middle of your 5 marks on the y-axis. Then mark the values on the other 4 marks on the y-axis, each value being double the value on the mark below it.

3   Calculate the amount by which x must increase in order to double any y-value. Place this value next to the first mark to the right of the origin on the x-axis, and place corresponding values next to the other marks.

4   Mark 5 points on the curve as shown on the figure above, and sketch the curve. To decide which of the two curves below to draw, examine what happens to the value of y as $x \rightarrow \infty$. This will depend on whether the exponent b is positive or negative.

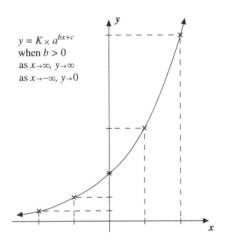

$y = K \times a^{bx+c}$
when $b > 0$
as $x \rightarrow \infty$, $y \rightarrow \infty$
as $x \rightarrow -\infty$, $y \rightarrow 0$

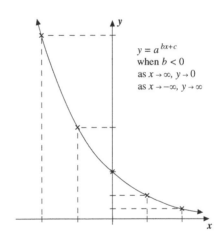

$y = a^{bx+c}$
when $b < 0$
as $x \rightarrow \infty$, $y \rightarrow 0$
as $x \rightarrow -\infty$, $y \rightarrow \infty$

All the above should become clear if you study the examples below. Note that in each case the axes are drawn first, then the scale-marks made on each axis as described above.

Example

Sketch the curve $y = 3.2^x$

- As $x \rightarrow \infty$, $y \rightarrow \infty$; as $x \rightarrow -\infty$, $y \rightarrow 0$
- When $x = 0$, $y = 1$
- y doubles when 3.2^x increases from 1 to 2.

    If $3.2^x = 2$

    then $\log 3.2^x = \log 2$

continued

continued

$$\therefore x \log 3.2 = \log 2$$

$$\therefore x = (\log 2) \div (\log 3.2)$$

$$\therefore x \approx 0.6$$

We now have the scales for both axes:

- On the x-axis the equally-spaced values 0.6 units apart.
- On the y-axis the value doubles with each increase of 0.6 on the x-axis.
- These 5 points are sufficient to show the shape of the curve:

  $(-1.2, 0.25), (-0.6, 0.5), (0, 1), (0.6, 2), (1.2, 4)$.

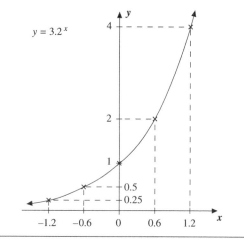

Example

Sketch the curve $y = 18^{3x + 2}$.

- As $x \to \infty$, $y \to \infty$; as $x \to -\infty$, $y \to 0$.
- When $x = 0$, $y = 18^2 = 324$.
- y doubles when $\log 18^{3x + 2} = 2 \times 324 = 648$

$$\therefore \log 18^{3x+2} = \log 648$$

$$\therefore (3x + 2) \log 18 = \log 648$$

$$\therefore 3x + 2 = (\log 648) \div (\log 18) \approx 2.24$$

$$\therefore x \approx 0.08$$

We now have the scales for both axes:

- On the x-axis the equally-spaced values 0.08 units apart.
- On the y-axis the value doubles with each increase of 0.08 on the x-axis.
- These 5 points are sufficient to show the shape of the curve:
  $(-0.16, 81), (-0.08, 162), (0, 324), (0.08, 648), (0.16, 1296)$.

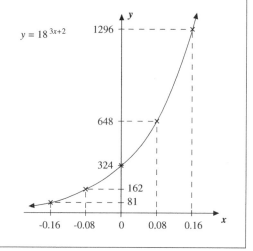

Example

Sketch the curve $y = 5.9^{2 - 3x}$.

- As $x \to \infty$, $y \to 0$; as $x \to -\infty$, $y \to \infty$
- When $x = 0$, $y = 5.9^2 \approx 34.8$,
- y doubles when $5.9^{2 - 3x}$ increases to 69.6

$$\therefore (2 - 3x) \log 5.9 = \log 69.6$$

$$\therefore 2 - 3x = (\log 69.6) \div (\log 5.9)$$

∴ $2 - 3x \approx 2.39$

∴ $3x \approx -0.39$

∴ $x \approx -0.13$

We now have the scales for both axes:

- On the x-axis the equally-spaced values 0.13 units apart
- On the y-axis the value doubles with each increase of -0.13 on the x-axis, i.e. with each *decrease* of 0.13.
- These 5 points are sufficient to show the shape of the curve:

    $(-0.26, 13.9)$, $(-0.18, 69.6)$, $(0, 34.8)$, $(0.13, 17.4)$, $(0.26, 8.7)$.

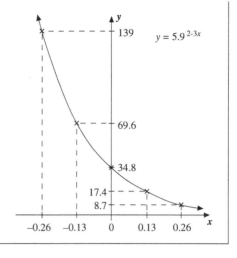

$y = 5.9^{\,2-3x}$

Exercises 16.4

Sketch each of the following curves. You may use the long method by setting up your own table of values for x and y or the quicker method described in the examples above. The answers provided are the coordinates of the 5 points obtained using the quick method, but if you use the long method the 5 points given in the answers should lie on your curve.

1 $y = 6.3^{\,1.25x + 0.83}$

2 $y = 0.78^{\,4.5x - 5.6}$

3 $y = 4.3^{\,2.6 - 2.5x}$

4 $y = 13^{\,0.9x + 1.3}$

5 $y = 81^{\,1.3x - 0.39}$

16.4 EXPONENTIAL RELATIONSHIPS

Three very common relationships are:

a

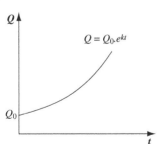

$Q = Q_0.e^{kt}$

b

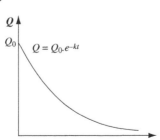

$Q = Q_0.e^{-kt}$

c

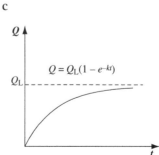

$Q = Q_L(1 - e^{-kt})$

The values of Q, Q_0, t and k were defined in the Summary in section 16.2 on page 276.

Examples of exponential growth

See the above figure **a** for this example.

- Population growth (human, animal, bacteria cultures—provided nothing such as famine interferes with the rate of growth).
- All 'chain reactions'.

Examples of exponential decay

See the above figure **b** for this example.

- The radioactive decay of a substance, Q being the mass of the substance remaining unchanged at time t.
- A heated body cooling, Q being the excess of the body's temperature above room temperature (*Newton's law of cooling*).
- The speed of a rotating flywheel when the power is disconnected, Q being measured in revolutions per minute, for example.
- Electrical quantities.

Examples of exponential growth towards an upper limit

See the above figure **c** for this example.

- The concentration of a substance during a chemical reaction, Q_L being the final limiting concentration.
- The temperature of a body after it is placed in an oven, Q being the excess of the body's temperature above its original temperature, and Q_L the temperature of the oven.
- The speed of an electrically driven flywheel after the power is connected.
- Electrical quantities.

Students of electrical engineering will easily be able to interpret the following diagrams. For a series R–C or a series L-C circuit, when charging or discharging, all the currents and voltages increase or decrease exponentially. (Current = rate of flow of charge.)

Charging	Discharging

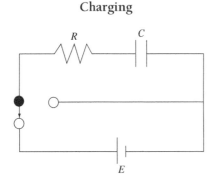

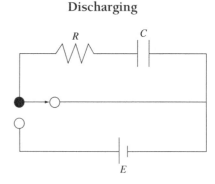

$$i = \frac{E}{R}e^{-\frac{1}{RC}t}$$

$$V_R = iR = Ee^{-\frac{1}{RC}t}$$

$$V_C = E - V_R = E(1 - e^{-\frac{1}{RC}t})$$

$$i = \frac{E}{R}e^{-\frac{1}{RC}t}$$

$$V_R = iR = Ee^{-\frac{1}{RC}t}$$

$$V_C = -V_R = -Ee^{\frac{1}{RC}t}$$

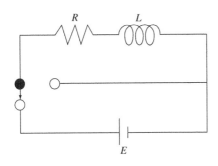
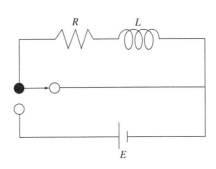

$$i = \frac{E}{R}(1 - e^{-\frac{R}{L}t})$$

$$V_R = iR = E(1 - e^{-\frac{R}{L}t})$$

$$V_L = E - V_R = Ee^{-\frac{R}{L}t}$$

$$i = \frac{E}{R}e^{-\frac{R}{L}t}$$

$$V_R = iR = Ee^{-\frac{R}{L}t}$$

$$V_L = -V_R = -Ee^{-\frac{R}{L}t}$$

Each of the above 12 equations is of the form $Q = Q_0\, e^{-kt}$ or $Q = Q_L\,(1 - e^{-kt})$, whose graphs are shown at the beginning of this section. In each case, quantity i, V_R, V_C or V_L either commences at an initial value Q_0 and decays towards zero, or commences at value zero and grows exponentially towards a final limiting value, Q_L.

For example, $I = \frac{E}{R}e^{-\frac{1}{RC}t}$ means that the current commences at value $\frac{E}{R}$ (when $t = 0$) and then decays exponentially with time towards zero as $t \to \infty$. In this case the continuous fractional rate of increase of i per unit of time is $k = -\frac{1}{RC}$.

Also, $i = \frac{E}{R}(1 - e^{-\frac{R}{L}t})$ means that the current commences at value zero (when $t = 0$) and grows exponentially towards a final limiting value of $\frac{E}{R}$ (as $t \to \infty$). In this case the continuous fractional rate of increase of i per unit of time is $k = \frac{R}{L}$.

Note how the values of the functions change with equal increases in the value of x:

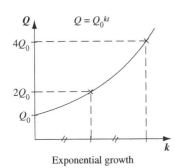
Exponential growth

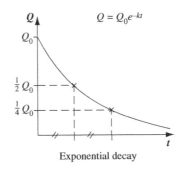
Exponential decay

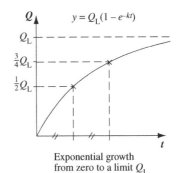
Exponential growth
from zero to a limit Q_L

The examples below show the steps in sketching such functions.

Remember: $\log_{10} 10 = 1$, $\log_e e = 1$ (i.e. $\ln e = 1$)

Examples

1 Sketch the graph of $V = 10e^{0.02t}$.
When $t = 0$, $V = 10e^{0.02t} = 10$
V doubles when $10 \times e^{0.02t} = 20$, i.e. $e^{0.02t} = 2$,
i.e. when $0.02t = \ln 2$, or $t \approx 35$

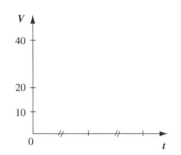 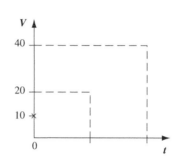 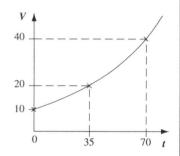

2 Sketch the graph of $M = 400(1 - e^{-1.25t})$.
When $t = 0$, $M = 400(1 - 1) = 0$
As $t \to \infty$, $M \to 400$
M halves when $400(1 - e^{-1.25t}) = 200$
i.e. when $e^{-1.25t} = 0.5$
i.e. when $-1.25t = \ln 0.5$
or $t \approx 0.56$.

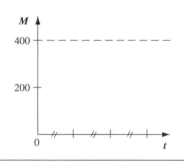 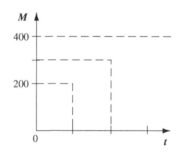 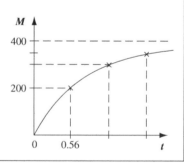

Exercises 16.5

1 Draw freehand, quick sketch-graphs of the functions:
   a $y = 2^x$, for $-3 \le x \le 3$ b $y = 2^{-x}$, for $-3 \le x \le 3$
   c $y = 1 - 2^{-x}$, for $0 \le x \le 4$ d $y = 327(1 - 2^{-x})$, for $0 \le x \le 4$
2 Draw accurate graphs of the functions:
   a $y = 5e^{0.2x}$ for $-6 \le x \le 10$ b $y = 24(1 - e^{-8x})$ for $0 \le x \le 0.5$
+++ 3 a Calculate the values of the function $y = 2^x$ when x has the values $-2, -1, 0, 1, 2$ and 3, and
         use these values to plot the graph of the function over this range.

b What numbers correct to 2 significant figures would replace the numbers 1, 2 and 3 on the x-axis of the graph in part **a** in order to convert it to the graph of the function:

   i $y = 2^{2x}$ ii $y = 16^x$ iii $y = e^x$ iv $y = e^{2x}$?

   (*Hint:* Replace 1 by the number that makes the new function equal to 2.)

c What numbers would replace the numbers 1, 2, 4 and 8 on the y-axis of the graph in part **a** in order to convert it to the graph of the function $y = 43 \times 2^x$?

4 Find the approximate values of x for which:

   a $e^x = 2, 4, 8$ and 16 b $e^x = 1, \frac{1}{2}, \frac{1}{4}$ and $\frac{1}{8}$

   Hence plot a graph of the function e^x over the above range of values.

5 The number of bacteria in a particular culture is given by $N = 1000 \times e^{2t}$, where the time t is in minutes. State how long it takes:

   a for the number to double

   b for the number to increase from 1000 to 4000

   c for the number to increase from 1000 to 8000.

   d Draw a graph showing the growth of the culture from 1000 to 8000. From your graph find approximately how long it takes for the number to increase from 1000 to 6000.

+++ 6 The electric current through an induction coil is given by $I = 200(1 - e^{-0.5t})$ mA, where the time t is in seconds.

   a Find the times taken for the current to become 100 mA, 150 mA and 175 mA.

   b Draw a graph showing the growth of the current during the first 5 seconds.

   c From your graph find the approximate size of the current after 1 second.

SELF-TEST

1 \$7000 is invested for 4 years at an interest rate of 5% p.a. What sum, to the nearest cent, does this grow to:

   a if the interest is compounded yearly?

   b if the interest is compounded each quarter?

   c if the interest is compounded daily?

   d if the interest is compounded continuously?

2 A certain bacterial culture grows so that each cell (on average) takes 9.70 minutes for mitosis (i.e. the 'doubling time'—the time for one cell to divide into two cells—is 9.70 minutes).

   a Calculate the continuous rate of growth k, correct to 3 significant figures.

   b If 3.20 mg of bacteria is originally present, how much will be present after 2 hours' growth?

+++ 3

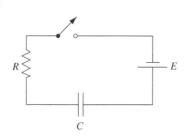

When the key is closed, the voltage E across the capacitor C increases exponentially towards a maximum value according to the equation $E = 85(1-e^{-kt})$, where t is in seconds.

a What is the voltage of the source?

b If E rises from 0 to 25 V in the first 730 ms, how long does it take for E to increase from 0 to 80 V?

4 Draw sketch-graphs of the following curves:

a $y = 5e^{14x}$ for $0 \le x \le 0.2$

b $i = 40e^{-100t}$ mA, where t is in seconds

c $Q = 16(1-e^{-500t})$ mC, where t is in seconds

5 A radioactive substance decomposes so that the mass present after t years is given by $M = 100e^{-0.1t}$ g.

a Find, to the nearest year, how long it will take for the mass present to reduce to:

    i   50 g   ii   25 g   iii   12.5 g   iv   3.125 g

b Draw a graph showing the decay from 100 g to 3.125 g and from your graph find approximately the mass present after 10 years.

CHAPTER 17

CIRCULAR FUNCTIONS

Learning objectives

- Form a mental picture of the unit circle and the projections that give the sine, cosine and tangent of an angle of any magnitude, and hence to state these approximate values for any given angle.
- Know the relationships between the trigonometric ratios of the four angles θ, $180° - \theta$, $180° + \theta$ and $360° - \theta$.
- State the relationships between the reciprocal ratios (cosecant, secant and cotangent) and the ratios sine, cosine and tangent respectively.
- State the relationships between the co-ratios (cosine, cotangent and cosecant) and the ratios sine, tangent and secant respectively.
- Express the size of an angle in circular (radian) measure and convert between degrees and radians.
- Calculate angular velocity and express it in various units of measurement.
- Calculate for a circle the length of an arc and the area of a sector.

17.1 ANGLES OF ANY MAGNITUDE

Sines

Our definition of $\sin \theta = \dfrac{\text{opposite}}{\text{hypotenuse}}$, where θ is an angle in a right-angled triangle, can apply only to *acute* angles (since no angle in a right-angled triangle can be greater than 90°). The more general definition of $\sin \theta$, which applies to angles of *any* magnitude, is as follows.

Draw a circle of unit radius with centre at the origin of a set of rectangular coordinate axes. Let the radius vector OP, commencing from the positive direction of the x-axis, rotate through *any* angle θ in the anticlockwise direction. Sin θ is defined as the *ordinate* (i.e. the y-value) of the point P.

Although this definition may seem rather complicated at first, the following examples should help you to understand it.

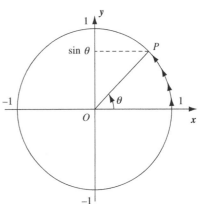

Examples

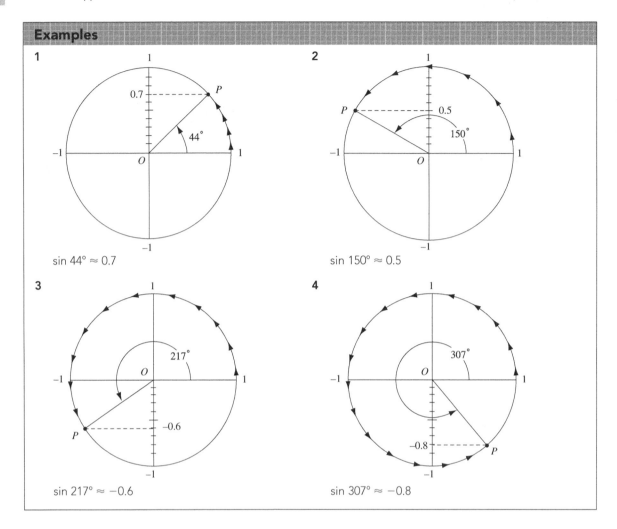

1   sin 44° ≈ 0.7

2   sin 150° ≈ 0.5

3   sin 217° ≈ −0.6

4   sin 307° ≈ −0.8

The sine of an angle is the 'height' of the point *P* after it has rotated through that angle, as described.

Note: ■ The sine of an angle can never lie outside the range −1 to 1.

■ Our previous definition of sin θ for an acute angle $\left(\dfrac{\text{opposite}}{\text{hypotenuse}}\right)$ gives the same result by our new definition. In example 1 above, sin 44° $= \left(\dfrac{\text{opposite}}{\text{hypotenuse}}\right) \approx \dfrac{0.7}{1} \approx 0.7$. We have merely *extended* the definition so as to cover angles of *any* magnitude.

If we number the quadrants of the circle 1, 2, 3, 4 as shown below:
■ the sine of an angle in the first quadrant (0 < θ < 90°) is positive
■ the sine of an angle in the second quadrant (90° < θ < 180°) is positive

- the sine of an angle in the third quadrant ($180° < \theta < 270°$) is negative
- the sine of an angle in the fourth quadrant ($270° < \theta < 360°$) is negative.

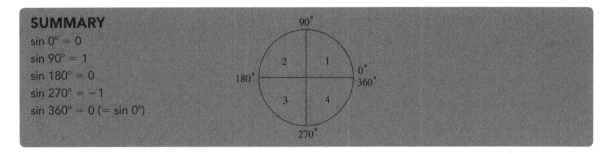

SUMMARY

$\sin 0° = 0$
$\sin 90° = 1$
$\sin 180° = 0$
$\sin 270° = -1$
$\sin 360° = 0 \,(= \sin 0°)$

Confusion can exist unless you realise that compass bearings (Chapter 10, section 10.8) commence from direction *north* with a clockwise rotation, but 'angles of any magnitude' commence from '*east*' (on the page) with an *anticlockwise* rotation.

Exercises 17.1

1 Given that each of the following is -1, -0.2, 0, 0.2 or 1 (correct to 1 significant figure), select the correct value for each. (Do this mentally, without a calculator.)

   a $\sin 12°$ b $\sin 180°$ c $\sin 360°$
   d $\sin 0°$ e $\sin 90°$ f $\sin 350°$

2 Evaluate mentally:

   a $\sin 90° + \sin 270°$ b $\sin 180° \times \sin 90°$
   c $\sin 0° - \sin 270° + \sin 90°$ d $\sin 270° + \sin 360° - \sin 90°$

Cosines

The cosine of an angle of any magnitude is defined by means of the same anticlockwise rotation around a circle of unit radius, but the value of the cosine of the angle is defined as the *abscissa* (i.e. the *x*-value) of the point *P*.

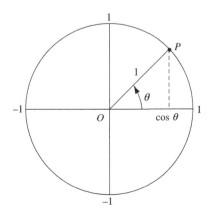

Examples

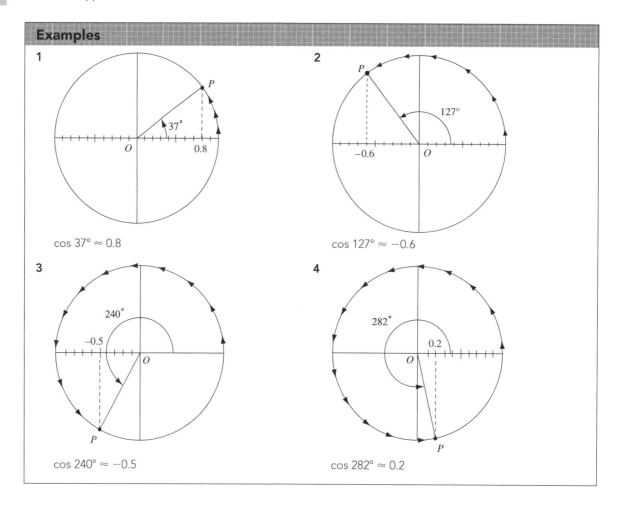

1 cos 37° ≈ 0.8

2 cos 127° ≈ −0.6

3 cos 240° ≈ −0.5

4 cos 282° ≈ 0.2

The cosine of an angle is the *horizontal* displacement of the point P after it has rotated through that angle, as described.

Note: The cosine is *positive* for angles in the first and fourth quadrants.

The cosine is *negative* for angles in the second and third quadrants.

SUMMARY

cos 0° = 1

cos 90° = 0

cos 180° = −1

cos 270° = 0

cos 360° (= cos 0°) = 1

Exercises 17.1 *continued*

3   Given that the value of each of the following is $-1, -0.3, 0, 0.3$ or 1 (correct to 1 significant figure), select the correct value for each. (Do this mentally, without a calculator.)

   a cos 73°   **b** cos 287°   **c** cos 252°   **d** cos 270°

   e cos 106°   **f** cos 90°   **g** cos 287°   **h** cos 360°

4   Evaluate mentally:

   a cos 0° + cos 360°   **b** cos 90° − cos 180°

   c cos 180° + cos 360°   **d** cos 0° × cos 180°

Tangents

The tangent of an angle of any magnitude is defined in the same way as the sine and cosine, but the value of the tangent is defined as $\dfrac{a}{b}$ (the *y*-value of *P* divided by its *x*-value). We have extended the result $\tan \theta = \dfrac{\sin \theta}{\cos \theta}$ to an angle of *any* magnitude.

However, from similar triangles, it can be seen that $\dfrac{a}{b} = \dfrac{d}{1} = d$. Hence, the value of tan θ can be obtained by drawing a scale *tangential* to the circle, as shown, graduated in the same units.

Tan θ is the value on this scale where *OP*, produced in either direction, intersects this scale.

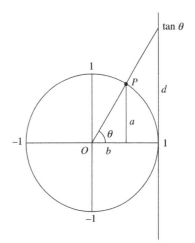

Examples

1

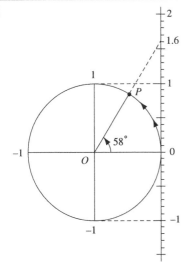

tan 58° ≈ 1.6

2

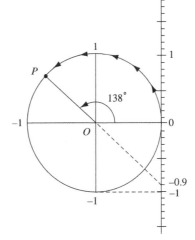

tan 138° ≈ −0.9

continued

continued

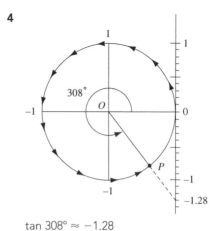

3

tan 215° ≈ 0.7

4

tan 308° ≈ −1.28

Note: The tangent is *positive* for angles in the first and third quadrants.
The tangent is *negative* for angles in the second and fourth quadrants.

SUMMARY

tan 0° = 0 (= tan 360°)

tan 180° = 0

tan 90° and tan 270° are *undefined* (i.e. they do not exist)

Exercises 17.1 *continued*

5 Given that the value of each of the following is −1.2, 0 or 1.2 (correct to 2 significant figures), select the correct value for each. (Do this mentally, without a calculator.)

a tan 130°   b tan 230°   c tan 180°

d tan 50°   e tan 0°   f tan 310°

A mnemonic (memory aid) to assist in remembering in which quadrants the ratios tan θ, sin θ and cos θ are positive, and in which they are negative, is:

All **S**tations **T**o **C**entral (ASTC):

In the first quadrant, they are *all* positive.   (A)

In the second quadrant, the *sine* is positive.   (S)

In the third quadrant, the *tangent* is positive.   (T)

In the fourth quadrant, the *cosine* is positive.   (C)

Exercises 17.1 *continued*

6 Given that in each case below the value is approximately $0.2, -0.2, 0.98, -0.98, 6$ or -6, state the approximate value in each case without using a calculator. (*Hint:* Imagine the angle on the unit circle and mentally picture the projections onto the *x*- and *y*-axes.)

a $\sin 80°$ b $\cos 170°$ c $\tan 280°$

d $\cos 10°$ e $\sin 170°$ f $\sin 350°$

g $\tan 80°$ h $\cos 260°$ i $\tan 100°$

j $\cos 190°$ k $\cos 280°$ l $\sin 260°$

m $\cos 350°$ n $\tan 260°$ o $\sin 100°$

Supplementary angles

From the diagram:

$$\sin(180° - \theta) = a = \sin\theta$$
$$\cos(180° - \theta) = -b = -\cos\theta$$
$$\tan(180° - \theta) = \frac{a}{-b} = -\tan\theta$$

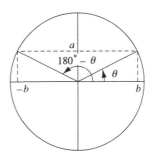

Note: Stress each 'S' when you say this mnemonic and you will remember it easily. *This result is very important:* The **S**ine of an angle i**S** the **S**ame a**S** the **S**ine of it**S** **S**upplement.

Negative angles

For negative angles we rotate the radius vector in the opposite direction (clockwise), e.g. $-2° = 358°$, $-90° = 270°$.

From the diagram:

$$\sin(-\theta) = -a = -\sin\theta$$
$$\cos(-\theta) = b = \cos\theta$$
$$\tan(-\theta) = \frac{-a}{b} = -\tan\theta$$

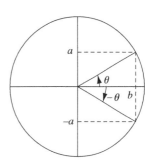

Remember: $\cos(-\theta) = \cos\theta$.

Exercises 17.1 *continued*

7 Simplify:

a $\dfrac{\sin(180° - \theta)}{\cos(-\theta)}$ b $\dfrac{\tan(-j)}{\tan(180° - j)}$ c $\dfrac{\cos(180° - i)}{\sin(-i)}$

The four angles θ, 180° − θ, 180° + θ and 360° − θ

The four angles marked are the angles θ, 180° − θ, 180° + θ and 360° − θ. (Notice that, in each case, the radius vector makes the same angle θ with the horizontal axis.)

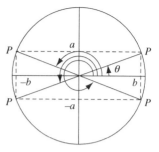

The sines of these four angles are respectively: a, a, −a and −a. The cosines of these four angles are respectively b, −b, −b and b.

> **Note:** ■ They all have the same value for their sines, except that two are positive and two are negative.
> ■ They all have the same value for their cosines, except that two are positive and two are negative.
> ■ It is also true that they all have the same value for their tangents, except that two are positive and two are negative.

Example

If we know that sin 20° = 0.3420, then:
  sin 160° [= sin (180° − 20°)] = 0.3420
  sin 200° [= sin (180° + 20°)] = −0.3420
  sin 340° [= sin (360° − 20°)] = −0.3420

We know which two are positive and which two are negative by visualising the unit circle, or by using our mnemonic: '**All S**tations **T**o **C**entral'. (**Note:** sin (−20°) = −0.3420 (fourth quadrant, hence negative).)

Example

From my calculator I find that $\sin^{-1} 0.682 = 43°$. Using this information, write down the sines of three other angles between 0° and 360°.
  (180 − 43)° = 137°, ∴ sin 137° = 0.682
  (180 + 43)° = 223°, ∴ sin 223° = −0.682
  (360 − 43)° = 317°, ∴ sin 317° = −0.682

Exercises 17.1 *continued*

8 Given that sin 30° = 0.5, write down the sines of three other angles between 0° and 360°.

9 Given that tan 60° = $\sqrt{3}$, write down the tangents of three other angles between 0° and 360°.

10 Given that sin 230° ≈ −0.766, write down the sines of three other angles between 0° and 360°.

11 From my calculator I find that $\cos^{-1}(−0.342) ≈ 110°$. From this information, write down the cosines of four angles between 0° and 360°.

The relationship between the sines and cosines of the angles $\pm\,\theta$, $90° \pm \theta$, $180° \pm \theta$, $270° \pm \theta$ and $360° - \theta$ (i.e. $-\theta$)

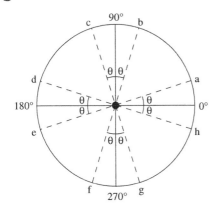

Rotate the radius vector so as to obtain, in turn, positions a, b, c, d, e, f, g and h.

The angles formed with the positive direction of the x-axis are, in turn:

θ, $90° - \theta$, $90° + \theta$, $180° - \theta$, $180° + \theta$,

$270° - \theta$, $270° + \theta$ and $360° - \theta$ (i.e. $-\theta$).

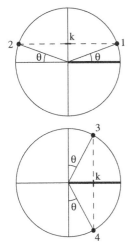

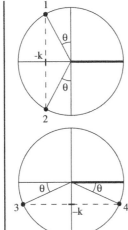

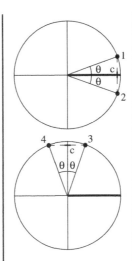

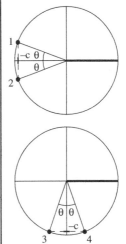

These values are equal:	These values are equal:	These values are equal:	These values are equal:
$\sin\theta$	$\cos(90° + \theta)$	$\cos\theta$	$\cos(180° - \theta)$
$\sin(180° - \theta)$	$\cos(270° - \theta)$	$\cos(360° - \theta)$	$\cos(180° + \theta)$
$\cos(90° - \theta)$	$\sin(180° + \theta)$	$\sin(90° - \theta)$	$\sin(270° - \theta)$
$\cos(270° + \theta)$	$\sin(360° - \theta)$	$\sin(90° + \theta)$	$\sin(270° + \theta)$
If $\theta = 20°$, they all equal 0.3420.	If $\theta = 20°$, they all equal -0.3420	If $\theta = 20°$, they all equal 0.9397	If $\theta = 20°$, they all equal -0.9397
If $\theta = 80°$, they all equal 0.9848	If $\theta = 80°$, they all equal $= -0.9848$	If $\theta = 80°$, they all equal 0.1736	If $\theta = 80°$, they all equal -0.1736
You can test using any other θ value you wish	You can test using any other θ value you wish	You can test using any other θ value you wish	You can test using any other θ value you wish

Note: Any of the above results can be obtained by picturing the above circle diagrams in your mind. The only ones worth memorising are:

$\sin\theta = \cos(90° - \theta)$ $\sin\theta = \sin(180° - \theta)$

$\cos\theta = \sin(90° + \theta)$ $\cos\theta = \sin(90° - \theta)$

Exercises 17.1 *continued*

12 State the trigonometrical ratio (sine or cosine) or the angle $\leq 360°$ required to complete the following statements:

a $\sin (90° - \theta) = \sin (\ldots + \theta)$

b $\cos (270° + \theta) = \ldots (180° - \theta)$

c $\sin (180° + \theta) = \cos (\ldots + \theta)$

d $\cos (90° + \theta) = \ldots (270° - \theta)$

e $\sin \theta = \ldots (270° + \theta)$

f $\sin (270° - \theta) = \cos (\ldots - \theta)$

g $\cos (180° - \theta) = \ldots (180° + \theta)$

h $\cos \theta = \cos (\ldots - \theta)$

i $\sin (270° + \theta) = \cos (\ldots - \theta)$

j $\cos (90° - \theta) = \ldots (270° + \theta)$

k $\sin \theta = \ldots (180° - \theta)$

l $\cos (180° + \theta) = \sin (\ldots - \theta)$

m $\sin (90° + \theta) = \ldots (90° - \theta)$

n $\cos \theta = \sin (\ldots + \theta)$

o $\sin (180° - \theta) = \cos (\ldots + \theta)$

p $\sin (360° - \theta) = \cos (\ldots - \theta)$

13 State the two angles $\leq 360°$ that could be used to complete the following statements:

a $\cos (270° + \theta) = \sin (\ldots)$

b $\sin (180° + \theta) = \cos (\ldots)$

c $\cos \theta = \sin (\ldots)$

d $\cos (180° - \theta) = \sin (\ldots)$

e $\sin (90° + \theta) = \cos (\ldots)$

f $\cos (90° - \theta) = \sin (\ldots)$

14 State whether each of these identities is true or false (state 'T' or 'F'):

a $\sin (90° - \theta) = \sin (90° + \theta)$

b $\cos (270° - \theta) = \sin (180° + \theta)$

c $\sin (90° - \theta) = \cos (90° + \theta)$

d $\cos (180° - \theta) = \cos (180° + \theta)$

e $\cos (90° - \theta) = \sin (180° + \theta)$

f $\sin \theta = \cos (90° - \theta)$

g $\cos (180° + \theta) = \sin (270° - \theta)$

h $\sin (-\theta) = \cos (90° + \theta)$

i $\cos (270° + \theta) = \sin (270° - \theta)$

j $\sin (180° + \theta) = \sin (180° - \theta)$

k $\cos (180° - \theta) = \sin (270° + \theta)$

l $\cos (90° - \theta) = \cos (90° + \theta)$

m $\sin (180° - \theta) = \cos (180° + \theta)$

n $\cos (270° - \theta) = \sin (180° + \theta)$

o $\cos (\theta) = \cos (180° + \theta)$

p $\sin (90° + \theta) = \cos \theta$

q $\sin (270° - \theta) = \sin (270° + \theta)$

r $\cos (270° + \theta) = \sin \theta$

s $\sin \theta = \cos (270° + \theta)$

t $\cos (270° - \theta) = \sin (90° - \theta)$

17.2 THE RECIPROCAL RATIOS

Definitions

The reciprocal of $\tan \theta$ is called the **cotangent** of θ ($\cot \theta$).

The reciprocal of $\sin \theta$ is called the **cosecant** of θ ($\operatorname{cosec} \theta$).

The reciprocal of $\cos \theta$ is called the **secant** of θ ($\sec \theta$).

Example

$$\cot 26.41° = \frac{1}{\tan 26.41°} \approx \frac{1}{0.49662} \approx 2.01$$

Exercises 17.2

1

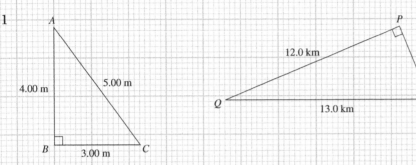

A, B, C, P, Q and R are the angles shown in the above triangles. Evaluate, as decimals, correct to 3 significant figures:

a   sec C   b   cot Q   c   cosec A   d   cot C   e   sec Q

f   cosec R   g   sec A   h   cot R   i   cosec Q   j   sec R

Your calculator does not provide a cosec, sec or cot key—use the sin, cos and tan keys respectively followed by the reciprocal key, $\boxed{x^{-1}}$ or $\boxed{\frac{1}{x}}$.

2   Using your calculator, evaluate:

a   cot 47°   b   sec 53°   c   sec 17.54°

d   cot 78.32°   e   cosec 56° 17′   f   sec 28° 46′

3   Using your calculator, evaluate:

a   23.7 sec 62°   b   cot 41° + cosec 53°

c $\dfrac{\text{cosec } 38.42°}{1.34}$ d   sec 27.4° − cot 63.7°

4   a   Given that sec θ = 1.642, evaluate θ in decimal degrees.

    b   Given that cosec A = 2.35, evaluate A in degrees and minutes.

5   a   Given that cosec G = 1.073, evaluate tan G.

    b   Given that sec K = 2.146, evaluate cos K.

17.3 THE CO-RATIOS

Exercises 17.3

1

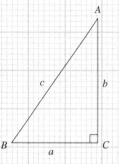

Let ABC be any right-angled triangle having sides of length a, b and c units.

A and B are complementary angles,
i.e. $A = 90° − B$ and $B = 90° − A$.
Write down the values (in terms of a, b and c) of:

a	**i**	$\sin A$	**ii**	$\cos B$	**b**	**i**	$\cos A$	**ii**	$\sin B$
c	**i**	$\tan A$	**ii**	$\cot B$	**d**	**i**	$\cot A$	**ii**	$\tan B$
e	**i**	$\sec A$	**ii**	$\operatorname{cosec} B$	**f**	**i**	$\operatorname{cosec} A$	**ii**	$\sec B$

If A and B are **c**omplementary angles:

- the sine of one = the **co**sine of the other
- the tangent of one = the **cot**angent of the other
- the secant of one = the **co**secant of the other.

> ## Examples
> **1** $\sin 10° = \cos 80°$ (since $10°$ and $80°$ are complementary)
> **2** $\operatorname{cosec} 85° = \sec 5°$ (since $85°$ and $5°$ are complementary)
> **3** $\cos x° = \sin (90° − x°)$ (since $x°$ and $90° − x°$ are complementary)

Exercises 17.3 continued

2 Write down the missing angles:

  a $\sin 20° = \cos \ldots$ **b** $\operatorname{cosec} 40° = \sec \ldots$ **c** $\sec 33° = \operatorname{cosec} \ldots$

  d $\cot 49° = \tan \ldots$ **e** $\operatorname{cosec} 48° = \sec \ldots$ **f** $\cot A° = \tan \ldots$

3 Write down the missing trigonometrical ratios:

  a $\cos 30° = \ldots 60°$ **b** $\cot 3° = \ldots 87°$ **c** $\tan \theta = \ldots (90 − \theta)°$

17.4 CIRCULAR MEASURE

Just as a time interval may be measured in different units (days, hours, minutes, etc.), an angle may be measured in several different units. The unit that we have been using is the **degree**, one degree being defined as $\frac{1}{360}$ of a revolution. This is called the 'degree system' or 'sexagesimal system' (since 1° = 60′ and 1′ = 60″). (When you worked through Chapter 10, did you investigate why the 'minute' and 'second' are used for both units of time and units of angle?)

An angle can also be measured in **radians.** One radian is defined as the angle subtended at the centre of a circle by an arc that is equal in length to the radius, as shown in the diagram.

Since the radius of a circle is contained 2π times in the circumference of a circle ($C = 2\pi r$), it follows that:

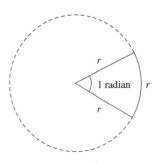

2π radians = 360° (1 revolution)

∴ π radians = 180°

Hence, 1 radian = $\dfrac{180°}{\pi} \approx 57.30°$.

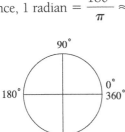

Degrees

Radians

Note: 360° = 2π radians   90° = $\dfrac{\pi}{2}$ radians

180° = π radians   270° = $\dfrac{3\pi}{2}$ radians

When an angle is measured in radians, the unit is usually omitted. For example, '$\angle A = 1.23$' means '$\angle A = 1.23$ radians' and '$\theta = \dfrac{\pi}{2}$' means '$\theta = \dfrac{\pi}{2}$ radians'.

Exercises 17.4

1 Convert the following angles into degrees, mentally:

a $\dfrac{\pi}{4}$ b π c $\dfrac{\pi}{2}$ d $\dfrac{\pi}{10}$ e $\dfrac{5\pi}{12}$

2 Express the following angles in circular measure in terms of π, mentally:

a 90°   b 60°   c 45°   d 360°   e 30°

In general, a calculator is used for conversions. Your calculator may provide a conversion key, but if not, use the fact that π radians = 180°.

To convert radians to degrees, multiply by $\dfrac{180}{\pi}$.

To convert degrees to radians, multiply by $\dfrac{\pi}{180}$.

Exercises 17.4 *continued*

3 Express the following angles in radians (correct to 5 significant figures):

a 47° b 1° c 342°

d 23.78° e 191.25° f 0.01°

4 Convert into degrees (correct to 4 decimal places):

a 3 radians b 1 radian c 2.0347 radians

d $\dfrac{\pi}{2.64}$ radians e 1.836π radians f $\dfrac{3\pi}{7}$ radians

5 Express each of the following angles in degrees and minutes (correct to the nearest minute):

a 2 radians b 1.384 92 radians

c 3.106 82 radians d $\dfrac{2\pi}{13}$ radians

6 Answer mentally (do not use a calculator or tables):

a $\cos\dfrac{\pi}{2}$ b $\tan\pi$ c $\sin\dfrac{\pi}{2}$ d $\tan 3\pi$

e $\cos\pi$ f $\sin\dfrac{3\pi}{2}$ g $\cos-\dfrac{\pi}{2}$ h $\sin-\dfrac{3\pi}{2}$

7 Evaluate correct to 4 significant figures:

a $\sin 2.5326$ b $\tan 0.682\ 09$ c $\tan\dfrac{\pi}{1.623\ 80}$ d $\sin-1.23\pi$

8 Evaluate $\sin\omega t$, when:

a $\omega = 419$ rad/s, $t = 0.002$ s b $\omega = 314$ rad/s, $t = 3.50$ ms

c $\omega = 6.25 \times 10^4$ rad/s, $t = 85\ \mu s$

9 Evaluate $12\cos\omega t$, when:

a $\omega = 265$ rad/s, $t = 4.35$ ms b $\omega = 8.14 \times 10^4$ rad/s, $t = 32.5\ \mu s$

17.5 ANGULAR VELOCITY (ROTATIONAL SPEED)

Angular velocity is a measure of angular rotation in unit time. The angular rotation may be measured in degrees, radians or revolutions.

e.g. $60°/s = \dfrac{\pi}{3}$ rad/s (≈ 1.05 rad/s) $= 10$ rpm

(Note the abbreviations: 'r' for 'revolution(s)' and 'rad' for 'radian(s)'.)

An angular velocity of ω radians per second for a time interval of t seconds produces a rotation ('angular displacement') of ωt radians.

Rule: $\theta = \omega t$ $\begin{cases} \theta = \text{angular displacement (in radians)} \\ \omega = \text{angular velocity (in radians per second)} \\ t = \text{time interval (in seconds)} \end{cases}$

Exercises 17.5

1 Express the following rotational speeds in radians per second (correct to 3 significant figures):

   a 4000 rpm   **b** $33\frac{1}{3}$ rpm   **c** 500 r/s

2 Express the following rotational speeds in revolutions per minute (correct to 3 significant figures):

   a 260 rad/s   **b** 400 rad/s

3 Express the rotational speed of $33\frac{1}{3}$ rpm in degrees per second.

+++ 4 How long does it take a wheel to make 10 revolutions if its angular velocity is:

   a 2.50°/ms?   **b** 100 rad/s?   **c** 850 rad/min?

+++ 5 What is the angular displacement (in degrees) made by a wheel in one millisecond, if its angular velocity is:

   a 650 rad/s?   **b** 850 rpm?

17.6 THE CIRCLE: LENGTH OF ARC, AREA OF SECTOR, AREA OF SEGMENT

Length of arc

If angle at centre $= 1$ rad, $l = r$

If angle at centre $= \theta$ rad, $l = r \times \theta$

> **Rule:** $l = r\theta$
>
> (θ in radians)

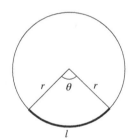

Area of sector

If angle at centre $= 2\pi$ rad, $A = \pi r^2$

If angle at centre $= 1$ rad, $A = \pi r^2 \div 2\pi = \frac{1}{2}r^2$

If angle at centre $= \theta$ rad, $A = \frac{1}{2}r^2\theta$

> **Rule:** $A = \frac{1}{2}r^2\theta$
>
> (θ in radians)

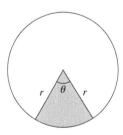

Examples

AB is a chord of a circle of radius 5.00 m, which subtends an angle of 110° at the centre C. Find:

1 the length of the minor arc AB

2 the area of the minor sector CAB

   angle at centre $= 110° \approx 1.92$ radians

1 $l = r\theta$ **2** $A = \frac{1}{2}r^2\theta$

      $= 5 \times 1.92$ $= \frac{1}{2} \times 25 \times 1.92$

      $= 9.60$ m $= 24.0$ m^2

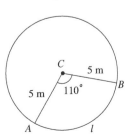

Exercises 17.6

1 In each case below, the radius of a circle (r) and the angle (θ) subtended at the centre C by a chord PQ are given. In each case find:

    i the length of the minor arc PQ **ii** the area of the minor sector CPQ

    a $r = 8.00$ m, $\theta = 30.0°$ **b** $r = 60.0$ mm, $\theta = 150°$

    c $r = 123$ mm, $\theta = 116°$ **d** $r = 38.7$ m, $\theta = 8° \, 47'$

2 PQ is an arc of a circle whose centre is C and whose radius is 73.4 mm. Given that the length of the minor arc PQ is 105 mm, find:

    a the area of the minor sector PCQ **b** the area of the major sector PCQ

3 AB is a chord of a circle whose centre is C. Given that $\angle ACB = 24° \, 47'$ and that the area of the minor sector ACB is 1.56 m^2, find:

    a the radius of the circle   **b** the area of the triangle ABC

SELF-TEST

1 Simplify:

    a $\tan(-x) \times \cos(180° - x)$ **b** $\sin(180° - \theta) \div \tan(-\theta)$

2 Given that $\cos 70° = 0.342$, write down the cosines of three other angles between $0°$ and $360°$.

3 From my calculator I find that $\cos^{-1}(-0.788) = 142°$. From this information, write down the cosines of four angles between $0°$ and $360°$.

4 Solve: $\tan 2K = 1.782$, where $0° \le K \le 360°$.

5 Solve: $\tan R = -0.2651$, where $90° \le R \le 180°$.

6 Use your calculator to evaluate $\dfrac{\sec 78° - 1.45}{5.63 - \cot 13°}$.

7 Through how many degrees does a wheel rotate in 5 ms if its angular velocity is:

    a 13 rad/s?   **b** 255 rpm?

+++ 8 KN is a chord of a circle whose centre is C and whose radius is 246 mm. Given that the length of the chord KN is 366 mm, find:

    a the length of the minor arc KN **b** the area of triangle KNC

    c the area of the minor sector CKN **d** the area of the major sector CKN

CHAPTER 18

TRIGONOMETRIC FUNCTIONS AND PHASE ANGLES

Learning objectives

- State the amplitude, period and frequency of wave motions given by the functions $a \sin b\theta$, $a \cos b\theta$ and $a \tan b\theta$, and to be able to draw sketch-graphs of these functions.
- State the equation for any sine or cosine wave when given the graph of the function.
- Draw sketch-graphs of functions of the forms:
  - $a \sin (\theta \pm \alpha)$, $a \cos (\theta \pm \alpha)$
  - $a \sin (b\theta + \alpha)$, $a \cos (b\theta + \alpha)$ and $a \tan (b\theta + \alpha)$.
- Determine the phase difference between two given sine or cosine functions.
- Find the resultant curve by superimposing the graphs of two waveforms.

 The amount of material in this chapter that is appropriate for study will depend upon the needs and interests of the student. This material is particularly relevant for students of electrical engineering.

18.1 GRAPHS OF $a \sin b\theta$, $a \cos b\theta$

A **periodic function** of x is one whose values periodically repeat themselves at regular intervals of x. When the function value is plotted against x, the graph consists of a pattern that repeats itself at regular intervals. The pattern that repeats itself in this way is called one **cycle**. The interval on the x-axis corresponding to one complete cycle is called the **period** of the function.

The largest excursions that the function value makes from zero into positive and negative values respectively are called its **peak positive amplitude** and its **peak negative amplitude.** The total excursion between the peak positive value and the peak negative value is called the **peak-to-peak amplitude.** (**Note:** Amplitudes have no sign; they are absolute values.)

Although the above description and definitions may be a little hard to understand, they should become clear after careful study of the following examples.

Examples

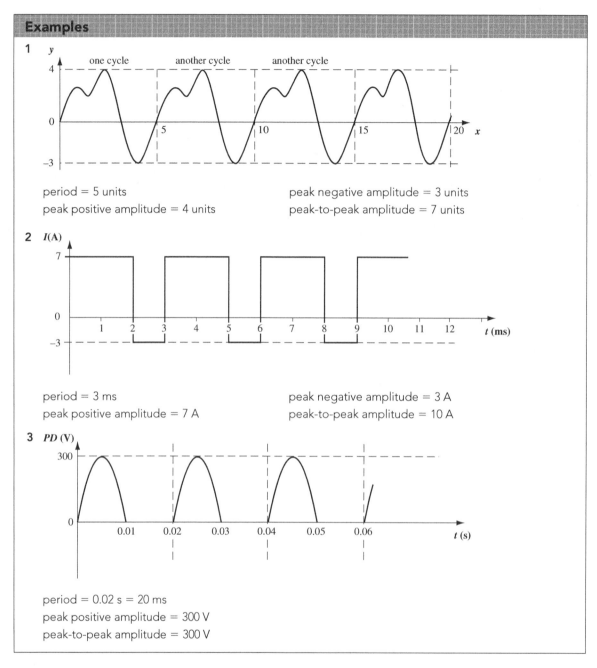

1

period = 5 units

peak positive amplitude = 4 units

peak negative amplitude = 3 units

peak-to-peak amplitude = 7 units

2

period = 3 ms

peak positive amplitude = 7 A

peak negative amplitude = 3 A

peak-to-peak amplitude = 10 A

3

period = 0.02 s = 20 ms

peak positive amplitude = 300 V

peak-to-peak amplitude = 300 V

At this stage of the course we are concerned with the functions $a \sin b\theta$ and $a \cos b\theta$ for any values of a and b, with their graphs, amplitudes and periods.

The graph of sin θ

For any value of θ, the value of $\sin \theta$ is given by the ordinate (i.e. the y-value) *or* the 'height' of P above the horizontal axis. By drawing a set of cartesian axes for θ and $\sin \theta$, as shown, and using any convenient scale for our degree intervals on the θ-axis, we can plot the graph of $\sin \theta$ (against θ) by projecting the height of point P onto the graph.

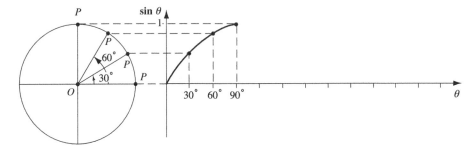

As *P* rotates anticlockwise from its standard position:

- from 0° to 90°: sin θ increases from 0 to 1
- from 90° to 180°: sin θ decreases from 1 to 0
- from 180° to 270°: sin θ decreases from 0 to −1
- from 270° to 360°: sin θ increases from −1 to 0.

The above graph shows the value of the sine plotted for 0°, 30°, 60° and 90°. Greater accuracy is achieved by plotting a greater number of values.

The complete graph of one cycle (obtained by rotating *P* through 360°) is shown below.

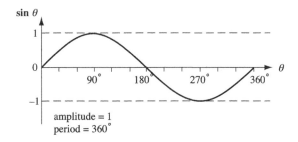

Note: Because the peak positive amplitude = the peak negative amplitude, we usually refer simply to 'the amplitude' (meaning the largest excursion the function value makes from zero, in either direction).

The graph of cos θ

Cos θ has its maximum value (= 1) when θ = 0°, 360°, 720°, ... Hence it also has amplitude 1 and period 360°. Since cos θ = sin (90° − θ), the graph of cos θ has the *same shape* as the sin θ graph. The cos θ graph is simply the sin θ graph moved along the θ-axis so that it peaks at θ = 0°, 360°, 720°, ...

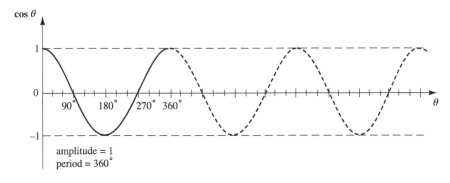

The graphs of sin $b\theta$

Sin $\theta = 0$ when $\theta = 0°, 180°, 360°, 540°$, etc.

$$\begin{cases} \text{amplitude} = 1 \\ \quad \text{period} = 360° \end{cases}$$

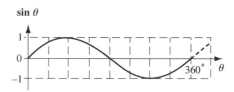

Sin $2\theta = 0$ when $2\theta = 0°, 180°, 360°, 540°$, etc.
i.e. when $\theta = 0°, 90°, 180°, 270°$, etc.

$$\begin{cases} \text{amplitude} = 1 \\ \quad \text{period} = 180° \end{cases}$$

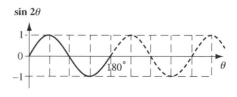

Sin $4\theta = 0$ when $4\theta = 0°, 180°, 360°, 540°$, etc.
i.e. when $\theta = 0°, 45°, 90°, 135°$, etc.

$$\begin{cases} \text{amplitude} = 1 \\ \quad \text{period} = 90° \end{cases}$$

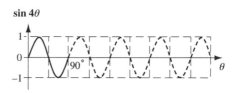

Note:
- The values of sin θ start to repeat when θ increases by 360°.
  ∴ period = 360°
- The values of sin 2θ start to repeat when 2θ increases by 360°, i.e. when θ increases by 180°.
  ∴ period = 180°
- The values of sin 4θ start to repeat when 4θ increases by 360°, i.e. when θ increases by 90°.
  ∴ period = 90°
- The values of sin $b\theta$ start to repeat when $b\theta$ increases by 360°, i.e. when θ increases by $\dfrac{360°}{b}$.
  ∴ period $= \dfrac{360°}{b}$

The amplitude factor

$y = 5 \times \sin\theta$: The effect of the '5' is to make all the y-values five times as large.
           (amplitude = 5)

$y = 0.2 \sin\theta$: The effect of the '0.2' is to make all the y-values 0.2 times as large.
           (amplitude = 0.2)

The function $a \sin b\theta$ has amplitude a,

period $\dfrac{360°}{b}$.

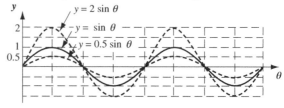

Sketching the graph of $y = a \sin b\theta$

Sketch a sine curve, y against θ, and label the axes with the correct amplitude and period.

Examples

1

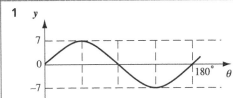

$y = 7 \sin 2\theta$

$\begin{cases} \text{amplitude} = 7 \\ \text{period} = \dfrac{360°}{2} = 180° \end{cases}$

2

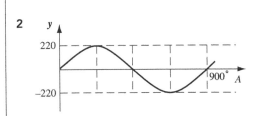

$y = 220 \sin 0.4 A$

$\begin{cases} \text{amplitude} = 220 \\ \text{period} = \dfrac{360°}{0.4} = 900° \end{cases}$

SUMMARY

$a \sin b\theta$ is a periodic function.

$\begin{cases} \text{amplitude} = a \\ \text{period} = \dfrac{360°}{b} \end{cases}$

$a \cos b\theta$ is a periodic function.

$\begin{cases} \text{amplitude} = a \\ \text{period} = \dfrac{360°}{b} \end{cases}$

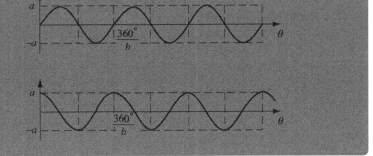

Exercises 18.1

1 Sketch the following curves. In each case state the amplitude and the period of the function.

   a $y = 13 \sin 5\theta$, one cycle only   **b** $y = 6 \sin 3\theta, 0° \le \theta \le 360°$

   c $y = 200 \sin 2A, 0° \le A \le 360°$ **d** $y = 24 \cos 4\theta, 0° \le \theta \le 360°$

2 For what values of θ $(0° \le \theta \le 180°)$ is the function $27 \sin 4\theta$ equal to zero?

3 For what values of A $(0° \le A \le 180°)$ is the function $12 \cos 3A$ equal to zero?

4 What is the maximum positive value of the function $7 \sin 6\theta$ and for what values of θ between $0°$ and $90°$ does it occur?

5 What is the greatest negative value of the function $31 \cos 5A$ and for what values of A between $0°$ and $90°$ does it occur?

+++ 6 How many cycles are there in the graph of $y = 24 \sin 6\theta$ between $\theta = 0°$ and $\theta = 180°$?

+++ 7 How many cycles are there in the graph of $y = 100 \cos 10A$ between $A = 0°$ and $A = 90°$?

18.2 THE TANGENT CURVES

In most branches of technology, the tangent functions are not nearly as important as the sinusoidal functions. However, you should be able to sketch simple tangent curves.

The tangent function is periodic. As θ increases from 0 to $\dfrac{\pi}{2}$ (90°), tan θ increases from 0 and approaches infinity. (Infinity is not a number, so tan $\dfrac{\pi}{2}$ (tan 90°) is not defined.)

Tan $\theta = 0$ when $\theta = 0$, π, 2π, 3π, 4π, . . .

and approaches ∞ or $-\infty$ at values midway between the above values

(i.e. at $\theta = \dfrac{\pi}{2}, \dfrac{3\pi}{2}, \dfrac{5\pi}{2}, \ldots$).

Note that the period $= \pi$, unlike the sinusoidal functions which have a period of 2π.

The amplitude of a periodic curve is the maximum excursion of the value of the function from the horizontal axis. Since tangent functions do not *have* a maximum or minimum value, amplitude is not defined for them.

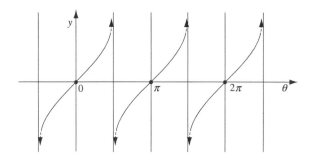

The curve $y = \tan b\theta$

Tan $b\theta = 0$ when $b\theta = 0$, π, 2π, 3π, . . .

that is, when $\theta = 0, \dfrac{\pi}{b}, \dfrac{2\pi}{b}, \dfrac{3\pi}{b}, \ldots$

The period, $T = \dfrac{\pi}{b}$

The vertical lines at . . . , $-\dfrac{\pi}{2b}, \dfrac{\pi}{2b}, \dfrac{3\pi}{2b}, \ldots$, which the curve approaches but never contacts, are called asymptotes.

18.3 SKETCH-GRAPHS OF $a \sin(\theta \pm \alpha)$, $a \cos(\theta \pm \alpha)$

To sketch the graph of $a \sin(\theta \pm \alpha)$:

1   Sketch the graph of $a \sin \theta$ in *pencil* (amplitude $= a$, period $= 2\pi$ or 360°).
2   Move the curve along the horizontal axis to give the correct phase angle α, thus sketching the curve $y = a \sin(\theta \pm \alpha)$ *in ink*.
3   Erase the pencil sketch of $a \sin \theta$.
    Mark the values where the curve crosses each axis.

Example

Sketch one cycle of the graph $y = 16.4 \sin(\theta + 45°)$.

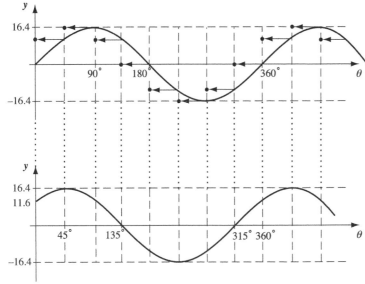

1   Sketch $y = 16.4 \sin \theta$ *in pencil* (amplitude 16.4, period 360°).
2   Since $16.4 \sin(\theta + 45°)$ leads $16.4 \sin \theta$ by 45°, now move the curve to the left by 45°, thus sketching the required curve *in ink*. **(In practice, it is easier to move the y-axis to the right by 45° and change the scale on the x-axis, which produces the same effect.)**
3   Then erase the pencil sketch of $16.4 \sin \theta$.
4   Mark the values where the curve crosses the θ-axis and the y-axis.

If the phase angle is expressed in degrees, it is customary to state the period in degrees and to plot y against θ where the θ scale is in degrees. However, if the phase angle is given in radians, the period is usually stated in radians also and the θ scale is marked in radians.

Example

Sketch one cycle of the graph

$$y = 12 \cos\left(\theta - \frac{\pi}{3}\right).$$

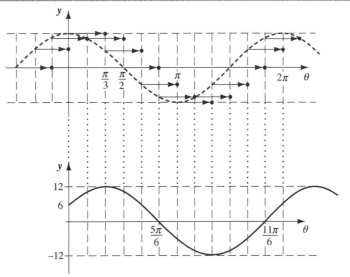

continued

continued

1   Sketch 12 cos θ *in pencil* (amplitude = 12, period = 2π).

2   Since $12 \cos\left(\theta - \dfrac{\pi}{3}\right)$ *lags* 12 cos θ by $\dfrac{\pi}{3}$, then move the curve to the *right* by $\dfrac{\pi}{3}$,

sketching $y = 12 \cos\left(\theta - \dfrac{\pi}{3}\right)$ *in ink*.

3   Then erase the pencil sketch of $y = 12 \cos \theta$.

4   Mark the values where the curve crosses the axes.

Exercises 18.2

For each of the following functions, state the amplitude and the period. Sketch two cycles of each function, marking the values where the curve crosses each axis:

1   $y = 18 \sin (\theta - 30°)$

2   $y = 200 \cos\left(\theta + \dfrac{\pi}{6}\right)$

3   $y = 0.6 \sin\left(\theta + \dfrac{\pi}{4}\right)$

4   $y = 50 \sin (\theta + 60°)$

5   $y = 2.3 \cos\left(\theta - \dfrac{\pi}{8}\right)$

6   $y = 6 \sin\left(\theta - \dfrac{\pi}{2}\right)$

18.4 THE FUNCTIONS $a \sin (b\theta + \alpha)$, $a \cos (b\theta + \alpha)$

$a \sin (b\theta + \alpha) = 0,$ when $(b\theta + \alpha) = 0, \pi, 2\pi, 3\pi, \ldots$

i.e. when $b\theta = -\alpha, \pi - \alpha, 2\pi - \alpha, 3\pi - \alpha, \ldots$

i.e. when $\theta = -\dfrac{\alpha}{b}, \dfrac{\pi}{b} - \dfrac{\alpha}{b}, \dfrac{2\pi}{b} - \dfrac{\alpha}{b}, \dfrac{3\pi}{b} - \dfrac{\alpha}{b}, \ldots$

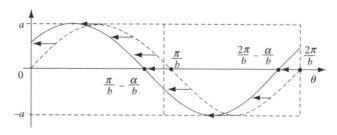

Hence, the introduction of the **phase angle** α has no effect on the amplitude or the period of the waveform, but shifts the curve $y = a \sin b\theta$ along the horizontal axis by $\dfrac{\alpha}{b}$ to the left.

Thus, **phase shift** $= \overset{\leftarrow}{\dfrac{\alpha}{b}}$ (A *negative* α shifts the curve to the *right*.)

The curve $y = a \sin (b\theta + \alpha)$ *leads* the curve $y = a \sin b\theta$ by $\dfrac{\alpha}{b}$.

All of the above also applies to the curve
$y = a \cos(b\theta + \alpha)$, except that the curve $y = a \cos b\theta$
(the dotted curve in this diagram) starts from $y = a$
instead of from $y = 0$.

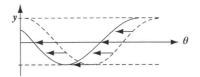

SUMMARY

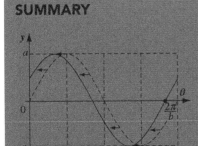   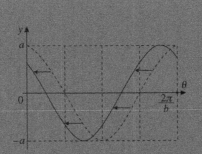

Dotted curve: $y = a \sin b\theta$ $y = a \cos b\theta$
Solid curve: $y = a \sin(b\theta + \alpha)$ $y = a \cos(b\theta + \alpha)$

Both functions have amplitude $= a$, period $= \dfrac{2\pi}{b}$.

The solid curve is obtained by shifting the dotted curve by $\dfrac{\alpha}{b}$ (to the left if α is positive; to the right if α is negative).

The curve $y = \tan(b\theta + \alpha)$

$\tan(b\theta + a) = 0$ when $b\theta + a = 0, \pi, 2\pi, 3\pi, \ldots$

i.e. when $b\theta = -\alpha, \pi - \alpha, 2\pi - \alpha, 3\pi - \alpha, \ldots$

i.e. when $\theta = -\dfrac{\alpha}{b}, \dfrac{\pi}{b} - \dfrac{\alpha}{b}, \dfrac{2\pi}{b} - \dfrac{\alpha}{b}, \dfrac{3\pi}{b} - \dfrac{\alpha}{b}, \ldots$

Hence, the phase angle α causes a phase shift of $\overset{\leftarrow}{\dfrac{\alpha}{b}}$.

As suggested for sketching sinusoidal curves, it is easiest to first sketch the basic curve with no phase angle, in pencil or with dots, and then to calculate the number of millimetres the curve should be moved on the diagram in order to achieve any required phase shift.

Exercises 18.3

1   For each of the following functions:

   i   state the period, in radians (per cycle)

   ii   state the period, in degrees (per cycle)

   iii state the frequency (in cycles per radian)

   iv   state the frequency (in cycles per degree)

   v   state the number of cycles which occur between $\theta = 0$ and $\theta = 2\pi$ or $360°$

continued

continued

    vi draw a sketch graph showing two cycles of the curve, with θ measured in both radians
    and in degrees.

    a $y = 24 \sin 3\theta$ **b** $y = 100 \sin 4\theta$

    c $y = 50 \cos 2\theta$ **d** $y = 75 \cos 5\theta$

Complete the following table. Two examples have been done for you.

The curve:	is the same as the curve:	but shifted by:	along the horizontal axis to the:
$y = 7 \sin (3\theta + 13°)$	$y = 7 \sin 3\theta$	$\dfrac{13°}{3}$	left
$y = 24 \cos \left(2\theta - \dfrac{\pi}{3}\right)$	$y = 24 \cos 2\theta$	$\dfrac{\pi}{6}$	right
2 $y = 12 \sin (6\theta - 7°)$	**a**	**b**	**c** . . . .
3 $V = 40 \cos (5\theta + 29°)$	**a**	**b**	**c** . . . .
4 $i = 17 \sin (4\theta + 1.2\pi)$	**a**	**b**	**c** . . . .
5 $V = 35 \cos (3\theta - 120°)$	**a**	**b**	**c** . . . .
6 $i = 23 \sin (8\theta - 4\pi)$	**a**	**b**	**c** . . . .
7 $V = 240 \cos (20\pi\theta + 0.2\pi)$	**a**	**b**	**c** . . . .

Example

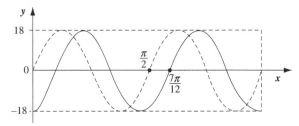

a State the equation for the dotted graph.   **b** State the equation for the solid graph.

Solution

a The dotted graph is a sine curve having amplitude 18 units and period $\dfrac{\pi}{2}$.

$$\therefore \frac{2\pi}{b} = \frac{\pi}{2}$$

$$\therefore b = 4$$

$\therefore$ Its equation is $y = 18 \sin 4x$.

b The solid curve has a phase shift to the *right* of $\dfrac{7\pi}{12} - \dfrac{\pi}{2}$, that is, $\dfrac{\pi}{12}$.

$\therefore \dfrac{\alpha}{b} = \dfrac{\pi}{12}$

$\therefore \dfrac{\alpha}{4} = \dfrac{\pi}{12}$

$\therefore \alpha = \dfrac{\pi}{3}$

$\therefore$ Its equation is $y = 18 \sin\left(4x - \dfrac{\pi}{3}\right)$.

Exercises 18.3 *continued*

8 For each sketch-graph below, state:

  i the equation for the dotted curve

  ii the equation for the solid curve.

a 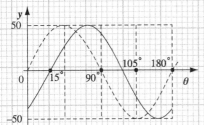 **b**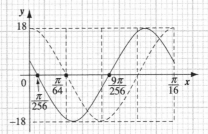

9 For each exercise below, sketch the two graphs on the same axes, showing two cycles for each. Sketch the first as a dotted curve and the second as a solid curve.

  a $y = 40 \sin 6\theta$

      $y = 40 \sin\left(6\theta - \dfrac{\pi}{4}\right)$

  b $y = 37 \cos 25\pi x$

      $y = 37 \cos(25\pi x - 0.375\pi)$

10 For each part of this question, sketch one cycle of each of the two curves on the same axes, the first as a dotted curve and the second as a solid curve. Mark the values on the horizontal axis where each curve crosses it.

  a $y = 93 \sin 4\theta$

      $y = 93 \sin\left(4\theta + \dfrac{\pi}{4}\right)$

  b $y = 19 \cos 2\theta$

      $y = 19 \cos\left(2\theta - \dfrac{\pi}{2}\right)$

18.5 PHASE ANGLES

Two periodic functions that have the same period are said to be **coherent**.

Examples

1 The functions $12 \sin \theta$ and $23 \sin \theta$ both have a period of $360°$ (i.e. 2π) and are hence coherent.

In this case the two functions have their peaks at the same values of θ. They are **in phase** with each other (i.e. 'in step')

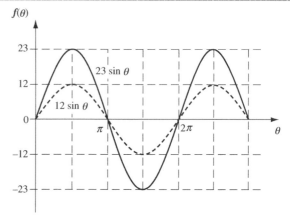

2 The functions $46 \sin \theta$ and $30 \cos \theta$ both have a period of 2π and are hence coherent.

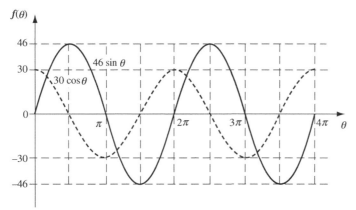

In this case there is a difference of $\dfrac{\pi}{2}$ between the angles at which the two functions have their peak values. We say there is a **phase difference** of $\dfrac{\pi}{2}$ between the two functions. Since as θ increases, the cosine function always peaks for an angle θ that is $\dfrac{\pi}{2}$ *less* than that for the corresponding peak of the sine curve, we say that the cosine function **leads** the sine function by $\dfrac{\pi}{2}$ (or 90°), and the sine function **lags** the cosine function by $\dfrac{\pi}{2}$ (or 90°). As we trace the curves from left to right, the cosine function peaks 'earlier' than the sine function by an angle of $\dfrac{\pi}{2}$ (or 90°).

If $y = A \sin \theta$, then $y = \pm A$ when $\theta = 90°, 270°, 450°, \ldots$

If $y = A \sin (\theta + 30°)$, then $y = \pm A$ when $\theta + 30° = 90°, 270°, 450°, \ldots$

i.e. when $\theta = 60°, 240°, 420°, \ldots$

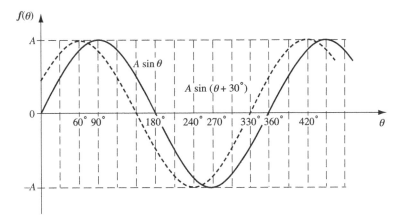

Hence, the function sin $(\theta + 30°)$ *leads* the function sin θ by 30°.

In general:

- function sin $(\theta + \alpha°)$ *leads* function sin θ by $\alpha°$

Also, since both of these functions have a period of 360°, a *lead* of θ is equivalent to a *lag* of $360° - \theta$, and a *lag* of θ is equivalent to a *lead* of $360° - \theta$.

Three facts that enable you to determine phase relationships:

1   For two sine functions or two cosine functions, the one with the larger phase angle leads the one with the smaller, the lead being the difference between the phase angles.

Examples

sin $(\theta + 80°)$ leads sin $(\theta + 30°)$ by 50° $\left.\begin{array}{l}\end{array}\right\}$

sin $(\theta + 20°)$ leads sin $(\theta - 50°)$ by 70°   If all these sine functions were cosine functions

$\sin\left(\theta - \dfrac{\pi}{2}\right)$ leads $\sin\left(\theta - \dfrac{\pi}{3}\right)$ by $\dfrac{\pi}{6}$ the results would be the same.

---

2   If one function is a sine function and the other a cosine function, the simplest method is to convert the cosine function to a sine function using the relationship

$$\cos \theta = \sin (90° + \theta) = \sin\left(\dfrac{\pi}{2} - \theta\right)$$

Examples

$\cos (\theta + 40°) = \sin (\theta + 130°)$
$\therefore \cos (\theta + 40°)$ leads sin $(\theta + 60°)$ by 70°

$\cos\left(\theta - \dfrac{\pi}{6}\right) = \sin\left(\dfrac{\pi}{2} + \left[\theta - \dfrac{\pi}{6}\right]\right) = \sin\left(\theta + \dfrac{\pi}{3}\right)$

$\therefore \cos\left(\theta - \dfrac{\pi}{6}\right)$ leads sin $\left(\theta + \dfrac{\pi}{12}\right)$ by $\dfrac{\pi}{4}$

3

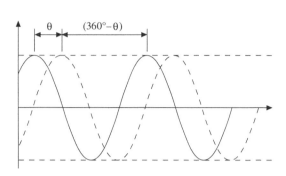

For two periodic functions each with period 360°, if one leads the other by θ the second leads the first by $(360° - \theta)$. It is convenient to state all phase differences θ such that $0° \leq \theta \leq 180°$ (i.e. $0 \leq \theta \leq \pi$). For example if one function leads another by 240°, it is more convenient to state that the *second* function leads the *first* by 120° or that the first *lags* the second by 120°.

Example

Cos $(\theta + 80°)$, which $= \sin (\theta + 170°)$, leads $\sin (\theta - 120°)$ by 290°, but it is more convenient to state that $\sin (\theta - 120°)$ leads $\cos (\theta + 80°)$ by 70° or that $\cos (\theta + 80°)$ lags $\sin (\theta - 120°)$ by 70°.

Exercises 18.4

State which function leads the other, (state 'first' or 'second'), and state the angle of lead (as an angle $\leq 180°$ or π):

1 $5 \sin (\theta + 50°)$ and $7 \sin (\theta + 20°)$

2 $8 \sin (\theta - 30°)$ and $10 \sin (\theta - 70°)$

3 $6 \sin (\theta - 60°)$ and $9 \sin (\theta + 40°)$

4 $3 \cos (\theta - 150°)$ and $\sin (\theta + 20°)$

5 $\cos (\theta + 40°)$ and $5 \sin (\theta - 70°)$

6 $4 \sin \left(\theta - \dfrac{\pi}{3} \right)$ and $8 \cos \left(\theta + \dfrac{2\pi}{3} \right)$

7 $12 \sin \left(\theta + \dfrac{5\pi}{6} \right)$ and $18 \sin \left(\theta - \dfrac{7\pi}{6} \right)$

+++ 8 $7 \cos \left(\theta - \dfrac{5\pi}{6} \right)$ and $12 \sin \left(\theta + \dfrac{\pi}{4} \right)$

+++ 9 $8 \cos \left(\theta - \dfrac{7\pi}{12} \right)$ and $9 \cos \left(\theta + \dfrac{11\pi}{12} \right)$

+++ 10 $4 \sin \left(\theta + \dfrac{5\pi}{12} \right)$ and $3 \cos \left(\theta + \dfrac{5\pi}{12} \right)$

18.6 ADDING SINE AND COSINE FUNCTIONS

If the graphs of two different functions of x, $f(x)$ and $g(x)$, are plotted on the same axes, and the ordinates (i.e. the y-values; the 'heights' of the curves) are added for each value of x, the resulting graph is that of $y = f(x) + g(x)$.

To add the ordinates at any particular value of x, add the 'height' PB of one curve to the 'height' PA of the other curve. A convenient method is to mark the distance PB on the edge of a sheet of paper and then move the paper vertically to add this distance to the 'height' of point A. (Adding a *negative* ordinate means *subtracting* the height.)

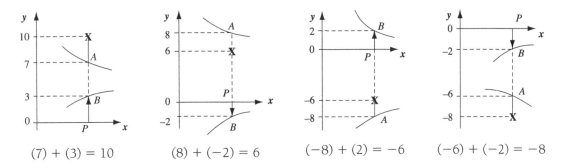

$$(7) + (3) = 10 \qquad (8) + (-2) = 6 \qquad (-8) + (2) = -6 \qquad (-6) + (-2) = -8$$

In the above examples, **X** marks the ordinate for $PA + PB$. A pair of dividers would save time for measuring the distances. A ruler could be used to measure the distances, or squares counted if working on graph paper, but these methods are more tedious.

Sine and cosine curves have the same shape and are called *sinusoidal*. If two sinusoidal waves that have the *same frequency* are added, the resulting curve is also sinusoidal and also has the same frequency.

To add the ordinates of two curves, the easiest points to plot are:

- those at places on the horizontal axis where one of the curves crosses the axis. At such points the sum of the heights is the height of the other curve. In the example below, such points are marked ⊙
- those at places where one of the curves is the same distance *below* the axis as the other curve is *above* it. The sum of the heights is then zero and the resulting curve crosses the axis. In the example below, such points are marked **X**.

Some other points will then need to be plotted to obtain the shape of the resulting curve, especially near the expected turning points.

Example

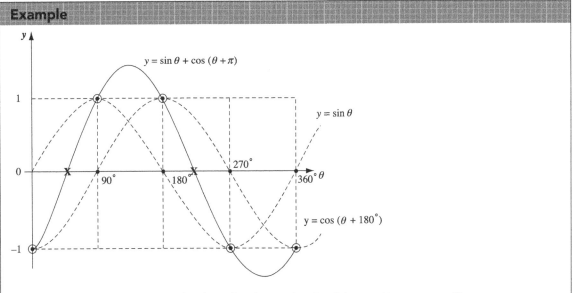

On the above diagram, measured with a ruler, the amplitude of the resulting curve ≈ 28 mm.
20 mm on the *y*-axis represents about 1 unit

continued

continued

∴ 28 mm represents about 1.4 units

∴ The amplitude ≈ 1.4 units

Measured with a ruler, the peak of the resulting curve is about 9 mm to the right of the peak of the $y = \sin \theta$ curve.

36 mm on the θ axis represents about 180°

∴ 9 mm represents about 45°

∴ The phase shift $\overrightarrow{=}$ 45°

Hence, the equation for the resulting curve is approximately $y = 1.4 \sin(\theta - 45°)$.

Exercises 18.5

1 On the same axes, sketch $y = 24 \cos \theta$ and $y = 12 \sin(\theta + 45°)$. Hence, sketch the curve $y = 24 \cos \theta + 12 \sin(\theta + 45°)$. Express the resulting curve in the form $A \cos(\theta + \alpha)$.

2 On the same axes, sketch $y = 9 \sin(x - \pi)$ and $y = 6 \sin\left(x - \dfrac{\pi}{2}\right)$. Hence, sketch the curve $y = 9 \sin(x - \pi) + 6 \sin\left(x - \dfrac{\pi}{2}\right)$. Express the resulting curve in the form $A \sin(x + \alpha)$.

SELF-TEST

1 How many cycles are there in the graph of $y = 12 \cos 50A$ between $A = 0°$ and $A = 180°$?

2 Sketch the graphs of the following functions and state the amplitude and the period of each:

   a $y = 12 \sin 4\theta$ ($0° \leq \theta \leq 360°$)   **b** $y = 6 \cos 30A$ (one cycle only)

3 What is the greatest possible value of $17 \cos 3\theta$ and for what values of θ does it occur, where $0° \leq \theta \leq 360°$?

4 Sketch one cycle of each of the following functions, showing the values where the curve crosses each axis:

   a $y = 12 \cos(\theta + 45°)$ **b** $y = 100 \sin(A - 60°)$

   c $V = 200 \sin(3.7 \times 10^3\, t)$ V   **d** $i = 90 \cos(8\pi \times 10^4\, t)$ mA

5 Sketch one cycle of each of the following functions, showing the values where the curve crosses each axis:

   a $y = 35 \sin(2\theta + 40°)$ **b** $v = 425 \sin(125t - 0.85)$ V

CHAPTER 19

TRIGONOMETRY OF OBLIQUE TRIANGLES

Learning objectives

- Solve triangles using the sine rule and the cosine rule and apply this skill to solving both pure (geometric) and applied (practical) problems.

19.1 THE SINE RULE

Right-angled triangles are solved by means of the formulae:

$$\tan \theta = \frac{\text{opposite side}}{\text{adjacent side}}$$

$$\sin \theta = \frac{\text{opposite side}}{\text{hypotenuse}}$$

$$\cos \theta = \frac{\text{adjacent side}}{\text{hypotenuse}}$$

But these formulae apply *only* to *right-angled triangles*; they cannot be used for an 'oblique' triangle (i.e. one in which there is no right angle).

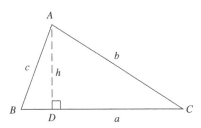

Let *ABC* be any triangle and let the perpendicular from *A* to *BC* have length *h* units.

From the right-angled triangle *ABD*: $\sin B = \dfrac{h}{c}$

$$\therefore h = c \sin B$$

From the right-angled triangle *ACD*: $\sin C = \dfrac{h}{b}$

$$\therefore h = b \sin C$$

$$\therefore b \sin C = c \sin B \text{ (both being equal to } h\text{)}$$

$$\therefore \frac{b}{\sin B} = \frac{c}{\sin C}$$

Similarly, by drawing a perpendicular from C to AB, it may be shown that:

$$\frac{a}{\sin A} = \frac{b}{\sin B}$$

(It is simple to show that these results are also true in any obtuse-angled triangle; a useful exercise if you care to do it!)

For any $\triangle ABC$:

Rule: $\dfrac{a}{\sin A} = \dfrac{b}{\sin B} = \dfrac{c}{\sin C}$ (This is called the **sine rule**.)

$$\frac{\sin A}{a} = \frac{\sin B}{b} = \frac{\sin C}{c}$$

For a $\triangle PTX$: $\dfrac{p}{\sin P} = \dfrac{t}{\sin T} = \dfrac{x}{\sin X}$

Example

In $\triangle KMP$, $m = 325.0$ mm

$p = 587.0$ mm

$\angle P = 124.0°$

Find the size of $\angle M$.

$$\frac{\sin M}{m} = \frac{\sin P}{p}$$

$$\frac{\sin M}{325} = \frac{\sin 124°}{587}$$

$$\sin M = \frac{325 \sin 124°}{587}$$

$$\approx 0.4590$$

$$\therefore \ \angle M \approx 27.32°\ (\approx 27°19')$$

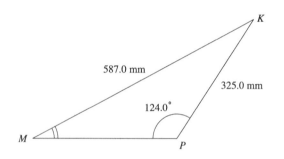

Exercises 19.1

1   In each triangle below, find the length of the side labelled S:

    a

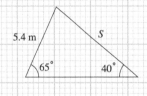

    b

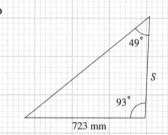

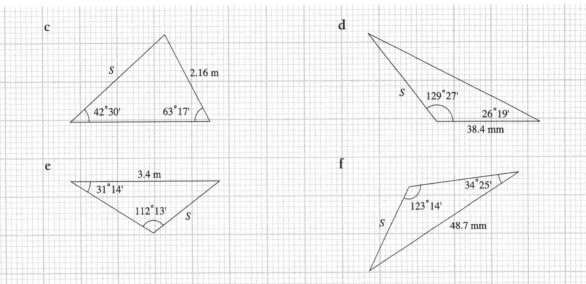

c

d

e

f

2   In each case below find the size of the angle labelled A in degrees and minutes correct to the
nearest minute:

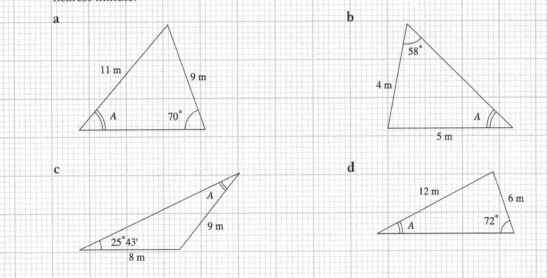

a

b

c

d

+++   3   In each case below find the size of any angle labelled A, in degrees, and the length of any side
labelled S:

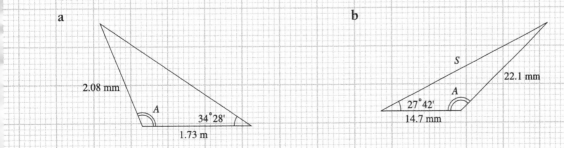

a

b

19.2 THE AMBIGUOUS CASE

Keep the following five facts in mind when using the sine rule to find the size of an angle in a triangle:

1   The sum of the three angles in any triangle is $180°$.

2   A side opposite a greater angle is always greater than a side opposite a smaller angle.

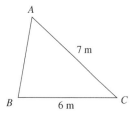

For example, $\angle B > \angle A$ because $b > a$.

3   No triangle can contain more than one obtuse angle (since there cannot be two angles greater than $90°$).

4   The only angle of a triangle that can possibly be obtuse is the greatest angle (i.e. the angle opposite the greatest side).

5   If A is an angle in a triangle and we know the value of sin A, there are *two* possible values for A.

   For example, if $\sin A = 0.9716$
$$A = 76.31°, \text{ or } 103.69° \,(180° - 76.31°)$$

In question 2 of Exercises 19.1 we knew that the required angle A could not be obtuse because it was not opposite the greatest side of the triangle. However, consider this case:

ABC is a triangle in which: $\angle A = 30.00°$
$$b = 55.00 \text{ mm}$$
$$a = 28.00 \text{ mm}$$

Find the magnitude of $\angle B$.

$$\frac{\sin B}{b} = \frac{\sin A}{a}$$

$$\frac{\sin B}{55} = \frac{\sin 30°}{28}$$

$$\sin B = \frac{55 \sin 30°}{28}$$

$$\approx 0.9821$$

$$\therefore \; \angle B = 79.16° \text{ or } 100.84°$$

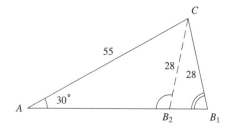

We know that $\angle B > \angle A$ (because $b > a$), but we do not know whether $\angle B = 79.16°$ or $100.84°$. There are two triangles that fit the data, AB_1C and AB_2C.

Examples

1 In $\triangle TKR$, $K = 73°$

$\qquad k = 8$ m

$\qquad t = 6$ m

The sine rule would give us $\angle T$:

$$\frac{\sin T}{t} = \frac{\sin K}{k}$$

But $\angle T < \angle K$ because $t < k$.

$\therefore$ $\angle T$ must be acute.

The triangle is not ambiguous.

2 In $\triangle NDP$, $P = 38°$

$\qquad d = 7$ m

$\qquad p = 5$ m

The sine rule would give us $\angle D$:

$$\frac{\sin D}{d} = \frac{\sin P}{p}$$

But $\angle D > \angle P$ because $d > p$.

$\therefore$ $\angle D > 38°$. It may be acute or obtuse.

The triangle is ambiguous.

Exercises 19.2

1 In each case below state **i** which angle you could find directly, using the sine rule, and **ii** whether or not the triangle is ambiguous:

a $\triangle PHR$ $\begin{cases} \angle H = 40° \\ h = 6 \text{ m} \\ r = 5 \text{ m} \end{cases}$

b $\triangle ATK$ $\begin{cases} \angle K = 50° \\ a = 18 \text{ mm} \\ k = 15 \text{ mm} \end{cases}$

c $\triangle DRH$ $\begin{cases} \angle D = 40° \\ d = 3 \text{ m} \\ h = 4 \text{ m} \end{cases}$

d $\triangle TPG$ $\begin{cases} \angle P = 30° \\ t = 44 \text{ mm} \\ p = 32 \text{ mm} \end{cases}$

e $\triangle KED$ $\begin{cases} \angle D = 57° \\ d = 19 \text{ mm} \\ k = 13 \text{ mm} \end{cases}$

f $\triangle FBN$ $\begin{cases} \angle F = 24° \\ n = 5 \text{ m} \\ f = 7 \text{ m} \end{cases}$

+++ 2 In $\triangle PKT$, $\angle PKT = 34° 26'$, $PK = 217$ mm and $PT = 154$ mm. Find the sizes of **a** $\angle KTP$, **b** $\angle TPK$, **c** side KT. If this triangle is ambiguous, state the two possible sets of values.

+++ 3 In $\triangle NAF$, $a = 21.9$ m, $n = 32.5$ m and $\angle A = 26.43°$. Find the sizes of **a** $\angle N$, **b** $\angle F$, **c** side f. If this triangle is ambiguous, state the two possible sets of values.

19.3 THE COSINE RULE

When we are given the lengths of the three sides of a triangle, or two sides and the included angle, the sine rule is of no assistance to us. In such cases we need another method for solving the triangle.

Let ABC be any triangle.

$$a^2 = h^2 + (b - x)^2$$
$$= h^2 + b^2 - 2bx + x^2$$
$$= b^2 + (h^2 + x^2) - 2bx$$

But $h^2 + x^2 = c^2$

and $2bx = 2b \times C \cos A$

(because $\cos A = \dfrac{x}{c}$)

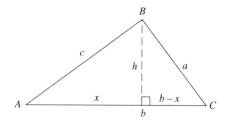

Hence $a^2 = b^2 + c^2 - 2bc \cos A$

Similarly $\left.\begin{array}{l}b^2 = c^2 + a^2 - 2ca \cos B \\ c^2 = a^2 + b^2 - 2ab \cos C\end{array}\right\}$ by symmetry (e.g. by naming the vertices different ways)

Rule: The **cosine rule** is:
$$a^2 = b^2 + c^2 - 2bc \cos A$$
$$b^2 = c^2 + a^2 - 2ca \cos B$$
$$c^2 = a^2 + b^2 - 2ab \cos C$$

This rule relates the magnitudes of three sides of a triangle and one of the angles.

Note: ■ The cosine of an obtuse angle is negative.

■ You should be able to use the cosine rule directly, regardless of the letters used to name the vertices of a triangle.

The cosine rule states that the square of any one side of a triangle equals the sum of the squares on the other two sides minus twice the product of these other two sides and the cosine of the angle between them.

Exercises 19.3

1 Applying the cosine rule to each of the following triangles, complete the given statement:

a

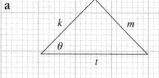

$m^2 = \ldots\ldots$

b

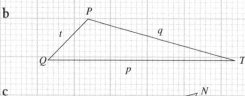

i $t^2 = \ldots\ldots$

ii $p^2 = \ldots\ldots$

c

i $x^2 = \ldots\ldots$

ii $v^2 = \ldots\ldots$

Examples

1 $x^2 = 7.24^2 + 13.8^2 - 2(7.24)(13.8)\cos 80°$

$ \approx 52.42 + 190.4 - 34.70$

$ \approx 208.2$

$\therefore x \approx 14.4$ mm

2 $124^2 = 307^2 + 346^2 - 2(307)(346)\cos\theta$

$\therefore \cos\theta = \dfrac{307^2 + 346^2 - 124^2}{2(307)(346)}$

$ \approx 0.935$

$\therefore \theta \approx 20.8°$

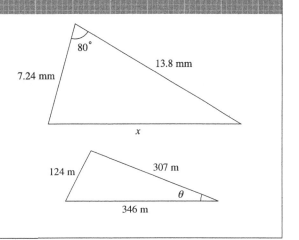

Exercises 19.3 *continued*

2 Evaluate the pronumerals in the following figures, stating lengths to 3 significant figures and angles to the nearest minute:

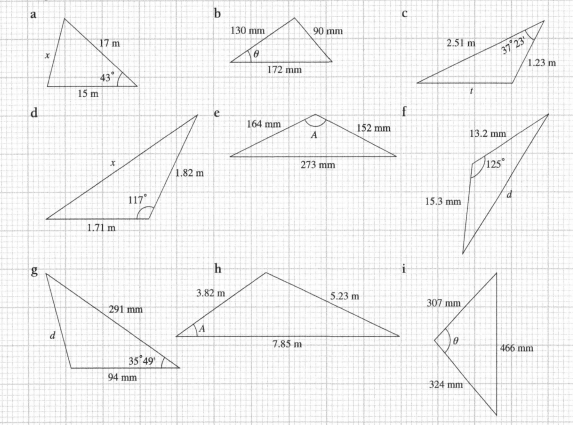

19.4 THE USE OF THE SINE AND COSINE RULES

Before commencing the following exercises you should make sure that you understand these three points:

1   **Which rule to use:** When calculating manually, using a scientific calculator, the sine rule is always easier to use than the cosine rule, so if a choice exists, the sine rule is to be preferred because it is a simpler formula. But to use the sine rule, you must know the *length of a side that is opposite a known angle.*

    When the data consists of the three sides (SSS), or two sides and the included angle (SAS), you are *forced* to use the *cosine* rule as the first step in solving the triangle.

    Triangle type

    SSS

    (Three sides)

    You are *forced* to use the *cosine* rule to find one of the angles. Then use *either* rule to find a second angle (but the sine rule is the simpler).

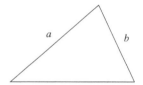

    SAS

    (Two sides and the included angle)

    You are *forced* to use the *cosine* rule to find the side opposite the angle θ. Then use *either* rule to find a second angle (but the sine rule is the simpler).

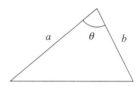

    AAS

    (Two angles and one side)

    The third angle is known $(180 - \alpha - \beta)$. You are forced to use the sine rule to find one of the unknown sides. Then use the sine rule again to find the third side (or use the cosine rule but this is more difficult as it gives a quadratic equation).

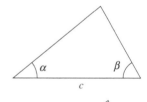

    SSA

    (Two sides and a non-included angle)

    Use the sine rule to find the angle opposite side b and then the sine rule again to find the third side. (You *could* use the cosine rule to find both of these but this is a more involved method.)

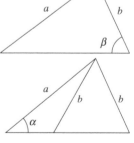

    Remember that the triangle is ambiguous if the *smaller* of the given sides is opposite the given angle.

2   Remember that the largest angle of a triangle (the angle that is opposite the largest side) *may* be obtuse. When using the sine rule it is always safer to find the other two angles first to avoid this difficulty.

3   There are two methods in common use for stating a direction or bearing (see section 10.8).

Exercises 19.4

Angles in these exercises are measured in degrees, as decimals, unless degrees and minutes are stated or requested.

Geometrical

1   The sides of a triangle have lengths in the ratio 2 : 3 : 4. Find the three angles in degrees and minutes correct to the nearest minute.

2   The sizes of the angles of a triangle are in the ratio 2 : 3 : 4. Given that the largest side has a length of 35.7 cm, find the length of the shortest side.

3   The adjacent sides of a parallelogram have lengths of 1.83 m and 792 mm, and the angle included between these sides is 52.8°. Find the lengths of the two diagonals.

+++ 4   The sides of a triangle have lengths 548 mm, 835 mm and 1024 mm.
Find:
  a   the length of the median drawn to the shortest side
  b   the angle that the above median makes with the longest side. ('Median' was defined in Chapter 7.)

+++ 5   A quadrilateral has sides of lengths 41.3 mm, 78.2 mm, 106 mm and 165 mm. Given that the two longest sides are adjacent and that the angle included between them is 37.2°, find the size of the angle between the two shortest sides.

Applications

6   Two steel beams, one of which is three times as long as the other, rest between a vertical wall and a horizontal floor, both reaching the same height on the wall. Given that the shorter beam makes an angle of 73.0° with the floor, find the angle that the longer beam makes with the floor. State the result in degrees and minutes correct to the nearest minute.

+++ 7   An aircraft flies at a constant ground speed of 537 km/h and at constant altitude around a triangular course starting from a position P vertically above a control tower. The pilot flies for 20.0 minutes on a bearing of 132.8°, then for 40.0 minutes on a bearing of 256.3° and finally on a new constant bearing, back to P. Find **a** the total length of the trip and **b** the direction bearing on the third leg of the trip.

8   An apprentice is instructed to cut out a triangular sheet of metal PQR such that PQ = 295 mm, QR = 153 mm and the angle RPQ = 28.5°. These instructions do not give sufficient information and the apprentice is uncertain as to the required length of the side PR. What are the two possibilities for the length of PR?

+++ 9   To determine the width of a river that has parallel banks, a surveyor observes an object P on the opposite bank from two points A and B that are 55.0 m apart. From A the surveyor measures the angle BAP to be 43.7° and from B the surveyor measures the angle ABP to be 127.3°. Find the width of the river.

continued

continued

+++ 10 A launch is situated due west of a ship that is moving in still water on a compass bearing of S37.5°E at a speed of 11.8 km/h. Given that the launch travels at a speed of 26.4 km/h, in what direction should the launch travel in order to intercept the ship?

11 A triangular steel frame has sides of lengths 9.32 m, 6.85 m and 5.47 m. Find the size of the largest angle in this frame.

+++ 12 The navigator on a ship observes two landmarks *P* and *Q*, knowing that the bearing of *P* from *Q* is 103.7° and that the distance separating *P* from *Q* is 4.68 km. From the ship, *P* bears 48.2° and *Q* bears 36.9°. Find the distance of the ship from *Q*.

SELF-TEST

1 In each of these triangles, find the length of the side labelled *S*:

a

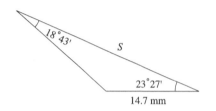

b

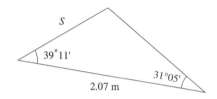

2 In △*ABC*, ∠*ABC* = 64°, *AB* = 35 mm and *CA* = 48 mm. Find the size of ∠*CAB*.

3 In △*KNT*, *N* = 120°, *n* = 273 m and *k* = 134 m. Find the length of side *t*.

4 In △*MTA*, *A* = 16°43′, *a* = 132 mm and *m* = 275 mm. Find the two possible sizes of ∠*M*.

5 For each of the following triangles, find the size of the side or angle marked by a pronumeral:

a

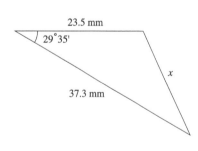

b

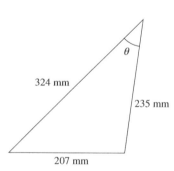

c

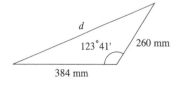

d

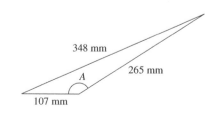

+++ 6 The navigator on a ship S sights two landmarks, P and Q. P is on a bearing of 156° and Q bears 219°. From a chart it is determined that the bearing of Q from P is 296° and the distance PQ is 15.9 km. Find the distances of the ship from P and from Q.

+++ 7 A ship sails 73.0 km on course S31°W and then 98.5 km on course N76°E. What is the final distance and bearing of the ship from its starting point?

CHAPTER 20

TRIGONOMETRIC IDENTITIES

Learning objectives

- Manipulate and simplify trigonometric expressions using memorised or given identities.
- Use trigonometric identities to solve practical problems.

20.1 DEFINITION

In general, an equation is true for only certain particular values of the variable. For example, the equation $3(x - 1) = 2x + 5$ is true only for $x = 8$ (which is called the **solution** of the equation). However, the equation $3(x - 1) = 3x - 3$ is true for *all* values of x because $3(x - 1)$ and $3x - 3$ are two different ways of writing the same expression. Such an equation is called an **identity.**

Exercises 20.1

1. Solve the following four equations and then state which one is an identity:

    a $x^3 + 1 = (x + 1)(x^2 + x + 1)$ **b** $x^3 + 1 = (x - 1)(x^2 + x - 1)$

    c $x^3 + 1 = (x + 1)(x^2 - x + 1)$ **d** $x^3 + 1 = (x - 1)(x^2 - x - 1)$

20.2 TRIGONOMETRIC IDENTITIES

In Chapter 17, $\sin \theta$ and $\cos \theta$ were defined as the ordinate and abscissa respectively of a point P when P has rotated around a circle of unit radius, through an angle θ, starting from the standard position (i.e. lying along the positive x-axis).

  a   **b**   **c**   **d**

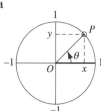 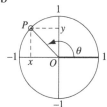 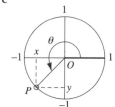 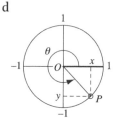

So in the above diagrams, *by definition,* for angle θ of any magnitude, $\sin \theta = y$ and $\cos \theta = x$.

Tan θ was defined as being $\dfrac{\sin \theta}{\cos \theta}$.

In the above diagrams the circle has unit radius, so $OP = 1$.

By Pythagoras' theorem, $y^2 + x^2 = OP^2$.

$$\therefore \sin^2 \theta + \cos^2 \theta = 1.$$

Note: $\sin^2 \theta$ is the abbreviation for $(\sin \theta)^2$, which is not the same as $\sin \theta^2$.

For example: $\sin \left(\dfrac{\pi}{3}\right)^2 = \sin \dfrac{\pi^2}{9} \approx 0.8897$

but $\qquad \sin^2 \dfrac{\pi}{3} = \left(\sin \dfrac{\pi}{3}\right)^2 = \left(\dfrac{\sqrt{3}}{2}\right)^2 = 0.75$

Exercises 20.2

1 Evaluate correct to 4 decimal places:

   a $\sin (0.47)^2$ **b** $\sin^2 0.47$ **c** $\cos^2 \dfrac{\pi}{5}$ **d** $\cos \left(\dfrac{\pi}{5}\right)^2$

2 Use your calculator to verify that $\sin^2 \theta + \cos^2 \theta = 1$ is an identity by substituting any value you like to choose for θ (in degrees or radians).

3 Verify that $\tan \theta = \dfrac{\sin \theta}{\cos \theta}$ is an identity, using your calculator and several values of θ.

Note: Regardless of how many times we test an equation by making substitutions, as in the last two exercises, we cannot *prove* that the equation is an *identity* because we have not shown that the equation is true for *every* value of the variable. No amount of verification can constitute a *proof*.

Exercises 20.2 *continued*

4 It has been proved above, from the definitions of $\sin \theta$ and $\cos \theta$ and by using Pythagoras' theorem, that $\sin^2 \theta + \cos^2 \theta = 1$ is an identity.

   a By dividing both sides of this equation by $\cos^2 \theta$, prove the identity $\sec^2 \theta = 1 + \tan^2 \theta$.

   b Prove the identity $\operatorname{cosec}^2 \theta = 1 + \cot^2 \theta$.

Note: $\sin^2 \theta + \cos^2 \theta = 1$

$\qquad\qquad \sec^2 \theta = 1 + \tan^2 \theta$

$\qquad \operatorname{cosec}^2 \theta = 1 + \cot^2 \theta$

These are called the 'Pythagorean identities'. They are very important and should be memorised.

20.3 PROVING AN IDENTITY

Consider the following argument: If $\quad 2 = 5$

then $\quad 5 = 2$

adding: $\therefore \underline{7 = 7}$

There is nothing wrong with the above proof, but it is important to understand just *what* has been proved. I have proved that *if* $2 = 5$, *then* $7 = 7$. The proof does not state that $2 = 5$ and the fact that 2 does *not* equal 5 is irrelevant. Note that the correctness of the conclusion ($7 = 7$) does not prove that $2 = 5$.

Rule: When proving an identity (or anything else), never argue from what you have to prove because if what you are attempting to prove is actually *wrong*, you may still end up with a correct conclusion. But this does not mean that your proof is valid.

There are three valid methods to prove that $A = B$:

1   Commence with A and by logical argument show that it is equal to B.
2   Commence with B and by logical argument show that it is equal to A.
3   Commence with A and B *separately* and by logical argument show that they are both equal to C.

Advice:

■   When simplifying trigonometric expressions it is usually helpful to express all the trigonometric ratios in terms of *sine* and *cosine*.

■   It is convenient to use the abbreviations LHS and RHS for 'left-hand side' and 'right-hand side'.

Example

Prove that $\operatorname{cosec} x - \sin x = \cos x \cot x$.

Proof 1

$$\text{LHS} = \frac{1}{\sin x} - \sin x$$

$$= \frac{1 - \sin^2 x}{\sin x}$$

$$= \frac{\cos^2 x}{\sin x}$$

$$= \cos x \times \frac{\cos x}{\sin x}$$

$$= \cos x \times \frac{1}{\tan x}$$

$$= \cos x \cot x$$

$$= \text{RHS}$$

Proof 2

$$\text{RHS} = \cos x \times \frac{1}{\tan x}$$

$$= \cos x \times \frac{\cos x}{\sin x}$$

$$= \frac{\cos^2 x}{\sin x}$$

$$= \frac{1 - \sin^2 x}{\sin x}$$

$$= \frac{1}{\sin x} - \frac{\sin^2 x}{\sin x}$$

$$= \operatorname{cosec} x - \sin x$$

$$= \text{LHS}$$

Proof 3

$$\text{LHS} = \frac{1}{\sin x} - \sin x$$

$$= \frac{1 - \sin^2 x}{\sin x}$$

$$= \frac{\cos^2 x}{\sin x}$$

$$\text{RHS} = \cos x \times \frac{1}{\tan x}$$

$$= \cos x \times \frac{\cos x}{\sin x}$$

$$= \frac{\cos^2 x}{\sin x}$$

$$\therefore \text{LHS} = \text{RHS} \left[= \frac{\cos^2 x}{\sin x} \right]$$

It would now be neater to link these two arguments together into a single argument, resulting in either proof 1 or proof 2.

20.4 SUMMARY: TRIGONOMETRIC IDENTITIES

These identities should be memorised:

By definition	By definition	The Pythagorean identities
$\sec x = \dfrac{1}{\cos x}$ $\text{cosec } x = \dfrac{1}{\sin x}$ $\cot x = \dfrac{1}{\tan x}$	$\tan x = \dfrac{\sin x}{\cos x}$ $\therefore \sin x = \tan x \cos x$ and $\cos x = \dfrac{\sin x}{\tan x}$	$\sin^2 x + \cos^2 x = 1$ $\sec^2 x = 1 + \tan^2 x$ $\text{cosec}^2 x = 1 + \cot^2 x$ $\therefore \sin^2 x = 1 - \cos^2 x$ $\cos^2 x = 1 - \sin^2 x$ $\tan^2 x = \sec^2 x - 1$ $\cot^2 x = \text{cosec}^2 x - 1$ $\sec^2 x - \tan^2 x = 1$ $\text{cosec}^2 x - \cot^2 x = 1$

Exercises 20.3

First revise section 17.3.

1 Simplify the following, expressing each as a single trigonometric ratio:

a $\dfrac{\sin 47°}{\cos 47°}$

b $\dfrac{1}{\sec 23°}$

c $\dfrac{\cos 38°}{\sin 38°}$

d $\dfrac{\text{cosec } \theta}{\sec \theta}$

e $\cos A \text{ cosec } A$

f $\cot x \sec x$

g $\sin \theta \div \tan \theta$

h $\dfrac{\sin \theta \sec \theta}{\tan \theta \cos \theta}$

i $\sin \theta \text{ cosec } \theta \cot \theta$

2 Write down the missing trigonometrical ratios:

a $\dfrac{\sin 20°}{\sin 70°} = \dots 20°$

b $\sin 38° \text{ cosec } 52° = \dots 38°$

c $\cot 26° \text{ cosec } 64° = \dots 64°$

d $\text{cosec } 32° \div \tan 58° = \dots 32°$

e $\cos 73° \div \sin 73° = \dots 17°$

f $\text{cosec } 48° \div \cot 48° = \dots 42°$

continued

continued

3 Simplify, expressing each as a single trigonometrical ratio:

  a $\dfrac{\cos \theta°}{\cos (90 - \theta)°}$

  b $\tan A° \sin (90 - A)°$

  c $\sin x° \operatorname{cosec} (90 - x)°$

  d $\cos (90 - \theta)° \div \tan \theta°$

4 Simplify:

  a $1 - \cos^2 \theta$

  b $\sec^2 \theta - \tan^2 \theta$

  c $1 - \sin^2 A$

  d $\cot^2 B - \operatorname{cosec}^2 B$

  e $(\sin \theta + \cos \theta)^2 - 2 \sin \theta \cos \theta$

  f $\cos^2 A - 1$

  g $\tan^2 \theta - \sec^2 \theta$

  h $3 - 3 \cos^2 K$

5 Prove the following identities:

  a $\sqrt{\operatorname{cosec}^2 \theta - \cot^2 \theta} = 1$

  b $\cot^2 \theta (1 - \cos^2 \theta) = \cos^2 \theta$

  c $\dfrac{1 - \sin^2 A}{1 - \cos^2 A} = \cot^2 A$

  d $(\sec^2 A - 1) \div (1 + \tan^2 A) = \sin^2 A$

  e $\sin^2 x + \tan^2 x + \cos^2 x = \sec^2 x$

  f $\dfrac{\cos^2 \theta - 1}{1 - \sec^2 \theta} = \dfrac{1}{\tan^2 \theta + 1}$

  g $\sqrt{\dfrac{1 - \cos^2 A}{1 - \sin^2 A}} = \tan A$

  h $\cos^2 x \tan^2 x + \cos^2 x = 1$

6 Prove the following identities:

  a $(\cos \theta + \sin \theta)(\cos \theta - \sin \theta) = 2 \cos^2 \theta - 1$

  b $\dfrac{\sin x}{1 - \cos x} - \dfrac{\sin x}{1 + \cos x} = 2 \cot x$

  c $\dfrac{1 + \cot \theta}{\operatorname{cosec} \theta} = \dfrac{\tan \theta + 1}{\sec \theta}$

  d $\sec x - \tan x = \dfrac{1 - \sin x}{\cos x}$

  e $\cot A \cos A = \operatorname{cosec} A - \sin A$

7 Simplify, expressing each in terms of a single trigonometrical ratio, or as a constant:

  a $1 - \sin^2 \theta$

  b $\tan^2 x - \sec^2 x$

  c $\dfrac{\sqrt{1 - \sin^2 \theta}}{\sin \theta}$

  d $\sqrt{1 - \cos^2 x}$

  e $\sec^2 \theta - \sin^2 \theta - \cos^2 \theta$

  f $(\cos^2 \theta - 1) \sec^2 \theta$

Exercises 20.4

Applications of $\tan \theta = \dfrac{\sin \theta}{\cos \theta}$

1 If a force of magnitude F is resolved into two components F_1 and F_2 which have directions at right angles to each other, then it may be shown that $F_1 = F \cos \theta$ and $F_2 = F \sin \theta$, where θ is the angle that the direction of F_1 makes with the direction of F.

  a Show that $\tan \theta = \dfrac{F_2}{F_1}$.

  b Evaluate F_2 when $\theta = 27°30'$ and $F_1 = 1.24$ kN.

2   A body of mass m kg is held in equilibrium by a force F on
    a smooth inclined plane whose angle of inclination to the
    horizontal is θ, the force F being parallel to the plane. It may
    be shown that $F = mg \sin \theta$ and $R = mg \cos \theta$ where g is the
    acceleration due to gravity and R is the normal reaction force.

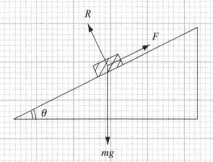

   a   Show that $\tan \theta = \dfrac{F}{R}$.

   b   Evaluate R when $\theta = 23.5°$ and $F = 235$ N.

3   A mass of m kg, suspended from a light cable, is acted upon
    by a horizontal force of P newtons, which holds the cable at
    angle θ to the vertical. It may be shown that $\sin \theta = \dfrac{P}{T}$ and
    $\cos \theta = \dfrac{mg}{T}$ where T newtons is the tension in the cable and g is
    the acceleration due to gravity, ≈ 9.81 m/s^2.

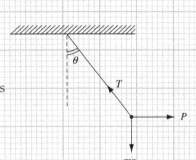

   a   Show that $\dfrac{P}{mg} = \tan \theta$.

   b   Find the force required to hold the cable at an angle of $40°$
    to the vertical when the mass is 1.20 t.

   c   Find the angle θ at which the cable is inclined to the vertical when a horizontal force of
    5.00 kN is applied, the mass being 750 kg.

4   If an AC voltage V supplied to a circuit produces a current
    of i amps which lags the supply voltage by angle ϕ, then
    $R = Z \cos \phi$ and $X = Z \sin \phi$, where R ohms and X ohms are
    the resistive and reactive components of the impedance Z ohms.

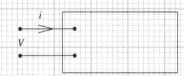

   a   Show that $\tan \phi = \dfrac{X}{R}$.

   b   Evaluate R when $X = 425 \ \Omega$ and $\phi = 7.00°$.

   c   Evaluate ϕ when $R = 1.00 \ k\Omega$ and $X = -300 \ \Omega$.

5   If $5 \sin \omega t + 3 \cos \omega t = A \sin (\omega t + \alpha)$, then it may be shown that $A \cos \alpha = 5$ and $A \sin \alpha = 3$.
    Evaluate:

   a $\tan \alpha$ **b** a **c** A

Exercises 20.5

Applications of the Pythagorean identities

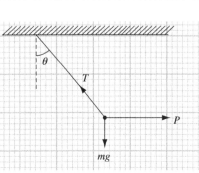

1   A horizontal force P holds a simple pendulum inclined at an
    angle θ to the vertical. Given that $T \cos \theta = mg$ and $T \sin \theta = P$,
    show that $(mg)^2 + P^2 = T^2$. Hence evaluate T, the tension in
    the string, when $m = 2.46$ kg and $P = 17.6$ N ($g = 9.80$ m/s^2).

continued

continued

2   A body is held at rest on a smooth inclined plane.
    Given that $mg \cos \theta = R$ and $mg \sin \theta = F$, show
    that $R^2 + F^2 = (mg)^2$, and hence evaluate the
    normal reaction force R when $m = 3.84$ kg and
    $F = 24.7$ N ($g = 9.80$ m/s^2).

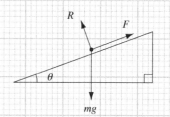

3   If a force of magnitude F is resolved into two components F_1 and F_2, which have directions at
    right angles to one another, it may be shown that:
    $$F_1 = F \cos \theta$$
    and $F_2 = F \sin \theta$
    where θ is the angle between the directions of F and F_1.
    Show that $F_1^2 + F_2^2 = F^2$, and hence evaluate F_2 when $F = 7.62$ kN and $F_1 = 5.34$ kN.

+++ 4   If $4 \sin \omega t + 7 \cos \omega t = A \sin (\omega t + \alpha)$, it may be shown that $A \cos \alpha = 4$ and $A \sin \alpha = 7$.
    Evaluate $(A \cos \alpha)^2 + (A \sin \alpha)^2$, and hence evaluate:
    a A b α

20.5 OTHER TRIGONOMETRIC IDENTITIES

There are many other identities that are useful when simplifying or otherwise manipulating
trigonometric expressions.

You are by now quite familiar with the fact that $a(x + y) = ax + ay$, but this 'distribution law' is
true only for *multiplication*.

For example, $\sqrt{(x + y)} \neq \sqrt{x} + \sqrt{y}$, and similarly, $\cos (\alpha - \beta) \neq \cos \alpha - \cos \beta$.

For example, $\cos (60° - 30°) = \cos 30° = \dfrac{\sqrt{3}}{2}$.

But $\cos 60° - \cos 30° = \dfrac{1}{2} - \dfrac{\sqrt{3}}{2} = \dfrac{1 - \sqrt{3}}{2}$.

Actually it can be shown that $\cos (\alpha - \beta) = \cos \alpha + \cos \beta + \sin \alpha \sin \beta$ and this identity can be
verified by substituting any values you care to choose for α and β. You will prove this identity in the
following exercises.

Exercises 20.6

1   Write down the exact coordinates of the point K,
    as an ordered pair, given that the circle has unit
    radius (i.e. radius $= 1$).

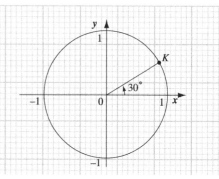

2

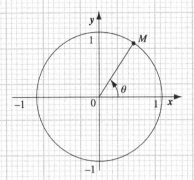

Write down the exact coordinates of the point M, as an ordered pair.

3 Given that the radius of this circle = 1 unit and that $\angle AOQ = \alpha$, $\angle AOP = \beta$:

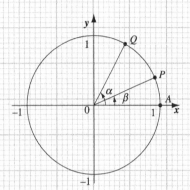

  a Write down the exact coordinates of the points P and Q.

  b Apply the cosine rule to $\triangle OPQ$ to find PQ^2.

  c Use the formula for the distance between two points to find PQ^2. (See section 9.1.)

  d By equating the above two expressions for PQ^2, find a formula for $\cos(\alpha - \beta)$.

Note: You have proved the identity $\cos(\alpha - \beta) = \cos\alpha\cos\beta + \sin\alpha\sin\beta$. Many other important identities follow from this result and the most famous and frequently used of these are included in the supplement to this chapter, where you will also find additional exercises. For present purposes only the four 'addition formulae' are included and throughout the following exercises, you can use these identities whenever necessary:

$$\sin(\alpha + \beta) = \sin\alpha\cos\beta + \cos\alpha\sin\beta$$
$$\sin(\alpha - \beta) = \sin\alpha\cos\beta - \cos\alpha\sin\beta$$
$$\cos(\alpha + \beta) = \cos\alpha\cos\beta - \sin\alpha\sin\beta$$
$$\cos(\alpha + \beta) = \cos\alpha\cos\beta + \sin\alpha\sin\beta$$

Exercises 20.6 *continued*

4 Express $\sin 2x$ in terms of $\sin x$ and $\cos x$. (*Hint:* Express $\sin 2x$ as $\sin(x + x)$.)

5 Express $\cos 2x$ in terms of $\cos^2 x$.

In the following exercises, always watch for:

- **factorising possibilities**

  e.g. $3\sin B\cos A - 6\cos A\cos B = 3\cos A(\sin B - 2\cos B)$

- **expressions of the form $\sin(\alpha \pm \beta)$ and $\cos(\alpha \pm \beta)$**

  e.g. $\cos 2x\sin 3y - \sin 2x\cos 3y = \sin 3y\cos 2x - \cos 3y\sin 2x$
  $$= \sin(3y - 2x)$$

- **expressions of the form $(x - y)(x + y)$**
  e.g. $(\sin x + 2\cos y)(\sin x - 2\cos y) = (\sin x)^2 - (2\cos y)^2$
  $$= \sin^2 x - 4\cos^2 y$$
- **conversions from $\sin^2 \theta$ to $\cos^2 \theta$, and vice versa, using $\sin^2 \theta + \cos^2 \theta = 1$**
  e.g. $\sin^2 A \cos B + \cos B \cos^2 A = \cos B (\sin^2 A + \cos^2 A)$
  $$= \cos B$$

Exercises 20.6 continued

6 Simplify $\cos x \cos 2x - \sin x \sin 2x$.

7 Simplify $\sin^2 x \sin y + \cos^2 x \sin y$.

8 Simplify $\sin (x + y) \sin y + \cos (x + y) \cos y$.

9 Express $\dfrac{\sin y (\sin x + \cos x)}{\cos x \cos y}$ in terms of $\tan x$ and $\tan y$.

10 Express $\dfrac{\sin (A + B)}{\cos (A - B)}$ in terms of $\tan A$ and $\tan B$. (*Hint:* After expanding, divide the numerator and denominator by $\cos A \cos B$.)

11 Simplify $\sin (A + B) \sin (A - B)$.

12 If $\tan (x + y) = A$ and $\tan x = \frac{1}{7}$:

  a express $\tan y$ in terms of A, using the identity $\tan (\alpha + \beta) = \dfrac{\tan \alpha + \tan \beta}{1 - \tan \alpha \tan \beta}$

  b evaluate $\tan y$ when $A = \frac{1}{7}$.

13 Simplify $\sin 5x \cos 4x - \cos 5x \sin 4x$.

14 Given that $\sin 2\theta = 2 \sin \theta \cos \theta$, express $\dfrac{2 \tan A}{\tan^2 A + 1}$ as the sine of an angle. (*Hint:* Use $\tan A = \dfrac{\sin A}{\cos A}$.)

15 Simplify:

  a $\dfrac{\sin^2 \theta}{(1 + \sin \theta)(1 - \sin \theta)}$

  b $\dfrac{1}{1 + \sin \theta} + \dfrac{1}{1 - \sin \theta}$

+++ 16 a Find the *exact* value of $\sin 15°$ thus: $\sin 15° = \sin (45° - 30°)$
  $$= \sin 45° \cos 30° - \cos 45° \sin 30°$$
  and then use *exact values*.

  b Find the *exact* values of

  i $\sin 75°$ ii $\cos 165°$ iii $\tan 75°$ iv $\sin 195°$

  (You can use your calculator to make an approximate check of your results.)

SELF-TEST

1 Simplify:

  a $\operatorname{cosec}^2 (90° - A) - (\tan^2 A + \sin^2 A)$

  b $\sec^2 \theta \cos^2 (90° - \theta) + 1$

  c $(1 + \tan^2 x) \tan^2 (90° - x)$

  d $\sqrt{\dfrac{1 + \cot^2 A}{\operatorname{cosec}^2 A - 1}}$

  e $\sin x \operatorname{cosec} (90° - x) \sqrt{\operatorname{cosec}^2 x - 1}$

2 Simplify:

a $\dfrac{\cos \theta}{\sec \theta} + \dfrac{\sin \theta}{\operatorname{cosec} \theta}$

b $\cos x \sec x - \dfrac{\sec x}{\cos x}$

c $(\tan x + \sec x)(\sec x - \tan x)$

d $\dfrac{\sec x - \tan x \sin x}{\sec x - \cos x}$

3 Prove the following identities:

a $\sec^2 x \operatorname{cosec}^2 x = \sec^2 x + \operatorname{cosec}^2 x$

b $\cos \theta + 1 = \dfrac{\sin^2 \theta}{1 - \cos \theta}$

c $\dfrac{1 + \tan A}{\sec A} = \dfrac{1 + \cot A}{\operatorname{cosec} A}$

d $\sin^2 x - \cos^2 x = \dfrac{1 - \cot^2 x}{1 + \cot^2 x}$

4 Solve the simultaneous equations $\begin{cases} A \sin \theta = 14.72 \\ A \cos \theta = -11.38 \end{cases}$ in two ways:

a by squaring both sides of each equation and then adding

b by equating the ratios of the left-hand sides and the right-hand sides.

5 Expand:

a $\cos (\theta + 60°)$

b $\tan \left(\dfrac{3\pi}{4} - x \right)$

+++ 6 a Given that $\cos (A + B) = \cos A \cos B - \sin A \sin B$, simplify $\cos (60° + \theta) + \cos (60° - \theta)$.

b Given that $\tan (A - B) = \dfrac{\tan A - \tan B}{1 + \tan A \tan B}$, simplify $\dfrac{\sin 7\theta - \cos 7\theta \tan 6\theta}{\cos 7\theta + \sin 7\theta \tan 6\theta}$.

+++ 7 a Given that $\sin (A + B) = \sin A \cos B + \cos A \sin B$, prove the identity
$\tan \alpha + \tan \beta = \sin (\alpha + \beta) \sec \alpha \sec \beta$.

b Given that $\cos (A - B) = \cos A \cos B + \sin A \sin B$, prove the identity
$\cos \dfrac{x + y}{2} \cos \dfrac{x - y}{2} - \sin \dfrac{x + y}{2} \sin \dfrac{x - y}{2} = \cos x$.

+++ 8 Use the triangle shown in the diagram to verify that
$\cos (\alpha + \beta) = \cos \alpha \cos \beta - \sin \alpha \sin \beta$. (*Hint:* First
evaluate $\cos (\alpha + \beta)$ using the cosine rule.)

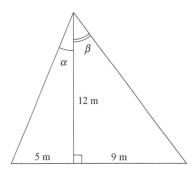

+++ 9 Express $2.74 \sin \theta + 4.13 \cos \theta$ in the form $A \sin (\theta + \alpha°)$, stating the values of A and θ correct
to 3 significant figures, given that $A > 0$. (*Hint:* Expand $A \sin (\theta + \alpha°)$ using the given identity
$\sin (A + B) = \sin A \cos B + \cos A \sin B$.)

CHAPTER 21

INTRODUCTION TO VECTORS

Learning objectives

- Resolve a vector into two components at right angles to each other.
- Determine the resultant of two or more vectors analytically and graphically.
- Solve two-dimensional problems by representing quantities as vectors.

21.1 A MATHEMATICAL VECTOR

A **vector** is a directed line segment. A vector has only two properties: its length (i.e. its magnitude) and its direction. A vector is completely specified (i.e. distinguished from every other vector) once these two quantities are stated.

On this diagram, only *one* vector is shown. This is the vector having length 18 mm and in the direction making an angle of 48° with the *x*-axis.

A vector that runs from point *A* to point *B* is designated as $\overrightarrow{AB}$. (This is not the same as vector $\overrightarrow{BA}$, which would have the same length but which would run in the opposite direction.)

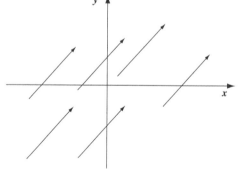

21.2 SCALAR AND VECTOR QUANTITIES

A **scalar quantity** is one that has magnitude (i.e. size) only, so only one number is required to specify (i.e. describe) it completely. These quantities have no direction. Examples of scalar quantities are length, distance, area, volume, mass, density, temperature, energy and electric charge. A length is completely specified, for example, by '2500 mm' or '250 cm' or '2.5 m'.

A **vector quantity**, however, has magnitude and direction, and both of these must be stated in order to specify it. 'Displacement' is a vector quantity. If movement occurs (of a 'point', a 'particle', a 'body') from point *P* to point *Q*, by any route, the 'displacement' is defined as the directed line segment from *P* to *Q*.

This diagram shows the movement of a particle
from point K to point M along a curved path. The
displacement of the particle is the directed line segment
$\overrightarrow{KM}$, i.e. the shortest possible path from K to M.
A displacement can be represented by a vector (a
directed line segment drawn to scale) and so it is said
to be a 'vector quantity'. Another example of a vector quantity is a
force. To know the effect of a force upon a body we must know not
only its size but also its direction. A force of 17 N acting upon a body in direction due north has
the opposite effect on the body to a force of 17 N acting in direction due south. Other examples of
vector quantities are velocity, acceleration, momentum,
electric field and magnetic field.

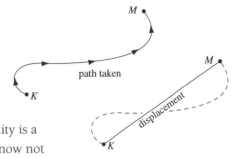

Any vector quantity may be represented by a vector
(i.e. by a directed line segment). The length of the vector
represents the magnitude of the quantity (to some stated
scale) and the arrow indicates its direction.
The vector $\overrightarrow{PQ}$, being 34 mm in length, represents a force
of 136 N acting in direction NE.

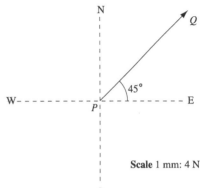

21.3 ADDITION OF VECTORS AND VECTOR QUANTITIES

If the displacement $\overrightarrow{AB}$ occurs, followed by the displacement $\overrightarrow{BC}$, the
net result (the **resultant displacement**) is the displacement $\overrightarrow{AC}$, i.e. the
single displacement that achieves the same effect. We say that when we
'add' the displacements $\overrightarrow{AB}$ and $\overrightarrow{BC}$, the 'sum' is the displacement $\overrightarrow{AC}$. We
write $\overrightarrow{AB} + \overrightarrow{BC} = \overrightarrow{AC}$. Note that adding displacements is not the same as
adding scalar quantities, since the sum of the lengths of AB and BC is not
the length of AC.

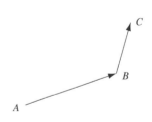

Example

If a body undergoes a displacement of 12 m due north, followed
by a displacement of 5 m due east, the resultant displacement is
13 m in the direction 22.62° E of N (since as shown in the diagram,
$SF = \sqrt{5^2 + 12^2} = 13$, and $\tan \theta = \frac{5}{12}$, $\theta \approx 22.62°$).

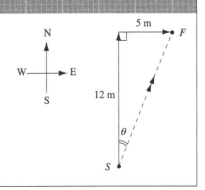

Successive displacements

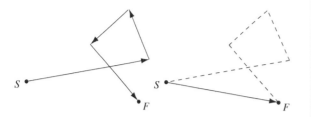

Note: If several displacements occur successively, the **resultant** (i.e. the net effect) is the displacement represented by the vector drawn from the starting point, S, to the finishing point, F. In the diagram, the resultant displacement is $\overrightarrow{SF}$.

Simultaneous displacements, velocities, etc.

It should be noted that a body may undergo several displacements *simultaneously* and that in this case the resultant is exactly the same as if the displacements had occurred in succession. For example, suppose that a sailor walks 10 m due north on the deck of a stationary ship, and then while the sailor is standing at attention, the ship moves 15 m due east. The final position of the sailor with respect to the ground will be exactly the same as if these two motions had occurred simultaneously. It makes no difference whether the sailor walks the 10 m while the ship is moving 15 m or whether one of these motions occurs first, followed by the other.

If the vectors in the diagram above represent simultaneous displacements, the resultant displacement is represented by the vector $\overrightarrow{SF}$. Now if these several simultaneous displacements all occurred with different constant speeds during a time interval of 1 second (or 1 minute, 1 hour, etc.), these vectors represent simultaneous *velocities* of the body and $\overrightarrow{SF}$ represents the resultant *velocity*. Hence, simultaneous *velocities* are added by the polygon law. If all these velocities are velocity *increases* of the body in unit time, these vectors represent simultaneous *accelerations* of the body, and $\overrightarrow{SF}$ represents the resultant *acceleration*.

Since $F = ma$, the acceleration being in the same direction as the force, these vectors may represent *forces* acting upon the body and $\overrightarrow{SF}$ represents the resultant *force* acting on the body (i.e. the single force that would have the same effect upon the body as the several forces acting simultaneously). Hence, *all* vector quantities are added in the same manner, by representing them by vectors placed end-to-end ('nose-to-tail'), and applying the polygon law for the addition of vectors.

Rule: The polygon law for the addition of vectors

The 'sum' (i.e. 'net effect' or 'resultant') of a number of vectors is the vector that runs from the starting point to the finishing point when the vectors are placed end-to-end ('nose-to-tail').

The triangle and the parallelogram laws

When we have only *two* vectors to add, the polygon becomes a triangle.

If two vectors are placed end-to-end, forming two sides of a triangle, the resultant of these vectors is the third side of the triangle in the direction from the starting point to the finishing point.

We may draw the vectors from a common point, and the resultant is then the diagonal of the parallelogram from this point, where the two vectors are the adjacent sides of the parallelogram. (The shaded triangle is the same triangle as in the first diagram.)

21.4 ANALYTICAL ADDITION OF TWO VECTORS AT RIGHT ANGLES TO EACH OTHER

In this particular case the resultant of the vectors may be found analytically (i.e. by *computation*), using Pythagoras' theorem and right-angled triangle trigonometry.

Example

Find the resultant (R) of two vectors of magnitudes 5.00 and 7.00 units, given that their directions are at right angles to each other.

$R = \sqrt{5^2 + 7^2}$ $\tan \theta = \frac{5}{7}$

$\quad = \sqrt{74}$ ≈ 0.7143

$\quad \approx 8.6023$ $\therefore \ \theta \approx 35.54°$

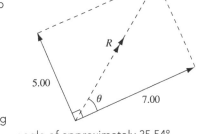

The resultant has magnitude 8.60 units, its direction being between the directions of the two given vectors, making an angle of approximately 35.54° with the larger of the two.

Exercises 21.1

1 A person walks 4.00 km due east, then 3.00 km due north. Find analytically the person's final position.

2 Two forces act at a point. The forces are 500 N acting vertically upwards and 1.20 kN acting horizontally. Find analytically the resultant force.

3 Four forces act at a point. The forces are 1.38 kN acting due north, 3.76 kN acting due east, 2.54 kN acting due south and 5.91 kN acting due west. Find analytically the resultant of these forces.

4 An aircraft has maximum airspeed 300 km h^{-1}. If this aircraft flies at maximum speed for 2 hours, its compass reading being 045°, while the air through which it is flying is moving at 50 km h^{-1} *from* the north-west, find analytically the displacement of the aircraft after this time. State distance correct to 3 significant figures and direction to nearest minute.

5 At the instant when a certain mass is released above the earth, two forces act upon it:

  i   gravitational attraction towards the earth (its 'weight'), which is a force of 49.0 N

  ii a horizontal force of 26.0 N due to the wind that is blowing.

continued

continued

   a Find analytically the direction in which the mass will initially move.

   b Given that the acceleration of a body equals the resultant force upon it divided by its mass, find the initial acceleration of the body, given that its mass is 5.0 kg.

6 While a railway car is moving at a speed of 9.3 m s^{-1}, a person inside it walks at a speed of 3.1 m s^{-1} in a direction perpendicular to the motion of the car. Find the person's velocity with respect to the ground.

21.5 RESOLUTION OF A VECTOR INTO TWO COMPONENTS AT RIGHT ANGLES

A vector quantity has no effect whatever in a direction at right angles to itself.

Example

1 If a body is given a displacement due north, it does not move any further east or west.

2 If a body has a velocity due west, it does not move any further north or south.

3 If a body is acted upon by a vertical force (e.g. gravitational attraction), this force produces no acceleration in a horizontal direction.

A force that is continuously applied to a moving body at right angles to the direction in which it is moving cannot change the speed of the body, i.e. cannot make it move any faster or slower. *However,* a vector quantity *does* have an effect in all directions that are *not* at right angles to that vector.

Any force acting upon a body in any direction *between* the vertical and the horizontal will have an effect in both the vertical and the horizontal directions. The trolley will both accelerate and press with a greater force on the ground.

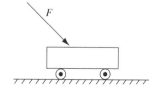

To find the effect of a force (i.e. its **component**) in any particular direction, we **resolve** (i.e. separate) the vector into two components, one in the required direction and the other at right angles to the required direction.

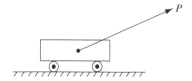
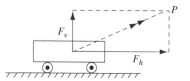

To find the horizontal (or vertical) component of the force P, we resolve it into the two components F_h and F_v in the horizontal and vertical directions respectively.

F_v has no effect in the horizontal direction, hence F_h is the total horizontal effect (the 'horizontal component') of force P.

F_h has no effect in the vertical direction, hence F_v is the total vertical effect (the 'vertical component') of force P.

> **Note:** If we were to separate the force P into two forces that are *not* at right angles, *neither* of these will represent the total force in that direction.

For example, if we separate P into the forces F_1 and F_2, F_1 does not represent the total force in the horizontal direction because F_2 *also* has an effect in the horizontal direction.

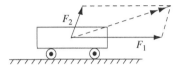

To find the component of a vector V in some other direction, we draw a **rectangle** with V as a diagonal and one of its sides in the required direction. The two sides of the rectangle will represent the components of the vector in these two directions.

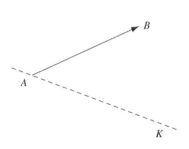

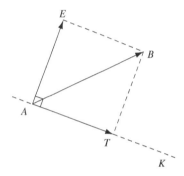

To find the component of vector $\vec{AB}$ in direction $\vec{AK}$, we construct the rectangle $AEBT$. $\vec{AT}$ is the component of $\vec{AB}$ in the required direction AK and AE is the component of $\vec{AB}$ in direction at right angles to AK.

It is usually quicker to construct the triangle ABT since the two components are then $\vec{AT}$ and $\vec{TB}$.

SUMMARY

- The component of vector $\vec{AB}$ in direction AK is $\vec{AT}$, where $BT \perp AK$.

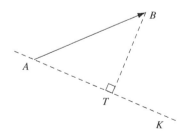

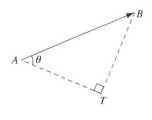

If the angle between the direction of the vector and the direction of the required component is θ, the magnitude of the component is the magnitude of the vector multiplied by $\cos\theta$ (since $AT = AB \times \cos\theta$).

SUMMARY

- The component (i.e. the effective part) of a vector $\vec{V}$ in some other direction is $V_m \cos \theta$, where V_m is the magnitude of $\vec{V}$ and θ is the angle between the two directions.
- The component in a direction at right angles to this is $V_m \sin \theta$.

Example

If F_v and F_h are the vertical and horizontal components of a vector, we have:

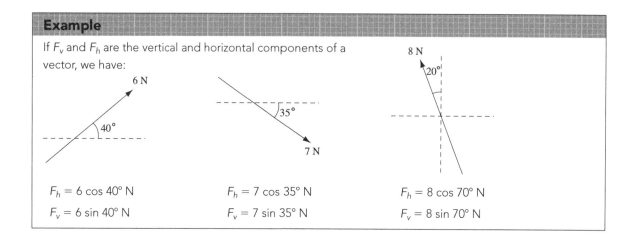

$F_h = 6 \cos 40° \text{ N}$

$F_v = 6 \sin 40° \text{ N}$

$F_h = 7 \cos 35° \text{ N}$

$F_v = 7 \sin 35° \text{ N}$

$F_h = 8 \cos 70° \text{ N}$

$F_v = 8 \sin 70° \text{ N}$

Note: In Appendix B there is a brief explanation of the meanings of the terms acceleration, force, mass and weight. If you do not understand the difference between the 'mass' and the 'weight' of a body and why a body of mass m has a weight of $m \times g$ where $g \approx 9.8 \text{ m/s}^2$, it is recommended that you study this appendix.

Exercises 21.2

1   A projectile is launched with a speed of 300 m s^{-1} in a direction making an angle of 70.0° with the horizontal. Find, correct to 3 significant figures:

  a its initial horizontal velocity component

  b its initial vertical velocity component

2   An object is moving at a constant speed of 20 km h^{-1}. How long will it take (in hours and minutes, correct to the nearest minute) to move a distance of 100 km further to the north, given that the direction of its motion is:

  a due north?   **b** 20° E of N?   **c** 40° E of N?

  d 60° E of N?   **e** 80° E of N?

3   An object is moving at a speed of 40.5 m s^{-1}. How long (in seconds correct to 3 significant figures) will it take to move 200 m further to the west given that the direction of its motion:

  a is 35.0° W of S?   **b** is N47.0° W?

  c has a bearing of 200°?   **d** has a bearing of 348°?

4   Each of the forces shown has a
    horizontal component. Find these
    three horizontal components and
    hence the net horizontal force acting
    upon the body.

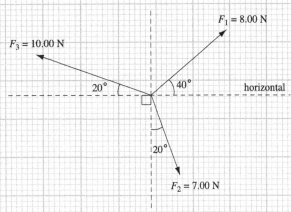

5   Find the net vertical force acting upon
    the body in question 4.

6

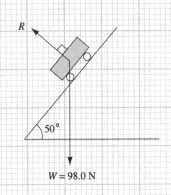

This diagram shows a trolley on a smooth inclined
plane. The only forces acting on this trolley are
(i) gravitational attraction (W, 98.0 N) and (ii) the force
with which the plane presses against the wheels (R).
Force R is at right angles to the plane and hence has
no component parallel to the plane. W, however, does
have a component parallel to the plane.

a   Find the force component acting on the trolley
    parallel to the plane, correct to 3 significant figures.

b   Given that the resultant force on a body in any
    particular direction equals the mass of the body
    multiplied by its acceleration in that direction, find
    the acceleration of this body down the plane, given
    that its mass is 10.0 kg.

7   While this body slides down the plane, the only forces
    acting upon it are (i) gravitational attraction (W, 242 N),
    (ii) a force (R) perpendicular to the plane, and (iii) a
    friction force parallel to the plane (F, 20 N).
    Find, correct to the nearest newton:

    a   the component of W parallel to the plane

    b   the total force acting on the body parallel to the plane

8   The force on an electric charge when moving through a magnetic field
    is given by the product of the size of the charge, the strength of the
    magnetic field and the velocity component of the charge in a direction perpendicular to the direction
    of the field. Find the force that acts upon an electron moving at a speed of 2.5×10^8 m s^{-1} through
    a magnetic field of strength 0.35 T if the angle between the direction of the motion and the direction
    of the field is:

    a   zero   b   30°   c   60°   d   90°

    (The charge on an electron = 1.6×10^{-19} C.)

21.6 RESULTANT OF A NUMBER OF VECTORS ANALYTICALLY

To find the resultant of a number of vectors analytically, we:

1   choose any two directions at right angles to one another (it is usually convenient to choose north and east for vectors in a horizontal plane, and vertical and horizontal for vectors in a vertical plane)
2   resolve each of the vectors into its components in each of these directions
3   add, to find the *total* of the components in each of these directions
4   find the resultant of these two total components.

Example

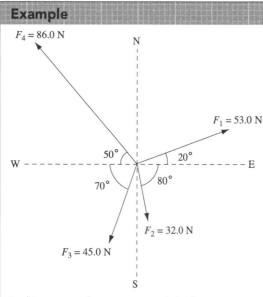

	→ East	↑ North
F_1	53 cos 20°	53 sin 20°
F_2	32 cos 80°	−32 sin 80°
F_3	−45 cos 70°	−45 sin 70°
F_4	−86 cos 50°	86 sin 50°
	−15.3102	10.2068

In this case we have computed the force components in directions east and north. Note that components to the west and south are *negative*.

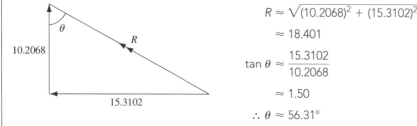

$$R \approx \sqrt{(10.2068)^2 + (15.3102)^2}$$

$$\approx 18.401$$

$$\tan \theta \approx \frac{15.3102}{10.2068}$$

$$\approx 1.50$$

$$\therefore \theta \approx 56.31°$$

The resultant is a force of 18.4 N in direction N56.3°W.

Note: ■ The working should be set out clearly, preferably tabulated as shown in the above example.
■ Results should be stated only to the degree of accuracy justified by the data. In the example above the forces and angles are given to 3 significant figures, hence the results should be expressed in the same manner.
■ For convenience it is advisable to use angles all measured from the one line (in the above example, from the east-west line). This means that all the components in one direction will be calculated using *cosines* and all the components in the direction at right angles to this will be calculated using *sines* of the same angles.

Example

An aircraft departs from point *A* and files 234 km on bearing 73.0°, then 317 km on bearing 146°, then 521 km on bearing 248°, then 133 km on bearing 305°. Find the final position of the aircraft.

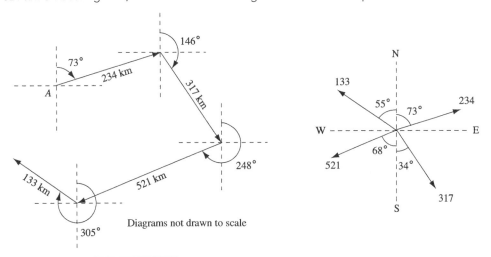

Diagrams not drawn to scale

North ↑	East →
234 cos 73°	234 sin 73°
−317 cos 34°	317 sin 34°
−521 cos 68°	−521 sin 68°
133 cos 55°	−133 sin 55°
−313.2743	−190.970 54

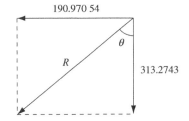

$$R \approx \sqrt{(313.2743)^2 + (190.970\ 54)^2}$$

$$\approx \sqrt{134\ 611}$$

$$\approx 367\ km$$

$$\tan \theta \approx \frac{190.970\ 54}{313.2743}$$

$$\approx 0.6096$$

$$\therefore \theta \approx 31.4°$$

The final position is approximately 367 km from *A* on a bearing of 211°.

21.7 ADDITION OF VECTORS GRAPHICALLY

Vectors may be added by representing them by directed line segments drawn to a suitable scale added nose-to-tail and finding the resultant by measurement. In the aircraft example (p. 349) a suitable scale would be to use 1 mm to represent 5 km. It is recommended that you do draw this diagram as accurately as possible. Your result should be within 1% of the correct result obtained analytically. The same can be done with the previous example on page 348 in which four forces were added analytically. In the diagram on page 348 the forces are shown all acting at a point but in order to add them graphically they would be drawn nose-to-tail in any desired order using a suitable scale.

Note: If more practice is required at the graphical addition of vectors, the approximate answers to Exercises 21.3 can be found by this method.

Exercises 21.3

1 Find the resultant of the forces shown in the diagram.
2 An aircraft departs from airport *P* and flies 123 km on bearing 286°, then 254 km on bearing 166°, then 318 km on bearing 28.0°, then 143 km on bearing 172°. Find the final position of the aircraft.
3 Three tugs exert forces on the bow of a ship. These forces are 153 kN, 174 kN and 197 kN in directions N10°E, N35°W and N40°E respectively. Find the resultant force on the ship.

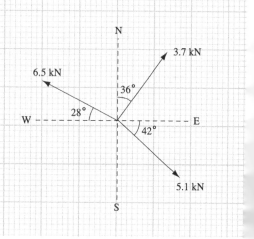

If we are not given the specific directions of the vectors to be added analytically, but only the angles *between* these directions (i.e. their *relative* directions), it is convenient to resolve each vector into two components at right angles, one of the directions chosen being the direction of one of the vectors.

Examples

1 Two forces, of 8.0 kN and 5.0 kN, act upon a body simultaneously, the angle between their directions being 53.0°. Find the resultant force.

Method A

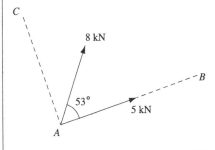

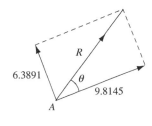

Force in direction $AB = 5 + 8 \cos 53°$

≈ 9.8145 kN

Force in direction $AC = 8 \sin 53°$

≈ 6.3891 kN

$R \approx \sqrt{(9.8145)^2 + (6.3891)^2}$

$\approx \sqrt{137.145}$

≈ 11.7 kN

$\tan \theta \approx \dfrac{6.3891}{9.8145}$

≈ 0.651

$\therefore \theta \approx 33.1°$

Method B

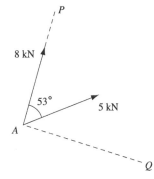

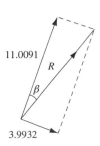

Force in direction $AP = 8 + 5 \cos 53°$

≈ 11.0091 kN

Force in direction $AQ = 5 \sin 53°$

≈ 3.9932 kN

$R \approx \sqrt{(11.0091)^2 + (3.9932)^2}$

$\approx \sqrt{137.146}$

≈ 11.7 kN

$\tan \beta \approx \dfrac{3.9932}{11.0091}$

≈ 0.3627

$\therefore \beta \approx 19.9°$

continued

continued

The resultant is a force of approximately 11.7 kN in a direction between that of the two given forces, making an angle of 33.1° with the 5 kN force (i.e. making an angle of 19.9° with the 8 kN force).

2

186 N

324 N

110°

150°

186 N

261 N

Find the resultant of the three forces shown in the diagram.

70°

30°

324 N

261 N

44.2829

R

θ

34.3515

→	↑
324	–
− 186 cos 70°	186 sin 70°
− 261 cos 30°	− 261 sin 30°
34.3516	44.2828

$$R \approx \sqrt{(44.2828)^2 + (34.3516)^2}$$
$$\approx \sqrt{3141}$$
$$\approx 56.0446$$

$$\tan \theta \approx \frac{44.2828}{34.3516}$$
$$\approx 1.2891$$
$$\theta \approx 52.2°$$

The resultant is a force of 56.0 N in direction between the 324 N force and the 186 N force making an angle of 52.2° with the 324 N force.

Exercises 21.3 *continued*

4 In each of these cases, find the resultant of the two given forces:

a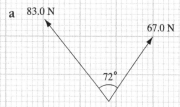
83.0 N

67.0 N

72°

b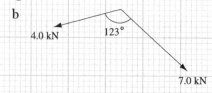

4.0 kN   123°

7.0 kN

+++ 5 A ship sails a distance of 13.0 km on a constant bearing. It then turns to starboard through an angle of 38.0° and travels on this new bearing for 7.0 km. It then turns to starboard through an angle of 27.0° and travels on this bearing for 9.0 km. Find the distance of the ship from its departure point.

21.8 EQUILIBRIUM

If a number of forces has a resultant of zero, we say that these forces are 'in **equilibrium**'. It follows that for forces in equilibrium, the total of their components (effective parts) in *any* direction must be zero. Any body that has no acceleration must have zero resultant force acting upon it and therefore the forces upon it must be in equilibrium.

Examples

1   Given that these forces are in equilibrium, find the magnitude of force *F*.

Method A

Find the resultant of the two given forces.

North	West
2.6 cos 25°	−2.6 sin 25°
−3.2 cos 70°	3.2 sin 70°
1.2619	1.9082

$$R \approx \sqrt{(1.2619)^2 + (1.9082)^2}$$

$$\approx \sqrt{5.2336}$$

$$\approx 2.2877$$

F must balance this resultant.

$$\therefore F \approx 2.2877 \text{ kN} \approx 2.3 \text{ kN}$$

Method B

North	East
2.6 cos 25°	2.6 sin 25°
−3.2 cos 70°	−3.2 sin 70°
−F cos θ	F sin θ
−F cos θ + 1.2619	F sin θ −1.9082

continued

continued

Since the total component in *any* direction must be zero:

$$F \cos \theta \approx 1.2619$$

$$F \sin \theta \approx 1.9082$$

$$\therefore F^2 (\sin^2 \theta + \cos^2 \theta) \approx (1.2619)^2 + (1.9082)^2$$

$$\therefore F^2 \approx 5.2336$$

$$\therefore F \approx 2.2877 \text{ kN} \approx 2.3 \text{ kN}$$

2

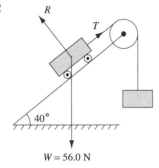

A body rests upon a smooth inclined plane. Given that the three forces shown are in equilibrium, find the magnitudes of the forces T and R. (T is the tension in the string and R is the force that the plane exerts on the wheels of the trolley.)

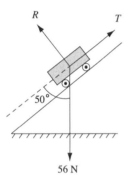

$T = 56 \cos 50°$ (resultant force parallel to plane = zero)

$\approx 36.0 \text{ N}$

$R = 56 \sin 50°$ (resultant force perpendicular to plane = zero)

$\approx 42.9 \text{ N}$

3

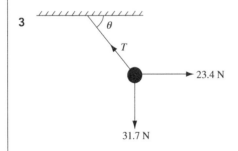

A mass hanging from a cable is drawn to one side and held in equilibrium by a horizontal force as shown. Find the magnitudes of T (the tension in the cable) and the angle θ.

$T \cos \theta = 23.4$ (resultant horizontal force = zero)

$T \sin \theta = 31.7$ (resultant vertical force = zero)

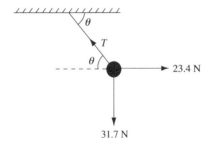

$$\frac{T \sin \theta}{T \cos \theta} = \frac{31.7}{23.4}$$

$$\therefore \tan \theta \approx 1.3547$$

$$\therefore \theta \approx 53.6°$$

$$T = \frac{23.4}{\cos \theta} \left(\text{or } T = \frac{31.7}{\sin \theta} \approx 39.4 \text{ N} \right)$$

$$\approx 39.4 \text{ N}$$

Exercises 21.4

1 In each of the diagrams below, given that the forces are in equilibrium, find the magnitudes of
   the forces and angles that are represented by pronumerals.

a
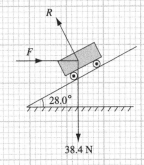

b
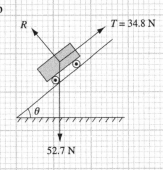

c
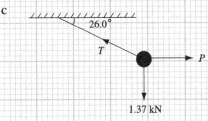

d
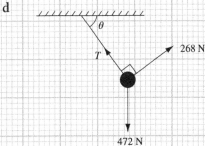

2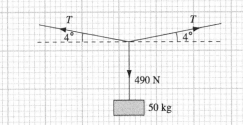

A mass of 50 kg is supported by two cables, both
of which make an angle of 4° with the horizontal.
Find:

a   the tension in the cables

b   the ratio of this tension to the load supported

+++ 3 A boom is used to support a mass of 400 kg.
      The force F must be such that the resultant
      of the forces T and F is directed along
      the boom, which is inclined at 70° to the
      horizontal. Find the size of force F.

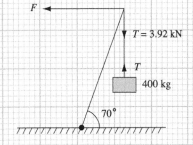

SELF-TEST

1 t seconds after a certain projectile is launched from the ground, it has a vertical upwards velocity component of $30.0 - 9.80t$ m s^{-1}. Given that it has a constant horizontal velocity component of 20.0 m s^{-1}, find its velocity (i.e. its speed and direction of motion):

   a 2 seconds after launching

   b 5 seconds after launching

   (State directions as angles made with the horizontal.)

2 Find the resultant of the four forces shown, expressing its direction in relation to the direction of force F_1.

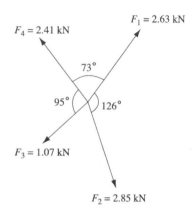

3

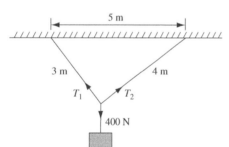

Find the tensions in the cables (T_1 and T_2).

+++ 4 P, Q, R and S are four landmarks. Q is 30.4 km from P on a bearing of 223°; R is 20.6 km from Q on a bearing of 112°; S is 41.0 km from R on a bearing of 335°. Find the bearing of S from P, to the nearest degree.

CHAPTER 22

ROTATIONAL EQUILIBRIUM AND FRAME ANALYSIS

Learning objectives

■ Solve problems involving moments of forces, including the analysis of the forces in simple frames.

22.1 A FEW REMINDERS

1   A mass of m kg experiences a weight force of mg N where $g \approx 9.81$ m/s^2.
2   The mass of a body may be considered to be concentrated at a point called the 'centre of mass' or the 'centre of gravity'. For a uniform symmetrical body this is the midpoint of the body. When considering the equilibrium of a body, its weight may be regarded as a vertical downwards force acting from the centre of gravity.
3

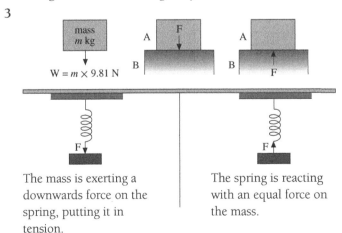

The mass is exerting a downwards force on the spring, putting it in tension.

The spring is reacting with an equal force on the mass.

When one body A exerts a force on another body B, body B always exerts an equal force on body A in the opposite direction, called the *reaction* force. This is Newton's third law of motion.

In the diagram above, body A exerts force F on the spring, stretching it and putting it in tension. The spring reacts with force F on the body, supporting its weight.

The tyres on an accelerating car exert a backward friction force on the road surface. The equal reaction force of the road on the tyres pushes the car forward.

4   The effect of a force in a direction making an angle θ with the direction of the force is called the *component* of the force in this other direction

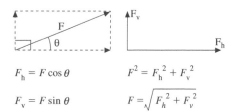

$F_h = F \cos \theta$

$F_v = F \sin \theta$

$F^2 = F_h^2 + F_v^2$

$F = \sqrt{F_h^2 + F_v^2}$

A force may be 'resolved' into two components at right angles to each other. It is usually convenient for these two directions to be the horizontal and the vertical. The force may be replaced by these two components.

5   A force is specified by stating both its size and its direction.

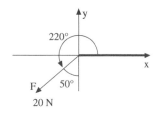

When specifying the direction in which a force acts, it is usual practice to state the angle between the direction of the force and the positive direction of the x-axis in an anticlockwise direction.

The force illustrated adjacent is a force of 20 N at 220°.

6   When performing calculations, always work with quantities expressed with more significant figures than are required in the final result.

22.2 THE MOMENT OF A FORCE ABOUT A POINT

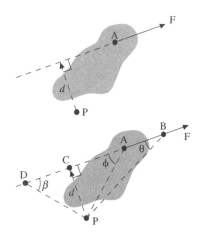

Using the principle of the conservation of energy, it can be shown that the *moment* (turning effect) of a force about a point is given by the product of the magnitude of the force and the perpendicular distance from the point to the line of action of the force. $\mathbf{M_P = F \times d.}$

Since $d = $ PB $\sin \theta = $ PA $\sin \phi = $ PD $\sin \beta$, it follows that the moment M_P is given by the product of the distance from P to *any* point on the line of action of the force and the component of the force perpendicular to the line joining P to the line of action.

A moment of a force is regarded as positive if it acts in a clockwise direction and negative if in an anticlockwise direction. A body will be in *rotational equilibrium* about a point if the sum of the moments of the forces acting upon it is zero, i.e. if the sum of the clockwise moments equals the sum of the anticlockwise moments.

Equilibrium

■   If there is a resultant force acting upon a body it will move (and accelerate) in that direction. If there is no resultant force acting upon a body it is said to be in 'translational equilibrium'.

■   If the forces acting upon a body have a resultant moment about some point (inside or outside the body), the body will rotate about that point (in a clockwise or anticlockwise direction).

- If a body is in translational equilibrium and there is a point about which the resultant forces exert no resultant moment, then they will exert no resultant moment about any other point (inside or outside the body). The body is said to be in 'rotational equilibrium'.
- We are concerned in this chapter only with bodies that are stationary. These bodies have no resultant force acting upon them and no resultant moment about any point (inside or outside the body). These bodies are in translational and rotational equilibrium.

Example

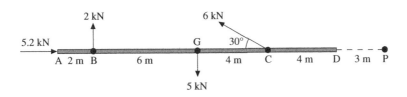

With clockwise moments regarded as positive and anticlockwise moments as negative:

Taking moments about point A: $M_A = (5 \times 8) - (2 \times 2) - (3 \times 12) = 0$.

So this body is in translational equilibrium ($F\uparrow = F\downarrow = 5$ kN and $\overrightarrow{F} = \overleftarrow{F} = 5.2\,kN$), and the forces are so arranged that there is no resultant moment about the point A. Hence the resultant moment about *any other point* will be zero.

$M_B = (5 \times 6) - (6 \sin 30° \times 10) = 0$ $M_D = (2 \times 14) - (6 \sin 30° \times 4) - (5 \times 8) = 0$
$M_P = (2 \times 17) - (5 \times 11) + (6 \sin 30° \times 7) = 0$ $M_C = (2 \times 10) - (5 \times 4) = 0$

Exercises 22.1

All systems in these exercises are in equilibrium.

All lengths are in metres.

State all results of forces correct to 3 significant figures and angles correct to 0.1°.

1

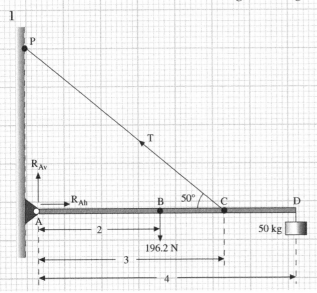

A beam of mass 20 kg (hence weight = 20 × 9.81 N), is held horizontal by a cable attached between a vertical wall and point C on the beam, as illustrated. The beam is hinged at the wall and a load of mass 50 kg is suspended from the other end. At the hinge there will be a horizontal reaction force R_{Ah} (resisting the pull towards the wall) and a vertical reaction force R_{Av} which we will assume to be directed upwards.

continued

continued

   a without performing any multiplications write down the equations expressing the following facts:

     i   the vertical forces on the beam are in equilibrium, $(F\uparrow = F\downarrow)$

     ii the horizontal forces on the beam are in equilibrium, $(\overrightarrow{F} = \overleftarrow{F})$

     iii the sum of the clockwise moments about A = the sum of the anticlockwise moments

     iv the sum of the clockwise moments about C = the sum of the anticlockwise moments

     v   the sum of the clockwise moments about D = the sum of the anticlockwise moments.

   b Use equation (iv) to evaluate R_{Av}.

   c Use equation (iii) to evaluate T, the tension in the cable.

   d Use equation (ii) to evaluate R_{Ah}.

   e Use equation (i) to check your results.

   f Use equation (v) to check your results.

   g Evaluate R_A, the resultant reaction force at the hinge.

   h Evaluate the reaction force at the pin P on the wall.

2

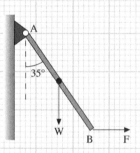

A beam hinged to a wall is held at an angle of 35° to the vertical by a horizontal force F as illustrated. Given that the length of the beam AB is 6.00 m and the mass of the beam is 20.0 kg, evaluate:

a the force F

b the reaction at the hinge.

3

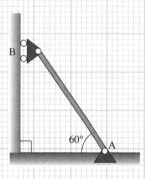

A uniform beam AB of length 5.00 m and mass 83.0 kg has a roller at end B where it leans against a wall and is hinged at the ground. Determine:

a the reaction force by the wall at B

b the reaction force by the ground at A.

4

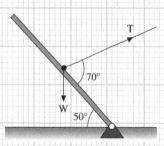

A beam is hinged at the ground and is held at an angle of 50° to the horizontal by a cable attached to its midpoint and making an angle of 70° with the beam as illustrated. Given that the beam has a length of 6.00 m and a mass of 125 kg, determine:

a the tension in the cable

b the reaction at the ground.

5

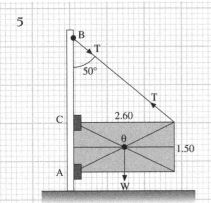

The illustration shows a gate of length 2.60 m, height 1.50 m and having a mass 50.0 kg. It is attached to a vertical post by hinges at A and C and with a cable attached between the post and the gate. The cable is tensioned so that it and the lower hinge support the whole weight of the gate, thus removing any force at the hinge C. Determine:

a the tension *T* in the cable

b the reaction force at the lower hinge.

+++ 6

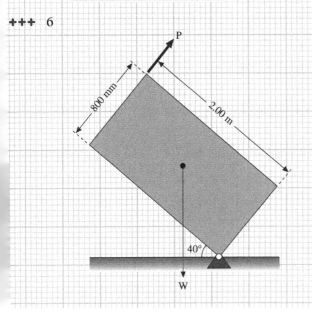

A rectangular block of wood of mass 380 kg is supported by a hinge and held at an angle of 40° to the horizontal by a force *P* as illustrated. Determine:

a the magnitude of the force *P*

b the reaction at the hinge.

22.3 AN INTRODUCTION TO THE ANALYSIS OF FRAMES

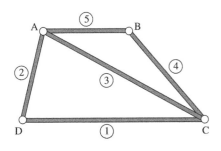

A frame consists of 'members' ('elements') connected by 'joints' at points called 'nodes'. This frame has 5 members and 4 joints (A, B, C, D).

Joints: All joints are hinged ('pinned'). The members may for example be connected by a single bolt or a single rivet.

Supports: We will consider only frames that are supported by (a) a hinge, or (b) a roller.

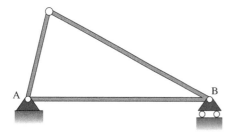

A **hinged support** is shown at joint A. This prevents translational movement in any direction, allowing only rotation.

A **roller support** is shown at B. This allows translational movement.

The hinge-roller combination shown in the diagram is commonly used.

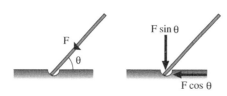

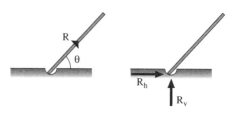

It may be helpful to think of a hinged support as a cavity in a solid immovable base. This may make it easier to understand the reaction forces R_h and R_v.

The reaction force R = the applied force F, but acts in the opposite direction. The component reaction forces R_h and R_v are the horizontal and vertical components of the applied force, $F \cos \theta$ and $F \sin \theta$ respectively, but acting in the opposite directions.

The forces F place this member in **compression**. The member has been 'squeezed' and may be considered as 'trying' to **expand** as shown by the reaction forces R.

The forces F place this member in **tension.** The member has been 'stretched' and may be considered as 'trying' to **contract,** as shown by the reaction forces R.

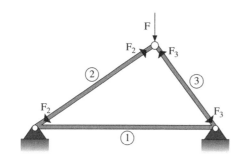

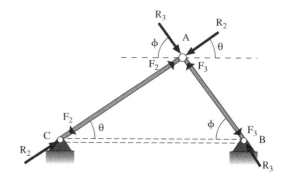

Force F puts members ② and ③ in compression. The forces in these members, F_2 and F_3, are the reaction forces in these members which are 'trying' to expand. Since the hinge supports are immovable, the forces F_2 and F_3 have no effect on member ①, so it will not experience any force and could be regarded as not being present; $F_1 = 0$. At A, the force R_2 is the reaction force to F_2; it is equal in magnitude to F_2 but acts in the opposite direction. (Similarly for the reaction force at B, $R_3 = F_3$).

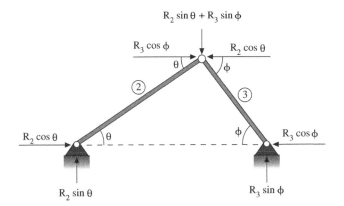

The resultants of both the horizontal and the vertical force components on the frame are zero. The frame is in translational equilibrium. R_2 is the resultant of the horizontal and the vertical reaction forces at A:

$$R_2 = \sqrt{(R_2 \sin \theta)^2 + (R_2 \cos \theta)^2}$$

(Similarly for R_3).

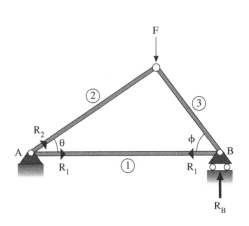

The support for this frame, being a hinge-roller combination, would not be stable without the presence of member ① and the frame would collapse. With member ① present we have a stable construction in equilibrium. There is no resultant force on the frame as a whole or on any section of it and there is no resultant moment when taken about any point, inside or outside of the frame. Force F_3 will have pushed joint B to the right until the tension produced in member ① equals the horizontal force on it, $F_3 \cos \phi$. The changes in the angles will be negligible.

Since there is no horizontal resultant force on joint B there will be no horizontal reaction. The reaction force $R_B = F_3 \sin \phi$ vertically upwards. The reaction force at a roller support is always perpendicular to the surface.

$F_1 = F_3 \cos \phi$

$R_1 = F_1$ (in opposite direction)

$R_2 = F_2$ (in opposite direction)

The reaction at A, R_A, is the resultant of R_1 and R_2. At B, R_1 balances the horizontal component of F_3. $R_B = F_3 \sin \phi$, vertically upwards.

Note: If any frame is scaled up or down in size, the forces are not affected; they depend only upon the *shape* of the frame, i.e. on the angles and not on the size of the frame. If a frame is enlarged or diminished in size all the triangles of the new frame are similar to the original triangles and the forces are unaffected. In the following example no lengths are given in **method B** so a length of 10 units is arbitrarily assigned to member AB. The length assigned could be x units but an actual value makes the calculations simpler.

Example

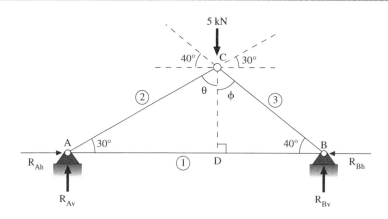

Method A

F_2 would equal 5 cos $\theta°$ and F_3 would equal 5 cos $\phi°$ *only if* F_2 and F_3 were at right angles to each other. In this case, $\theta = 60°$ and $\phi = 50°$

At C the resultant of F_2 and F_3 must be 5 kN↑, so we can solve the parallelogram of forces to determine their values:

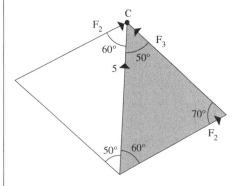

$$\frac{F_2}{\sin 50°} = \frac{5}{\sin 70°}$$

$$F_2 = \frac{5 \sin 50°}{\sin 70°}$$

$$= 4.08 \text{ kN}$$

$$\frac{F_3}{\sin 60°} = \frac{5}{\sin 70°}$$

$$F_3 = \frac{5 \sin 60°}{\sin 70°}$$

$$= 4.61 \text{ kN}$$

$F_2 = 4.08$ kN at 30° (compression) $R_{Av} = F_2 \sin 30° = 2.04$ kN $R_{Ah} = F_2 \cos 30° = 2.53$ kN
$F_3 = 4.61$ kN at 140° (compression) $R_{Bv} = F_3 \sin 40° = 2.96$ kN $R_{Bh} = F_3 \cos 40° = 3.53$ kN

Method B

Forces acting on a frame = forces supporting the frame
On the frame:

$F↑ = F↓: \therefore R_{Av} + R_{Bv} = 5$ (1)

$\overrightarrow{F} = \overleftarrow{F}: \therefore R_{Ah} - R_{Bh} = 0$ (2)

$\therefore F_2 \sin 30° + F_3 \sin 40° = 5$ (1)

$F_2 \cos 30° + F_3 \cos 40° = 0$ (2)

Solving these simultaneous equations gives F_1, F_2 and the reactions, as in method A

Method C

First some trigonometry. Let AB = 10 m (see the note before the previous example):

$\triangle$ ABC: $\dfrac{10}{\sin 110°} = \dfrac{AC}{\sin 40°}$

$\therefore$ AC = 6.840

M_A: $(R_{Bv} \times 10) - (5 \times AD) = 0$

$\triangle$ ACD: AD = AC cos 30°

= 5.924

$\therefore$ BD = 4.076

$R_{Bv} = \dfrac{5 \times 5.924}{10}$

= 2.96 kN

CD = AC sin 30°

= 3.420

$\therefore R_{Av} = 5 - 2.962$

= 2.04 kN

At A:

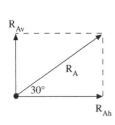

$R_{Av} = R_A \sin 30°$

= 2.04 kN

$R_A = \dfrac{2.04}{\sin 30°}$

= 4.08 kN

$R_{Ah} = R_A \cos 30°$

= 3.53 kN

At B:

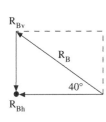

$R_{Bv} = R_B \sin 40°$

= 2.962 kN

$R_B = \dfrac{2.962}{\sin 40°}$

= 4.61 kN

$R_{Bh} = R_B \cos 40°$

= 3.53 kN

Hence the results as shown on the previous page.

If the fixed support at B is replaced by a roller, joint B will be pushed to the right (a very small distance), stretching member AB until the tension produced in it equals the horizontal component of F_3 which is pushing it. There will then be no horizontal reaction at support B. Hence all the above results will be the same except that now $F_1 = F_3 \cos 40° = 3.53$ kN and $R_B = 2.04$ kN at 90°.

Note: To solve this particular example, method A is the simplest and quickest to use, but for many other frames it is necessary or desirable to take moments. If you are familiar with all three methods above, you will have all the tools you need to analyse most simple frames.

Exercises 22.2

1

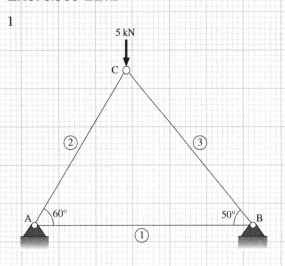

Study method A in the example in section 22.3 and then, without further reference to this, use this method to determine the reaction forces and the force in each member of the frame shown in the adjacent diagram. Then repeat using methods B and C. As in the example, state any changes to the forces that would occur if the fixed support at B were replaced by a roller.

2

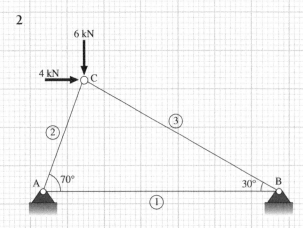

For the frame shown in the adjacent diagram, first determine the reactions at A and B and then the force in each of the members.

+++ 3 For each of the frames shown in the diagrams below determine the reaction at the supports, the tension in the cable CD and the force in each of the members.

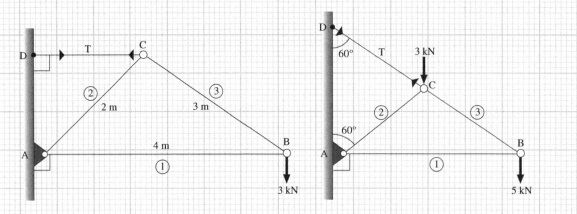

DETERMINANTS AND MATRICES

Learning objectives

- Evaluate determinants.
- Solve simultaneous equations using determinants.
- Multiply matrices.
- Understand the algebra of matrices.
- Calculate the inverse of a matrix.
- Express simultaneous equations in matrix form.
- Solve simultaneous linear equations using matrices.

There is another method for solving simultaneous linear equations that is applicable to *all* such equations and often saves a lot of time. Once learnt, this method is very concise, very simple to use and provides less opportunity for careless error.

However (there is always a catch, of course), there are a few facts that you will have to learn first.

23.1 DEFINITION AND EVALUATION OF A 2 × 2 DETERMINANT

$\begin{vmatrix} a & c \\ b & d \end{vmatrix}$ is called a **two-by-two determinant.** It is an array of numbers having two rows and two columns enclosed between vertical bars. It is shorthand notation for $ad - bc$.

Note: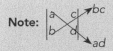

Examples

1 $\begin{vmatrix} 2 & 3 \\ 4 & 5 \end{vmatrix} = (2 \times 5) - (4 \times 3) = -2$ **2** $\begin{vmatrix} -2 & 3 \\ -4 & -5 \end{vmatrix} = (-2 \times -5) - (-4 \times 3) = 10 + 12 = 22$

Exercises 23.1

Evaluate the following 2 × 2 determinants:

1 $\begin{vmatrix} 3 & 4 \\ 2 & 5 \end{vmatrix}$ **2** $\begin{vmatrix} 2 & 0 \\ 6 & 1 \end{vmatrix}$

continued

continued

3   $\begin{vmatrix} 2 & 6 \\ 5 & 3 \end{vmatrix}$

4   $\begin{vmatrix} 3 & -2 \\ 5 & 4 \end{vmatrix}$

5   $\begin{vmatrix} 3 & -1 \\ -4 & -2 \end{vmatrix}$

6   $\begin{vmatrix} 6.72 & 3.91 \\ 4.84 & 5.06 \end{vmatrix}$

23.2 SOLUTION OF SIMULTANEOUS EQUATIONS USING 2 × 2 DETERMINANTS

If we have two simultaneous equations, for example:

$$\begin{cases} 2x + 3y = 4 \\ 5x + 6y = 7 \end{cases}$$

$\triangle$ is the determinant we obtain from the coefficients on the left-hand sides, in the same order and arrangement as they appear in the equations.

In this case, $\triangle = \begin{vmatrix} 2 & 3 \\ 5 & 6 \end{vmatrix} = 12 - 15 = -3$

$\triangle_x$ is the same determinant as above, except that the x-coefficients are replaced by the right-hand numbers of the equations.

In this case, $\triangle_x = \begin{vmatrix} 4 & 3 \\ 7 & 6 \end{vmatrix} = 24 - 21 = 3$

$\triangle_y$ is the same determinant as $\triangle$ but with the y-coefficients replaced by the right-hand numbers of the equations.

In this case, $\triangle_y = \begin{vmatrix} 2 & 4 \\ 5 & 7 \end{vmatrix} = 14 - 20 = -6$

Examples

1   If $\begin{cases} 3x + 5y = 2 \\ 6x + 4y = 7 \end{cases}$

then $\triangle = \begin{vmatrix} 3 & 5 \\ 6 & 4 \end{vmatrix}$ $\triangle_x = \begin{vmatrix} 2 & 5 \\ 7 & 4 \end{vmatrix}$ $\triangle_y = \begin{vmatrix} 3 & 2 \\ 6 & 7 \end{vmatrix}$

$= 12 - 30$ $= 8 - 35$ $= 21 - 12$

$= -18$ $= -27$ $= 9$

2   If $\begin{cases} 4x - 3y = -5 \\ -2x - y = 6 \end{cases}$

then $\triangle = \begin{vmatrix} 4 & -3 \\ -2 & -1 \end{vmatrix}$ $\triangle_x = \begin{vmatrix} -5 & -3 \\ 6 & -1 \end{vmatrix}$ $\triangle_y = \begin{vmatrix} 4 & -5 \\ -2 & 6 \end{vmatrix}$

$= (-4) - (6)$ $= (5) - (-18)$ $= 24 - 10$

$= -4 - 6$ $= 5 + 18$ $= 14$

$= -10$ $= 23$

Exercises 23.2

1 Given that $\begin{cases} 4x + 3y = 1 \\ x + 2y = 5 \end{cases}$

   evaluate: **a** $\triangle$ **b** $\triangle_x$ **c** $\triangle_y$

2 Given that $\begin{cases} m - 5t = 3 \\ -2m + 4t = -1 \end{cases}$

   evaluate: **a** $\triangle$ **b** $\triangle_m$ **c** $\triangle_t$

3 Given that

   $\begin{cases} 3.6V + 2.8e = 1.8 \\ -1.2V - 5.7e = -3.3 \end{cases}$

   evaluate: **a** $\triangle$ **b** $\triangle_v$ **c** $\triangle_e$

Solution of simultaneous equations

Consider any two simultaneous linear equations:

$$\begin{cases} a_1x + b_1y = c_1 \quad \underline{\hspace{2cm}} \quad ① \\ a_2x + b_2y = c_2 \quad \underline{\hspace{2cm}} \quad ② \end{cases}$$

$$\therefore \quad a_1b_2x - b_1b_2y = b_2c_1 \quad \underline{\hspace{2cm}} \quad ① \times b_2$$
$$a_2b_1x + b_1b_2y = b_1c_2 \quad \underline{\hspace{2cm}} \quad ② \times b_1$$

Subtracting: $a_1b_2x - a_2b_1x = b_2c_1 - b_1c_2$

$$\therefore x(a_1b_2 - a_2b_1) = b_2c_1 - b_1c_2$$

$$\therefore x = \frac{b_2c_1 - b_1c_2}{a_1b_2 - a_2b_1}$$

$$= \frac{\begin{vmatrix} c_1 & b_1 \\ c_2 & b_2 \end{vmatrix}}{\begin{vmatrix} a_1 & b_1 \\ a_2 & b_2 \end{vmatrix}}$$

$$= \frac{\triangle_x}{\triangle}$$

Similarly, it may be shown that $y = \dfrac{\triangle_y}{\triangle}$.

SUMMARY

The solution of two simultaneous linear equations:

$$\begin{cases} a_1x + b_1y = c_1 \\ a_2x + b_2y = c_2 \end{cases}$$

is: $x = \dfrac{\triangle_x}{\triangle}, y = \dfrac{\triangle_y}{\triangle}$

This is known as 'Cramer's Rule'. To apply this rule, both equations must be expressed with the constants c_1 and c_2 on the right-hand sides.

Example

Solve:

$$\begin{cases} 23.5I_2 + 16.7I_1 - 58.7 = 0 \\ 81.2I_1 - 34.2I_2 + 13.9 = 0 \end{cases}$$

We first rearrange the equations: $\begin{cases} 16.7I_1 + 23.5I_2 = 58.7 \\ 81.2I_1 - 34.2I_2 = -13.9 \end{cases}$

$$\triangle = \begin{vmatrix} 16.7 & 23.5 \\ 81.2 & -34.2 \end{vmatrix} = -571.14 - 1908.2 = -2479.34$$

$$\triangle_{I_1} = \begin{vmatrix} 58.7 & 23.5 \\ -13.9 & -34.2 \end{vmatrix} = -2007.54 + 326.65 = -1680.89$$

$$\triangle_{I_2} = \begin{vmatrix} 16.7 & 58.7 \\ 81.2 & -13.9 \end{vmatrix} = -232.13 - 4766.44 = -4998.57$$

$$I_1 = \frac{\triangle_{I_1}}{\triangle} = \frac{-1680.89}{-2479.34} \approx 0.678, \quad I_2 = \frac{\triangle_{I_2}}{\triangle} = \frac{-4998.57}{-2479.34} \approx 2.02$$

Answer: $I_1 \approx 0.678$, $I_2 \approx 2.02$

Note the advantage of this method. Applying the substitution or elimination method would be very tedious for such equations.

Exercises 23.2 *continued*

Solve the following simultaneous equations correct to 3 significant figures:

4 $\begin{cases} 7E + 19V = 24 \\ 13E - 5V = 9 \end{cases}$ 5 $\begin{cases} 13L - 15W = 87 \\ 17L - 11W = 231 \end{cases}$

6 $\begin{cases} 2.70x - 1.40y = 3.40 \\ 3.80x + 4.60y = 1.30 \end{cases}$ 7 $\begin{cases} 29.3I_1 - 37.8I_2 = 83.7 \\ 41.2I_1 - 26.4I_2 = -16.3 \end{cases}$

23.3 MATRICES: INTRODUCTION

Matrices can be used, among other purposes, to solve simultaneous equations, provided we define their operations (e.g. addition and multiplication) in special ways. The interest in and use of matrices has increased greatly since the introduction of computers because their operations are easy to program on a computer. Manually, for example, it is usually quicker to solve simultaneous equations using determinants. However, with a computer, matrices are much easier to use, regardless of how many variables are involved.

Definitions

A *matrix* is a set of numbers (called *elements*) arranged in a rectangular pattern (or *array*) of rows and columns. A *determinant,* as we saw in section 23.1, is this kind of array distinguished by a vertical bar at each side. To distinguish a *matrix,* the array is enclosed in *parentheses,* either round or square.

For example, $\begin{vmatrix} 2 & 1 \\ 3 & 4 \end{vmatrix}$ is a *determinant*, which has the value 5, but $\begin{pmatrix} 2 & 1 \\ 3 & 4 \end{pmatrix}$ or $\begin{bmatrix} 2 & 1 \\ 3 & 4 \end{bmatrix}$ is a *matrix,* which does not have a value. **A matrix does not represent a number.**

If a matrix has *a* rows and *b* columns, it is said to be an '*a* × *b* matrix' or to 'have an *order* of *a* × *b*'. A matrix of order '*a* × 1' is called a *column matrix*. A matrix of order '1 × *b*' is called a *row matrix*.

Examples

1 $A = \begin{pmatrix} 3 & 0 \\ -2 & 4 \\ 5 & 3 \end{pmatrix}$ is a matrix of order 3 × 2.

2 $D = \begin{pmatrix} 5 \\ 3 \\ -1 \\ 2 \end{pmatrix}$ is a matrix of order 4 × 1, a *column matrix.*

3 $K = \begin{pmatrix} 5 & -3 & 2 \end{pmatrix}$ is a matrix of order 1 × 3, a *row* matrix.

4 $P = \begin{pmatrix} 4 & -2 \\ 0 & 3 \end{pmatrix}$ is a 2 × 2 matrix, a *square* matrix of order 2.

5 $T = \begin{pmatrix} -1 & 0 & 3 \\ 2 & 3 & 1 \\ 4 & 3 & 0 \end{pmatrix}$ is a 3 × 3 matrix, a *square* matrix of order 3.

We identify a matrix by a capital letter and its order can be shown under this letter. For example, $\underset{3\times2}{K}$ is a matrix that we call **K** and its order is 3 × 2 (i.e. 3 rows and 2 columns).

Exercises 23.3

Below is a set of matrices that are also referred to in following exercises.

$A = \begin{pmatrix} 1 & 0 & 2 \\ 3 & 2 & 1 \end{pmatrix}$ $B = \begin{pmatrix} 2 & 1 \\ 3 & 2 \end{pmatrix}$ $C = \begin{pmatrix} 3 \\ 1 \\ 2 \end{pmatrix}$

$D = \begin{pmatrix} 2 & 1 & 3 \\ 1 & 4 & 2 \end{pmatrix}$ $E = \begin{pmatrix} 1 & 3 & 4 \\ 2 & 5 & 2 \\ 4 & 1 & 3 \end{pmatrix}$ $F = \begin{pmatrix} 3 & 1 & 2 \end{pmatrix}$

$G = \begin{pmatrix} 2 & 1 \\ 1 & 3 \\ 3 & 2 \end{pmatrix}$ $H = \begin{pmatrix} 1 & 3 \\ 2 & 1 \end{pmatrix}$ $K = \begin{pmatrix} 3 & 1 & 2 \\ 0 & 2 & 3 \\ 1 & 0 & 2 \end{pmatrix}$

1 Using the above set of matrices, state the order of: **a** D, **b** G, **c** C, **d** A, **e** F.

2 Which of the matrices in the above set is: **a** a 2 × 3 matrix, **b** a 3 × 2 matrix, **c** a square matrix, **d** a row matrix, **e** a column matrix?

23.4 SOME DEFINITIONS AND LAWS

Equal matrices

Two matrices are said to be *equal* if, and only if, they are identical in every respect—that is, the elements of each are the same and in the same positions.

Example

If $\begin{pmatrix} a & b \\ c & d \end{pmatrix} = \begin{pmatrix} 2 & 4 \\ 1 & 3 \end{pmatrix}$, then $a = 2$, $b = 4$, $c = 1$, $d = 3$.

The sum or difference of two matrices

These are found by adding or subtracting the corresponding elements of each matrix.

Example

$$\begin{pmatrix} a & b & c \\ d & e & f \end{pmatrix} + \begin{pmatrix} g & h & i \\ j & k & l \end{pmatrix} = \begin{pmatrix} a+g & b+h & c+i \\ d+j & e+k & f+l \end{pmatrix}$$

$$\begin{pmatrix} 5 & 3 & 1 \\ 2 & 4 & 0 \end{pmatrix} - \begin{pmatrix} 3 & 3 & 0 \\ 5 & 1 & 2 \end{pmatrix} = \begin{pmatrix} 2 & 0 & 1 \\ -3 & 3 & -2 \end{pmatrix}$$

Note: Two matrices can be added or subtracted only when they have the same order (i.e. they must have the same number of rows and the same number of columns; they must have the 'same shape'). The resulting sum or difference will also have the same order.

Note: From the definition, it can be seen that **A + B = B + A** (the commutative law for addition).

The zero matrix

Note: For matrix **A** the zero matrix, **O**, is defined to be the matrix such that **A + O = O + A = A** (the law of addition of zero). The letter **O** is used to denote a zero matrix.

Example

The zero matrix of $\begin{pmatrix} a & b & c \\ d & e & f \end{pmatrix}$ is $\begin{pmatrix} 0 & 0 & 0 \\ 0 & 0 & 0 \end{pmatrix}$.   The zero matrix of $\begin{pmatrix} 2 & -3 \\ 0 & 5 \end{pmatrix}$ is $\begin{pmatrix} 0 & 0 \\ 0 & 0 \end{pmatrix}$.

Note: The matrix **O** is not the *number* zero but is the matrix of the same order as **A**, which has the number 0 for each of its elements.

Multiplication by a constant

By definition, a matrix is multiplied by a constant by multiplying *every element* of the matrix by that constant.

Example

$$3 \times \begin{pmatrix} 2 & -1 & 0 \\ 0 & 3 & -2 \end{pmatrix} = \begin{pmatrix} 6 & -3 & 0 \\ 0 & 9 & -6 \end{pmatrix}$$

Exercises 23.4

1 Solve the following matrix equations:

   a $\begin{pmatrix} x \\ y \end{pmatrix} = \begin{pmatrix} 7 \\ 3 \end{pmatrix}$
              b $\begin{pmatrix} x + 2 \\ y - 3 \end{pmatrix} = \begin{pmatrix} 5 \\ 6 \end{pmatrix}$

   c $\begin{pmatrix} x - 1 \\ 2y + 3 \end{pmatrix} = \begin{pmatrix} 3 - x \\ y + 9 \end{pmatrix}$
        d $\begin{pmatrix} x + y \\ x - y \end{pmatrix} = \begin{pmatrix} 3 \\ 5 \end{pmatrix}$

2 Write the single matrix equation

$$\begin{pmatrix} 3x + 2y - 5z \\ 4x - 3y + 2z \\ 5x + 5y - 3z \end{pmatrix} = \begin{pmatrix} 8 \\ 7 \\ 9 \end{pmatrix}$$

   as three separate simultaneous equations.

3 Using the set of matrices (**A, B, C, . . ., K**) in Exercises 23.3:

   a state which pairs of those matrices can be added or subtracted

   b write down the matrix **K + E**

   c state the zero matrix of **D**

   d state the zero matrix of **E**

   e write down the matrix $3 \times$ **A**

   f write down the matrix **2E + 3K**

23.5 MULTIPLICATION OF MATRICES

Since a matrix is simply an array (arrangement) of numbers in a rectangular pattern and does not have a value, we can define the product of two matrices in any way we choose.

For the purposes of explanation, the rows and columns of a matrix will be designated as shown in the matrix below: the rows being called $R_1, R_2, R_3, \ldots$ and the columns $C_1, C_2, C_3, \ldots$.

Any particular element of a matrix can be identified by stating its row and its column.

For example, in the matrix $\begin{array}{c} \\ R_1 \rightarrow \\ R_2 \rightarrow \end{array} \overset{\begin{array}{ccc} C_1 & C_2 & C_3 \\ \downarrow & \downarrow & \downarrow \end{array}}{\begin{pmatrix} 3 & 6 & 7 \\ 2 & 5 & 4 \end{pmatrix}}$, $R_1C_3 = 7$, $R_1C_1 = 3$ and $R_2C_2 = 5$.

By definition, when two matrices are multiplied, the product is another matrix and regardless of how many rows and columns the matrices may possess, when the elements in R_n of the first matrix are multiplied in succession by the elements of C_m of the second matrix and these products are added, this gives element R_nC_m of the product matrix.

When put into words this definition seems very complicated but some illustrations and some practice should enable you to gain facility with this process.

Ignoring all the other rows and columns that may be present:

$$
1 \quad R_3 \rightarrow \begin{pmatrix} \times & \times & \times \\ \times & \times & \times \\ 5 & 2 & 7 \\ \times & \times & \times \end{pmatrix} \times \begin{pmatrix} \times & \overset{C_2}{\underset{\downarrow}{4}} & \times \\ \times & 0 & \times \\ \times & 1 & \times \end{pmatrix} = R_3 \rightarrow \begin{pmatrix} \times & \times & \times \\ \times & \times & \times \\ \times & \underset{\uparrow\ C_2}{27} & \times \\ \times & \times & \times \end{pmatrix}
$$

In the product matrix, element $R_3C_2 = (5 \times 4) + (2 \times 0) + (7 \times 1) = 27$

$$
2 \quad \begin{pmatrix} 1 & 2 \\ 3 & 4 \end{pmatrix} \times \begin{pmatrix} 5 & 7 & 9 \\ 6 & 8 & 10 \end{pmatrix} = \begin{pmatrix} (1 \times 5) + (2 \times 6) & (1 \times 7) + (2 \times 8) & (1 \times 9) + (2 \times 10) \\ (3 \times 5) \times (4 \times 6) & (3 \times 7) + (4 \times 8) & (3 \times 9) + (4 \times 10) \end{pmatrix}
$$

$$
= \begin{pmatrix} 5 + 12 & 7 + 16 & 9 + 20 \\ 15 + 24 & 21 + 32 & 27 + 40 \end{pmatrix}
$$

$$
= \begin{pmatrix} 17 & 23 & 29 \\ 39 & 53 & 67 \end{pmatrix}
$$

You are advised to practise this process until it becomes quite familiar to you. Below are some exercises to enable you to practise the multiplication of two matrices. You will quickly discover the benefit of using fingers or a pen to obscure the rows and columns not being used to obtain a particular element of the product matrix.

Example

$$
\begin{pmatrix} 2 & 3 \\ 4 & 5 \end{pmatrix} \times \begin{pmatrix} 4 & 2 \\ 1 & 3 \end{pmatrix} = \begin{pmatrix} (2 \times 4) + (3 \times 1) & (2 \times 2) + (3 \times 3) \\ (4 \times 4) + (5 \times 1) & (4 \times 2) + (5 \times 3) \end{pmatrix}
$$

$$
= \begin{pmatrix} 8 + 3 & 4 + 9 \\ 16 + 5 & 8 + 15 \end{pmatrix}
$$

$$
= \begin{pmatrix} 11 & 13 \\ 21 & 23 \end{pmatrix}
$$

When the elements are small you should be able to obtain the product matrix without needing to write down the intermediate steps.

$$
\text{e.g.} \quad \begin{pmatrix} 3 & 1 \\ 2 & 4 \end{pmatrix} \times \begin{pmatrix} 2 & 0 \\ 1 & 3 \end{pmatrix} = \begin{pmatrix} 7 & 3 \\ 8 & 12 \end{pmatrix}
$$

Exercises 23.5

1 a You are given the following matrices:

$$A \begin{pmatrix} 3 & 1 \\ 2 & 4 \end{pmatrix}, \quad B \begin{pmatrix} 5 & 1 \\ 3 & 2 \end{pmatrix}, \quad C \begin{pmatrix} 1 & 2 \\ 4 & 3 \end{pmatrix}, \quad D \begin{pmatrix} 5 & 3 \\ 2 & 1 \end{pmatrix}, \quad E \begin{pmatrix} 2 & 0 \\ 0 & 3 \end{pmatrix}$$

Write down the following matrices:

i $A \times B$	ii $A \times C$	iii $A \times D$	iv $A \times E$
v $B \times C$	vi $B \times D$	vii $B \times E$	viii $C \times D$
ix $C \times E$	x $D \times E$		

b You are given the following matrices:

$$P \begin{pmatrix} -1 & 3 \\ -2 & 4 \end{pmatrix}, \quad Q \begin{pmatrix} 3 & -2 \\ 6 & -4 \end{pmatrix}, \quad R \begin{pmatrix} 4 & -3 \\ 2 & -1 \end{pmatrix}, \quad S \begin{pmatrix} -2 & 0 \\ -3 & 0 \end{pmatrix}, \quad T \begin{pmatrix} -4 & 2 \\ -6 & 3 \end{pmatrix}$$

Write down the following matrices:

i $P \times Q$	ii $P \times R$	iii $P \times S$	iv $P \times T$
v $Q \times R$	vi $Q \times S$	vii $Q \times T$	viii $R \times S$
ix $R \times T$	x $S \times T$		

2 As always in algebra, if there is no operation sign between two terms, multiplication is to be assumed. **AK** means $A \times K$, i.e. matrix **A** × matrix **K**.

Find the product of the matrices in each case below:

a $\begin{pmatrix} 1 & 3 & 0 \\ 0 & 2 & 1 \\ 2 & 0 & 3 \end{pmatrix} \begin{pmatrix} 1 & 2 & 3 \\ 2 & 0 & 2 \\ 0 & 3 & 1 \end{pmatrix}$

b $\begin{pmatrix} -1 & 2 & -1 \\ 0 & 0 & 3 \\ -2 & 0 & 1 \end{pmatrix} \begin{pmatrix} -2 & -1 & 0 \\ 2 & -3 & 0 \\ 1 & 0 & -1 \end{pmatrix}$

c $\begin{pmatrix} 2 & 1 & 0 \\ 1 & 0 & 3 \\ 3 & 1 & 1 \end{pmatrix} \begin{pmatrix} 0 & 1 & 2 \\ 1 & 1 & 0 \\ 2 & 3 & 1 \end{pmatrix}$

d $\begin{pmatrix} 1 & 0 \\ -1 & 2 \end{pmatrix} \begin{pmatrix} 3 & -2 \\ 1 & 5 \end{pmatrix}$

e $\begin{pmatrix} 1 & 0 \\ 3 & 2 \end{pmatrix} \begin{pmatrix} 5 & 4 & 8 \\ 6 & 7 & 9 \end{pmatrix}$

f $\begin{pmatrix} 0 & 1 & 1 \\ 1 & 3 & -3 \end{pmatrix} \begin{pmatrix} 2 & 1 & -1 \\ 4 & 3 & 0 \\ 0 & -1 & 5 \end{pmatrix}$

g $\begin{pmatrix} 1 & 2 \\ 0 & 3 \end{pmatrix} \begin{pmatrix} 1 & 2 & 4 \\ 3 & 0 & 1 \end{pmatrix}$

h $\begin{pmatrix} 3 & 0 & 1 \\ 0 & 2 & -1 \end{pmatrix} \begin{pmatrix} 1 & 5 \\ 4 & 2 \\ 6 & 0 \end{pmatrix}$

i $(0 \quad 3 \quad 2) \begin{pmatrix} 1 \\ -1 \\ 4 \end{pmatrix}$

j $\begin{pmatrix} 2 \\ -1 \\ 3 \end{pmatrix} (-2 \quad 3 \quad -1)$

continued

continued

k $\begin{pmatrix} 2 & 1 \\ 4 & 3 \end{pmatrix}\begin{pmatrix} 5 \\ 6 \end{pmatrix}$

l $(7 \quad -2 \quad 5)\begin{pmatrix} 6 \\ -2 \\ 8 \end{pmatrix}$

m $\begin{pmatrix} 2 & 0 \\ -1 & 3 \\ -2 & 4 \end{pmatrix}\begin{pmatrix} -3 \\ 0 \end{pmatrix}$

n $(4 \quad 5 \quad 1)\begin{pmatrix} 0 \\ -2 \\ 3 \end{pmatrix}$

o $\begin{pmatrix} -2 & -1 \\ -1 & -3 \end{pmatrix}\begin{pmatrix} -1 & 3 \\ 2 & 1 \end{pmatrix}$

p $\begin{pmatrix} 5 & 0 & -1 \\ 1 & 4 & 2 \\ -3 & -2 & 0 \end{pmatrix}\begin{pmatrix} -4 & 0 \\ 3 & -2 \\ 0 & 1 \end{pmatrix}$

3 If $\mathbf{A} = \begin{pmatrix} 3 & -1 & 2 \\ 0 & 2 & 1 \\ 1 & 3 & 0 \end{pmatrix}$ and $\mathbf{B} = \begin{pmatrix} -1 & 2 \\ 2 & 0 \end{pmatrix}$, find: **a** $\mathbf{A}^2$, **b** $\mathbf{B}^3$.

23.6 COMPATIBILITY

By now it has probably become clear to you that multiplication of two matrices is only possible when the number of columns in the first matrix equals the number of rows in the second matrix. i.e. the product $\underset{m \times n}{\mathbf{A}} \times \underset{p \times q}{\mathbf{B}} = \mathbf{C}$ exists only when $n = p$. When, and only when, this is so, the matrices are said to be 'compatible' for this multiplication. If $n \neq p$, the matrices cannot be multiplied and are said to be 'incompatible' for this operation.

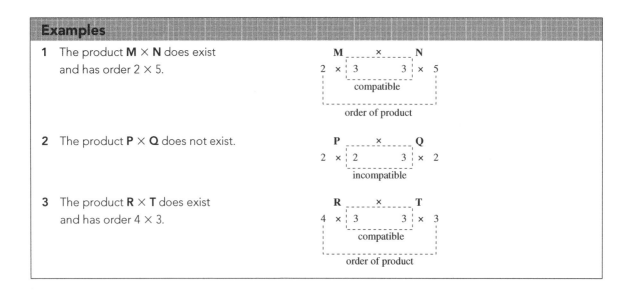

Examples

1 The product **M** × **N** does exist and has order 2 × 5.

$$\begin{array}{ccccc} \mathbf{M} & & \times & & \mathbf{N} \\ 2 & \times & 3 & \qquad 3 & \times & 5 \\ & & \text{compatible} & & \\ & & \text{order of product} & & \end{array}$$

2 The product **P** × **Q** does not exist.

$$\begin{array}{ccccc} \mathbf{P} & & \times & & \mathbf{Q} \\ 2 & \times & 2 & \qquad 3 & \times & 2 \\ & & \text{incompatible} & & \end{array}$$

3 The product **R** × **T** does exist and has order 4 × 3.

$$\begin{array}{ccccc} \mathbf{R} & & \times & & \mathbf{T} \\ 4 & \times & 3 & \qquad 3 & \times & 3 \\ & & \text{compatible} & & \\ & & \text{order of product} & & \end{array}$$

It is quite common for a product $\mathbf{A} \times \mathbf{B}$ to exist but for the product $\mathbf{B} \times \mathbf{A}$ *not* to exist.

For example:

$$\underset{2 \times 3}{\mathbf{V}} \quad \times \quad \underset{3 \times 3}{\mathbf{W}} \quad \text{exists (and has order } 2 \times 3\text{), but}$$

$$\underset{3 \times 3}{\mathbf{W}} \quad \times \quad \underset{2 \times 3}{\mathbf{V}} \quad \text{does } not \text{ exist.}$$

It is easy to show that $\mathbf{A} \times \mathbf{B}$ and $\mathbf{B} \times \mathbf{A}$ both exist only for $\underset{m \times n}{\mathbf{A}} \times \underset{n \times m}{\mathbf{B}}$:

e.g. $\quad \underset{2 \times 3}{\mathbf{A}} \quad \times \quad \underset{3 \times 2}{\mathbf{B}} \quad$ and $\quad \underset{3 \times 2}{\mathbf{B}} \quad \times \quad \underset{2 \times 3}{\mathbf{A}} \quad$ both exist.

SUMMARY

Facts about the product of two matrices:

- $\mathbf{A} \times \mathbf{B}$ may not *exist*.
- Even if $\mathbf{A} \times \mathbf{B}$ does exist, it is possible that $\mathbf{B} \times \mathbf{A}$ does *not* exist.
- Even when both products exist, in general, $\mathbf{A} \times \mathbf{B} \neq \mathbf{B} \times \mathbf{A}$.
- It is possible that, $\mathbf{A} \times \mathbf{B} = \mathbf{0}$ even though neither $\mathbf{A}$ nor $\mathbf{B}$ is the zero matrix $\mathbf{O}$.

Example

$$\begin{pmatrix} 3 & 2 \\ 6 & 4 \end{pmatrix} \times \begin{pmatrix} 4 & -2 \\ -6 & 3 \end{pmatrix} = \begin{pmatrix} 0 & 0 \\ 0 & 0 \end{pmatrix} = 0$$

You can verify this after studying the next section.

Exercises 23.6

1. Using the set of matrices in Exercises 23.3, state whether the given product exists in each case (answering 'yes' or 'no') and, if it *does* exist, state its order.

a AB	b BA	c AC	d DA	e FE	f EF
g CF	h AE	i BD	j CD	k DC	l CE

23.7 THE IDENTITY MATRIX, I

Exercises 23.7

1 Write down the product matrix:

$$\text{a} \quad \begin{pmatrix} a & b \\ c & d \end{pmatrix}\begin{pmatrix} 1 & 0 \\ 0 & 1 \end{pmatrix} \qquad\qquad \text{b} \quad \begin{pmatrix} 1 & 0 \\ 0 & 1 \end{pmatrix}\begin{pmatrix} a & b \\ c & d \end{pmatrix}$$

continued

continued

2 Write down the product matrix:

a $\begin{pmatrix} a & b & c \\ d & e & f \\ g & h & i \end{pmatrix} \begin{pmatrix} 1 & 0 & 0 \\ 0 & 1 & 0 \\ 0 & 0 & 1 \end{pmatrix}$ 　　　**b** $\begin{pmatrix} 1 & 0 & 0 \\ 0 & 1 & 0 \\ 0 & 0 & 1 \end{pmatrix} \begin{pmatrix} a & b & c \\ d & e & f \\ g & h & i \end{pmatrix}$

The *principal diagonal* of a square matrix is the diagonal that runs from the top left-hand corner to the bottom right-hand corner. The square matrix, which has the number 1 for each element on the principal diagonal and all other elements zero, plays a very special role in the theory of matrices.

$\begin{pmatrix} 1 & 0 \\ 0 & 1 \end{pmatrix}$ is called the *identity matrix* of order 2, and is specified as I_2.

$\begin{pmatrix} 1 & 0 & 0 \\ 0 & 1 & 0 \\ 0 & 0 & 1 \end{pmatrix}$ is called the *identity matrix* of order 3, and is specified as I_3.

For any square matrix A_n (i.e. of order $n \times n$), $A_n \times I_n = I_n \times A_n$.

I_n plays the same role in matrix theory as unity does in arithmetic (e.g. $7 \times 1 = 1 \times 7 = 7$), and so it is called the **unit matrix, U_n** or the **identity matrix, I_n**. We use the latter name and symbol in this book.

We use capital letters to identify matrices, but the capital letter I is reserved for the *identity matrix* and the capital letter O is reserved for the *zero matrix*.

Note: For a matrix $\underset{n \times m}{A}$ which is *not* square, $\underset{n \times m}{A} \times I_m = \underset{n \times m}{A}$ (but $I_m \times \underset{n \times m}{A}$ does not exist) and $I_n \times \underset{n \times m}{A} = \underset{n \times m}{A}$ (but $\underset{n \times m}{A} \times I_n$ does not exist).

Remember: I stands for the Identity matrix, not the numeral 1.

Therefore, non-square matrices do not have an identity matrix.

Exercises 23.7 *continued*

3 For the following, write down the product matrix if it exists. If it does not exist, write 'incompatible'.

a $\begin{pmatrix} 1 & 2 & 3 \\ 4 & 5 & 6 \end{pmatrix} \times \begin{pmatrix} 1 & 0 & 0 \\ 0 & 1 & 0 \\ 0 & 0 & 1 \end{pmatrix}$ 　　　**b** $\begin{pmatrix} 1 & 0 & 0 \\ 0 & 1 & 0 \\ 0 & 0 & 1 \end{pmatrix} \times \begin{pmatrix} 2 & 1 & 3 \\ 1 & 3 & 2 \end{pmatrix}$

c $\begin{pmatrix} 1 & 0 \\ 0 & 1 \end{pmatrix} \begin{pmatrix} 1 & 2 & 3 \\ 4 & 5 & 6 \end{pmatrix}$ 　　　**d** $\begin{pmatrix} 1 & 2 & 3 \\ 4 & 5 & 6 \end{pmatrix} \begin{pmatrix} 1 & 0 \\ 0 & 1 \end{pmatrix}$

4 **a** If $\begin{pmatrix} 17 & -13 & 19 \\ 34 & 28 & -15 \end{pmatrix} \times A = \begin{pmatrix} 17 & -13 & 19 \\ 34 & 28 & -15 \end{pmatrix}$, write down the matrix **A**.

Remember: An identity matrix is always *square*.

b If $\mathbf{B} \times \begin{pmatrix} 17 & -13 & 19 \\ 34 & 28 & -15 \end{pmatrix} = \begin{pmatrix} 17 & -13 & 19 \\ 34 & 28 & -15 \end{pmatrix}$, write down the matrix **B**.

c Does $\begin{pmatrix} 17 & -13 & 19 \\ 34 & 28 & -15 \end{pmatrix}$ have an identity matrix? State the reason for your answer.

5 For each of the following matrices, state whether an identity matrix exists (answering 'yes' or 'no') and, if it *does* exist, write it down.

a $\begin{pmatrix} -3 & 2 \\ 7 & 5 \end{pmatrix}$

b $\begin{pmatrix} -2 & 8 \\ 7 & 3 \\ 1 & 0 \end{pmatrix}$

c $\begin{pmatrix} 3 & -4 & 2 \\ 6 & 1 & -5 \end{pmatrix}$

d $\begin{pmatrix} 3 & -2 & 5 \\ 4 & 1 & 6 \\ -7 & 9 & 8 \end{pmatrix}$

6 If $\mathbf{M} = \begin{pmatrix} 0 & -1 \\ 1 & 0 \end{pmatrix}$, show that: **a** $\mathbf{M}^2 = -\mathbf{I}$,   **b** $\mathbf{M}^3 = -\mathbf{M}$.

23.8 THE INVERSE MATRIX, A⁻¹

The numbers 13 and $\frac{1}{13}$ are said to be multiplicative *inverses* of one another because in multiplication one *undoes* what the other *does*.

For example, $957 \times 13 \times \frac{1}{13} = 957$, $\sqrt{7} \times \frac{1}{13} \times 13 = \sqrt{7}$.

This occurs because $13 \times \frac{1}{13} = 1$ and $\frac{1}{13} \times 13 = 1$.

Now we will see how this applies to matrices.

Exercises 23.8

1 If $\mathbf{A} = \begin{pmatrix} 5 & -3 \\ 2 & -1 \end{pmatrix}$ and $\mathbf{B} = \begin{pmatrix} -1 & 3 \\ -2 & 5 \end{pmatrix}$, write down: **a** the matrix **AB**, **b** the matrix **BA**.

Since both products in the exercise above are the identity matrix, you can probably guess that **A** and **B** are said to be *inverses* of each other.

The inverse of matrix **M** is written as $\mathbf{M}^{-1}$, so you have proved for the above matrices **A** and **B** that $\mathbf{AB} = \mathbf{BA} = \mathbf{I}$, that is, that $\mathbf{B} = \mathbf{A}^{-1}$ and $\mathbf{A} = \mathbf{B}^{-1}$.

Definition: Two matrices, **A** and **B**, are said to be inverses of one another (i.e. $\mathbf{A} = \mathbf{B}^{-1}$ and $\mathbf{B} = \mathbf{A}^{-1}$) if $\mathbf{AB} = \mathbf{BA} = \mathbf{I}$.

Exercises 23.8 *continued*

2   Given that $C = \begin{pmatrix} 1 & 1 & 1 \\ 1 & 1 & 2 \\ 1 & 2 & 1 \end{pmatrix}$ and $D = \begin{pmatrix} 3 & -1 & -1 \\ -1 & 0 & 1 \\ -1 & 1 & 0 \end{pmatrix}$:

   a   write down the matrix **CD**

   b   write down the matrix **DC**

   c   what have we proved about the matrices **C** and **D**?

3   Given that matrix $\underset{p \times q}{A}$ has an inverse $\underset{r \times s}{B}$:

   a   what is the order of matrix **AB**?

   b   what is the order of matrix **BA**?

   c   Since these matrices are inverses of each other, by definition $AB = BA = \underset{n \times n}{I}$. Therefore, the orders of **AB** and **BA** are both $n \times n$. What can you deduce about the values of p, q, r, s and n?

   d   Hence, if matrices **A** and **B** are inverses of each other, what can you deduce about the *shapes* of the matrices **A** and **B**?

Most square matrices have an inverse, but not *all* of them. (Actually, it can be shown that all square matrices have an inverse **except those for which the determinant |A|= 0.**)

A matrix that has an inverse is said to be **invertible.**

SUMMARY

- If $A \times B = B \times A = I$, then A and B are inverses of each other (i.e. $A = B^{-1}$ and $B = A^{-1}$).
- A non-square matrix cannot have an inverse.
- Most, but not all, square matrices have an inverse (i.e. they are *invertible*).

Note: You are not required to be able to *find* the inverse A^{-1} of a given matrix **A**, but you should be able to determine whether or not two given matrices **A** and **B** are inverses of each other by testing whether $AB = BA = I$.

Exercises 23.8 *continued*

4   If $A = \begin{pmatrix} 2 & 1 & 1 \\ 1 & 1 & 1 \\ 2 & 2 & 1 \end{pmatrix}$ $B = \begin{pmatrix} 1 & 2 & -2 \\ 2 & 5 & -4 \\ 3 & 7 & -5 \end{pmatrix}$ $C = \begin{pmatrix} 1 & -1 & 0 \\ -1 & 0 & 1 \\ 0 & 2 & -1 \end{pmatrix}$

$D = \begin{pmatrix} 1 & 2 & 0 \\ 0 & 1 & 2 \\ 1 & 0 & 2 \end{pmatrix}$ $E = \begin{pmatrix} 3 & -4 & 2 \\ -2 & 1 & 0 \\ -1 & -1 & 1 \end{pmatrix}$:

a write down the products:

i AB	**ii** AC	**iii** AD	**iv** AE
v BC	**vi** BD	**vii** BE	

b hence, write down: (i) matrix A^{-1} (ii) matrix B^{-1}

5 Given that $J = \begin{pmatrix} 1 & 2 & -1 \\ 0 & 3 & -2 \\ 0 & 0 & -1 \end{pmatrix}$ and $K = \begin{pmatrix} 3 & -2 & 1 \\ 0 & 1 & -2 \\ 0 & 0 & -3 \end{pmatrix}$:

    a write down the matrix JK

    b write down the matrix J^{-1}

6 Given that $P = \begin{pmatrix} 1 & 2 & -2 \\ 1 & 3 & -1 \\ 3 & 2 & 0 \end{pmatrix}$ and $Q = \begin{pmatrix} 2 & -4 & 4 \\ -3 & 6 & -1 \\ -7 & 4 & 1 \end{pmatrix}$:

    a write down the matrix PQ

    b write down the matrix Q^{-1}

23.9 THE ALGEBRA OF MATRICES

As a result of the definitions of the operations of matrices, the laws for the algebra of matrices are mostly the same as the laws for the algebra of real numbers.

Name of law	Real numbers	Matrices
Commutative law for addition	$x + y = y + x$	$\mathbf{A} + \mathbf{B} = \mathbf{B} + \mathbf{A}$
Associative law for addition	$(x + y) + z = x + (y + z)$	$(\mathbf{A} + \mathbf{B}) + \mathbf{C} = \mathbf{A} + (\mathbf{B} + \mathbf{C})$
Identity law for addition	$x + 0 = 0 + x = x$ (0 is the identity element for addition)	$\mathbf{A} + \mathbf{O} = \mathbf{O} + \mathbf{A} = \mathbf{A}$ ($\mathbf{O}$ is the identity matrix for addition, the *zero* matrix of $\mathbf{A}$)
Identity law for multiplication	$x \times 1 = 1 \times x = x$ (1 is the identity element for multiplication)	$\mathbf{A} \times \mathbf{I} = \mathbf{I} \times \mathbf{A} = \mathbf{A}$ ($\mathbf{I}$ is the identity matrix for multiplication for matrix $\mathbf{A}$)
Law of multiplicative inverse	$x \times x^{-1} = x^{-1} \times x = 1$ (x^{-1} is the multiplicative inverse of x)	$\mathbf{A} \times \mathbf{A}^{-1} = \mathbf{A}^{-1} \times \mathbf{A} = \mathbf{I}$ ($\mathbf{A}^{-1}$ is the multiplicative inverse of $\mathbf{A}$)
Distributive law for multiplication	$C(x + y) = Cx + Cy$ $(x + y)C = xC + yC$ $x(y + z) = xy + xz$ $(y + z)x = yx + zx$	$k(\mathbf{A} + \mathbf{B}) = k\mathbf{A} + k\mathbf{B}$ $(\mathbf{A} + \mathbf{B})k = \mathbf{A}k + \mathbf{B}k$ $\mathbf{A}(\mathbf{B} + \mathbf{C}) = \mathbf{AB} + \mathbf{AC}$ $(\mathbf{B} + \mathbf{C})\mathbf{A} = \mathbf{BA} + \mathbf{CA}$

Remember:

- **O, the zero matrix for matrix A,** has been defined as the matrix with the same order as **A** but having all its elements zeros.
- **I, the identity matrix for matrix A,** has been defined for square matrices only, being the matrix having the same order as **A**, with all the elements on the principal diagonal being 1 and all the other elements being zero.
- **A^{-1}, the inverse of matrix A,** has been defined for *square* matrices only, being the matrix such that $AA^{-1} = A^{-1}A = I$. The only matrices that have an inverse are *square matrices whose determinant $\neq 0$.*

Hence, most algebraic operations with matrices are already quite familiar to us.

Examples

1 If $3A + 4B = 5C$

then $3A = 5C - 4B$

$\therefore\ A = \frac{1}{3}(5C - 4B)$

2 $(A + B)(C + D) = AC + AD + BC + BD$

3 $(A + B)(A + I) = A^2 + AI + BA + BI$

$= A^2 + A + BA + B$

4 $A(A^{-1} + I) = AA^{-1} + AI$

$= I + A,$ or $A + I$

5 $A(A^{-1}B) = (AA^{-1})B = IB = B$

6 $AB + A = AB + AI$

$= A(B + I),$ *not* $A(B + 1)$ because we cannot add a real number to a matrix

7 $A^{-1}(A + I) = A^{-1}A + A^{-1}I$

$= I + A^{-1},$ or $A^{-1} + I$

However, there is a difference between the algebra for matrices and the algebra for real numbers when we are *multiplying* because, as we have already noted, in general:

$A \times B \neq B \times A.$

(Don't forget that this statement does not mean that they can *never* be equal but that we cannot *assume* they *are* equal, because *usually* they are not equal.)

Hence, care must be taken to maintain the *correct order* of matrices when multiplying.

For example, $A(B + C) \neq (B + C)A$ and $ABA^{-1} \neq A^{-1}AB$ (which would equal **B**).

The only occasion when multiplying matrices is commutative is when they are inverses of one another ($AA^{-1} = A^{-1}A$ [$=I$]) or when one of them is the identity matrix ($AI = IA$ [$= A$]).

Example

Make **K** the subject of the equation $AK = B$.

We proceed thus: $AK = B$

$\therefore\ A^{-1}(AK) = A^{-1}B$ (*not* BA^{-1})

$\therefore\ (A^{-1}A)K = A^{-1}B$

$\therefore\ K = A^{-1}B$

Note: We cannot add a matrix and a real number (e.g. **A** + 2 makes no sense). But we can always replace **A** by **AI** or by **IA**.

Example

Solve the matrix equation   **AF** + 2**F** = **B**, for **F**.

$$\mathbf{AF} + 2\mathbf{F} = \mathbf{B}$$

$$\therefore \ \mathbf{AF} + 2\mathbf{IF} = \mathbf{B}$$

$$(\mathbf{A} + 2\mathbf{I})\mathbf{F} = \mathbf{B}$$

$$\therefore \ (\mathbf{A} + 2\mathbf{I})^{-1}(\mathbf{A} + 2\mathbf{I})\mathbf{F} = (\mathbf{A} + 2\mathbf{I})^{-1}\mathbf{B}$$

$$\therefore \ \mathbf{F} = (\mathbf{A} + 2\mathbf{I})^{-1}\mathbf{B}$$

Exercises 23.9

1   Simplify, removing all brackets:

   a $\mathbf{A}(\mathbf{A}^{-1}\mathbf{B})$

   b $\mathbf{A}(\mathbf{A}^{-1}\mathbf{I})$

   c $\mathbf{A}(\mathbf{B}\mathbf{A}^{-1}) + \mathbf{A}(\mathbf{A}^{-1}\mathbf{B})$

   d $\mathbf{AB}(\mathbf{A}^{-1}\mathbf{B} + \mathbf{B}^{-1}\mathbf{A})$

   e $\mathbf{I}^2$

   f $\mathbf{I}^2 + \mathbf{I}^3$

   g $(\mathbf{A} + \mathbf{I})^2$

   h $(\mathbf{A} + \mathbf{I})(\mathbf{B} + \mathbf{A})$

   i $\mathbf{A}(\mathbf{I} - \mathbf{A}^{-1}) + \mathbf{I}^2$

   j $\mathbf{A}(\mathbf{A}^{-1} + \mathbf{I}) - \mathbf{B}^{-1}\mathbf{B} - \mathbf{IA}$

2   Solve the following matrix equation for **X**:

   a $2\mathbf{X} - \mathbf{A} = \mathbf{B}$

   b $\mathbf{C} - 3\mathbf{X} = \mathbf{D}$

   c $3(\mathbf{A} - 2\mathbf{X}) = 2(3\mathbf{A} - \mathbf{X})$

   d $\dfrac{\mathbf{X} + \mathbf{A}}{3} = \dfrac{\mathbf{X} - 3\mathbf{B}}{2}$

3   Solve the following matrix equations for **X**:

   a $\mathbf{AX} = \mathbf{B}$

   b $\mathbf{XA} = \mathbf{B}$

   c $2\mathbf{A} + \mathbf{AX} = \mathbf{B}$

   d $\mathbf{AX} + \mathbf{I} = \mathbf{B}$

   e $\mathbf{X}^{-1} = \mathbf{A}$

   f $3\mathbf{X} - \mathbf{AX} = \mathbf{B}$

   g $\mathbf{AX} + \mathbf{X} = \mathbf{B}$

   h $\mathbf{AX} + \mathbf{B} = \mathbf{C} - \mathbf{X}$

   i $\mathbf{X} + \mathbf{XA} = \mathbf{B}$

   j $2\mathbf{X} + \mathbf{B} = \mathbf{C} - 2\mathbf{XA}$

4   If **AB** = **AC**:

   a   does it follow that **B** = **C** ('yes' or 'no')?

   b   and **B** ≠ **C**, what *is* **B** equal to?

5   If **AB** = **CA**:

   a   does it follow that **B** = **C** ('yes' or 'no')?

   b   and **B** ≠ **C**, what *is* **B** equal to?

Note:

- The matrix, $A \times B \times C = (A \times B) \times C = A \times (B \times C)$, the associative law, **but we must not change the order of the matrices when they are being multiplied.**
- The matrices **ABC, ACB, BCA, BAC, CAB** and **CBA** are probably all different matrices.
- However, consideration will show you that if k is a constant (i.e. a real number), then $k \times (A \times B) = (k \times A) \times B = A \times (k \times B)$. Which matrix we multiply by k, either $A \times B$ or A or $B,$ makes no difference to the final result. Although we must not change the positions of the *matrices*, we may change the position of a constant, k.
- Note also that $k \times A = k \times AI = A \times kI$.

Exercises 23.9 *continued*

6 Solve the following matrix equations for X:

a $3X + 2XA = C - B$

b $A + 2X = B + 3XC$

23.10 EXPRESSING SIMULTANEOUS EQUATIONS IN MATRIX FORM

Exercises 23.10

1 Find the following products and state their order:

a $\begin{pmatrix} 2 & 3 \\ 5 & 4 \end{pmatrix} \begin{pmatrix} x \\ y \end{pmatrix}$

b $\begin{pmatrix} 3 & 5 & -2 \\ 1 & 4 & 3 \\ 2 & -3 & 4 \end{pmatrix} \begin{pmatrix} x \\ y \\ z \end{pmatrix}$

2 Express as the product of two matrices:

a $\begin{pmatrix} 3x + 4y \\ 2x - 5y \end{pmatrix}$

b $\begin{pmatrix} 2x - 3y + 4z \\ x + 2y - 5z \\ 3x - y + 2z \end{pmatrix}$

c $\begin{pmatrix} 3a - 5b + 2c \\ 2a + 4b \\ 5a - 7c \end{pmatrix}$

3 Solve for x and y:

a $\begin{pmatrix} 2x \\ 3y \end{pmatrix} = \begin{pmatrix} 8 \\ -9 \end{pmatrix}$

b $\begin{pmatrix} 13 & 1 \\ 12 & 1 \end{pmatrix} \begin{pmatrix} x \\ y \end{pmatrix} = \begin{pmatrix} 29 \\ 27 \end{pmatrix}$

4 Express as a system of simultaneous equations:

a $\begin{pmatrix} 2x + 3y \\ 5x - 2y \end{pmatrix} = \begin{pmatrix} 3 \\ 4 \end{pmatrix}$

b $\begin{pmatrix} 3 & 5 \\ 2 & -4 \end{pmatrix} \begin{pmatrix} x \\ y \end{pmatrix} = \begin{pmatrix} 6 \\ 7 \end{pmatrix}$

c $\begin{pmatrix} 3x - 2y + 4z \\ 2x + 3y - 5z \\ x - 4y + 2z \end{pmatrix} = \begin{pmatrix} 6 \\ 7 \\ 8 \end{pmatrix}$

d $\begin{pmatrix} 1 & 2 & 3 \\ 0 & 1 & 2 \\ 1 & -2 & 0 \end{pmatrix} \begin{pmatrix} x \\ y \\ z \end{pmatrix} = \begin{pmatrix} 4 \\ 5 \\ 6 \end{pmatrix}$

5 Express each of the following systems of simultaneous equations as a single matrix equation:

a $\begin{cases} 5a + 7b - 3c = 17 \\ 7a - 2b - 8c = 13 \\ 3a + 5b + 5c = 19 \end{cases}$

b $\begin{cases} 4x - 5y + 6z = 9 \\ 7x + 3y = 5 \\ 3x - 8z = 7 \end{cases}$

23.11 SOLVING SIMULTANEOUS LINEAR EQUATIONS USING MATRICES

If a set of simultaneous equations is represented in matrix form

$$\mathbf{C} \times \mathbf{U} = \mathbf{K}$$

where:

C is the matrix formed by the coefficients

U is the matrix formed by the unknowns

K is the matrix formed by the constants

then the solution to the matrix of unknown constants **U** can be solved by multiplying both sides of the matrix equation by the inverse of **C**, the matrix of coefficients.

$$\mathbf{C}^{-1} \times \mathbf{C} \times \mathbf{U} = \mathbf{C}^{-1} \times \mathbf{K}$$

We know that:

$$\mathbf{C}^{-1} \times \mathbf{C} = \mathbf{I} \text{ (the identity matrix)}$$

Therefore,

$$\mathbf{I} \times \mathbf{U} = \mathbf{C}^{-1} \times \mathbf{K}$$

We also know that:

$$\mathbf{I} \times \mathbf{U} = \mathbf{U}$$

Therefore the solution to **U**, the matrix of unknown constants, becomes:

$$\mathbf{U} = \mathbf{C}^{-1} \times \mathbf{K}$$

This is a simple but powerful equation. In order to find the solution to a matrix of unknowns we only need to determine the inverse of the matrix of coefficients.

Solving two equations with two unknowns

For the general equation

$$\begin{cases} a_1 x + b_1 y = c_1 \\ a_2 x + b_2 y = c_2 \end{cases}$$

represented in matrix equation form by

$$\begin{pmatrix} a_1 & b_1 \\ a_2 & b_2 \end{pmatrix} \begin{pmatrix} x \\ y \end{pmatrix} = \begin{pmatrix} c_1 \\ c_2 \end{pmatrix},$$

$$\mathbf{C} = \begin{pmatrix} a_1 & b_1 \\ a_2 & b_2 \end{pmatrix}, \mathbf{U} = \begin{pmatrix} x \\ y \end{pmatrix}, \mathbf{K} = \begin{pmatrix} c_1 \\ c_2 \end{pmatrix}$$

$$\mathbf{C} \times \mathbf{U} = \mathbf{K} \text{ or } \mathbf{U} = \mathbf{C}^{-1} \times \mathbf{K}$$

Example

Solve the simultaneous equations

$$\begin{cases} 5x - 2y = 6 \\ 3x - y = 5 \end{cases}$$

given that the inverse of $\begin{pmatrix} 5 & -2 \\ 3 & -1 \end{pmatrix}$ is $\begin{pmatrix} -1 & 2 \\ -3 & 5 \end{pmatrix}$.

continued

continued

Steps

1 Express the equations as a single matrix equation:

$$\begin{pmatrix} 5 & -2 \\ 3 & -1 \end{pmatrix}\begin{pmatrix} x \\ y \end{pmatrix} = \begin{pmatrix} 6 \\ 5 \end{pmatrix}$$

2 Multiply both sides by the inverse matrix:

$$\begin{pmatrix} -1 & 2 \\ -3 & 5 \end{pmatrix}\begin{pmatrix} 5 & -2 \\ 3 & -1 \end{pmatrix}\begin{pmatrix} x \\ y \end{pmatrix} = \begin{pmatrix} -1 & 2 \\ -3 & 5 \end{pmatrix}\begin{pmatrix} 6 \\ 5 \end{pmatrix}$$

3 Multiply out the matrices on each side—we *know* the product on the left-hand side because

$$\mathbf{A}^{-1}\mathbf{A} = \mathbf{I} = \begin{pmatrix} 1 & 0 \\ 0 & 1 \end{pmatrix}.$$

$$\begin{pmatrix} 1 & 0 \\ 0 & 1 \end{pmatrix}\begin{pmatrix} x \\ y \end{pmatrix} = \begin{pmatrix} 4 \\ 7 \end{pmatrix}$$

$$\therefore \begin{pmatrix} x \\ y \end{pmatrix} = \begin{pmatrix} 4 \\ 7 \end{pmatrix}$$

$$\therefore x = 4, y = 7$$

Warning:

The error most often made is in step 3. Remember that $\begin{pmatrix} -1 & 2 \\ -3 & 5 \end{pmatrix}\begin{pmatrix} 6 \\ 5 \end{pmatrix}$ is not the same as $\begin{pmatrix} 6 \\ 5 \end{pmatrix}\begin{pmatrix} -1 & 2 \\ -3 & 5 \end{pmatrix}$ because for matrices $\mathbf{A} \times \mathbf{B} \neq \mathbf{B} \times \mathbf{A}$.

Finding the inverse of a 2 × 2 matrix

In order to find the solution to the matrix of unknowns $\mathbf{U}$ we need to find the inverse of the matrix of coefficients $\mathbf{C}^{-1}$.

To solve for $\mathbf{C}^{-1}$ we manipulate the matrix of coefficients $\mathbf{C}$ using the following steps:

1 Interchange the elements on the principal diagonal.
2 Reverse the signs of the elements on the secondary diagonal.
3 Divide by the determinant of the original matrix.

$$\text{For } \mathbf{C} = \begin{pmatrix} a_1 & b_1 \\ a_2 & b_2 \end{pmatrix}, \mathbf{C}^{-1} = \frac{1}{|\mathbf{C}|}\begin{pmatrix} b_2 & -b_1 \\ -a_2 & a_1 \end{pmatrix}$$

Example

Solve the following set of linear equations:

$$-4x + 3y = 2$$
$$-2x + y = 4$$

Representing in matrix form:

$$\begin{pmatrix} -4 & 3 \\ -2 & 1 \end{pmatrix} \times \begin{pmatrix} x \\ y \end{pmatrix} = \begin{pmatrix} 2 \\ 4 \end{pmatrix}$$

$$\mathbf{C} \times \mathbf{U} = \mathbf{K}$$

Manipulate the matrix of the coefficients **C**:

$$\mathbf{C} = \begin{pmatrix} -4 & 3 \\ -2 & 1 \end{pmatrix}$$

Interchange the elements on the principal diagonal and reverse the signs of the elements on the secondary diagonal.

$$\mathbf{C}^{-1} = \frac{1}{|\mathbf{C}|} \begin{pmatrix} 1 & -3 \\ 2 & -4 \end{pmatrix}$$

Calculate the determinant of the matrix of coefficients.

$$|\mathbf{C}| = \begin{vmatrix} -4 & 3 \\ -2 & 1 \end{vmatrix} = (-4) - (-6) = 2$$

Divide by the determinant of the matrix of coefficients.

$$\mathbf{C}^{-1} = \frac{1}{2} \begin{pmatrix} 1 & -3 \\ 2 & -4 \end{pmatrix}$$

Now that we have determined $\mathbf{C}^{-1}$ we can continue and solve for the matrix of unknowns **U**. Multiply both sides of the matrix equation by $\mathbf{C}^{-1}$:

$$\mathbf{C}^{-1} \times \mathbf{C} \times \mathbf{U} = \mathbf{C}^{-1} \times \mathbf{K}$$
$$\mathbf{U} = \mathbf{C}^{-1} \times \mathbf{K}$$

$$\begin{pmatrix} x \\ y \end{pmatrix} = \frac{1}{2} \begin{pmatrix} 1 & -3 \\ 2 & -4 \end{pmatrix} \begin{pmatrix} 2 \\ 4 \end{pmatrix}$$

$$\begin{pmatrix} x \\ y \end{pmatrix} = \begin{pmatrix} \frac{1}{2} \times (2 - 12) \\ \frac{1}{2} \times (4 - 16) \end{pmatrix} = \begin{pmatrix} -5 \\ -6 \end{pmatrix}$$

$$\therefore x = -5, y = -6$$

Note: If $|\mathbf{A}| = 0$, the matrix **A** has no inverse (because division by zero is not defined). Therefore, the matrix is not invertible.

Solving three equations with three unknowns

The general equations are:

$$\begin{cases} a_1x + b_1y + c_1z = d_1 \\ a_2x + b_2y + c_2z = d_2 \\ a_3x + b_3y + c_3z = d_3 \end{cases}$$

Representing in matrix form:

$$\begin{pmatrix} a_1 & b_1 & c_1 \\ a_2 & b_2 & c_2 \\ a_3 & b_3 & c_3 \end{pmatrix} \begin{pmatrix} x \\ y \\ z \end{pmatrix} = \begin{pmatrix} d_1 \\ d_2 \\ d_e \end{pmatrix}$$

$$\mathbf{C} = \begin{pmatrix} a_1 & b_1 & c_1 \\ a_2 & b_2 & c_2 \\ a_3 & b_3 & c_3 \end{pmatrix}, \mathbf{U} = \begin{pmatrix} x \\ y \\ z \end{pmatrix}, \mathbf{K} = \begin{pmatrix} d_1 \\ d_2 \\ d_e \end{pmatrix}$$

$\mathbf{C} \times \mathbf{U} = \mathbf{K}$, where **U** is the matrix of unknowns, given by $\mathbf{U} = \mathbf{C}^{-1} \times \mathbf{K}$

Example

If $\mathbf{M} = \begin{pmatrix} 1 & 2 & -1 \\ 3 & 5 & -1 \\ -2 & -1 & -2 \end{pmatrix}$ and $\mathbf{N} = \dfrac{1}{2}\begin{pmatrix} 11 & -5 & -3 \\ -8 & 4 & 2 \\ -7 & 3 & 1 \end{pmatrix}$:

a find the matrix **MN**

b write down the matrix $\mathbf{M}^{-1}$

c express the given system linear equations as a single matrix equation and use the result of **b** above to solve these simultaneous equations:

$$\begin{cases} x + 2y - z = -1 \\ 3x + 5y - z = 2 \\ -2x - y - 2z = -9 \end{cases}$$

Solutions

a $\mathbf{MN} = \dfrac{1}{2}\begin{pmatrix} 2 & 0 & 0 \\ 0 & 2 & 0 \\ 0 & 0 & 2 \end{pmatrix} = \begin{pmatrix} 1 & 0 & 0 \\ 0 & 1 & 0 \\ 0 & 0 & 1 \end{pmatrix}$

b $\mathbf{M}^{-1} = \dfrac{1}{2}\begin{pmatrix} 11 & -5 & -3 \\ -8 & 4 & 2 \\ -7 & 3 & 1 \end{pmatrix}$

c $\begin{pmatrix} 1 & 2 & -1 \\ 3 & 5 & -1 \\ -2 & -1 & -2 \end{pmatrix}\begin{pmatrix} x \\ y \\ z \end{pmatrix} = \begin{pmatrix} -1 \\ 2 \\ -9 \end{pmatrix}$

$$\mathbf{M} \times \begin{pmatrix} x \\ y \\ z \end{pmatrix} = \begin{pmatrix} -1 \\ 2 \\ -9 \end{pmatrix}$$

$$\mathbf{M}^{-1}\mathbf{M} \times \begin{pmatrix} x \\ y \\ z \end{pmatrix} = \dfrac{1}{2}\begin{pmatrix} 11 & -5 & -3 \\ -8 & 4 & 2 \\ -7 & 3 & 1 \end{pmatrix}\begin{pmatrix} -1 \\ 2 \\ -9 \end{pmatrix}$$

$$= \dfrac{1}{2}\begin{pmatrix} 6 \\ -2 \\ -4 \end{pmatrix}$$

$$\therefore \begin{pmatrix} x \\ y \\ z \end{pmatrix} = \begin{pmatrix} 3 \\ -1 \\ 2 \end{pmatrix}$$

$$\therefore x = 3,\ y = -1,\ z = 2$$

Finding the inverse of a 3 × 3 matrix

As with a 2 × 2 matrix, the solution to $\mathbf{U}$, the matrix of unknowns, first requires us to be able to determine $\mathbf{C}^{-1}$. Again, to find $\mathbf{C}^{-1}$ we manipulate the matrix $\mathbf{C}$, but in the case of a 3 × 3 matrix the manipulation is more complex. The solution however is highly methodical and lends itself to the use of spreadsheets or computer programs. The inverse of a 3 × 3 matrix of coefficients, $\mathbf{C}^{-1}$, = the matrix of minors ÷ the determinant of the matrix of coefficients.

$$\mathbf{C}^{-1} = \frac{1}{|\mathbf{C}|} \times \begin{pmatrix} \begin{vmatrix} b_2 & c_2 \\ b_3 & c_3 \end{vmatrix} & -\begin{vmatrix} b_1 & c_1 \\ b_3 & c_3 \end{vmatrix} & \begin{vmatrix} b_1 & c_1 \\ b_2 & c_2 \end{vmatrix} \\ -\begin{vmatrix} a_2 & c_2 \\ a_3 & c_3 \end{vmatrix} & \begin{vmatrix} a_1 & c_1 \\ a_3 & c_3 \end{vmatrix} & -\begin{vmatrix} a_1 & c_1 \\ a_2 & c_2 \end{vmatrix} \\ \begin{vmatrix} a_2 & b_2 \\ a_3 & b_3 \end{vmatrix} & -\begin{vmatrix} a_1 & b_1 \\ a_3 & b_3 \end{vmatrix} & \begin{vmatrix} a_1 & b_1 \\ a_2 & b_2 \end{vmatrix} \end{pmatrix}$$

⋯⋯ The matrix of minors

where $|\mathbf{C}|$ is the determinant of the matrix of coefficients

and $|\mathbf{C}| = a_1 \begin{vmatrix} b_2 & c_2 \\ b_3 & c_3 \end{vmatrix} - b_1 \begin{vmatrix} a_2 & c_2 \\ a_3 & c_3 \end{vmatrix} + c_1 \begin{vmatrix} a_2 & b_2 \\ a_3 & b_3 \end{vmatrix}$

For computer applications such as spreadsheets and programs the above can be used directly and the calculations are straightforward. For hand calculations however it becomes difficult to remember the position of the determinants in the matrix and the order of the elements in the determinants, so a stepwise approach for hand calculations is the best approach as follows.

1   Calculate the matrix of minors. To develop the matrix of minors, each element in the matrix is replaced with the determinant of corresponding matrix elements. The corresponding elements are the elements that are not in the same row or the same column as the element being replaced.

Element replaced with the determinant of corresponding elements

Element a_3 is replaced with the determinant of elements b_1, c_1, b_2, c_2.
Element c_2 is replaced with the determinant of elements a_1, b_1, a_3, b_3.

2   The sign of each element in the matrix of minors is modified according to the following matrix of coefficients.

$$
\begin{array}{c}
 \begin{array}{ccc} C_1 & C_2 & C_3 \end{array} \\
\begin{array}{c} R_1 \\ R_2 \\ R_3 \end{array}
\left(
\begin{array}{ccc}
+ & - & + \\
- & + & - \\
+ & - & +
\end{array}
\right)
\end{array}
$$

This means that the element in R_1C_1 remains unchanged, i.e. it is multiplied by $+1$. The element in R_1C_2 changes sign, i.e. it is multiplied by -1.

$$
\left(
\begin{array}{ccc}
\begin{vmatrix} b_2 & c_2 \\ b_3 & c_3 \end{vmatrix} & -\begin{vmatrix} a_2 & c_2 \\ a_3 & c_3 \end{vmatrix} & \begin{vmatrix} a_2 & b_2 \\ a_3 & b_3 \end{vmatrix} \\
-\begin{vmatrix} b_1 & c_1 \\ b_3 & c_3 \end{vmatrix} & \begin{vmatrix} a_1 & c_1 \\ a_3 & c_3 \end{vmatrix} & -\begin{vmatrix} a_1 & b_1 \\ a_3 & b_3 \end{vmatrix} \\
\begin{vmatrix} b_1 & c_1 \\ b_2 & c_2 \end{vmatrix} & -\begin{vmatrix} a_1 & c_1 \\ a_2 & c_2 \end{vmatrix} & \begin{vmatrix} a_1 & b_1 \\ a_2 & b_2 \end{vmatrix}
\end{array}
\right)
$$

3   The matrix is transposed. This means that the elements in the columns of the matrix become the elements in the rows and vice versa.

$$
\left(
\begin{array}{ccc}
\begin{vmatrix} b_2 & c_2 \\ b_3 & c_3 \end{vmatrix} & -\begin{vmatrix} b_1 & c_1 \\ b_3 & c_3 \end{vmatrix} & \begin{vmatrix} b_1 & c_1 \\ b_2 & c_2 \end{vmatrix} \\
-\begin{vmatrix} a_2 & c_2 \\ a_3 & c_3 \end{vmatrix} & \begin{vmatrix} a_1 & c_1 \\ a_3 & c_3 \end{vmatrix} & -\begin{vmatrix} a_1 & c_1 \\ a_2 & c_2 \end{vmatrix} \\
\begin{vmatrix} a_2 & b_2 \\ a_3 & b_3 \end{vmatrix} & -\begin{vmatrix} a_1 & b_1 \\ a_3 & b_3 \end{vmatrix} & \begin{vmatrix} a_1 & b_1 \\ a_2 & b_2 \end{vmatrix}
\end{array}
\right)
$$

4   The fourth and final step is to divide this manipulated matrix by the determinant of the original matrix. This provides us with the inverse of the matrix.

$$
\mathbf{C}^{-1} = \frac{1}{|\mathbf{C}|}
\left(
\begin{array}{ccc}
\begin{vmatrix} b_2 & c_2 \\ b_3 & c_3 \end{vmatrix} & -\begin{vmatrix} b_1 & c_1 \\ b_3 & c_3 \end{vmatrix} & \begin{vmatrix} b_1 & c_1 \\ b_2 & c_2 \end{vmatrix} \\
-\begin{vmatrix} a_2 & c_2 \\ a_3 & c_3 \end{vmatrix} & \begin{vmatrix} a_1 & c_1 \\ a_3 & c_3 \end{vmatrix} & -\begin{vmatrix} a_1 & c_1 \\ a_2 & c_2 \end{vmatrix} \\
\begin{vmatrix} a_2 & b_2 \\ a_3 & b_3 \end{vmatrix} & -\begin{vmatrix} a_1 & b_1 \\ a_3 & b_3 \end{vmatrix} & \begin{vmatrix} a_1 & b_1 \\ a_2 & b_2 \end{vmatrix}
\end{array}
\right)
$$

where:

$$
|\mathbf{C}| = a_1 \begin{vmatrix} b_2 & c_2 \\ b_3 & c_3 \end{vmatrix} - b_1 \begin{vmatrix} a_2 & c_2 \\ a_3 & c_3 \end{vmatrix} + c_1 \begin{vmatrix} a_2 & b_2 \\ a_3 & b_3 \end{vmatrix}
$$

Example

Solving the following set of linear equations:

$$
\begin{cases}
x - 2y + 3z = 4 \\
3x + y - 2z = -7 \\
4x - 4y + 3z = -3
\end{cases}
$$

Representing in matrix form

$$
\begin{pmatrix} 1 & -2 & 3 \\ 3 & 1 & -2 \\ 4 & -4 & 3 \end{pmatrix}
\begin{pmatrix} x \\ y \\ z \end{pmatrix}
=
\begin{pmatrix} 4 \\ -7 \\ -3 \end{pmatrix}
$$

$$\mathbf{C} \times \mathbf{U} = \mathbf{K}$$

$$\mathbf{C} = \begin{pmatrix} 1 & -2 & 3 \\ 3 & 1 & -2 \\ 4 & -4 & 3 \end{pmatrix}$$

Step 1 Solve for $\mathbf{C}^{-1}$.

Find the matrix of minors.

$$\begin{pmatrix} \begin{vmatrix} 1 & -2 \\ -4 & 3 \end{vmatrix} & \begin{vmatrix} 3 & -2 \\ 4 & 3 \end{vmatrix} & \begin{vmatrix} 3 & 1 \\ 4 & -4 \end{vmatrix} \\ \begin{vmatrix} -2 & 3 \\ -4 & 3 \end{vmatrix} & \begin{vmatrix} 1 & 3 \\ 4 & 3 \end{vmatrix} & \begin{vmatrix} 1 & -2 \\ 4 & -4 \end{vmatrix} \\ \begin{vmatrix} -2 & 3 \\ 1 & -2 \end{vmatrix} & \begin{vmatrix} 1 & 3 \\ 3 & -2 \end{vmatrix} & \begin{vmatrix} 1 & 1 \\ 3 & 1 \end{vmatrix} \end{pmatrix}$$

$$\begin{pmatrix} (1 \times 3) - (-2 \times -4) & (3 \times 3) - (-2 \times 4) & (3 \times -4) - (1 \times 4) \\ (-2 \times 3) - (3 \times -4) & (1 \times 3) - (3 \times 4) & (1 \times -4) - (-2 \times 4) \\ (-2 \times -2) - (3 \times 1) & (1 \times -2) - (3 \times 3) & (1 \times 1) - (-2 \times 3) \end{pmatrix}$$

$$\begin{pmatrix} -5 & 17 & -16 \\ 6 & -9 & 4 \\ 1 & -11 & 7 \end{pmatrix}$$

Step 2 Modify according to the matrix of coefficients.

$$\begin{pmatrix} + & - & + \\ - & + & - \\ + & - & + \end{pmatrix} \begin{pmatrix} -5 & -17 & -16 \\ -6 & -9 & -4 \\ 1 & 11 & 7 \end{pmatrix}$$

Step 3 Transpose the matrix.

$$\begin{pmatrix} -5 & -6 & 1 \\ -17 & -9 & 11 \\ -16 & -4 & 7 \end{pmatrix}$$

Step 4 Divide this matrix by the determinant |**C**| of the original matrix to obtain $\mathbf{C}^{-1}$.

$$|\mathbf{C}| = 1 \begin{vmatrix} 1 & -2 \\ -4 & 3 \end{vmatrix} - (-2) \begin{vmatrix} 3 & -2 \\ 4 & 3 \end{vmatrix} + 3 \begin{vmatrix} 3 & 1 \\ 4 & -4 \end{vmatrix}$$

$$= 1 \times (1 \times 3) - (-2 \times -4) - (-2) \times (3 \times 3) - (-2 \times 4) + 3 \times (3 \times -4) - (1 \times 4)$$

$$= -19$$

$$\mathbf{C}^{-1} = \tfrac{1}{-19} \begin{pmatrix} -5 & -6 & 1 \\ -17 & -9 & 11 \\ -16 & -4 & 7 \end{pmatrix}$$

Step 5 Multiply both sides of the matrix equation by the inverse matrix.

$$\mathbf{C}^{-1} \times \mathbf{C} \times \mathbf{U} = \mathbf{C}^{-1} \times \mathbf{K}$$

$$\mathbf{U} = \mathbf{C}^{-1} \times \mathbf{K}$$

$$\begin{pmatrix} x \\ y \\ z \end{pmatrix} = \overset{\mathbf{C}^{-1}}{\begin{pmatrix} 4 \\ -7 \\ -3 \end{pmatrix}}$$

$$\begin{pmatrix} x \\ y \\ z \end{pmatrix} = \tfrac{1}{-19} \begin{pmatrix} -5 & -6 & 1 \\ -17 & -9 & 11 \\ -16 & -4 & 7 \end{pmatrix} \begin{pmatrix} 4 \\ -7 \\ -3 \end{pmatrix}$$

continued

continued

$$= \frac{1}{-19} \begin{pmatrix} (-5 \times 4) & + & (-6 \times -7) & + & (1 \times -3) \\ (-17 \times 4) & + & (-9 \times -7) & + & (11 \times -3) \\ (-16 \times 4) & + & (-4 \times -7) & + & (7 \times -3) \end{pmatrix}$$

$$= \frac{1}{-19} \begin{pmatrix} 19 \\ -38 \\ -57 \end{pmatrix}$$

$$= \begin{pmatrix} -1 \\ 2 \\ 3 \end{pmatrix}$$

$$x = -1, y = 2, z = 3$$

Exercises 23.11

1 **a** Express the simultaneous equations

$$2x - 3y = 9$$
$$5x - 7y = 22$$

as a single matrix equation.

  b Solve the above equations using matrices, given that the inverse of the matrix

$$\begin{pmatrix} 2 & -3 \\ 5 & -7 \end{pmatrix} \text{ is } \begin{pmatrix} -7 & 3 \\ -5 & 2 \end{pmatrix}.$$

2 You are given that if $\mathbf{A} = \begin{pmatrix} a & b \\ c & d \end{pmatrix}$ then $\mathbf{A}^{-1} = \frac{1}{|\mathbf{A}|} = \begin{pmatrix} d & -b \\ -c & a \end{pmatrix}.$

  a If $\mathbf{P} = \begin{pmatrix} 3 & 2 \\ 5 & 4 \end{pmatrix}$:

    i evaluate $|\mathbf{P}|$

    ii write down the matrix $\mathbf{P}^{-1}$

  b Use the result for $\mathbf{P}^{-1}$ above to solve the system of equations

$$\begin{cases} 3x + 2y = 4 \\ 5x + 4y = 10 \end{cases}$$

3 If $\mathbf{M} = \begin{pmatrix} 4 & 3 \\ 1 & -2 \end{pmatrix}$ and $\mathbf{N} = \begin{pmatrix} 2 & 3 \\ 1 & -4 \end{pmatrix}$:

  a find the product $\mathbf{MN}$

  b write down the matrix $\mathbf{M}^{-1}$

  c use the result of **a** above to solve the simultaneous equations below, showing each step of the working:

$$\begin{cases} 4x + 3y = 7 \\ x - 2y = 10 \end{cases}$$

4 For the following systems of simultaneous equations:

  i express the equations in matrix form   **ii** find the determinant

  iii find the inverse matrix   **iv** solve the equations

a $a - 3b = -1$
 $4a - 11b = -2$

b $a - b = 6$
 $2a + 3b = 2$

c $3x - y = 6$
 $-2x + y = 2$

d $4p + 4q = 8$
 $5p - 3q = 2$

e $x - 3y = 4$
 $4x + 8y = 6$

5 The inverse of $\begin{pmatrix} 4 & -1 & 1 \\ 3 & -2 & -1 \\ 1 & -1 & -1 \end{pmatrix}$ is $\begin{pmatrix} 1 & -2 & 3 \\ 2 & -5 & 7 \\ -1 & 3 & -5 \end{pmatrix}$. Use this result to solve the system

of equations $\begin{cases} 4a - b + c = 15 \\ 3a - 2b - c = 13 \\ a - b - c = 3 \end{cases}$

6 If $\mathbf{P} = \begin{pmatrix} 11 & -5 & -3 \\ -8 & 4 & 2 \\ -7 & 3 & 1 \end{pmatrix}$ and $\mathbf{Q} = \frac{1}{2}\begin{pmatrix} 1 & 2 & -1 \\ 3 & 5 & -1 \\ -2 & -1 & -2 \end{pmatrix}$:

a write down the matrix **PQ**

b write the system of equations below as a single matrix equation

$\begin{cases} 11x - 5y - 3z = -12 \\ -8x + 4y + 2z = 10 \\ -7x + 3y + z = 10 \end{cases}$

c use the result of **a** above to solve the simultaneous equations in **b** using matrices.

7 If $\mathbf{P} = \begin{pmatrix} 3 & -4 & 2 \\ -2 & 1 & 0 \\ -1 & 1 & 0 \end{pmatrix}$, $\mathbf{Q} = \begin{pmatrix} 1 & 2 & -2 \\ 2 & 5 & -4 \\ 3 & 7 & -5 \end{pmatrix}$, $\mathbf{R} = \begin{pmatrix} 3 & -4 & 2 \\ -2 & 1 & 0 \\ -1 & -1 & 1 \end{pmatrix}$:

a write down the matrices (i) **PQ** (ii) **PR** (iii) **QR**

b which two of these matrices are inverses of each other?

c express the simultaneous equations $\begin{cases} a + 2b - 2c = 3 \\ 2a + 5b - 4c = 7 \\ 3a + 7b - 5c = 8 \end{cases}$ as a single matrix equation

d solve the equations in **c** using matrices.

8 Solve the following systems of simultaneous equations using matrices:

a $\begin{cases} 3x + 2y - 2z = -3 \\ 2x + 2y - z = 1 \\ -4x - 3y + 2z = 0 \end{cases}$ given that the inverse of $\begin{pmatrix} 3 & 2 & -2 \\ 2 & 2 & -1 \\ -4 & -3 & 2 \end{pmatrix}$ is $\begin{pmatrix} -1 & -2 & -2 \\ 0 & 2 & 1 \\ -2 & -1 & -2 \end{pmatrix}$

b $\begin{cases} 2a - 4b - c = 0 \\ 5a - 10b - 3c = 1 \\ 15a - 29b - 9c = 5 \end{cases}$ given that if $\mathbf{K} = \begin{pmatrix} 2 & -4 & -1 \\ 5 & -10 & -3 \\ 15 & -29 & -9 \end{pmatrix}$, then $\mathbf{K}^{-1} = \begin{pmatrix} 3 & -7 & 2 \\ 0 & -3 & 1 \\ 5 & -2 & 0 \end{pmatrix}$

c $\begin{cases} 2p + 3q + 3r = -2 \\ 3p + 5q + 5r = -4 \\ 5p + 3q + 4r = 0 \end{cases}$ given that $\begin{pmatrix} 2 & 3 & 3 \\ 3 & 5 & 5 \\ 5 & 3 & 4 \end{pmatrix}\begin{pmatrix} 5 & -3 & 0 \\ 13 & -7 & -1 \\ -16 & 9 & 1 \end{pmatrix} = \begin{pmatrix} 1 & 0 & 0 \\ 0 & 1 & 0 \\ 0 & 0 & 1 \end{pmatrix}$

continued

continued

$$\text{d} \quad \begin{cases} 2x + 3y + 2z = 1 \\ 3x + 4y + 3z = 1 \\ 5x + y + 4z = -1 \end{cases} \text{ given that } \begin{pmatrix} 2 & 3 & 2 \\ 3 & 4 & 3 \\ 5 & 1 & 4 \end{pmatrix}^{-1} = \begin{pmatrix} 13 & -10 & 1 \\ 3 & -2 & 0 \\ -17 & 13 & -1 \end{pmatrix}$$

$$\text{e} \quad \begin{cases} 23x + 5y - 35z = -2 \\ 13x + 3y - 20z = -1 \\ 19x + 4y - 29z = -2 \end{cases} \text{ given that } \begin{pmatrix} 7 & -5 & -5 \\ 3 & 2 & -5 \\ 5 & -3 & -4 \end{pmatrix} \begin{pmatrix} 23 & 5 & -35 \\ 13 & 3 & -20 \\ 19 & 4 & -29 \end{pmatrix} = \begin{pmatrix} 1 & 0 & 0 \\ 0 & 1 & 0 \\ 0 & 0 & 1 \end{pmatrix}$$

$$\text{f} \quad \begin{cases} a + 2b + 3c = -2 \\ 2a + 3b + 4c = 0 \\ 3a + b - 2c = 17 \end{cases} \text{ given that } \begin{pmatrix} 1 & 2 & 3 \\ 2 & 3 & 4 \\ 3 & 1 & -2 \end{pmatrix} \begin{pmatrix} -10 & 7 & -1 \\ 16 & -11 & 2 \\ -7 & 5 & -1 \end{pmatrix} = \mathbf{I}$$

$$\text{g} \quad \begin{cases} 2a + 2b = -6 \\ 3a + 2b + c = 0 \\ 7a + 5b + 2c = -1 \end{cases} \text{ given that if } \mathbf{M} = \begin{pmatrix} -1 & -6 & 3 \\ 1 & 4 & -2 \\ 1 & 11 & -5 \end{pmatrix}, \text{ then } \mathbf{M}^{-1} = \begin{pmatrix} 2 & 3 & 0 \\ 3 & 2 & 1 \\ 7 & 5 & 2 \end{pmatrix}$$

+++ 9 For the following systems of simultaneous equations:

i   express the equations in matrix form

ii find the determinant

iii find the inverse matrix

iv solve the equations

$$\text{a} \quad \begin{cases} 2x - 2y - z = 1 \\ x - 5y - 3z = 2 \\ x + 6y + 4z = -3 \end{cases} \qquad \text{b} \quad \begin{cases} 2a + 3b + 2c = -4 \\ 2a + 2b + 2c = -2 \\ 5a + b + 4c = -6 \end{cases} \qquad \text{c} \quad \begin{cases} p + q + r = 4 \\ 2p - 2q + r = 5 \\ 2p + 2q + 4r = 6 \end{cases}$$

$$\text{d} \quad \begin{cases} 2x + 2y - 2z = -2 \\ -2x + 2y - z = 1 \\ -4x - 3y + 3z = 0 \end{cases} \qquad \text{e} \quad \begin{cases} 4a + 2b - 4c = -4 \\ 4a - 5b + 2c = -2 \\ 5a - 2b - c = -5 \end{cases}$$

23.12 3 × 3 DETERMINANTS: DEFINITION AND EVALUATION

$\begin{vmatrix} a & b & c \\ d & e & f \\ g & h & i \end{vmatrix}$ is a square array of 'elements' having three rows and three columns. It is called a '3 × 3 determinant' or a 'determinant of third order'.

The value of a **second-order** determinant may be defined to provide a shorthand method of solving two simultaneous equations in two unknowns. The value of a **third-order** determinant is defined so that it provides a shorthand method of solving **three** simultaneous equations in three unknowns.

The value of the above determinant is defined as $aei + bfg + cdh - gec - hfa - idb$. There are several ways of obtaining this result without having the very difficult task of committing it to memory. We will use the method called the *Rule of Sarrus*. This method is the simplest but it *applies only to determinants of order* **three.** (Later, when you study determinants of higher orders, you will learn other methods and ones that are easier to program for a computer.)

The Rule of Sarrus

1 Write down the determinant, repeating the first two columns.
2 Obtain the products on the diagonals as shown below.
3 Add the lower products and add the upper products.
4 Subtract the sum of the upper products from the sum of the lower products.

Follow the application of this rule as we apply it to the general determinant $\begin{vmatrix} a & b & c \\ d & e & f \\ g & h & i \end{vmatrix}$

$$
1 \quad \begin{matrix} a & b & c & a & b \\ d & e & f & d & e \\ g & h & i & g & h \end{matrix}
$$

2, 3 $(Sum = gec + hfa + idb)$

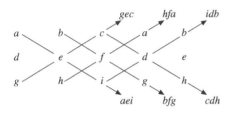

$(Sum = aei + bfg + cdh)$

4 Value: $(aei + bfg + cdh) - (gec + hfa + idb)$

Example

Evaluate: $\begin{vmatrix} 2 & 0 & -3 \\ -1 & 4 & 0 \\ 3 & 1 & -2 \end{vmatrix}$

Method: $(Sum = -36)$

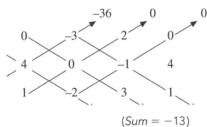

$(Sum = -13)$

Value: $= (-13) - (-36)$

$= 23$

Note: When copying down a determinant be very careful to include any negative signs. *Check* your copy before working on it. If you copied it row by row, check it column by column. It is very annoying to work on data that is later discovered to have been copied down incorrectly.

Exercises 23.12

1 Evaluate: (**Note:** All the elements are exact numbers.)

a $\begin{vmatrix} 2 & 1 & -2 \\ 3 & 0 & -1 \\ 5 & 0 & 1 \end{vmatrix}$ b $\begin{vmatrix} 4 & 3 & 5 \\ 0 & 5 & 0 \\ -4 & 4 & -5 \end{vmatrix}$ c $\begin{vmatrix} 1 & 1 & 0 \\ 1 & -1 & 2 \\ -1 & 1 & 2 \end{vmatrix}$ d $\begin{vmatrix} 1 & 2 & 3 \\ 2 & -3 & -8 \\ -5 & -2 & 1 \end{vmatrix}$

e $\begin{vmatrix} 3 & 1 & 2 \\ 0 & 4 & 1 \\ 2 & 3 & 2 \end{vmatrix}$ f $\begin{vmatrix} 0 & t & n \\ -t & 0 & x \\ -n & -x & 0 \end{vmatrix}$ g $\begin{vmatrix} 42 & 28 & 62 \\ 36 & 29 & 91 \\ 37 & 30 & 47 \end{vmatrix}$ h $\begin{vmatrix} 20 & -24 & 28 \\ -31 & 47 & -64 \\ 83 & -51 & -86 \end{vmatrix}$

2 In each case below, evaluate the pronumeral:

a $\begin{vmatrix} x & 2x & 3x \\ 1 & 2 & 0 \\ 0 & 1 & 3 \end{vmatrix} = 9$ b $\begin{vmatrix} 1 & x & 2 \\ 2 & 0 & x \\ 3 & 1 & 1 \end{vmatrix} = 10$ c $\begin{vmatrix} 2 & n & 1 \\ 1 & n & 2 \\ 1 & n & 1 \end{vmatrix} = 5$

d $\begin{vmatrix} t & t & 1 \\ 4 & 1 & 0 \\ -2 & t & t \end{vmatrix} = 3$ e $\begin{vmatrix} 1 & 0 & k \\ -1 & 2 & 2 \\ k & 2 & 0 \end{vmatrix} = \begin{vmatrix} -1 & k & 6 \\ 1 & 0 & -2 \\ -2 & 1 & k \end{vmatrix}$

23.13 SOLUTIONS OF SIMULTANEOUS LINEAR EQUATIONS USING 3 × 3 DETERMINANTS

A system of three linear equations in three unknowns may be solved algebraically by the elimination method.

However, this method is usually very tedious and the equations are more easily solved using determinants.

The general equations are: $\begin{cases} a_1x + b_1y + c_1z = d_1 \\ a_2x + b_2y + c_2z = d_2 \\ a_3x + b_3y + c_3z = d_3 \end{cases}$

If we solve these equations by elimination we obtain the solution:

$$x = \frac{\triangle x}{\triangle}, \ y = \frac{\triangle y}{\triangle}, \ z = \frac{\triangle z}{\triangle}$$

where: $\triangle = \begin{vmatrix} a_1 & b_1 & c_1 \\ a_2 & b_2 & c_2 \\ a_3 & b_3 & c_3 \end{vmatrix}$, which is the determinant formed by using the coefficients on the left-hand side of the equation

$\triangle_x = \begin{vmatrix} d_1 & b_1 & c_1 \\ d_2 & b_2 & c_2 \\ d_3 & b_3 & c_3 \end{vmatrix}$, which is the same determinant but with the coefficients of x replaced by the constants (i.e. the numbers on the right-hand side)

$$\triangle_y = \begin{vmatrix} a_1 & d_1 & c_1 \\ a_2 & d_2 & c_2 \\ a_3 & d_3 & c_3 \end{vmatrix}, \text{ which is } \triangle \text{ again but with the coefficients of } y \text{ replaced by the constants}$$

$$\triangle_z = \begin{vmatrix} a_1 & b_1 & d_1 \\ a_2 & b_2 & d_2 \\ a_3 & b_3 & d_3 \end{vmatrix}, \text{ which is } \triangle \text{ again but with the coefficients of } z \text{ replaced by the constants.}$$

Example

Given: $\begin{cases} x - 2y + 3z = 4 \\ 3x + y - 2z = -7 \\ 4x - 4y + 3z = -3 \end{cases}$

$$\triangle = \begin{vmatrix} 1 & -2 & 3 \\ 3 & 1 & -2 \\ 4 & -4 & 3 \end{vmatrix} = (-17) - 2 = -19$$

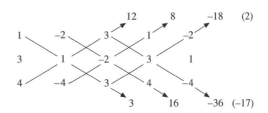

$$\triangle_x = \begin{vmatrix} 4 & -2 & 3 \\ -7 & 1 & -2 \\ -3 & -4 & 3 \end{vmatrix} = (84) - (65) = 19$$

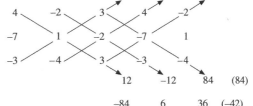

$$\triangle_y = \begin{vmatrix} 1 & 4 & 3 \\ 3 & -7 & -2 \\ 4 & -3 & 3 \end{vmatrix} = (-80) - (-42) = -38$$

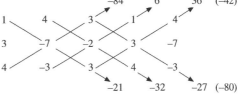

$$\triangle_z = \begin{vmatrix} 1 & -2 & 4 \\ 3 & 1 & -7 \\ 4 & -4 & -3 \end{vmatrix} = (5) - (62) = -57$$

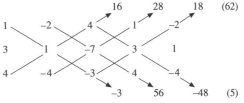

Solution

$$x = \frac{\triangle_x}{\triangle} = \frac{19}{-19} = -1$$

$$y = \frac{\triangle_y}{\triangle} = \frac{-38}{-19} = 2$$

$$z = \frac{\triangle_z}{\triangle} = \frac{-57}{-19} = 3$$

Note:

■ After much practice, the value of a determinant can be found on a calculator, using the *Rule of Sarrus*, showing no intermediate results. The diagonal products can be summed using the M^+ and M^- keys. However, for the time being, you are advised to *write down* the product of each diagonal before adding it to the calculator memory, as in the examples given. This allows you to check your work more easily and also allows for credit to be given in an examination for knowledge of the *method* even if an error is made during the computation.

■ When you have solved a set of simultaneous equations, check your result by substituting your values back into the original equations. You will be surprised how often careless errors are made during a long series of calculations.

■ When using a calculator, always work with *all* the significant figures in the data. State your result to the *appropriate* number of significant figures, that is, the number required or justified.

Remember: **a** Before using the *Rule of Sarrus* to solve a set of simultaneous equations, make sure that:

i all the equations have the pronumerals in the *same order*

ii all the pronumerals are on the left-hand sides of the equations and all the constants are on the right-hand sides

iii if a pronumeral is *absent* from an equation, write it in with a zero coefficient—for example, if there is no y in an equation, write it in as $0y$.

b When copying down an array to work on, be careful to place the elements in a neat rectangle so that the diagonal elements are approximately in straight lines. An untidy array leads to errors in computation.

c Before working on your arrays, check that you have made no errors, especially by omitting any negative signs. Discover any errors *before* you start to multiply.

d When multiplying out diagonals, either mentally or with a calculator, ignore any negative signs present. Decide *after* the multiplication whether the product should be positive or negative.

Exercises 23.13

1   Solve for the pronumerals using determinants (do not use a calculator):

a $\begin{cases} 2x + y + 3z = 4 \\ 3x - y - 4z = 5 \\ 4x + 3y + 2z = 1 \end{cases}$

b $\begin{cases} 2p + 4q + 3r + 8 = 0 \\ p + 3q - 2r + 2 = 0 \\ 3p - 5q - 4r - 4 = 0 \end{cases}$

c $\begin{cases} 3x + 2y + 4z = -7 \\ 2y - 3z + 4x = 6 \\ 4z - 5x - 2y = -7 \end{cases}$

d $\begin{cases} 4n - 3p + 5t = -3 \\ 3n + 4p = -6 \\ 5p - 4t = -4 \end{cases}$

2 Solve for the pronumerals using a calculator:

a $\begin{cases} 37x + 49y + 76z = 98 \\ 68y - 34x - 93z = 136 \\ 82z - 36y + 29x = -72 \end{cases}$ b $\begin{cases} 0.002x + 0.003y + 0.001z = 0.01 \\ 0.3x - 0.4y + 0.2z = 0.8 \\ 4000x + 5000y - 3000z = 4000 \end{cases}$

(*Hint:* In **b**, multiply or divide both sides of each equation by a constant so as to obtain simpler numbers. State results correct to 3 significant figures.)

c $\begin{cases} 2(3.7k - 1.4t) + 3(4.3t + 2.6w) + 8.7 = 0 \\ 4(2.3k + 1.7w) - 5.3w - 6.1 = 0 \\ 3(3.9k + 4.2t) = 0 \end{cases}$

3

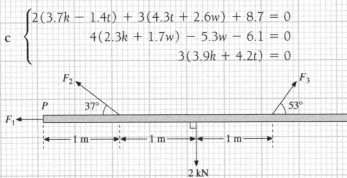

The equations for equilibrium of this beam are:

Vertical forces: $\quad\quad\quad 0.6F_2 - 2 + 0.8F_3 = 0$

Horizontal forces: $\quad\quad F_1 + 0.8F_2 - 0.6F_3 = 0$

Moments about point P: $\quad 0.6F_2 - 4 + 2.4F_3 = 0$

Using determinants, solve for F_1, F_2 and F_3, stating the results correct to 3 significant figures. Explain the meaning of any negative values.

4

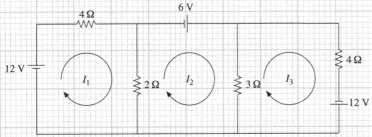

Applying Kirchhoff's laws to this network:

$$4I_1 + 2(I_1 - I_2) = 12$$
$$2(I_2 - I_1) + 3(I_2 - I_3) = 6$$
$$3(I_3 - I_2) + 4I_3 = 12$$

Use determinants to solve for I_1, I_2 and I_3, stating their values correct to 3 significant figures.

continued

continued

5   A developer receives council permission to divide his land of area 44 ha into 100 blocks, the only areas allowed for a block being 2 ha, 0.5 ha and 0.2 ha. He prices the 2 ha blocks at $100 000 each, the 0.5 ha blocks at $40 000 each and the 0.2 ha blocks at $20 000 each. He sells all the blocks, the total gross income from the sales being $3 200 000. How many blocks of each size did he sell?

6   A firm has a stock of three different bronze alloys. Alloy A consists of 95% copper, 3% tin and 2% zinc. Alloy B consists of 90% copper, 9% tin and 1% zinc. Alloy C consists of 80% copper, 15% tin and 5% zinc. How many kilograms of each of these alloys must be melted and mixed in order to produce 100 kg of a new alloy that consists of 87% copper, 9.6% tin and 3.4% zinc?

CHAPTER 24

STATISTICS AND PROBABILITY

Learning objectives

■ Calculate the median, mode, mean, range, variance and standard deviation of data sets.
■ Understand the difference between a population and a sample.
■ Understand what a binomial event is and be able to calculate the probability of a binomial event.
■ Calculate the mean and standard deviation of a binomial event.
■ Understand the normal curve and be able to use it to calculate the probability of a binomial event.

24.1 STATISTICS

Statistics is the field of mathematics involved in collecting, analysing, organising and representing data. There is virtually no limit to subjects that the data could be related to. By allowing the data to be organised and represented in a meaningful and easy-to-understand manner, statistics can help us to make decisions or infer conclusions from data that would otherwise be confusing and overwhelmingly complex.

A *population* is the entire set of data. For instance, if we are applying statistical analysis techniques to data about students in a school then the population would include all the students enrolled at that school.

A *sample* is a part or portion of the population, for example one hundred students randomly chosen from the school population.

24.2 MEDIAN, MODE AND MEAN

In statistics there are three parameters that we use to explore the middle or average of the data. These three measurements are called the Median, the Mode and the Mean.

Median

The median is the middle value in the ordered data set. This means that there is the same number of values in the data set that are larger and smaller than the median. If a sample consists of five numbers and they are ordered from lowest to highest, then the third number is the median.

Example

The median of the ordered set of numbers (3, 6, 8, 9, 10) is 8.

If the data set contains an even number of elements then the median is the average of the two middle values.

Example

The median of the ordered set of numbers (3, 6, 6, 8, 9, 10) is 7.

The median is based on the order that the numbers appear in the data set and is not affected by extremes in the data set—for example, in the ordered set of numbers (3, 6, 8, 9, 22) the median is still 8.

Mode

While the median defines the *middle* number in the data set, the **mode** is the value that occurs most often.

Example

For the data set (3, 6, 6, 8, 9, 10) the mode is 6.

Again the mode is not affected by extremes in the data set. For the data set (3, 6, 6, 8, 9, 22) the mode is still 6.

Mean

The mean, μ, is the average value in the data set. It is defined as the sum of all the values divided by the number of values in the data set. Thus, unlike the mode and median, the mean considers all the values in the data set.

For a population with individual variables, the mean of the **population**, is defined as:

$$\mu = \frac{x_1 + x_2 + x_3 + \ldots + x_N}{N}$$

where N is the number of variables in the population.

The mean of a **sample** is designated by the symbol $\bar{x}$.

For a sample with n individual variables $x_1, x_2, x_3, \ldots, x_n$, the mean is defined as

$$\bar{x} = \frac{x_1 + x_2 + x_3 + \ldots + x_n}{n}.$$

Examples

For the data set: (3, 6, 6, 8, 9, 10), $\mu = \dfrac{3 + 6 + 6 + 8 + 9 + 10}{6} = 7$

For the data set: (3, 6, 6, 8, 9, 22), $\mu = \dfrac{3 + 6 + 6 + 8 + 9 + 22}{6} = 9$

Even though the mean considers all the values in a particular data set, different data sets can still have the same mean.

Example

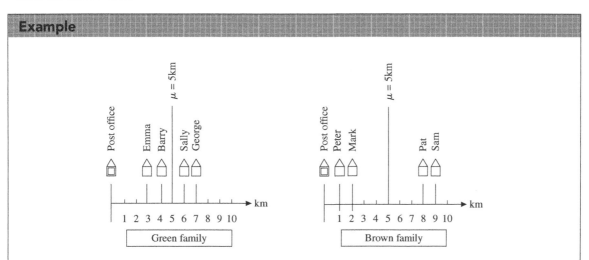

The members of the Green and Brown families all live in separate houses. The diagram shows how far each member of each family lives from the town's post office.

The mean distance that the Green family lives from the post office is:

$$\frac{3 + 4 + 6 + 7}{4} = 5 \text{ km; i.e. } \mu = 5 \text{ km.}$$

The mean distance that the Brown family lives from the post office is:

$$\frac{1 + 2 + 8 + 9}{4} = 5 \text{ km; i.e. } \mu = 5 \text{ km.}$$

24.3 RANGE, VARIANCE AND STANDARD DEVIATION

So far we have been learning about statistical parameters that explore the middle of the data. To further understand a data set we need to also explore the spread of the data and the range of possible values. In the above example we saw that both families live the same distance, on average, from the town post office. However, in the case of the Green family, a family member of might live anywhere between 3 km and 7 km from the post office, and a Brown family member might live anywhere between 1 km and 9 km from the post office. Thus, while the mean of the data is the same, the spread or variance of the data is different in each case.

Range

The range of a sample is obtained by subtracting the smallest amount in the data set from the largest amount. In the above case, the range of the Green family data is $7 - 3 = 4$ km, but the range of the Brown family data is $9 - 1 = 8$ km.

Variance

Variance is a parameter that quantifies how individual data values vary from the mean of the data. Variance of a population is defined as the sum of the squares of the differences between the data and the mean divided by the number of data items.

It is calculated differently for population data than for sample data.

The variance of a population is denoted by the symbol σ^2, and is calculated by:

$$\sigma^2 = \frac{(x_1 - \mu)^2 + (x_2 - \mu)^2 + \ldots + (x_N - \mu)^2}{N}$$

The variance of a sample is denoted by the symbol s^2, and is calculated by:

$$s^2 = \frac{(x_1 - \bar{x})^2 + (x_2 - \bar{x})^2 + \ldots + (x_n - \bar{x})^2}{n - 1}$$

Standard deviation

The standard deviation is the square root of the variance.

For a population the standard deviation $= \sigma$.

For a sample the standard deviation $= s$.

Example

The data for the Green and Brown families is for all members of the Green family and all members of the Brown family. Thus we have gathered data about the populations.

For the Green family data:

The population variance is $\sigma^2 = \dfrac{(3 - 5)^2 + (4 - 5)^2 + (6 - 5)^2 + (7 - 5)^2}{4} = 2.5$ km.

The standard deviation, $\sigma = \sqrt{2.5} \approx 1.58$ km.

For the Brown family data:

The population variance is $\sigma^2 = \dfrac{(1 - 5)^2 + (2 - 5)^2 + (8 - 5)^2 + (9 - 5)^2}{4} = 12.5$ km.

The standard deviation, $\sigma = \sqrt{12.5} \approx 3.54$ km.

So, while the mean of both data sets is 5 km, the standard deviation for the Green family data is 1.58 km and for the Brown family data the standard deviation is 3.54 km. We immediately see that a data value for the Brown family is likely to vary more from the mean that a data value for the Green family.

Example

Five water tanks are randomly removed from a production line and measured as part of a quality control program. The heights and diameters of the five tanks are measured and tabulated below. These measurements constitute a sample taken from the population.

Tank No	Height (mm)	Diameter (mm)
1	2200	5200
2	2180	5220
3	2160	5220
4	2170	5170
5	2180	5190

The mean height, $\bar{x} = \dfrac{2200 + 2180 + 2160 + 2170 + 2180}{5} = 2178$ mm.

The mean diameter, $\bar{x} = \dfrac{5200 + 5220 + 5220 + 5170 + 5190}{5} = 5200$ mm.

The standard deviation of the height data is therefore:

$$s = \sqrt{\frac{(2200 - 2178)^2 + (2180 - 2178)^2 + (2160 - 2178)^2 + (2170 - 2178)^2 + (2180 - 2178)^2}{5 - 1}}$$

$$= \sqrt{\frac{484 + 4 + 324 + 64 + 4}{5 - 1}} \approx 14.83 \text{ mm.}$$

The standard deviation of the diameter data is therefore:

$$s = \sqrt{\frac{(5200 - 5200)^2 + (5220 - 5200)^2 + (5220 - 5200)^2 + (5170 - 5200)^2 + (5190 - 5200)^2}{5 - 1}}$$

$$= \sqrt{\frac{0 + 400 + 400 + 900 + 100}{5 - 1}} \approx 21.21 \text{ mm.}$$

24.4 EXPLORING THE LOCATION OF THE DATA

Now that we have statistical parameters for the middle and the extremes of the data, we can quantify where a data element is located within a data set.

We describe the location of a data element within a data set by the number of standard deviations it is from the mean.

We refer to this parameter as Z.

For a variable x_1

$$Z = \frac{|x_1 - \mu|}{\sigma}$$

So if $Z = 1$, then $x_1 = \mu \pm \sigma$.

That means that when $Z = 1$, the corresponding data value x_1 varies from the mean by one standard deviation. Put another way, x_1 is one standard deviation either side of the mean when $Z = 1$.

Example

This example again refers back to the data for the Green and Brown families.
For the Green family:

$\mu = 5$ km $\sigma = 1.58$ km

For the Brown family

$\mu = 5$ km $\sigma = 3.54$ km

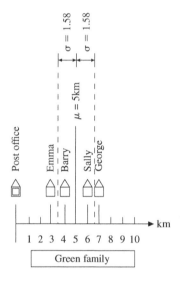 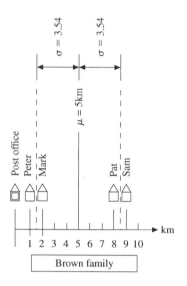

For Emma, who lives 3 km from the post office, we can calculate the value of Z:

$$Z = \frac{|3 - 5|}{1.58} = 1.27$$

i.e. Emma lives 1.27 standard deviations away from the mean distance.

Example

How far from the post office would a Green family member need to live to have a corresponding Z value of 2?

$$2 = \frac{x - \mu}{\sigma}$$

$\therefore 2\sigma + \mu = x$

$\therefore x = (2 \times 1.58) + 5 = 8.16$ km.

Exercise 24.1

For the following data sets calculate the mean, mode, range and median. Calculate your answer to one decimal place.

a   3, 9, 3, 6, 7, 8

b   6, 2, 5, 7, 7, 9

c 19, 22, 21, 26, 21, 20, 18

d 5, 3, 8, 12, 56, 16, 34, 12

e 2, 5, 8, 5, 12, 34, 4, 6

Exercises 24.2

1 Twelve people apply to a position vacant advertisement. As part of the selection criteria the applicants agree to undergo an IQ test. The results of the IQ tests are:

   120, 110, 92, 90, 130, 92, 87, 95, 99, 108, 77, 118

   For the population of data calculate the mean, range and standard deviation.

2 The following data sets describe a population. Calculate the population mean, variance and the standard deviation to two decimal places.

   a 4, 5, 4, 4, 8

   b 1, 2, 11, 5, 8, 9

   c 12, 8, 4

   d 7, 3, 7, 4, 5, 2

   e 9, 1, 12, 3, 2

3 If the above data sets were for a sample taken from a larger population, calculate the sample variance and standard deviation to two decimal places.

Exercises 24.3

+++ 1 A teacher sets a test for a class of students. He decides to analyse the test results using statistical analysis techniques and then grade the students accordingly.

A student scoring between the mean and one standard deviation above the mean is assigned a grade of B.

A student scoring between the mean and one standard deviation below the mean is assigned a grade of C.

A student scoring between one standard deviation above the mean and two standard deviations above the mean is assigned a grade of A.

continued

continued

A student scoring between one standard deviation below the mean and two standard deviations below the mean is assigned a grade of D.

A student scoring greater than two standard deviations above the mean is assigned a grade of A+.

A student scoring more than two standard deviations below the mean fails the test.

Class results	
Student	**Test score (%)**
Thomas	74
Charles	55
Sarah	81
Mathew	68
James	71
Jessica	74
Daniel	86
Jack	66
Emma	73
Laura	72
Joshua	84
Alice	68
Samantha	70

    a Calculate the mean and range of the class test scores.

    b Calculate the population variance and standard deviation.

    c What grade will Joshua, Alice and Jessica receive?

    d Does anyone fail the test, if so whom?

    e Does anyone receive a grade of A+, if so who?

2 From the data sets listed in Exercises 24.2, question 2, calculate Z for the data element listed below—i.e. calculate how many standard deviations from the mean the nominated data element is. Give your answer to two decimal places.

    a data element 4

    b data element 9

    c data element 12

    d data element 7

    e data element 9

24.5 GRAPHS

An important graph used in the field of probability and statistics is called a histogram. A histogram is a graphical representation of the frequency of an event occurring. The *x*-axis provides a description of the event, and the *y*-axis shows the frequency of the event occurring.

> **Rule:** The frequency of an event can be converted to the relative frequency of the event using the equation.
>
> $$\text{Relative frequency} = \frac{\text{frequency of events}}{\text{number of events}}$$

Examples

1   Children standing by the side of the road record how many cars of different colours they see. They record a total of 39 cars.

Frequency

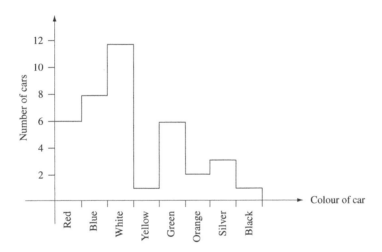

Relative frequency

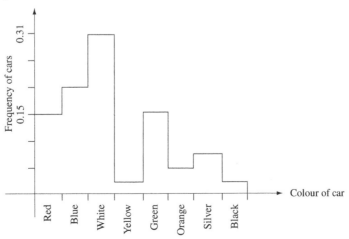

continued

continued

2   The total length of bolts produced by a machine is monitored to determine when the machine requires maintenance. The following histogram describes the length of bolts produced by the machine. A total of 2550 bolts are measured.

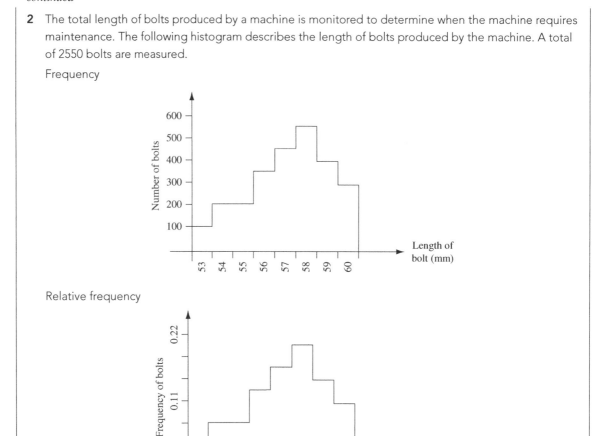

24.6 PROBABILITY

The probability of an event is a measure of the likelihood of the event occurring. If it will definitely happen then it has a probability of 1, and if an event will definitely *not* happen it is said to have a probability of 0. The range of probabilities therefore lies between 0 and 1.

The 'probability' of the occurrence of an event is defined as the value at which the relative frequency of occurrence stabilises in a long series of trials.

The statement that the probability of obtaining 'heads' on the toss of a coin is 50% means that in a long series of trials, as the number of tosses increases, the relative frequency (fraction) of heads will stabilise at 50%. The relative frequency will almost certainly not steadily approach 50% but at times will drift away from the 50% value. It is quite possible for example that there will be six heads after 10 tosses (60%) and seven heads after 11 tosses (approx. 63.6%), or there may be 48 heads after 100 tosses (48%) and still only 48 heads after 101 tosses (approx. 47.5%). However as the number of tosses increases these 'wanderings' from the 50% value will decrease in size and there will be a definite tendency toward the 50% value. When an unbiased coin is tossed there are only two possible outcomes to this experiment—heads or tails.

If an experiment has n possible outcomes, all equally likely, it follows that the probability of occurrence of each outcome is $\dfrac{1}{n}$.

> **SUMMARY**
> The probability of tossing a coin and its landing heads up is 0.5
> The probability of rolling a six on a die is 1/6
> The probability of rolling an eight on a standard six-sided die is 0

Binomial events

A binomial event is an event with two possible outcomes. The first outcome is that a result with a particular probability does occur and the second outcome is that a result with a particular probability does not occur. These events are complementary—if the probability of the first event is p then the probability of the second event is $1 - p$.

The probability of tossing a coin and its landing heads up is 0.5. The probability that it will *not* land heads up is $1 - 0.5 = 0.5$.

On a six-sided die, the probability of rolling a six is $\dfrac{1}{6}$. The probability of *not* rolling a six is therefore $1 - \dfrac{1}{6} = \dfrac{5}{6}$.

Factorial

'N factorial', written as $N!$, is defined as N multiplied by each positive integer less than N:

$$N! = N \times (N - 1) \times (N - 2) \times \ldots \times 3 \times 2 \times 1, \text{ with the special definition that } 0! = 1.$$

For example: $5! = 5 \times 4 \times 3 \times 2 \times 1 = 120$.

> **Rule:** ^nC_r is defined as $\dfrac{n!}{r!(n - r)!}$
> A scientific calculator has special keys for the calculation of $n!$ and ^nC_r

The binomial distribution

We can calculate the probability of a binomial event occurring an exact number of times from a set number of trials using the following formula.

> **Rule:** The probability of the event occurring r times out of n trials is $^nC_r p^r q^{n-r}$
> where:
> p = probability of the event occurring; q = probability of the event not occurring

For example, examining the probability of rolling exactly 5 sixes out of 10 rolls of a die:

$$p = 1/6, q = 5/6, r = 5, n = 10$$

$$\text{The probability} = {}^{10}C_5 \times \left(\frac{1}{6}\right)^5 \times \left(\frac{5}{6}\right)^{10-5}$$

$$= 252 \times \left(\frac{1}{6}\right)^5 \times \left(\frac{5}{6}\right)^5$$

$$= 252 \times \left(\frac{1}{7776}\right) \times \left(\frac{3125}{7776}\right)$$

$$\approx 0.013$$

Example

The results of a large number of trials show that a particular medication cures a particular disease in 73% of cases. If 8 patients are given the medication, find the probability of each of the following results both correct to 4 decimal places and as a percentage correct to 1 decimal place:

a that the following number of patients will be cured: 1, 2, 3, 4, 5, 6, 7, 8.

b that fewer than 7 patients will be cured

c that more than 2 patients will be cured.

Solutions

$$\left.\begin{array}{l} p = 0.73 \\ \therefore\ q = 0.27 \\ n = 8 \end{array}\right\} P(r) = {}^nC_r p^r q^{n-r}$$

a $P(0) = {}^8C_0(0.73)^0(0.27)^8 \approx 0.0000 \approx 0.0\%$
$P(1) = {}^8C_1(0.73)^1(0.27)^7 \approx 0.0006 \approx 0.1\%$
$P(2) = {}^8C_2(0.73)^2(0.27)^6 \approx 0.0058 \approx 0.6\%$
$P(3) = {}^8C_3(0.73)^3(0.27)^5 \approx 0.0313 \approx 3.1\%$
$P(4) = {}^8C_4(0.73)^4(0.27)^4 \approx 0.1056 \approx 10.6\%$
$P(5) = {}^8C_5(0.73)^5(0.27)^3 \approx 0.2285 \approx 22.9\%$
$P(6) = {}^8C_6(0.73)^6(0.27)^2 \approx 0.3089 \approx 30.9\%$
$P(7) = {}^8C_7(0.73)^7(0.27)^1 \approx 0.2386 \approx 23.9\%$
$P(8) = {}^8C_8(0.73)^8(0.27)^0 \approx 0.0806 \approx 8.1\%$

b $P(7) + P(8) = 0.2386 + 0.0806 = 0.3192$
or
$P(0) + P(1) + P(2) + P(3) + P(4) + P(5) + P(6)$
$\therefore\ P(<7) = 1 - 0.3192$
$\approx 0.6808\,(68.08\%)$

c $P(0) + P(1) + P(2) = 0.0064$
or
$P(>2) = P(3) + P(4) + P(5) + P(6) + P(7) + P(8)$
$\therefore\ P(>2) = 1 - 0.0064 = 0.9936$
$\therefore\ P(\geq 3) = 1 - 0.0064$
$\approx 0.9936\,(99.36\%)$

Example

Past records over a long period of time show that 4.0% of the screws manufactured by a particular machine are defective. Find the probability of each of the following results for a random sample of 50 of these screws, both as a decimal expressed correct to 3 significant figures, and also as a percentage correct to 1 decimal place:

a that exactly 4% of them will be defective
b that at least 2% of them will be defective

$$p = 0.04$$
$$\therefore q = 0.96$$
$$n = 50$$

Solutions

a 4% of 50 = 2
$$P(2) = {}^{50}C_2 \times (0.04)^2 \times (0.96)^{48} = 0.276 \ (27.6\%)$$
$$P(0) = {}^{50}C_0 \times (0.04)^0 \times (0.96)^{50} = 0.129\ 9$$
$$P(1) = {}^{50}C_1 \times (0.04)^1 \times (0.96)^{49} = 0.270\ 6$$

b $P(\text{at least 2}) = P(\geq 2)$
$$= 1 - P(0) - P(1)$$
$$= 1 - 0.129\ 9 - 0.270\ 6$$
$$\approx 0.600 \ (60.0\%)$$

24.7 MEAN AND STANDARD DEVIATION OF BINOMIAL EVENTS

For statistically binomial events the mean (average) outcome is:

$$\mu = n \times p$$

The variance is:

$$\sigma^2 = n \times p(1 - p)$$

The standard deviation is:

$$\sigma = \sqrt{(n \times p(1 - p))}$$

For the values in Example 1 above,

$$p = 0.73$$
$$n = 8$$

giving the mean and standard deviation as:

$$\mu = 8 \times 0.73 = 5.84$$
$$\sigma = \sqrt{(8 \times 0.73 (1 - 0.73))} = 1.26$$

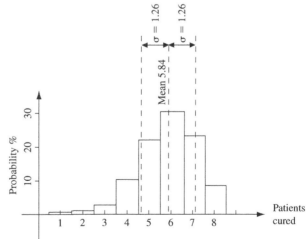

This provides us with useful information about the experiment and data set. We can use a histogram to graphically display the data to gain a better understanding.

The graph charts Example 1 on a histogram and overlaying the mean and standard deviation.

Example

If a fair coin is tossed 10 times, the calculation of the probability of obtaining exactly all the possible number of successes, X, is shown below:

$P(X = 0) = (0.5)^{10} \approx 0.00098$

$P(X = 1) = {}^{10}C_1(0.5)^1(0.5)^9 \approx 0.00987$

$P(X = 2) = {}^{10}C_2(0.5)^2(0.5)^8 \approx 0.04394$

$P(X = 3) = {}^{10}C_3(0.5)^3(0.5)^7 \approx 0.11719$

$P(X = 4) = {}^{10}C_4(0.5)^4(0.5)^6 \approx 0.20508$

$P(X = 5) = {}^{10}C_5(0.5)^5(0.5)^5 \approx 0.24609$

$P(X = 6) = P(X = 4) \approx 0.20508$

$P(X = 7) = P(X = 3) \approx 0.11719$

$P(X = 8) = P(X = 2) \approx 0.04394$

$P(X = 9) = P(X = 1) \approx 0.00987$

$P(X = 10) = P(X = 0) \approx \underline{0.00098}$

TOTAL = 1.0000

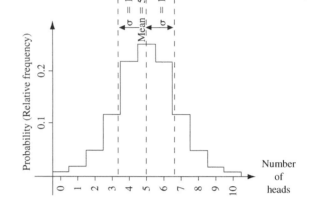

These are the probabilities—that is, the values at which the relative frequencies stabilise in a long series of trials.

All these frequencies can be made to occur as closely as we like to the given values by making the number of trials larger and larger.

Hence, for example, from the above results we can calculate that:

P(throw 5 heads) ≈ 0.246 = 24.6%

P(throw 3 or 4 heads) ≈ 0.11719 + 0.20508 ≈ 0.322 = 32.2%

P(throw 3 to 6 heads, inclusive) ≈ 0.11719 + 0.20508 + 0.24609 + 0.20508 ≈ 0.773 = 77.3%

For the above:

$\mu = n \times p$

$\quad = 10 \times 0.5$

$\quad = 5$

$\sigma = \sqrt{(n \times p(1 - p))}$

$\quad = \sqrt{(10 \times 0.5 \times (1 - 0.5))}$

$\quad = \sqrt{2.5}$

$\quad \approx 1.6$

The 'mean' (average) P(X) = 5

This is also called the 'expected' result, meaning not that the result is *actually* expected to occur but that it is the *most likely* result. In a long series of trials this result will be the one that occurs most frequently (in approximately 0.246 (24.6%) of the trials). Again, we can display this information on a histogram to gain a visual representation of the data. Referring to the histogram for this example, we see that the probability of throwing three heads is equal to the area between 2.5 and 3.5. The probability of throwing five heads is equal to the area between 4.5 and 5.5.

24.8 NORMAL CURVE

The 'normal curve' is a symmetrical bell-shaped curve. If a histogram is symmetrical and forms the shape of a bell curve the function can be approximated by the normal distribution function.

The diagrams illustrate how the above histogram approximates to a normal curve. We see that the probability of throwing any number of heads from 0 to 10 is given by the areas of the corresponding rectangles and also approximately by the corresponding area under the normal curve. On the normal curve we see that the probability of throwing three heads is equal to the area between 2.5 and 3.5. The probability of throwing five heads is equal to the area between 4.5 and 5.5.

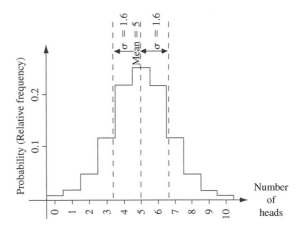

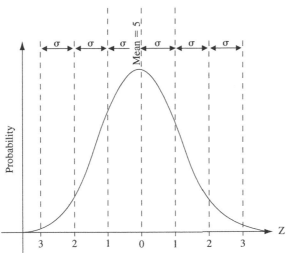

The normal curve is a standard curve where Z (the number of standard deviations from the mean) is plotted on the *x*-axis. The area under a normal curve is constant and is tabulated below.

z	0	1	2	3	4	5	6	7	8	9
0.0	.0000	.0040	.0080	.0120	.0160	.0199	.0239	.0279	.0319	.0359
0.1	.0398	.0438	.0478	.0517	.0557	.0596	.0636	.0675	.0714	.0754
0.2	.0793	.0832	.0871	.0910	.0948	.0987	.1026	.1064	.1103	.1141
0.3	.1179	.1217	.1255	.1293	.1331	.1368	.1406	.1443	.1480	.1517
0.4	.1554	.1591	.1628	.1664	.1700	.1736	.1772	.1808	.1844	.1879
0.5	.1915	.1950	.1985	.2019	.2054	.2088	.2123	.2157	.2190	.2224
0.6	.2258	.2291	.2324	.2357	.2389	.2422	.2454	.2486	.2518	.2549
0.7	.2580	.2612	.2642	.2673	.2704	.2734	.2764	.2794	.2823	.2852
0.8	.2881	.2910	.2939	.2967	.2996	.3023	.3051	.3078	.3106	.3133
0.9	.3159	.3186	.3212	.3238	.3264	.3289	.3315	.3340	.3365	.3389
1.0	.3413	.3438	.3461	.3485	.3508	.3531	.3554	.3577	.3599	.3621
1.1	.3643	.3665	.3686	.3708	.3729	.3749	.3770	.3790	.3810	.3830
1.2	.3849	.3869	.3888	.3907	.3925	.3944	.3962	.3980	.3997	.4015
1.3	.4032	.4049	.4066	.4082	.4099	.4115	.4131	.4147	.4162	.4177
1.4	.4192	.4207	.4222	.4236	.4251	.4265	.4279	.4292	.4306	.4319
1.5	.4332	.4345	.4357	.4370	.4382	.4394	.4406	.4418	.4429	.4441
1.6	.4452	.4463	.4474	.4484	.4495	.4505	.4515	.4525	.4535	.4545
1.7	.4554	.4564	.4573	.4582	.4591	.4599	.4608	.4616	.4625	.4633
1.8	.4641	.4649	.4656	.4664	.4671	.4678	.4686	.4693	.4699	.4706
1.9	.4713	.4719	.4726	.4732	.4738	.4744	.4750	.4756	.4761	.4767
2.0	.4772	.4778	.4783	.4788	.4793	.4798	.4803	.4808	.4812	.4817
2.1	.4821	.4826	.4830	.4834	.4838	.4842	.4846	.4850	.4854	.4857
2.2	.4861	.4864	.4868	.4871	.4875	.4878	.4881	.4884	.4887	.4890
2.3	.4893	.4896	.4898	.4901	.4904	.4906	.4909	.4911	.4913	.4916
2.4	.4918	.4920	.4922	.4925	.4927	.4929	.4931	.4932	.4934	.4936
2.5	.4938	.4940	.4941	.4943	.4945	.4946	.4948	.4949	.4951	.4952
2.6	.4953	.4955	.4956	.4957	.4959	.4960	.4961	.4962	.4963	.4964
2.7	.4965	.4966	.4967	.4968	.4969	.4970	.4971	.4972	.4973	.4974
2.8	.4974	.4975	.4976	.4977	.4977	.4978	.4979	.4979	.4980	.4981
2.9	.4981	.4982	.4982	.4983	.4984	.4984	.4985	.4985	.4986	.4986
3.0	.4987	.4987	.4987	.4988	.4988	.4989	.4989	.4989	.4990	.4990
3.1	.4990	.4991	.4991	.4991	.4992	.4992	.4992	.4992	.4993	.4993
3.2	.4993	.4993	.4994	.4994	.4994	.4994	.4994	.4995	.4995	.4995
3.3	.4995	.4995	.4995	.4996	.4996	.4996	.4996	.4996	.4996	.4997
3.4	.4997	.4997	.4997	.4997	.4997	.4997	.4997	.4997	.4997	.4998
3.5	.4998	.4998	.4998	.4998	.4998	.4998	.4998	.4998	.4998	.4998
3.6	.4998	.4998	.4999	.4999	.4999	.4999	.4999	.4999	.4999	.4999
3.7	.4999	.4999	.4999	.4999	.4999	.4999	.4999	.4999	.4999	.4999
3.8	.4999	.4999	.4999	.4999	.4999	.4999	.4999	.4999	.4999	.4999
3.9	.5000	.5000	.5000	.5000	.5000	.5000	.5000	.5000	.5000	.5000

We can use the values of the areas below the
normal curve instead of using the formulas to
solve functions that can be approximated with
the normal curve.

Areas under the Standard Normal Curve from 0 to Z

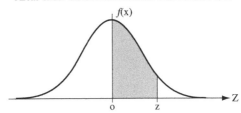

In the example on page 414, a fair coin is tossed 10 times. It was seen that the probability of
throwing between 3 and 8 heads is:

$P(X = 3) = {}^{10}C_3(0.5)^3(0.5)^7 \approx 0.117\ 19$

$P(X = 4) = {}^{10}C_4(0.5)^4(0.5)^6 \approx 0.205\ 08$

$P(X = 5) = {}^{10}C_5(0.5)^5(0.5)^5 \approx 0.246\ 09$

$P(X = 6) = {}^{10}C_6(0.5)^6(0.5)^4 \approx 0.205\ 08$

$P(X = 7) = {}^{10}C_7(0.5)^7(0.5)^3 \approx 0.117\ 19$

$P(X = 8) = {}^{10}C_8(0.5)^8(0.5)^2 \approx \underline{0.043\ 94}$

$\text{TOTAL} = \ \approx \underline{0.935}$

The above results are plotted on the original histogram.

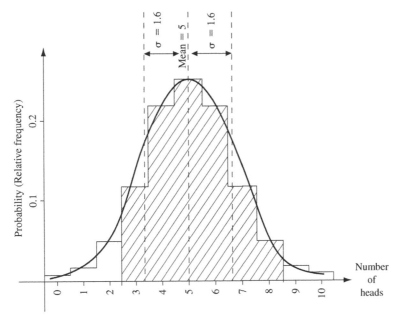

Instead of these manual calculations the normal curve can be used.

We have already calculated μ and σ

$\mu = 5$

$\sigma = 1.6$

For the probability of throwing between 3 and 8 heads we need to find the area between 2.5 and 8.5

We need to calculate Z for the upper value and the lower value.

$Z = \dfrac{|x_1 - \mu|}{\sigma}$

for $x = 2.5$, $Z = \dfrac{5 - 2.5}{1.6} = 1.5625$

for $x = 8.5$, $Z = \dfrac{8.5 - 5}{1.6} = 2.1875$

From the tables:

The area under the normal curve from 0 to 1.5625 = 0.4409

The area under the normal curve from 0 to 2.1875 = 0.4856

0.4409 + 0.4856 = 0.927 (92.7%)

Thus, using the ^nC_r formula we obtain a result of 93.5%, compared to the normal curve approximation of 92.7%.

Note: The total area under any normal curve = 1. The curve is symmetrical about the mean, therefore the area under the curve to the right of the mean = the area under the curve to the left of the mean = 0.5.

It can be seen from the areas table that the area under the standard normal curve between the mean and $Z \geq 3.9 = 0.5000$ correct to 4 significant figures. Hence the area to the left or right of any point on the horizontal axis that is 3.9 or more standard deviations from the mean is negligible and may be regarded as being zero. We may, for all practical purposes, assume that any curve meets the horizontal axis at points that are 3.9 standard deviations from the mean.

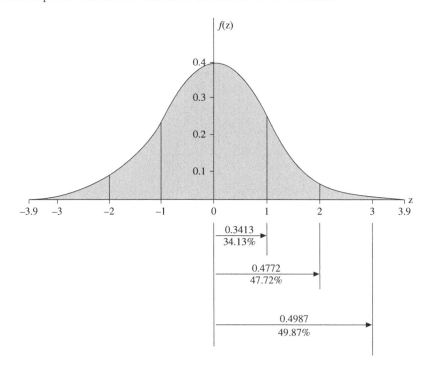

The areas table gives the area under the curve between the mean and:

- $Z = 1$ standard deviation as 0.3413
- $Z = 2$ standard deviations as 0.4772
- $Z = 3$ standard deviations as 0.4987
- $Z \geq 3.9$ standard deviations as 0.5000 (correct to four decimal places).

Note: The area under the normal curve between $x = x_1$ and $x = x_2$.

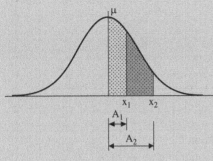

If the two x-values are on the same side of the mean (i.e. both $> \mu$ or both $< \mu$), the probability of the outcome between x_1 and $x_2 = A_2 - A_1$.

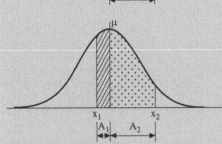

If the two x-values are on opposite sides of the mean the probability of the outcome between x_1 and $x_2 = A_2 + A_1$.

In all the following examples no attempt has been made to draw the diagrams to scale. It is always helpful to draw such diagrams.

Example

Referring to the third example on page 414, when a fair coin is tossed 10 times the probability of obtaining any number of heads, from 0 to 10, is given by the areas of the corresponding rectangles and also given approximately by the corresponding areas under the normal curve.

For this distribution, the mean, $\mu = np = 10 \times 0.5 = 5$ and the standard deviation,
$\sigma = \sqrt{npq} = \sqrt{10 \times 0.5 \times 0.5} \approx 1.58$

To find the probability of throwing 7 heads

P(7 heads) $\approx$ the area under the normal curve between the points where $x_1 = 6.5$ and $x_2 = 7.5$

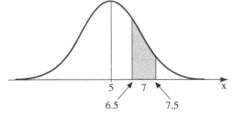

$Z = \dfrac{|x - \mu|}{\sigma}$, hence for $x_1 = 6.5$, $Z_1 = \dfrac{6.5 - 5}{1.58} \approx 0.95$

and for $x_2 = 7.5$, $Z_2 = \dfrac{7.5 - 5}{1.58} \approx 1.58$

From the areas table, the area between the mean and Z_1, $A_1 \approx 0.3289$, and the area between the mean and Z_2, $A_2 \approx 0.4429$

$\therefore$ the required area $\approx A_2 - A_1$

$$= 0.4429 - 0.3289$$

$$= 0.114$$

$\therefore$ the required probability ≈ 0.114 (11.4%)

The correct value (from the example on p. 414) is P(7) = 0.117 19 ($\approx$11.7%)

To find the probability of throwing more than 6 heads

P($>$6 heads) $\approx$ the area under the normal curve between the points where $x_1 = 6.5$ and $x_2 = 10.5$

continued

continued

$$Z_1 = \frac{6.5 - 5}{1.58} \approx 0.95 \text{ and } Z_2 = \frac{10.5 - 5}{1.58} \approx 3.48$$

From the areas table: $A_1 \approx 0.3289$ and $A_2 \approx 0.4997$

$A_2 - A_1 \approx 0.1708$

$\therefore$ the required probability ≈ 0.171 (17.1%)

The correct value (from the example on p. 414) is

$P(7) + P(8) + P(9) + P(10) = 0.171\,98 \approx 17.2\,\%$

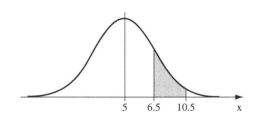

To find the probability of throwing 2 to 7 heads, inclusive

$P(2 \leq \text{no. heads} \leq 7) = (\text{area between } \mu \text{ and } x_1 = 1.5) + (\text{area between } \mu \text{ and } x_2 = 7.5)$

$$Z_1 \approx \frac{5 - 1.5}{1.58} = 2.22 \text{ and } Z_2 \approx \frac{7.5 - 5}{1.58} = 1.58$$

From the areas table: $A_1 \approx 0.4868$ and $A_2 \approx 0.4429$

Required area $= A_1 + A_2 = 0.4868 + 0.4429 = 0.9297$

$\therefore$ the required probability ≈ 0.930 (93.0%)

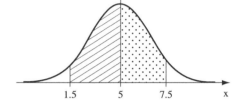

The correct value from the example on page 414 is $P(2) + P(3) + P(4) + P(5) + P(6) + P(7) = 0.934\,57 \approx 0.93$ but this would be a tedious method of calculation.

As has been shown, all the above problems could be, and have been, solved using the $^nC_r p^r q^{n-r}$ formula, without resort to the normal curve. However, when a large number of such calculations is involved it is much more convenient to use the normal curve approximation. When the value of ^nC_r is too large for the available calculator to evaluate, the normal curve approximation *has* to be used.

Exercises 24.4

1   A die is rolled 180 times. The probability of throwing a six is $\frac{1}{6}$.

   a   Find the probability of throwing exactly 30 sixes using the $^nC_r p^r q^{n-r}$ formula.

   b   From the result of part **a** deduce the probabilities of throwing fewer than 30 sixes and more than 30 sixes. *Hint:* $P(<30) + P(30) + P(>30) = 1$, and $P(<30) = P(>30)$.

   c   Find the probabilities of throwing fewer than 30 sixes and of throwing more than 30 sixes using the table of areas below the normal curve.

2   Past records show that the chance of a driver surviving a serious car accident if an air bag protection system is fitted to the vehicle is 95%.

   Find the probability that of the next 200 major car accidents:

   a   exactly 95% of drivers will survive

   b   more than 95% of drivers will survive

   c   fewer than 95% of drivers will survive. Use the same method of calculation as used in part **b**

   d   more than 185 of drivers will survive

   e   check that $P(<95\%) + P(\text{exactly } 95\%) + P(>95\%) = 1$

3 In a multiple-choice questionnaire a student encounters 40 questions for which he must choose between the 4 given answers. His complete ignorance forces him to guess the answer to all questions. What is the probability that for these 40 questions he will, by sheer luck, guess the correct answer to:

a 9 or more of them?

b 15 or more of them?

c fewer than 8 of them?

SELF-TEST

1 The following data sets describe a population. Calculate the mean, mode, range, median, variance and standard deviation. State your answers to one decimal place.

a 22.1, 22, 19.5, 19.2, 20, 22

b 7, 14, 6, 2, 20, 8, 2

c 10.3, 8.6, 8.6, 9.2

2 If the data sets described in question 1 were a sample, rather than a population, calculate the variance and the standard deviation.

+++ 3 A sample of historical rainfall data for the month of May is listed below for a remote township. A local member of the community has started recording rainfall data and publishing his measurements. Rather than only publishing the recorded measurement he has decided to also provide a description of the monthly rainfall. If the recorded rainfall data for the month is within one standard deviation of the mean, then the month is described as 'business as usual'; more than one standard deviation from the mean is described as 'wetter than a sock in a puddle'; and less than one standard deviation is described as 'dry as a chip'. For the sample of rainfall data:

a Calculate the mean.

b Calculate the standard deviation.

c Say how the month of May would be described if 99 mm of rainfall were recorded.

State your answers to one decimal place.

Sample of historical rainfall data: 77, 80, 88, 99, 101, 68, 88, 98, 88, 80, 99.

4 Further analysis of the historical rainfall data for the remote township described above indicates that for the month of June the probability of rainfall on any day is 25%. Using the equation $P(r) = {}^nC_r\, p^r\, q^{n-r}$:

a Find the probability that for the month of June less than 5 days will have rainfall.

b Find the probability that more than 7 days will have rainfall.

State your answers as percentages to one decimal place.

5 Use the normal curve to calculate an approximate answer to question 4.

+++ 6 Two decks of cards are shuffled together, totalling 104 cards. Four cards are randomly selected from the deck.

    a Using the equation $P(r) = {}^nC_r\, p^r\, q^{n-r}$ calculate the probability that exactly one queen is chosen.

    b Using the equation $P(r) = {}^nC_r\, p^r\, q^{n-r}$ calculate the probability that at least one queen is chosen.

    c Using the normal distribution curve approximate the probability that exactly one queen is chosen.

    d Using the normal distribution curve approximate the probability that at least one queen is chosen.

    e If 40 cards are chosen from the deck rather than 4, recalculate **a**, **b**, **c** and **d**.

    State your answers as percentages to one decimal place.

APPENDIX A: EULER'S CONSTANT, *e*, AND THE EXPONENTIAL GROWTH FORMULA

A1 Euler's constant, *e*

The proof that the expression $\left(1 + \dfrac{1}{h}\right)^h$ approaches a limiting value as h becomes larger and larger

(i.e. as $h \to \infty$) is beyond the scope of this course. However, you can *verify* the truth of this statement by evaluating the expression for larger and larger values of h. You should understand that no amount of *verification* of a statement constitutes a *proof* because you have not tested the statement for *all* possible cases. No matter how many tests you perform for different values of h, there is still the possibility that for some untested value the statement does not hold true.

For example, using your calculator, evaluate the expression $\left(1 + \dfrac{1}{h}\right)^h$ when $h = 1, 10, 100, 1000, \ldots$

In the eighteenth century, the Swiss mathematician Leonhard Euler discovered that although the value of this expression becomes larger and larger as h increases, this value does not increase without limit but approaches a value that is known as e. We say that the limit of the value of $\left(1 + \dfrac{1}{h}\right)^h$ as h becomes larger and larger is e. Mathematicians abbreviate this statement to $\lim\limits_{h \to \infty} \left(1 + \dfrac{1}{h}\right)^h = e$.

This value, like the value of π, has been calculated to thousands of decimal places and ≈ 2.718. Its value is such a useful number in mathematics that your calculator has keys on it that enable you to find the value of e^x, e^{-x} and logarithms to the base e.

A2 Proof of the formula for exponential growth: $Q = Q_0 \times e^{kt}$

The amount after 1 year of compound interest at $r\%$ per year when the interest is paid and compounded n times per year is given by:

$$A = P \times \left(1 + \frac{\frac{r}{n}}{100}\right)^n$$

$$\text{which} = P \times \left(1 + \frac{r}{100n}\right)^n = P \times \left(1 + \frac{1}{\frac{100n}{r}}\right)^n$$

$$= P \times \left(1 + \frac{1}{\frac{100n}{r}}\right)^{\frac{100n}{r} \times \frac{r}{100}} = P \times \left[\left(1 + \frac{1}{\frac{100n}{r}}\right)^{\frac{100n}{r}}\right]^{\frac{r}{100}}$$

$$= P \times \left[\left(1 + \frac{1}{h}\right)^h\right]^{\frac{r}{100}}, \text{ where } h = \frac{100n}{r}$$

As $n \to \infty$, $h \to \infty$ and so $\left(1 + \dfrac{1}{h}\right)^h \to e$.

$\therefore A \to P \times e^{\frac{r}{100}}$

i.e. $A \to P \times e^k$, where $k = \dfrac{r}{100}$, the interest rate expressed as a decimal.

During the second year we have $\$(P \times e^k)$ invested under the same conditions.

$\therefore$ at the end of the second year: $\quad A = \$(P \times e^k) \times e^k = \$P \times e^{2k}$

at the end of the third year: $\quad A = \$(P \times e^{2k}) \times e^k = \$P \times e^{3k}$

At the end of t years:

$$A = P \times e^{kt}$$

This formula does not, of course, apply to only a sum of money but to *any* quantity that grows exponentially.

The general formula for exponential growth is:

$$Q = Q_0 \times e^{kt}$$

where $\begin{cases} Q \text{ is the quantity originally present (i.e. the value of } Q \text{ when } t = 0) \\ k \text{ is the percentage rate of increase per some specified period of time expressed as} \\ \quad \text{a decimal} \\ t \text{ is the number of these growth periods} \end{cases}$

APPENDIX B: MASS AND WEIGHT

An acceleration of a m/s per second means that the velocity increases by a m/s each second. For example, a body that starts from rest and has an acceleration of 7 m/s per second (i.e. 7 m/s/s, 7 m/s^2) will have velocities on each successive second of 0 m/s, 7 m/s, 14 m/s, 21 m/s, etc.

A body that has no resultant force acting upon it will continue to remain at rest or will continue to move with a constant speed in a straight line (Newton's first law of motion). In order to give a body an acceleration (i.e. a change in speed or direction of motion, or both), a resultant force must act upon it. The force required to give a body a particular acceleration is proportional to the mass ('inertia') of the body and to the magnitude of the acceleration.

The force required to give a mass of 1 kg an acceleration of 1 m/s^2 is called a 'newton' (N).

The force required to give a mass of m kg an acceleration of a m/s^2 is given by the formula $F = ma$.

When a body is in free fall in a vacuum (i.e. with no resistance from air, water etc.), the only force acting upon it is the gravitational attraction towards the centre of the earth, which will give it an acceleration that is independent of its mass and is known as 'g'. Because the earth is not a perfect sphere this force and acceleration depend on its location on the earth (i.e. on how far it is from the centre of gravity of the earth). The value of g throughout Australasia has the value of 9.80 m/s^2 correct to 3 significant figures but in other places can vary from this value by about 0.2%. In our work we will take the value of g to be 9.80 m/s^2.

The force of gravitational attraction that produces this acceleration in free fall is commonly called the 'weight' of the body and is given by given by $F = mg$. For example, a body of mass 13.7 kg has a weight of approximately 13.7×9.80 N $\approx$ 134 N. A mass of 1.649 t (1649 kg) has a weight of approximately 1649×9.80 N $\approx$ 16 200 N (i.e. 16.2 kN).

Remember: 'Weight' is a *force*—the force with which the earth attracts the body (for a reason still undiscovered by scientists). This is the force required to support the body or to lift the body.

Isaac Newton (1642–1727) is commonly associated with the concept of the falling apple and gravity so it is appropriate that the unit of force is called the 'newton' (N) and that 1 N is approximately the weight of a medium-size apple. Remember when you hold an apple in your hand that the force on your hand is approximately 1 newton (1 N).

Note: In some exercises you will need to use the fact that on any body of mass m kg, there is a vertically downward force acting upon it of $m \times 9.80$ N (correct to 3 significant figures). This is the 'weight' of the body—the vertical force required to lift it or to prevent it from falling.

ANSWERS

1 Fractions and decimals

Exercises 1.1
1 a -8 b 4 c -2 d -2 e -4 f 2 2 a -3 b 10 c -3 3 a 0 b 0 c -2 4 a $-2°C$ b $-6°C$ 5 a contraction of 3 mm b contraction of 5 mm 6 a loss of 2 dB b loss of 8 dB c gain of 4 dB 7 a 2 b 14 8 a -6 b -8 c -4 d 4 e 6 f 12 9 a 4 b -2 c 6 d -4 e -8 f -1

Exercises 1.2
1 a 14 b -1 c 26 2 a 12 b 7 c 29 3 a 10 b 5 4 a 40 b 6 c 9 d 2

Exercises 1.3
1 a 2, 3, 5 b 3, 9 c 2, 3, 4 d 3, 5, 9 e 2, 3, 9 f 3, 5 g 2, 4 h 3, 9 2 a $\frac{13}{15}$ b $\frac{16}{17}$ c $\frac{3}{4}$ d $\frac{7}{19}$ e $\frac{3}{4}$ f $\frac{4}{7}$ g $\frac{2}{3}$ h $\frac{3}{5}$ 3 a $\frac{8}{13}$ b $\frac{3}{5}$ c $\frac{997}{7000}$ d 40 e $\frac{10}{99}$ 4 a $2\frac{1}{2}$ b $4\frac{2}{7}$ c $11\frac{7}{11}$ d $4\frac{1}{2}$ e $4\frac{1}{4}$ 5 a $\frac{7}{2}$ b $\frac{11}{4}$ c $\frac{37}{8}$ d $\frac{103}{8}$ e $\frac{124}{11}$ f $\frac{137}{12}$ 6 a $\frac{11}{12}$ b $1\frac{7}{12}$ c $\frac{31}{200}$ d $\frac{17}{20}$ e $\frac{7}{36}$ f $1\frac{11}{40}$ g $1\frac{1}{2}$ h $\frac{53}{200}$ 7 a $\frac{1}{12}$ b $\frac{3}{4}$ c $\frac{11}{18}$ d $-\frac{13}{100}$ 8 a $\frac{5}{6}$ b $1\frac{1}{18}$ c $\frac{1}{18}$ d $\frac{23}{30}$ e $\frac{1}{50}$ f $\frac{2}{3}$ 9 a 58 b $96\frac{1}{3}$ c $64\frac{2}{5}$ d $83\frac{1}{2}$ 10 a $60\frac{1}{2}$ b $31\frac{4}{5}$ c $13\frac{4}{7}$ d $16\frac{1}{4}$ 11 a $67\frac{1}{4}$ b $21\frac{11}{12}$ c $86\frac{3}{10}$ d $29\frac{9}{14}$ 12 a $\frac{3}{4}$ b $2\frac{1}{7}$ c $31\frac{1}{2}$ d $3\frac{3}{7}$ e $103\frac{1}{3}$ 13 a 19 b $\frac{10}{21}$ c 16 d 33 e $\frac{2}{3}$ f $\frac{1}{8}$ g 26 h $\frac{2}{9}$ i $\frac{3}{20}$ j 50 k $\frac{1}{2}$ l 21 14 a $1\frac{1}{5}$ b $4\frac{2}{3}$ c 1 d $2\frac{1}{3}$ 15 a 12 b $1\frac{1}{8}$ c $\frac{3}{7}$ d $\frac{1}{6}$ e $1\frac{1}{8}$ 16 a $2\frac{3}{13}$ b 7 c $2\frac{2}{9}$ d $2\frac{1}{3}$ e $1\frac{7}{8}$ 17 a $\frac{1}{5}$ b 2 c $7\frac{3}{5}$ d 21 e 11 18 a $\frac{4}{9}$ b $\frac{1}{49}$ c $\frac{9}{25}$ d $\frac{1}{4}$ e $1\frac{7}{9}$ f $20\frac{1}{4}$ g $5\frac{5}{16}$ 19 a $\frac{1}{9}$ b $5\frac{4}{9}$ c $2\frac{7}{9}$ d $12\frac{1}{4}$ e $30\frac{1}{4}$ 20 a $\frac{2}{3}$ b $\frac{1}{2}$ c $\frac{5}{9}$ d $\frac{1}{3}$ e $\frac{3}{4}$ f $1\frac{1}{9}$ 21 a $\frac{3}{4}$ b $\frac{1}{5}$ c $1\frac{1}{2}$ d $1\frac{2}{3}$ e $2\frac{1}{3}$ 22 a $\frac{9}{10}$ b $\frac{1}{10}$ c $\frac{1}{2}A$ 23 a $\frac{2}{3}$ b $1\frac{1}{3}$ c $4\frac{7}{8}$ m 24 a $13\frac{3}{4}$ Ω b $\frac{3}{10}$ m 25 $\frac{11}{18}$ h 26 $\frac{7}{12}$ 27 $\frac{5}{6}$ h 28 $\frac{2}{3}$ h 29 a 0 b U c U d 0 e U f 0 g U h 0 i U j 0 k U l 0

Exercises 1.4
1 a 21.124 b 239.58 c 30.008 2 a 63.68 b 3.05 c 11.618 3 a 0.043 b 0.63 c 2500 d 0.006 4 a 0.23 b 69 c 60 d 0.02 5 a 22.78 b 16.38 c 4.046 d 7.752 6 a 0.006 b 1.15 c 25.2 d 0.1 e 0.02 f 0.003 7 a 0.024 b 0.0006 c 0.002 88 d 1.8 e 0.6 f 0.000 06 8 a 0.09 b 0.0004 c 0.36 d 1.44 e 0.0144 f 0.000 121 g 0.0009 h 1.21 9 a 0.2 b 0.9 c 0.07 d 1.2 10 a 26.08 b 1.203 c 230.7 d 0.0521 11 a 2.03 b 732.8 c 6.003 d 1.703 e 0.0732 f 0.0039 12 a 0.68 b 5.33 c 62.15 d 234.20 e 0.04 f 0.00 g 0.01 h 0.00 13 a 400 b 0.567 c 0.234 14 a 30 b 0.4 c 0.2 d 0.03 15 a 20 b 20 c 0.3 16 a 3.6 Ω b 200 m 17 a 0.12 Ω b 0.16 Ω c 0.024 Ω d 0.48 Ω 18 a 150 b 0.06 s 19 a 1.46 mm b 43.8 mm c 58.4 mm d 7.3 mm 20 3066 mm^2 21 3.4 m/s 22 4000 W 23 19.8 J 24 20 s 25 a 1.5 s b 45 s 26 4.8 min 27 a 7000 kg b 0.002 m^3

Exercises 1.5
1 a 2500 b 2400 c 2500 d 600 e 700 f 800 g 1000 h 1000 2 a 74 b 73 c 74 d 99 e 100 f 1000 3 a 1.36 b 1.36 c 1.35 d 0.03 e 0.06 f 0.01

Exercises 1.6
1 a 3 b 2 c 3 d 4 e 2 f 1 2 a 0.0346 kg b 34.0 m c 0.0490 g d 601 m e 4.01 t f 781 km 3 a 520 g b 52 mm c 0.542 kg d 16.1 t e 650 m f 0.024 m^2

Exercises 1.7
1 a 1.00 b 6.45 c 0.02 d 0.92 e 1.34 f 10.01 g 21.26 h 1.00 i 2.14 j 19.67 k -18.83 l 16.21 m 1.81 n 2.45 o -8.25 p 4.03 q 1.62 r 11.31 s -16.21 t 1.40 u 9.36 v 2.78 w 8.27 x 6.72 2 a $4\frac{1}{9}$ b $2\frac{1}{35}$ c $3\frac{7}{8}$ d $2\frac{2}{5}$ e $25\frac{222}{245}$ f $10\frac{113}{120}$

Exercises 1.8
2 9.37 3 6.39 4 0.28 5 -0.72 6 -3.88 7 0.37 8 6.25 9 7.30 10 0.02 11 0.77 12 3.23 13 3.04

Self-test
1 a 2 b -2 2 a 0 b -8 c 0 d 2 e -4 3 a 5°C rise b 2°C drop 4 a -4 b -4 5 a 0 b 9 6 a -5 b -3 7 a 4 b 2 8 a 7

b 11 **9 a** $6\frac{1}{2}$ **b** $16\frac{11}{12}$ **10 a** $43\frac{3}{7}$ **b** $7\frac{1}{9}$ **11 a** $4\frac{3}{8}$
b $\frac{4}{9}$ **12 a** $\frac{7}{15}$ **b** $\frac{3}{5}$ **13 a** $\frac{13}{15}$ **b** $\frac{2}{15}$ **14 a** 8.10 **b** 4.37
c 3.216 **d** 3.17 **15 a** 0.06 **b** 0.0252 **c** 0.132
d 30 **e** 3.2 **f** 0.2 **g** 0.000 144 **h** 0.11 **i** 1.21
16 a 1.06 **b** 0.99 **c** 1.81 **d** 0.97 **e** 3.65 **f** 0.17
g 4.36 **h** 1.53 **i** 0.69 **j** 9.20 **k** 2.54 **l** 1.87
17 a 62.1 **b** 55.4 **c** 27.5 **d** 1291.4 **e** 5.1 **f** 1.1
g 82.9 **h** 44.7 **i** 199.9 **j** 189.3 **k** 0.8 **l** 3.3

2 Ratio, proportion and percentage

Exercises 2.1

1 15:17 **2** 17:48 **3** 30:1 **4** 7:10 **5** 9:40 **6** 3:25
7 1.54:1 **8** 1:2.80 **9 a** 0.810:1 **b** 1:1.24 **10** 6 m
11 10 L, $17\frac{1}{2}$L, $22\frac{1}{2}$L **12 a** 14 m **b** 2.8 m

Exercises 2.2

1 $1.60 **2** 1400 Ω **3** 70 km **4** 340 **5** 39 km

Exercises 2.3

1 yes **2** no **3 a** no **b** no **4** yes **5 a** no
b yes **c** no **d** yes **e** no **6 a** no **b** yes **c** no
7 a no **b** yes **8** 12.1 m/s **9** 500 g **10** 18.8 mN
11 84 Ω **12 a i** 32 **ii** 5 **b i** 27.4 **ii** 199
c i 66.8 **ii** 154 **13 a** 0.56 t **b** 20 mm
14 a 18.7 m/s **b** 19.5 N

Exercises 2.4

1 a 30.9 **b** 30.5 **c** 2.53 **2 a** 21 s **b** 3.0 A
3 a 0.506 revs/hour **b** 119 min **c** 42 300 km
d 384 000 km

Exercises 2.5

1 a 0.244 **b** 24.2 **2** 7.9 mm **3 a** 3.50
b 358 mm **c** 29.3 s

Exercises 2.6

1 a 0.38 **b** 0.02 **c** 0.0163 **d** 0.000 14 **2 a** 0.501
b 2.46 **c** 9.78 **d** 0.152 **e** 426 **3 a** 0.09 **b** 0.6
c 0.21 **d** 0.2 **4 a** 3.80 **b** 90 **c** 2.03 **d** 0.050
e 0.273 **5 a** 35 c **b** 50.0 Ω **c** 5 min 39 s **d** 3.70 t
e 47.3 m **6 a** 43.1% **b** 55.5% **c** 236% **d** 83.3%
e 400% **f** 66.7% **g** 0.318% **h** 37.5% **i** 2680%
j 91.7% **7 a** 21.1% **b** 27.5% **8** 140 kg **9** $2\frac{1}{2}$%
10 4.3% **11 a** 7.21 **b** 11.61 **c** 20.52 **d** 2.79
e 5.17 **f** 13.6 **12 a** 19.9 **b** 6.37 **c** 101 **d** 0.730

Self-test

1 $102, $170 **2** 82% **3 a** 327 Hz **b** 12.9%
decrease **4 a** 2.96 mL **b** $2.82 **5** 0.05% **6** 20%
7 a 128.5 kg **b** 0.7 kg **c** 147 kg

3 Measurement and mensuration

Exercises 3.1

1 a 0.001 kg **b** 1000 kg **c** 0.001 m **d** 0.000 001 m
e 0.01 m **f** 0.000 001 kg **2 a** 2.63 g **b** 75.4 km
c 0.5 t **d** 350 mm **e** 81.6 μm **f** 800 g

Exercises 3.2

1 1600 **2** 50 km

Exercises 3.3

1 30 **2** 600

Exercises 3.4

1 a 0.1 kg (100 g) **b** 40 mg **2 a** 3% **b** 0.7% **c** 1%

Exercises 3.5

1 a 0.5 km, 0.64%, 77.5 – 78.5 km **b** 0.0005 kg,
0.6%, 0.0825 – 0.0835 kg **c** 0.0005 m, 0.013%,
3.9995 – 4.0005 m **d** 0.000 000 5 t, 0.8%, 0.000 062
5 – 0.000 063 5 t **e** 0.005 t, 1.1%, 0.465 – 0.475 t
f 0.5 km, 0.076%, 653.5 – 654.5 km **2 a** 354.11
b 0.01 **c** 20.00 **d** 0.15 **e** 0.00 **f** 6.30

Exercises 3.6

1 9.14 **2** 6.99 **3** 4.84 **4** 15.8 **5** 6.90 **6** 7.82
7 6.35 **8** 2.45 **9** 24.8 **10** 270 **11** 1.01 **12** 13.5

Exercises 3.7

1 5.39 m **2** 2.65 m **3** 1.43 km **4** 1.987 km
5 983 mm **6** 1.46 km **7** 17.4 mm **8** 1180 mm
9 24.5 mm **10** 13.1 m **11** 1.8 m

Exercises 3.8

1 a 5 **b** 16 **c** 1 **d** 13 **e** 36 **2** $\sqrt{2}$:1
3 a $k = 13$ **b** $t = 15$ **c** $n = 26$

Exercises 3.9

1 a 0.211 m² **b** 211 000 mm² **2** $1.60

Exercises 3.10

1 30 m² **2** 12 m² **3** 6 m² **4** 6.02 m² **5 a** 6 m²
b 2.4 m **6 a** 6 m² **b** 2.4 m **7** 8 m² **8** 6 m²
9 20 m² **10** 12.25 m² **11** 25 m² **12** 21 cm²

Exercises 3.11

1 a 126 mm, 1260 mm^2 **b** 7.54 m, 4.52 m^2
c 400 mm, 12 700 mm^2 **d** 2.59 m, 0.536 m^2
e 1.02 m, 83 500 mm^2 **2 a** 15.7 m^2 **b** 93.9 mm^2
c 9570 mm^2 **d** 505 mm^2 **e** 5.74 m^2 **3** 549 mm
4 165 mm   **5** 1.39 m   **6** 174 mm   **7** 0.209 m^2
8 a 24.6 m^2 **b** 2340 m^2 **c** 57.9 m^2 **d** 16.1 m^2
e 20.6 m^2 **f** 10.3 m^2

Exercises 3.12

1 a 0.000 008 63 m^3 **b** 69 mL   **c** 4.5 cm^3
d 0.027 m^3 **e** 23 000 mm^3 **2 a** 376 000 mm^3
b 376 mL   **c** 0.000 376 m^3 **d** 0.376 L   **e** 376 cm^3
3 a 156 cm^3 **b** 156 mL   **4** 1.75 m^3 **5** 1560 cm^3
6 1.9 m^3 **7** 302 mL   **8** 47.2 cm^3 **9 a** 1.26 kL
b 0.707 m^3 **10 a** 4.81 m   **b** 9.08 m^2 **c** 5.78 m^2
d 83.2 m^3

Exercises 3.13

1 0.0177 m^2 **2** 62.7 m^2 **3** 176 mm^2 **4** 126 cm^2
5 283 m^2 **6** 110 m^2

Self-test

1 a 0.0567 km   **b** 2.700 g   **c** 54 cm^2 **d** 120 cm^3
e 0.086 00 t   **f** 4.83 mL   **2 a** 64.0 cm   **b** 144 cm^2
3 a 70.7 mm   **b** 2500 mm^2 **c** 3930 mm^2
d 55 300 cm^3 **e** 2.57 m^2 **4** 3530 cm^3, 27.6 kg
5 11.9 mm, 446 mm^2 **6 a** 7.23   **b** 4.64   **c** 1.91
d 29.3   **e** 195   **f** 17.0   **g** 4.83   **h** 2.39   **i** 0.263
j 4.89

4 Introduction to algebra

Exercises 4.1

1 a 6   **b** 18   **c** 11   **d** −1   **e** −5   **2 a** 9   **b** −5
c 18   **d** 36   **e** 12   **3 a** 64   **b** 49   **c** 4   **d** 16
4 a 9   **b** 9   **c** −9   **d** 25   **e** 1   **f** −5   **g** 10
h −3   **i** 18   **j** 12   **k** 36   **l** 36   **5 a** 0   **b** −5
c 5   **d** 12   **6 a** 1   **b** $2\frac{1}{6}$ **c** $-\frac{1}{6}$ **7 a** −8   **b** −2
c −1   **d** −6   **e** 1   **f** $\sqrt{5}$

Exercises 4.2

1 a i 6   **ii** 7   **iii** 6   **iv** 7   **b i** 9   **ii** 3   **iii** 9
iv 3   **c i** 3   **ii** 8   **iii** 3   **iv** 8   **d i** 6   **ii** 2
iii 6   **iv** 2   **2 a** $3a − 2b$ **b** $3x − 7y$ **c** $2g − 4f$
d $−2ab^2 − 5a^2b$ **e** $13xy − 4x − 6y$ **f** $2a^2b$

Exercises 4.3

1 a $a + b + c$ **b** $a + b + c$ **c** $a − b − c$
d $a − b + c$ **e** $a + b − c$ **f** a **2 a** $x + 1$
b $m + 3$ **c** $3k − 6$ **d** $−4$ **e** $8 − 5b$ **f** $2t$
3 a $2a^2 − b^2$ **b** $− x$ **c** $− 2q^2$

Exercises 4.4

1 a $12pq$ **b** $42lm$ **c** $−32a$ **d** $−15tx$ **e** $5x^2$
f $8x^2$ **g** $6x^2$ **h** $3a^2b$ **i** $8a^2b$ **j** $2ax^2$ **k** $15k^2m$
l $15x^2y$ **2 a** x **b** $3x$ **c** $3t^2$ **d** $18t$ **e** $4sy$
f $a + b$ **g** $3(x + y)$ **h** $a(2b − c)$ **i** $3a$

Exercises 4.5

1 a $3x + 6y$ **b** $2a − 2b$ **c** $6x + 12y$
d $2ab + 3ax$ **e** $a^2 + 2ab$ **f** $3m^2 − 2m$
g $8a − 8b$ **h** $2mp − 3pq$ **i** $−5$
2 a $ax + ay + bx + by$ **b** $at + 2a − 3t − 6$
c $mt − mf − kt + fk$ **d** $x^2 − 5x + 6$
e $6x^2 + 13x + 6$ **f** $2x^2 − 7x + 3$
g $mk + 3m + 2k + 6$ **h** $ab − ax − bx + x^2$
i $x^2 − a^2$ **3 a** $x^2 + 6x + 9$ **b** $4a^2 + 16a + 16$
c $k^2 − 6k + 9$ **d** $x^2 − 9$ **e** $4x^2 + 12x + 9$
f $h^2 − 1$ **g** $9t^2 − 12t + 4$ **h** $25b^2 − 10b + 1$

Exercises 4.6

1 a $3x$ **b** 3 **c** $2x$ **d** xy **e** $2pq^2$ **f** $2ab^2$
2 a $a + b$ **b** $3(x^2 − y)$ **c** $2ab(p − 3q)$
d $p(m − 2k)$ **e** $a(b − 2x)$ **f** $a(a + b)$ **3 a** $\dfrac{3x}{y}$
b $\dfrac{x}{y}$ **c** $\dfrac{3}{2a^2b}$ **d** $\dfrac{x(p − q)}{2(p + q)}$ **e** $\dfrac{c − d}{3}$ **f** $\dfrac{3(x + y)}{4k}$
4 a $6x$ **b** $6xy$ **c** $6xy$ **d** x^2y^2 **e** $24\,p^2q^2$ **f** $12\,a^2b^3$
5 a $6(a + b)$ **b** $18(x^2 − y)$ **c** $12a^2b(p − 3q)$
d $3p(m − 2k)$ **e** $6a(b − 2x)$ **f** $a^2(a + b)(p + q)$

Exercises 4.7

1 a $15x$ **b** $1 + 3x$ **c** $1 + 3a$ **d** $3ax$
2 a $\dfrac{1 + 3y}{4y}$ **b** $\dfrac{3x}{2}$ **c** $\dfrac{9k^2}{t}$ **d** $\dfrac{3 − 2k}{2t}$
3 a $4x − 1$ **b** $3b + 1$ **c** $\dfrac{2t − 1}{t}$ **4 a** $\dfrac{5}{7}$ **b** $\dfrac{5}{x}$
c $\dfrac{3}{m}$ **d** $\dfrac{a − b}{m}$ **e** $\dfrac{4}{x}$ **f** $\dfrac{7}{a + b}$ **5 a** $\dfrac{7x}{6}$ **b** $\dfrac{5R}{12}$
c $\dfrac{5C}{8}$ **6 a** $\dfrac{3m + 4x}{mx}$ **b** $\dfrac{4 − C^2}{2C}$ **c** $\dfrac{2M + 9}{6M^2}$
7 a $\dfrac{5m + 1}{6}$ **b** $\dfrac{1}{2}$ **c** $\dfrac{13}{6}$ **d** $\dfrac{x^2 + 6}{3x^2}$ **8 a** $\dfrac{ax}{by}$ **b** $\dfrac{a}{3}$

$c \dfrac{2t}{3m}$ **9 a** 6396 **b** $\dfrac{3t}{2b}$ **c** $\dfrac{5a^2}{3t}$ **10 a** $\dfrac{k}{3y}$ **b** $\dfrac{15}{b^2}$

$c \dfrac{9}{10m}$ **11 a** $\dfrac{n}{k^2}$ **b** $\dfrac{a}{2b}$ **c** $\dfrac{3k^2}{10t^2}$ **12 a** $\dfrac{a}{3b}$

$b \dfrac{5}{2m}$ $c \dfrac{a+b}{2a+3b}$ **d** $\dfrac{k}{3bc}$ **13 a** $\dfrac{x}{y}$ **b** $\dfrac{2a}{a-b}$

$c \dfrac{t^2}{t+1}$ **14 a** $\dfrac{x-2}{1-2x}$ **b** $\dfrac{8-2t^2}{3}$ $c \dfrac{y-x}{y+x}$

Exercises 4.8

1 a $2x+2$ **b** $3-T$ **c** $-3-3R$ **d** $7-2R$

e $2V+3$ **f** $-6t$ **2 a** $\dfrac{x-2}{6}$ **b** $\dfrac{16E+1}{12}$

$c \dfrac{5-3C}{6}$ **d** $\dfrac{3V-4}{12}$ **e** $-\dfrac{9L+5}{12}$ **f** $\dfrac{5k+15}{18}$

3 a $\dfrac{5-x}{6}$ **b** $\dfrac{1-5x}{6}$ **c** $\dfrac{8R-5}{6}$ **d** $\dfrac{10n-1}{12}$

e $\dfrac{25-26G}{24}$ **f** $\dfrac{5R-11}{18}$ **4 a** $\dfrac{2x}{1+x}$ **b** $\dfrac{4-3m}{2-2m}$

Exercises 4.9

1 a 1 **b** -1 **c** -1 **d** 1 **e** 1 **f** -1

2 a $\dfrac{1}{y-x}$ **b** $\dfrac{3+k}{k-3}$ **c** $\dfrac{3-2x}{x+2}$ **3 a** $\dfrac{b-a}{3}$

b $\dfrac{R}{6-R}$ **c** 1 **d** -2 **4 a** $3n+1$ **b** $8-9n$

c $n-7$ **d** $3-13n$ **e** $\dfrac{13-16n}{6}$ **f** $\dfrac{8n-23}{6}$

5 a $t+2$ **b** $4t+3$ **c** $12-9t$ **d** $\dfrac{31-17t}{6}$

e $\dfrac{15t-5}{4}$ **f** $\dfrac{19-13t}{15}$

Exercises 4.10

1 a $x=4$ **b** $x=2$ **c** $x=10$ **d** $x=1$

e $x=1\frac{1}{3}$ **f** $W=1$ **g** $x=0$ **h** $R=-2$

i $t=-1\frac{1}{2}$ **j** $x=\frac{1}{5}$ **k** $x=-\frac{2}{7}$ **l** $x=-3$

m $E=\frac{4}{7}$ **n** $l=-1$ **2 a** $x=3$ **b** $x=1\frac{3}{5}$

c $x=2\frac{4}{7}$ **d** $x=-\frac{1}{9}$ **e** $x=2\frac{1}{4}$ **f** $d=9$

g $W=1\frac{1}{2}$ **h** $x=1$ **i** $L=\frac{3}{4}$ **j** $x=3$ **3 a** $x=6$

b $x=-1\frac{1}{5}$ **c** $a=-6$ **d** $k=4\frac{1}{2}$ **e** $m=-7$

f $x=\frac{3}{4}$ **g** $n=-2$ **h** $W=-\frac{2}{3}$ **i** $y=6$ **j** $x=-2$

4 a $x=\frac{1}{3}$ **b** $x=\frac{1}{5}$ **c** $a=-2$ **d** $a=1$

e $x=-\frac{1}{7}$ **f** $m=\frac{5}{27}$ **5 a** $x=1$ **b** $x=-2\frac{1}{4}$

c $x=\frac{3}{5}$ **d** $x=1\frac{2}{11}$ **e** $x=-4\frac{1}{2}$ **f** $x=1\frac{2}{5}$

6 a $x=2$ **b** $x=-5$ **c** $x=0$

Exercises 4.11

1 a $x=-7$ **b** $n=1.1$ **c** $R=2$ **d** $L=3$

e $R=-2\frac{8}{11}$ **f** $M=1$ **2 a** $x=-3$ **b** $L=3$

c $V=6$ **d** $n=-10$ **3 a** $x=4.6$ **b** $t=-\frac{3}{10}$

c $k=-2$ **d** $E=-\frac{1}{2}$ **e** $x=\frac{1}{3}$ **f** $W=3\frac{2}{5}$

Exercises 4.12

1 Let n be the number

$$3n-(n-5)=28$$
$$2n+5=28$$
$$2n=23$$
$$n=11\frac{1}{2}$$

2 a 2.50 m **b** 18.8 m² **3** 20.5 m **4** 2.05 Ω

5 $BC=27$ mm **6 a** 5 m **b** 12 m² **7** 36 km/h

and 44 km/h **8** 3.5 km/h **9 a** at rate A: \$18; at

rate B: \$25 **b** 286

Exercises 4.13

1 a yes **b** yes **c** yes **2 a** yes **b** no **c** no

Exercises 4.14

1 a $x=2, y=3$ **b** $R=-3, r=-1$ **c** $F=-\frac{2}{3}$,

$W=\frac{1}{3}$ **2 a** $x=1, y=5$ **b** $E=3, V=2$

c $C_1=4, C_2=8$ **3 a** $x=2, y=5$

b $T=\frac{1}{3}, l=-2$ **c** $V=-3, E=-2$

Exercises 4.15

1 a $x=2, y=3$ **b** $W=4, d=-2$ **c** $I_1=-3$,

$I_2=-5$ **d** $t=2, x=-\frac{2}{3}$ **e** $L_1=3, L_2=\frac{5}{7}$

f $i=3, V=\frac{1}{5}$ **2 a** $a=5, b=-2\frac{1}{2}$ **b** $E=-1$,

$R=2$ **c** $v=-3\frac{1}{2}, d=-1$ **3 a** $x=1, b=3$

b $V_1=2, V_2=-3$ **c** $S=-5, v=-6$

4 a $a=1.35, b=0.435$ **b** $x=0.857$,

$y=2.71$

Exercises 4.16

1 a $k=0.0120, d_e=6.30$ mm **b** 7.86 mm

2 a $k=0.250, b=0.200$ kN **b** 2.70 kN

3 $m=20.0$ kg, $F=8.00$ N **4 a** $k=8.00$,

$C=200$ mm **b** 272 mm **c** 12.5 kg **5 a** 7.50 m/s²

b 4.50 m/s **c** 57.0 m/s **6** $R_1=17\,500$ Ω,

$R_2=2500$ Ω **7 a** $R=1.50$ Ω, $E=5.50$ V

b $R=235$ Ω, $E=53.1$ V **8 a** $R=10.0$ Ω,

$r=2.00$ Ω **b** $R=4.58$ Ω, $r=0.387$ Ω **9** 250 mL

of the weaker solution and 150 mL of the stronger.

10 a 8.25 km/h **b** 3.75 km/h **11** length

85 mm, breadth 35 mm   **12** iron 7.87 t/m^3 (g/cm^3), nickel 8.91 t/m^3 **13** 4.58 kg of A, 3.75 kg of B.

Self-test

1 a -10 **b** $-3\frac{1}{2}$ **c** 9   **d** 6   **e** 15   **f** -11

2 a $4x^2 - 9$ **b** $4t^2 - 12t + 9$ **c** $x^2 + 6tx + 9t^2$
d $4a^2 - 9b^2$ **e** $25k^2 - 20km + 4m^2$

3 a $\dfrac{4a + 1}{2a^2b}$ **b** $\dfrac{3LV - 2}{3RL}$ **c** $\dfrac{9x + y}{6x^2y}$ **d** $\dfrac{4tx - 3p}{12p^2t^2}$

4 a $\dfrac{2}{3t - 1}$ **b** $\dfrac{1}{x}$ **5 a** $\dfrac{7L - 7}{24}$ **b** $\dfrac{7V + 5}{24}$

6 a $a = -3$ **b** $b = 4$ **c** $k = \frac{1}{6}$ **d** $E = 2$
e $t = -3$ **f** $x = \frac{1}{3}$ **g** $a = 57$ **h** $W = -1\frac{1}{3}$
i $x = \frac{2}{3}$ **j** $t = 3\frac{1}{2}$ **7 a** 850 kg/m^3 **b** 5.1 kg
8 \$62 500   **9 a** $k = \frac{1}{6}$, $m = -\frac{1}{2}$ **b** $I_1 = -1\frac{5}{12}$,
$I_2 = -3\frac{1}{2}$ **c** $V = -5$, $t = 2$ **10 a** $F = -2$,
$m = 5$ **b** $a = -2$, $k = 0$ **11 a** $a = 1$, $t = 2$
b $S = 10.6$, $t = 1.85$ **12** 41 L of *Full-cream*,
59 L of *Slim*   **13** 21 kg/min and 14 kg/min

5 Formulae: evaluation and transposition

Exercises 5.1

1 0.386   **2** 0.398   **3** 14.1   **4** 0.988   **5** 93.8
6 -1.01 **7** -15.8 **8** 6.90   **9** 0.128 **10** 26.1
11 13.5   **12** -25.8 **13** 4.25   **14** 0.903
15 -8.39 **16** -1.49

Exercises 5.2

1 a 56.3 kJ   **b** 459 kJ   **2 a** 10.3 N   **b** 43.0 N
3 a 259 mm   **b** 560 mm   **4 a** 16.6 m/s   **b** 186 m/s
5 a 799 mF   **b** 366 μF   **6 a** 51.6 mHz
b 1.16 kHz   **7 a** 1.84 Hz   **b** 478 Hz   **8** 4.24 s

Exercises 5.3

1 a $x = 13$ **b** $x = \dfrac{3kn + ay}{a}$ or $\dfrac{3kn}{a} + y$

2 a $x = 15$ **b** $x = \dfrac{b(m + t)}{a}$

3 a $x = 8.5$ **b** $x = \dfrac{nbt - na}{b}$ or $nt - \dfrac{na}{b}$

4 a $x = 3.5$ **b** $x = \dfrac{h + mt}{t}$ or $\dfrac{h}{t} + m$

5 $x = (m - t)^2$ **6** $x = \pm\sqrt{\dfrac{3a - m}{k}}$

7 $x = \pm\sqrt{\dfrac{2a}{3ky}}$ **8** $x = a \pm 3\sqrt{t}$ **9** $x = \dfrac{3 \pm 5t}{2}$

10 $x = \dfrac{a - 2b}{(k - t)^2}$ **11 a** $T_b = \dfrac{n_aT_a}{n_b}$ **b** $n_b = \dfrac{T_an_a}{T_b}$

12 $d_2 = \dfrac{d_1(V - 2)}{V}$ **13 a** $T_1 = \dfrac{T_2v + 1000P}{v}$

b $T_2 = \dfrac{T_1v - 1000P}{v}$ **14** $v = \pm\sqrt{\dfrac{2FS}{m} + u^2}$

15 $F = VE - W$ **16** $m = \dfrac{m_s(1 - q)}{q}$

17 $R = \dfrac{V}{I}$ **18** $W = Pt$ **19** $V = \dfrac{Q}{C}$ **20** $I = \dfrac{R}{Bl}$

21 $l = \dfrac{IN}{H}$ **22** $l = \dfrac{Fd}{CI_1I_2}$ **23** $Q - \sqrt{2WC}$

24 $L_1 = \dfrac{M^2}{k^2L_2}$ **25** $H = \sqrt{\dfrac{2W}{\mu}}$ **26** $L = \dfrac{1}{4\pi^2f^2C}$

27 $a = l\rho G$ **28** $h = \dfrac{2W - mv^2}{2mg}$

29 $R_1 = \dfrac{R_2(I - I_1)}{I_1}$ **30** $R = \dfrac{R_1R_2}{R_1 + R_2}$

31 $L_3 = \dfrac{LL_1}{L_1 - L} - L_2$

Exercises 5.4

1 $x = \dfrac{5}{a + 4}$ **2** $x = \dfrac{k}{3t - m}$ **3** $x = \dfrac{t}{a + m}$

4 $n = \dfrac{5}{b + k}$ **5** $w = \dfrac{k}{3m + 1}$

6 $t = \dfrac{ma - b}{m - 1}$ or $\dfrac{b - ma}{1 - m}$

Exercises 5.5

1 $t = \dfrac{v - u}{a}$ **2 a** $m = \dfrac{PVM}{RT}$ **b** $M = \dfrac{mRT}{PV}$

3 a $Q_2 = Q_1(1 - \eta)$ **b** $Q_1 = \dfrac{Q_2}{1 - \eta}$

4 a $T_1 = \dfrac{T_2v + 1000P}{v}$ **b** $T_2 = \dfrac{T_1v - 1000P}{v}$

5 $m = \dfrac{2FS}{v^2 - u^2}$ **6 a** $F = VE - W$ **b** $V = \dfrac{W + F}{E}$

7 a $v = \dfrac{Ft + mu}{m}$ **b** $m = \dfrac{Ft}{v - u}$ **8** $W = \dfrac{E - b}{a}$

9 a $q = \dfrac{P}{2t(l + b)}$ **b** $b = \dfrac{P - 2qlt}{2qt}$

10 a $c = \dfrac{Q}{m(t_2 - t_1)}$ **b** $t_1 = \dfrac{mct_2 - Q}{mc}$

$$11 \ t = \frac{273(V - V_o)}{V_o} \qquad 12 \ \mathbf{a} \ V = \frac{W}{\eta(aw + b)}$$

$$\mathbf{b} \ a = \frac{1}{\eta V} - \frac{b}{W} = \frac{W - \eta bV}{\eta WV} \qquad \mathbf{c} \ W = \frac{\eta bV}{1 - \eta aV}$$

$$13 \ I_F = \frac{IR_s}{R_s + r} \qquad 14 \ V = \frac{R_i E}{R_x + R_i}$$

$$15 \ R_1 = \frac{R_a(R_2 + R_3)}{R_2 - R_a} \qquad 16 \ r = \frac{eR}{E - e}$$

$$17 \ R_1 = \frac{I_2 R_2}{I - I_2} \qquad 18 \ R_2 = \frac{R_3(R - R_1)}{R_1 + R_3 - R}$$

$$19 \ I_1 = \frac{E - I_2 R_b}{R_a + R_b} \qquad 20 \ C_3 = \frac{C_2(C_1 - C)}{C - C_1 - C_2} \ \text{or}$$

$$\frac{C_2(C - C_1)}{C_1 + C_2 - C} \qquad 21 \ r = \frac{8R}{4 - R}$$

Exercises 5.6

1 a 259 mW **b** $W = Pt$ **c** 1.36 kJ   **2 a** 36.3 J

$$\mathbf{b} \ s = \frac{W}{F} \quad \mathbf{c} \ 2.80 \text{ km} \quad \mathbf{3 \ a} \ 1.22 \text{ kV} \quad \mathbf{b} \ I = \frac{E}{R}$$

$$\mathbf{c} \ 725 \ \mu\text{A} \quad \mathbf{4 \ a} \ 260 \text{ mA} \quad \mathbf{b} \ R_1 = \frac{R_2(I - I_1)}{I_1}$$

$$\mathbf{c} \ 2.57 \text{ k}\Omega \quad \mathbf{5 \ a} \ 43.0 \text{ N} \quad \mathbf{b} \ v = \sqrt{\frac{2Fs}{M} + u^2}$$

$$\mathbf{c} \ 39.5 \text{ m/s} \quad \mathbf{6 \ a} \ 186 \text{ m/s} \quad \mathbf{b} \ h = \frac{2W - mv^2}{2mg}$$

$$\mathbf{c} \ 98.2 \text{ mm} \quad \mathbf{7 \ a} \ 773 \text{ mW} \quad \mathbf{b} \ V = \sqrt{\frac{P(R_1 + R_2)^2}{R_1}}$$

$$\mathbf{c} \ 875 \text{ mV} \quad \mathbf{8 \ a} \ 1.09 \text{ kHz} \quad \mathbf{b} \ C = \frac{1}{4\pi^2 Lf^2}$$

$$\mathbf{c} \ 16.0 \ \mu\text{F} \quad \mathbf{9 \ a} \ 845 \text{ mJ} \quad \mathbf{b} \ s = \frac{W}{F + P} \quad \mathbf{c} \ 833 \text{ mm}$$

$$\mathbf{10 \ a} \ 855 \text{ mN} \quad \mathbf{b} \ h = \frac{W}{F + mg} \quad \mathbf{c} \ 954 \text{ mm}$$

$$\mathbf{11 \ a} \ 832 \ \mu\text{A} \quad \mathbf{b} \ R = \frac{I_1 r}{I_2 - I_1} \quad \mathbf{c} \ 618 \text{ m}\Omega$$

$$\mathbf{12 \ a} \ 1.93 \text{ mF} \quad \mathbf{b} \ C_3 = \frac{C_2(C - C_1)}{C_1 + C_2 - C} \quad \mathbf{c} \ 3.12 \text{ mF}$$

Self-test

1 102 V   **2 a** 63.6 Ω **b** 27.4 Ω **3** 0.065 4

$$\mathbf{4 \ a} \ t = \frac{273(V - V_o)}{V_o} \quad \mathbf{b} \ R = \frac{V^2 t}{E} \quad \mathbf{c} \ l = \frac{Ra}{\rho}$$

$$\mathbf{d} \ L_s = L_A - 4M \quad \mathbf{e} \ R_2 = \frac{R_1 R_s}{R_x} \quad \mathbf{f} \ C_2 = \frac{CC_1}{C_1 - C}$$

$$\mathbf{g} \ i_2 = \frac{i_1 R_1 - V}{R_2} \quad \mathbf{h} \ n = \frac{IR}{E - Ir} \quad \mathbf{5 \ a} \ 1.32 \text{ kN}$$

$$\mathbf{b} \ t = \frac{mv}{F} \quad \mathbf{c} \ 541 \text{ ms} \quad \mathbf{6 \ a} \ 520 \ \mu\text{C} \quad \mathbf{b} \ C = \frac{Q}{V}$$

$$\mathbf{c} \ 474 \ \mu\text{F} \quad \mathbf{7 \ a} \ 6.47 \text{ m/s} \quad \mathbf{b} \ s = \frac{v^2 - u^2}{2a}$$

c 390 mm **8 a** 103 ms **b** $s = ut + \frac{1}{2}at^2$

c 2.36 km **9 a** 5.39 kΩ **b** 3.30 kΩ **c** 49.8 kΩ

6 Introduction to geometry

Exercises 6.1

1 a 159°51′, obtuse **b** 76°38′, acute **c** 88°45′, acute **d** 189°41′, reflex **e** 60°11′, acute
f 90°, right **g** 172°48′, obtuse **h** 180°40′, reflex **i** 180°, straight **j** 78°21′, acute

Exercises 6.2

1 a 61°14′ **b** 151°14′ **c** 16°32′ **d** 170°36′
e 41°43′ **f** 41°35′ **2 a** $(90 - \theta)°$ **b** $(180 - x)°$
c 38°26′ **d** 67°18′ **e** $(133 - t)°$ **f** $(94 + \theta)°$
3 a $x = 30$ **b** $x = 18$ **c** $m = 36$ **d** $y = 15$
e $k = 36$ **f** $t = 26$

Exercises 6.3

1 $t = 80$ **2** $k = 50$ **3** $m = 70$

Exercises 6.4

1 $a = 30, b = 80, c = 70$ **2** $p = 46, q = 72,$
$r = 62$ **3** $k = 76, l = 50, m = 54$ **4** $x = 70$
5 $b = 107, g = 128, p = 73$ **6** $x = 80$

Exercise 6.5

1 a 65 mm **b** 81°
2 An interval joining two sides of a triangle and parallel to the third side divides those two sides in equal ratios.
3 The opposite sides of a parallelogram are equal in length.

Exercises 6.6

2 a 60° **b** 30° **c** 1.15 ± 0.02

Self-test

1 a 122°32′ **b** 46°09′ **c** $(200 - k)°$
d $(t - 90)°$ **e** $(270 - 2x)°$ **2 a** $h = 64$
b $y = 36$ **c** $u = 208$ **d** $x = 132$ **3 a** $x = 70$
b $p = 49, q = 15$ **c** $x = 23$ **d** $x = 36$

7 Geometry of triangles and quadrilaterals

Exercises 7.1

1 $t = 30$ 2 $k = 70$ 3 $y = 15$ 4 $h = 115$
5 $x = 35, y = 30$ 6 $x = 20$

Exercises 7.2

1 $b = d, c = e$ 2 $a = b, d = f$ 3 $b = c, a = d$
4 $a = f, b = e, c = d$ 5 $c = f, b = e, a = d$
6 $b = c, f = d$

Exercises 7.3

2 206 mm ± 5 mm

Exercises 7.4

1 a yes, SSS b no c yes, AAS d no e yes,
SAS f yes, RHS g no h no 2 a yes, SSS
b no c no d yes, AAS

Exercises 7.5

1 $x = 8, y = 6$ 2 $d = 7.5$ 3 $t = 10$ 4 $x = 2.8$
5 $n = 39.9$ 6 $y = 0.49$ 7 $x = 4.33$ 8 $x = 1.6$
9 $p = 3.6, q = 1.8$ 10 $x = 4.5$ 11 $d = 6.92$
12 $x = 6$ 13 $x = 4.3$ 14 $x = 0.5$ 15 8.57 m
16 2.4 m

Exercises 7.6

1 a AAS b The opposite sides are equal lengths
and the opposite angles are equal. 2 a SSS
b Alternate angles are equal (or, corresponding
angles are equal), ∴ the opposite sides are parallel.
3 a Each is a pair of supplementary angles
(because the sum of the 4 angles is 360°)
b Cointerior angles are supplementary, ∴ the
opposite sides of ABCD are parallel, ∴ ABCD is
a parallelogram. 4 a parallelogram b rhombus
c no special type d trapezium e no special type
f square g parallelogram h rhombus i rectangle
j parallelogram k rectangle l parallelogram
5 a yes; PQBA will always be a parallelogram
because both pairs of opposite sides must
always remain equal in length b 2 m
c extremely small (approaching zero) as rope QB
is made longer and longer d no
6 a parallelogram b yes c no d 70°

Exercises 7.7

1 a 9.80 m^2 b 13.4 m^2 c 128 000 mm^2
d 1380 mm^2 2 168 cm^2 3 19.1 m^2

Exercises 7.8

1 28.8 m^2 2 a 7.00 m b 12.2 m c 84.0 m^2
3 a 10.0 m b 7.07 m c 50.0 m^2 4 a 21.6 m
b 14.4 m c 156 m^2 5 a 80.0 mm b 215 mm
c 113 mm 6 a 6.45 m b 15.7 m c 7.07 m
7 a 1.73 m b 3.46 m^2 c 3.46 m 8 a 12 m
b 60 m^2 c 60 m^2 9 a 24 m^2 b 6 m
10 a trapezium b 324 cm^2

Self-test

1 a $p = 120$ b $m = 30$ c $x = 53$ 2 a $b = d$,
$a = e = c$ b $c = d$ c $b = e, c = g$ d $a = d$,
$b = c$ 3 a yes, SAS b no c no d yes, AAS e no
f yes, SAS 4 a rectangle b yes c no d 8 m
e 28 m 5 a yes; ADPQ always remains a
parallelogram since both pairs of its opposite
sides remain equal, hence AD remains horizontal
(parallel to PQ) and hence AB remains vertical,
since ADCB is a rectangle b 424 mm c 88 mm
6 a 32.00 m^2 b 11.70 m c 6.403 m 7 a 641 mm
b 276 000 mm^2 8 a 11.6 m^2 b 272 mm^2

8 Geometry of the circle

Exercises 8.1

1 17.1 m 2 60°

Exercises 8.2

1 a 207 mm b 80.9 mm^2 2 a 117° b 0.237 m

Exercises 8.3

1 a $p = 40, q = 100, r = 80$ b $p = 25, q = 130$,
$r = 50$ c $p = x, q = 180 - 2x, r = 2x$
2 a i 40° ii 60° iii 100° b i $2x°$ ii $2y°$
iii $2(x + y)°$ c The angle subtended at the
centre of a circle is twice the angle subtended at
the circumference by the same arc. 3 a i 50°
ii 50° b i $\frac{1}{2}x°$ ii $\frac{1}{2}x°$ c All angles subtended
by the same arc at the circumference of a circle
are equal, i.e. angles in the same segment are
equal. 4 a i 40° ii 60° iii 80° iv 90°
b Any angle subtended by a diameter at the

circumference of a circle is a right angle, i.e. an angle in a semicircle is a *right angle*. **5 a** 75°
b 90° **c** 100° **d** 120° **6 a** **i** 65° **ii** 115°
iii 180° **b** **i** $\frac{1}{2}y°$ **ii** $\frac{1}{2}x°$ **iii** $\frac{1}{2}(x + y)° =$
$\frac{1}{2}(360°) = 180°$ **c** **i** $\frac{1}{2}\theta°$ **ii** $360° - \theta°$
iii $\frac{1}{2}(360° - \theta)° = 180° - \frac{1}{2}\theta°$ **iv** 180°
d Opposite angles in a cyclic quadrilateral are
supplementary. **7** $m = 74, n = 42$ **8** $g = 31$
9 $k = 45$ **10** $p = 55$ **11** $m = 52$ **12** $k = 36$
13 $x = 100, y = 92$ **14** $m = 25$ **15** $h = 67$
16 a $x = 55, y = 40$ **b** $x = 24$ **17** $t = 23$
18 $x = 105, y = 105$ **19** $q = 20, r = 30$
20 $x = 52, y = 38$ **21** $m = 136, t = 22$
22 $p = 30, q = 70$ **23** $x = 28$

Exercises 8.4

1 7.5 m **2** 9.5 m, 10.5 m

Exercises 8.5

1 a 90° **b** 50° **2 a** 24° **b** 132° **c** 66° **3** 28°
4 a 102° **b** 129° **5** 10.58 m **6** 12 m **7** 12 m
8 4.90 m

Exercises 8.6

1 $x = 74$ **2** $m = 40, n = 62$

Exercises 8.7

1 7.73 m **2** 273 mm

Exercises 8.8

1 3.08 mm **2** 9.17 mm **3** 93.4 mm **4** 9.80 mm
5 15.3 mm **6 a** **i** 433 mm **ii** 60° **iii** 2.23 m
b 2.36 m

Self-test

1 a 83.7 mm **b** 119 mm² **2 a** $p = x + 32$
b $q = x + 40$ **c** $x = 54$ **3 a** 40 **b** 64 **c** 50
d 40 **e** 100 **4 a** 50 **b** 54 **c** 47 **d** 67
5 6.5 m **6** 48 mm **7 a** $p = 106$ **b** $p = 60$,
$q = 30, r = 80, s = 70, t = 100$

9 Straight line coordinate geometry

Exercises 9.1

1 a (x_1, y_1) **b** $x_2 - x_1$ **c** $y_2 - y_1$
d $\sqrt{(x_2 - x_1)^2 + (y_2 - y_1)^2}$ **2 a** the AAS test
b $AQ = MP, QM = PB, MA = BM$ **c** they are the

opposite sides of a rectangle. **d** Q is the midpoint
of AK, P is the midpoint of BK **e** Q is the point
$\left(\dfrac{x_1 + x_2}{2}, y_1\right)$, P is the point $\left(x_2, \dfrac{y_1 + y_2}{2}\right)$
f $\left(\dfrac{x_1 + x_2}{2}, \dfrac{y_1 + y_2}{2}\right)$ **3 a** **i** 5 units **ii** (3.5, 5)
b **i** 13 units **ii** (5.5, 10) **c** **i** 10 units **ii** (1, 5)
d **i** 5 units **ii** (1, 0.5) **e** **i** 13 units
ii (−9, −10.5) **f** **i** $\sqrt{85}$ units **ii** (−2.5, −4)
4 a (3, 9) **b** (11, 0)

Exercises 9.2

1 a 3 **b** 10 **c** $3\frac{1}{3}$ **2 a** 3 **b** 6 **c** 2 **3 a** 3
b −6 **c** −2 **4 a** 4 **b** −12 **c** −3 **5 a** 10
b 3 **c** 0.3 **6 a** 2 **b** −10 **c** −5 **7 a** 3 **b** 6
c 4 **d** 2 **e** −2 **f** 7

Exercises 9.3

1 a $y = 8$ **b** $y = 11$ **c** $y = b + 20$
d $y = 4x + b$ **e** $y = mx + b$ **2 a** **i** 3 **ii** 5
iii $-1\frac{2}{3}$ **b** **i** 2 **ii** −7 **iii** $3\frac{1}{2}$ **c** **i** −5 **ii** −3
iii $-\frac{3}{5}$ **d** **i** 2 **ii** 4 **iii** −2 **e** **i** 3 **ii** 6
iii −2 **f** **i** $\frac{2}{3}$ **ii** −3 **iii** $4\frac{1}{2}$ **3 a** P, Q, S, V
b M, P, Q **c** $b = -1$ **d** $k = 1$ **e** $m = -3$
f $a = 2\frac{1}{2}$

Exercises 9.4

1 a $y = 2x + 3$ **b** $y = -3x + 5$
c $y = -3x + 3$ **d** $y = \frac{2}{3}x - 2$ **e** $y = -3x - 5$
2 a $y = 3x - 1$ **b** $y = -2\frac{1}{2}x + 6$
c $y = -3x - 5$ **3 1** $y = 3\frac{1}{3}x + 6\frac{2}{3}$ **2** $y = 2x + 2$
3 $y = -2x + 12$ **4** $y = -3x + 12$
5 $y = 0.3x$ **6** $y = -5x + 15$

Exercises 9.5

(The $\pm$ values indicate roughly the expected accuracy.)
1 a 12.1 ($\pm$ 0.2) **b** 6.3 ($\pm$ 0.1) **c** 2.90 ($\pm$0.05)
2 a 50.88 m/s ($\pm$0.02) **b** 1.77 s ($\pm$0.01)
c 2.32 m/s² ($\pm$0.05) **3 a** 15.53 mm ($\pm$0.02)
b 170 mm ($\pm$2) **c** 6.4 μm/mm ($\pm$0.1)
4 a 1.44 L ($\pm$0.01) **b** 479 K ($\pm$2)
c 3.35 mL/K ($\pm$0.05) **5 a** 31.0 mA ($\pm$0.5)
b 82 V ($\pm$2) **c** 300 μA/V ($\pm$10)

Exercises 9.6

1 b 106.5   c 3.35   2 a 29.4 s   4 67.4 kW
5 a $9100   b 32   c 96 articles

Exercises 9.7

1 a T b P c Q d U e S f V g R h R
2 a $x = 4$ b $x = 6$ c $x = -4$ d $x = 4$
e $x = 2$ f $x = 2$ 3 a i R ii Q iii T iv S v U
vi Q b U c V d Q e Q f $(3, 42)$ 4   c $(2, 5)$
d $x = 2, y = 5$ e '. . . read off the coordinates of
the point of intersection.'   5 a $x = 1\frac{1}{2}, y = 3$
b $x = -0.6, y = -7.2$ 6   b $t = 1.28$ h (± 0.05),
$D = 83$ km (± 1) c $D = 62t + 2, D = -49t + 145$

Self-test

1 a i 3   ii 2   iii -6 b i -2.5 ii -0.8 iii -2
c i 2.5   ii 1.4   iii -3.5 d i 0.2   ii -6 iii 1.2
2 a $y = 2x + 5$ b $y = -3x + 12$
c $y = -1\frac{1}{2}x - 2$ d $y = -x + 1$ e $y = -3x - 6$
f $y = 2.5x + 5$ 3 a 4.23 mA (± 0.01)
b 0.3% (± 0.1) c 28 μA/V (± 2)
4 a $-244°$C (± 10) b 0.49 Ω/°C (± 0.01)
6 283 W   7 $x = 2.14, y = 45.7$
8 b $t = 2.4$ h (± 0.1), $\theta = 43°$C (± 1)
c $\theta_A = 9.4t + 21, \theta_B = -15t + 80$
9 a 233 Hz (± 1) b 49.4 mA (± 0.2)

10 Introduction to trigonometry

Exercises 10.2

1 1.38   4 a 0.6009   b 1.428   c 0.7265   6 a 0.5048
b 3.149   c 0.2852   7 a 2.70   b 1.44   c 19.9   d 1.21

Exercises 10.3

1 a 35.4 m   b 7.48 m   c 448 mm   2 a 99.2 mm
b 8.61 m   c 36.7 mm   d 31.3 m

Exercises 10.4

2 a 35.70°   b 7.102°   c 2.090°   3 a 73.87°
b 57.55°   c 0.3679°   4 a 32°16′   b 86°48′
c 79°32′   5 $\theta = 53.23°$ (53°14′)   6 a 36.07°,
36°04′   b 53.13°, 53°08′   7 a 61°52′   b 57°59′
8 a 4.51 m   b 47°41′   c 3.79 mm   d 34°01′

Exercises 10.5

1 a 0.8387   b 0.7268   c 0.9781   d 0.5931
e 0.2275   2 a 5.109   b 0.079 76   c 58.35

d 1.676   3 a 40.26°, 40°16′   b 20.04°, 20°02′
c 87.50°, 87°30′   d 66.04°, 66°02′   4 a $\frac{3}{4}$ b $\frac{12}{13}$
c $\frac{3}{5}$ d $\frac{5}{13}$ e $\frac{12}{5}$ f $\frac{12}{13}$ g $\frac{4}{5}$ h $\frac{4}{5}$ 5 a 6.16 m
b 7.88 m   c 6.57 m   d 18.7 mm   e 228 m
f 11.6 m   6 a 25°50′   b 42°16′   7 a 24.7 m
b 4.11 m   8 a 37°27′   b 44°29′   9 63°26′
10 90°, 36°52′, 53°08′   11 14.8 m

Exercises 10.6

1 a 6.86 kN   b 28°14′   2 a 96.2 km/h
b 18°34′   3 9°23′   4 41°07′   5 196°00′
6 131 m   7 81.6 mm   8 42.0 mm   9 a 2.75 r
b 1.73r c 7.81 mm   10 a 77.5 V   b 7.10 V
c 26.8 V   d 44°52′   11 a 1.39 A   b 17.2 mA
c 23°36′   d 33°57′

Exercises 10.7

1 a $\sqrt{1 - t^2}$ b $\dfrac{\sqrt{1 - t^2}}{t}$ c $\dfrac{1 - t^2}{t}$ 2 $\dfrac{\sqrt{5}}{3}$
3 $\sqrt{15}$ 4 a 0.8808   b 0.6446   5 a 0.9454
b 0.95849   6 29.6 N   7 380 Ω

Exercises 10.8

1 a 160°   b 350°   c 225°   d 220°   e 070°
f 315°   2 31.1 km   3 94.8 km   4 82.0 km
5 a 64.7 m   b 43°38′   c 76.0 m   6 147.8 m

Exercises 10.9

1 a 84.4 m^2 b 5.86 m^2 c 1.13 km^2

Self-test

1 a 0.5543   b 0.5095   c 2.356   2 a 47.0 m
b 12.3 mm   3 a 0.9413   b 0.3375   c 0.5416
4 51.7 mm   5 a 750 mm   b 56°55′   6 5.27
7 a 654 m   b 3.46 m   8 a 23°32′   b 33°44′
9 4.31 km   10 3.706 m

11 Indices and radicals

Exercises 11.1

1 a 9   b 7   c 163.72   d 893.4   e 6   f 35
2 a 9x b ab c a^2b d ab e $\frac{1}{4}k^2n$ f 48ab
3 a 4   b 6   c 9   d 12   4 a $\dfrac{tx}{mk}$ b $\dfrac{a}{b}$ c $\dfrac{a}{b}$
d $\dfrac{a}{b}$ 5 a 3   b $\frac{1}{3}$ c 4   d 2   e 6   f 9   g 6   h 2
i 4   6 a $x^2 + 2x\sqrt{x} + x$ b $9m + 12\sqrt{m} + 4$

c $25 - 10\sqrt{a} + a$ d $1 - 4\sqrt{x} + 4x$
e $4n^2 + 12n\sqrt{n} + 9n$ f $x^3 - 2x\sqrt{x} + 1$

7 a 3 b $x^2 - 3$ c $b^2 - \dfrac{x}{y^2}$ d $\dfrac{c}{a}$ e c

Exercises 11.2

1 a 81 b 1.44 c 0.008 d $\frac{4}{9}$ e $\frac{1}{16}$ f -9 g 9
h -8 i -8 j $\frac{9}{16}$ 2 a 4 b -8 c 16 d -32
e 64 3 a P b N c N d P e P 4 a 1 b 1
c 1 d -1 5 a 0.09 b -0.008 c 0.01
d -0.09 e 0.09

Exercises 11.3

1 a 18 b 36 c 72 d 1440 e 4 f 25 g 49
h -5 2 a 4 b 36 c 15 d 216 e -2 f 3
g 16 h 72 3 a $8x^3$ b $\frac{1}{4}x^2$ c $12x^2$ d $9x^2$
e $-8x^3$ f $12x^2$ g $7x^2$ h $-3x^2$ i $8x^4$

Exercises 11.4

1 a $x^{16}y^3$ b $12ak^4m^6$ c $24x^8$ d $36x^6$ 2 a a^5b^3
b k^5m^2t c $6a^5x^4y^2$ d $2k^2m$ 3 a $a^8 + a^5$
b $6x^3 + 3x^2$ c $8k^4 - 11k^3$ d $7m^2 - 5m^3$
4 a $a^4 - b^2$ b $-a^8 + a^5 + 3a^3 - 3$
c $k - 5k^2 + 6$ d $25 - m^6$ 5 a $a^5 - a$
b $k^7 - 3k^4 + 2k$ c $6m^3 - 3m^2 - 2m + 1$ d $t^4 - t^6$
6 a m^4 b k^2 c $6a^4$ d $\frac{1}{16}x^6$ 7 a $3ab$ b mt
c $8k^2$ d $3m^2$ e $a + b$ f $k - x$ 8 a x^2 b $2\,m^7$
c $2a^4bc$ d m^2t^3 9 a x^{15} b $2a^6$ c $8a^6$ d a^2b^6
e m^6b^{15} f $-x^{12}$ g m^{18} h $-8m^6t^9w^{12}$ i $-2x^3y^6$
10 a $4x^6$ b $30x^{12}$ c 0 d $2\,m^{18}x^6$
11 a $p^6 + 6p^3 + 9$ b $4t^6 - 12t^5 + 9t^4$

c $1 - 6a^2b + 9a^4b^2$ 12 a $\frac{16}{81}$ b $\frac{1}{32}$ c $\dfrac{9n^2}{25}$

d $\dfrac{8t^3}{27n^6}$ e $\dfrac{16a^6}{b^4}$ f $\dfrac{t^4 - 4t^3 + 4t^2}{9k^6}$ 13 a 24^n
b 4^k c $16^x \times 15^t$ d $(2x + 3y)^t$ e 7^x f $(a - b)^7$
g $(x + y)^5$ 14 a 6^n b n^5 c $-$ d $8^k xy^2$ e a^{2n} f $-$

Exercises 11.5

1 a k^{-5} b $8t^{-1}$ c $3m^6$ d a^2 e $m^{-\frac{3}{4}}$ f $4b^{\frac{1}{8}}$
2 a $x^5 - x^2$ b $a^{-2} + a^{-5}$ c $6k^{\frac{1}{2}} + 2k^0$
d $e^{-x} + e^{2x}$ e $e^{-x} + e^{-x}$ f $2k^{\frac{1}{2}} - 5k + 3k^{\frac{1}{2}}$
3 a $a^6 + 2a^3$ b $1 - k^2$ c $e^x - e^{-x}$ d $x^{\frac{3}{2}} - x$
e $t^{-\frac{1}{2}} + t^{-\frac{1}{2}}$ f $3m^{\frac{1}{2}} + 2m^{\frac{3}{2}}$ 4 a 1 b 3 c 1
d 1 e 1 f 1 g 12 h 1 i 0 j 72 k 5 l 4

5 a 4 b 3 c 1 d -2 e -2 f 2 g -1 h 8
i 243 j 4 k -4 l 1000 6 a 9 b 2 c 5 d 4
e 3 f 2 g 3 h 27 7 a 2 b 7 c 1 d 9 e 3
8 a $\frac{1}{8}$ b $\frac{1}{16}$ c $\frac{1}{5}$ d $\frac{1}{81}$ e $\frac{1}{32}$ 9 a $\frac{1}{2}$ b $\frac{1}{2}$ c $\frac{1}{4}$ d $\frac{1}{3}$
e 1 10 a $\frac{1}{4}$ b $\frac{1}{27}$ c 1 d $\frac{1}{64}$ e $\frac{1}{8}$ 11 a 8 b 3

c 9 d 8 e 81 12 a $\dfrac{19}{9x^2}$ b $\dfrac{8}{3t}$ c $\dfrac{w}{(w - 1)^2}$

d $\dfrac{3x - 2}{x^3}$ e $\dfrac{1}{2}$ f $\dfrac{2x + 1}{2x}$

Exercises 11.6

1 a 1.23×10^4 b 5.91×10^{-1} c 8.36×10^7
d 5.47×10^{-2} e 4.38×10 f 6.39×10^{-8}
g 5.61×10^7 h 9.63×10^3 i 4×10^{10}
j 5.13×10^{-3} k 7×10^{-6} l 6.9×10^{-1}
2 a 10^6 b 10^5 c 10^7 3 a 10^{-4} b 10^{-12}
c 10^{-8} d 10^{-2} 4 a 8×10^{-12} b 10^{-2}
c 3×10^{-2} d 10^{-6} 5 a $10^5\,\text{m}^2$ b $10^3\,\text{m}^2$
c $10^{-5}\,\text{m}^2$ d $10^{-2}\,\text{m}^2$ 6 a $3 \times 10^{-2}\,\text{m/s}$
b $3 \times 10^6\,\text{m/s}$ c $2 \times 10^{-5}\,\text{km/h}$ d $5 \times 10^4\,\text{mm/s}$
7 a $3 \times 10^8\,\text{J}$ b $10^{-6}\,\text{J}$ c $6 \times 10^{-4}\,\text{J}$
d $4.8 \times 10^{-8}\,\text{J}$ 8 a $10^5\,\Omega$ b $10\,\Omega$ c $10^3\,\Omega$
d $10^{-6}\,\Omega$ 9 a $10^8\,\text{W}$ b $10^{-4}\,\text{W}$ c $10^2\,\text{W}$
d $10^{-8}\,\text{W}$ 10 a $10^4\,\text{W}$ b $10^{-5}\,\text{W}$ c $10^{-4}\,\text{W}$
d $10^3\,\text{W}$ 11 a 10^2 b $10^{-1}\,\text{A}$ c $10^{-3}\,\text{A}$
d $10^2\,\text{A}$ 12 a 1.36×10^{24} b 1.77×10^{16}
c -1.68×10^{15} d 7.60×10^{16} e 9.33×10^{-12}
f -2.00×10^{-7} g 9.11×10^{-5} h 1.50×10^{17}
i 4.29×10^{-1} j 4.42×10^{-2}

Exercises 11.7

1 a i $26.8 \times 10^3\,\mu\text{g}$ ii $26.8 \times 10^{-6}\,\text{kg}$ iii $26.8\,\text{mg}$
b i $26.8 \times 10^3\,\mu\text{A}$ ii $26.8 \times 10^{-3}\,\text{A}$ iii $26.8\,\text{mA}$
c i $82.5 \times 10^{-3}\,\text{mm}$ ii $82.5 \times 10^{-6}\,\text{m}$ iii $82.5\,\mu\text{m}$
d i $6.84 \times 10^3\,\text{kW}$ ii $6.84 \times 10^6\,\text{W}$ iii $6.84\,\text{MW}$
e i $637 \times 10^{-3}\,\text{MHz}$ ii $637 \times 10^3\,\text{Hz}$ iii $637\,\text{kHz}$
f i $3.8 \times 10^3\,\mu\text{H}$ ii $3.8 \times 10^{-3}\,\text{H}$ iii $3.8\,\text{mH}$
g i $74.5 \times 10^{-3}\,\text{MV}$ ii $74.5 \times 10^3\,\text{V}$ iii $74.5\,\text{kV}$
h i $62 \times 10^3\,\text{nm}$ ii $62 \times 10^{-6}\,\text{m}$ iii $62\,\mu\text{m}$
i i $13.8 \times 10^3\,\text{g}$ ii $13.8\,\text{kg}$ iii $13.8\,\text{kg}$
j i $4.25 \times 10^3\,\text{kHz}$ ii $4.25 \times 10^6\,\text{Hz}$ iii $4.25\,\text{MHz}$
k i $830 \times 10^{-3}\,\text{kg}$ ii $830 \times 10^{-3}\,\text{kg}$ iii $830\,\text{g}$
l i $68 \times 10^3\,\text{mm}$ ii $68\,\text{m}$ iii $68\,\text{m}$

2 a 11.32×10^6 **b** 94.65×10^{-6} **c** 181.3×10^3
d 429.5×10^{-6} **3 a** **i** 5.61×10^4 **ii** 56.1×10^3
b **i** 1.818×10^{-1} **ii** 181.8×10^{-3}
c **i** 2.346×10^7 **ii** 23.46×10^6
d **i** 9.152×10^5 **ii** 915.2×10^3 **e** **i** 3.71×10
ii 37.1 **f** **i** 5.636×10^{-7} **ii** 563.6×10^{-9}

Exercises 11.8

1 a $l = (hd^2)^{\frac{1}{3}}$ **b** 173 mm **2 a** $S = (2\sqrt{\pi V})^{\frac{2}{3}}$

b 1.37 m^2 **3 a** $R = 100\left[\left(\dfrac{P}{A}\right)^n - 1\right]$ **b** 14.5

4 a $r = \left(\dfrac{T^2 GM}{4\pi^2}\right)^{\frac{1}{3}}$ **b** 384 Mm

5 a $C_2 = \dfrac{CC_1}{C_1 - C}$ **b** 1.91 F

Self-test

1 a $6a^{\frac{1}{2}} - 2a$ **b** $x^2 - 2x^{\frac{3}{2}} + x$ **c** $1 + 2m^{\frac{3}{2}} + m^3$
d $2a^{\frac{1}{2}} - a - 1$ **2 a** $\frac{1}{81}$ **b** $\frac{1}{3}$ **c** 1 **d** 3 **e** $\frac{1}{4}$
3 a $\frac{1}{8}$ **b** 1 **c** $\frac{1}{5}$ **d** 1 **e** 4 **4 a** 6.69 **b** 0.996
c 0.292 **d** -1.81 **e** 7.17 **5 a** $x = 2\frac{1}{2}$ **b** $n = 0$
c $k = 6$ **d** $n = -1\frac{1}{2}$ **e** $t = -1$ **f** $x = -1\frac{1}{2}$
6 a 6 **b** 25 **c** $\frac{4}{9}$ **d** -2 **7 a** $2\frac{1}{4}$ **b** -1 **c** 9 **d** 8
e 5 **f** 144 **g** -6 **h** 6 **i** 10 **8 a** $3t^4$ **b** $2m^{-1}$
c $4a^{\frac{1}{6}}$ **d** k **e** $3t + 4t^4$ **f** $2n^{-6}$ **g** $(a + b)^9$
h $a^2 - b^2$ **9 a** $mt^{\frac{1}{2}} + 3m^2t$ **b** $a - a^{-1} - 1$
c $n^3 - 4n + 4n^{-1}$ **d** $3y^2 - y$ **e** $y^4 - y^2$

f $b - b^{\frac{1}{3}}$ **10** $V = \sqrt{\dfrac{ER}{t}}$, 83.5 mV

11 $t_2 = (t_1 + 273)\left(\dfrac{V_1}{V_2}\right)^{\gamma - 1} - 273$, $t_2 = 107°C$

12 Polynomials

Exercises 12.1

1 a $2x^3 - 11x^2 + 16x - 6$ **b** $x^6 - 2x^5 - x^4 + 2x^3$
$- 2x^2 + x - 1$ **c** $3x^4 - 2x^3 - 4x^2 + 2x + 1$
d $x^3 - 1$ **2 a** 4 **b** 8 **3** 51 **4** -7.01 **5** 4.088

Exercises 12.2

1 a $2(2x + 3)$ **b** $3(3 - 4a)$ **c** $a(b + 1)$
d $3x(2y - 3x)$ **e** $5(2a - 1)$ **f** $8a(2 - 3ax)$
g $3a^3(1 - 2b^2)$ **h** $5x^2(y - 2a + 1)$ **i** $3ab^2(a + 2)$
2 a $(p + q)(a + b)$ **b** $(b - c)(a + 2)$
c $(2a - y)(x + 3b)$ **d** $(3 - x)(2l - m)$
3 a $(b + c)(a + 1)$ **b** $(m + k)(l + 1)$

c $(x + y)(1 + m)$ **d** $(2 - y)(x + 1)$
4 a $(q + k)(p + 1)$ **b** $(q + k)(p - 1)$
c $(b - 3)(a - 1)$ **d** $(x - y)(y - 1)$
e $(t - m)(1 + k)$ **f** $(2x - 1)(k + m)$
5 a $a^k(1 - a^2)$ **b** $e^x(1 + e^{-2x})$ **c** $e^{-x}(e^{2x} + 1)$
d $2e^x(3e^x - 2)$ **6 a** e^{-x} **b** e^{2x}

Exercises 12.3

1 a $(E + V)(E - V)$ **b** $(L + 3)(L - 3)$
c $(3Q + 5)(3Q - 5)$ **d** $(2a + 3b)(2a - 3b)$
e $x(x + 2y)$ **f** $(m - 1)(m - 9)$ **g** $3(2G - 3)$
h $3t(2m + t)$ **i** $3e(2V - e)$ **2 a** 89 **b** 157
c 140 **d** 297 **e** 1920 **f** $0.175\ 1$
3 a $(V + 3)(V - 3)$ **b** $(2C + 5)(2C - 5)$
c $(3L + M)(3L - M)$ **d** $(4 + 3Q)(4 - 3Q)$
e $(7R + 1)(7R - 1)$ **f** $(ab + 2)(ab - 2)$
g $(1 + r)(1 - r)$ **h** $(5E + 4)(5E - 4)$
i $(4a + 5bc)(4a - 5bc)$

Exercises 12.4

1 a $x^2 + 5x + 6$ **b** $x^2 + 7x + 12$
c $x^2 + 8x + 15$ **d** $x^2 + 9x + 20$ **e** $x^2 + 9x + 8$
f $x^2 + 12x + 35$ **g** $x^2 + 10x + 21$
h $x^2 + 16x + 60$ **i** $x^2 + 13x + 40$
2 a $x^2 + 11x + 24$ **b** $x^2 + 14x + 24$
c $x^2 + 10x + 24$ **3 a** $(x + 1)(x + 3)$
b $(x + 3)(x + 5)$ **c** $(R + 1)(R + 7)$
d $(Q + 2)(Q + 6)$ **e** $(L + 2)(L + 3)$
f $(E + 1)(E + 4)$ **g** $(Z + 2)(Z + 5)$
h $(F + 1)(F + 16)$ **i** $(Q + 4)(Q + 8)$
j $(V + 3)(V + 4)$ **k** $(E + 3)(E + 8)$
l $(R + 2)(R + 8)$ **4 a** $x^2 - 5x + 6$
b $x^2 - 9x + 20$ **c** $x^2 - 7x + 12$
d $x^2 - 8x + 7$ **5 a** $(x - 3)(x - 8)$
b $(R - 2)(R - 4)$ **c** $(C - 1)(C - 9)$
d $(F - 1)(F - 36)$ **e** $(r - 3)(r - 3)$
f $(Q - 2)(Q - 50)$ **6 a** $x^2 + 3x - 10$
b $x + 4x - 21$ **c** $x - 3x - 4$ **d** $x + 6x - 16$
7 a $(x + 2)(x - 5)$ **b** $(y - 2)(y + 6)$
c $(R - 3)(R + 4)$ **d** $(C + 3)(C - 4)$
e $(E + 4)(E - 6)$ **f** $(Q - 2)(Q + 3)$
g $(t + 2)(t - 3)$ **h** $(L - 4)(L + 6)$
i $(V + 3)(V - 5)$ **j** $(r - 2)(r + 5)$

k $(Z - 4)(Z + 9)$ **l** $(F + 5)(F - 6)$

8 a $(2x + 1)(x + 1)$ **b** $(2x + 3)(x + 1)$

c $(2x + 1)(x + 3)$ **d** $(3a + 1)(2a + 1)$

e $(6a + 1)(a + 1)$ **f** $(4m + 5)(m + 1)$

9 a $(2x - 1)(x - 2)$ **b** $(2x - 3)(x - 1)$

c $(2a - 1)(a - 5)$ **d** $(3k - 1)(2k - 2)$

e $(3k - 2)(2k - 1)$ **f** $(6k - 1)(k - 2)$

10 a $(3x - 1)(2x + 3)$ **b** $(6x - 1)(x + 2)$

c $(3x - 2)(2x + 1)$ **d** $(18k + 5)(2k - 1)$

e $(6m + 5)(4m - 1)$ **f** $(2m + 1)(12m - 5)$

11 a $(3b + 4)(2b + 1)$ **b** $(6b + 1)(b + 4)$

c $(4a + 3)(a + 3)$ **d** $(6x + 1)(x + 6)$

e $(2x + 1)(4x + 9)$ **f** $(8x + 1)(x + 9)$

12 a $(3x - 2)(2x - 3)$ **b** $(3x - 8)(2x - 1)$

c $(3x - 1)(3x - 2)$ **d** $(9x - 2)(x - 1)$

e $(10k - 3)(k - 1)$ **f** $(5k - 1)(2k - 3)$

13 a $(2a + 5)(2a - 3)$ **b** $(2a - 5)(2a + 3)$

c $(4a - 5)(a + 3)$ **d** $(4a + 5)(a - 3)$

e $(3n - 2)(2n + 1)$ **f** $(6n + 1)(n - 2)$

14 a $(3n - 2)(2n - 1)$ **b** $(6a - 5)(a + 2)$

c $(4x + 5)(x - 1)$ **d** $(4n + 1)(n + 5)$

e $(8x - 2)(x - 3)$ **f** $(5t - 2)(2t + 5)$

Exercises 12.5

1 $3(1 + x)(1 - x)$ **2** $5(x + 2)(x + 4)4$

3 $w(t + 2)(t - 2)$ **4** $4(E + 3)(E - 3)$

5 $6(Q + 3)(Q - 2)$ **6** $5(L - 5)(L + 2)$

7 $7(E + 4)(E + 2)$ **8** $3(Z - 1)(Z - 4)$

9 $2(v + 6)(v - 3)$ **10** $3x(2x - 1)(2x + 1)$

11 $10t(t - 2)(t - 3)$ **12** $5k^2(3k + 4)(2k - 1)$

Exercises 12.6

1 $\dfrac{x + 2}{3}$ **2** $\dfrac{2}{x + 8}$ **3** $E + 1$ **4** $2 - C$ **5** $R + 2$

6 $R - 2$ **7** $\dfrac{1}{2Q - 1}$ **8** $\dfrac{6 - k}{k + 4}$ **9** $\dfrac{3t - 1}{2t - 1}$

Exercises 12.7

1 $x = 0, -2$ **2** $x = 0, 6$ **3** $k = -2, 1$

4 $t = -3, -4$ **5** $E = 0, 2.5$ **6** $C = 0, -3.5$

7 $Q = -1, -1.5$ **8** $x = 1.25, 3.5$

9 $V = -2.5, -1.2$

Exercises 12.8

1 a $V = \pm 5$ **b** $R = \pm 7$ **c** $R = \pm 5$ **d** $L = \pm 4$

e $x = \pm 2$ **f** $Q = \pm 6$ **2 a** $L = \pm 5.22$

b $R = \pm 3.73$ **c** $x = \pm 1.25$ **d** $R = \pm 0.938$

e $x = \pm 1.07$ **f** $x = 11.5$ **3 a** $x = \pm 6$

b $Q = \pm 9$ **c** $x = \pm 9$ **d** $L = \pm 4$ **e** $x = \pm 4$

f $t = \pm 8$ **4 a** $m = \pm 3.87$ **b** $x = \pm 1.73$

c $t = \pm 1.50$ **d** $k = \pm 4.58$ **e** $x = \pm 2.65$

f $n = \pm 3.61$ **5 a** $x = -1, 5$ **b** $R = -4, -2$

c $C = -4, 6$ **d** $R = -8, 2$ **e** $k = -72.9, 74.9$

f $t = -5.41, 1.41$ **6 a** $k = -7.12, 1.12$

b $C = -9.58, 13.6$ **c** $x = -0.962, 6.96$

d $n = -0.876, 2.21$ **e** $k = -4.45, -1.55$

f $x = -1.88, 5.88$

Exercises 12.9

1 a $V = 0, 6$ **b** $W = -2, 0$ **c** $M = 0, 2.5$

d $L = -1.5, 0$ **e** $Q = -0.5, 0$ **f** $R = 0, 5$

2 a $R = 0, 5$ **b** $V = 0, 0.6$ **c** $x = 0, 0.3$

d $K = 0, -0.2$ **e** $R = 0, 0.5$ **f** $t = -0.2, 0$

Exercises 12.10

1 a $R = -2, 3$ **b** $t = 2, 3$ **c** $W = -12, 2$

d $d = 4$ **2 a** $C = -6, -2$ **b** $T = -4, 9$

c $x = 6$ **d** $R = -8, 3$ **3 a** $k = 1\frac{1}{2}, 2\frac{1}{2}$

b $V = -\frac{3}{4}, -\frac{1}{2}$ **c** $C = \frac{1}{2}, \frac{2}{3}$ **d** $x = -1\frac{1}{2}, \frac{1}{6}$

4 a $x = 2, 6$ **b** $W = 2, 18$ **c** $x = -1, -5$

d $x = -18, 0$ **e** $R = 2, 3$ **5 a** $x = -4, 2$

b $R = 2, 10$ **c** $x = -2, 12$ **d** $x = 2, 3$

6 a $V = -1\frac{1}{2}, 0, 1\frac{1}{3}$ **b** $C = -2\frac{1}{2}, 0, \frac{3}{4}$

c $a = -\frac{1}{2}, 0, 2\frac{1}{4}$ **d** $x = -\frac{1}{3}, 0, 1\frac{1}{2}$

Exercises 12.11

1 $-1, 2$ **2** $-2, 6$ **3** $-2, -3$

4 $x^2 - 7x + 10 = 0$

Exercises 12.12

1 a $x = -3.45, 1.45$ **b** $t = -0.303, 3.30$

c $x = -2.35, 0.851$ **d** $Q = -0.434, 0.768$

2 a $x = -2.69, 0.186$ **b** $R = -0.425, 1.18$

c $C = -2.62, -0.382$ **d** $E = -3.56, 0.562$

3 a $x = -0.732, 2.73$ **b** $C = -3.00, -0.500$

4 a $x = 1.30, 7.70$ **b** $d = -0.303, 3.30$

c $V = -1.78, 0.281$ **d** $E = -3.19, 2.19$

Exercises 12.13

1 a $4\,\text{m}, 6\,\text{m}$ **b** $0\,\text{m}, 10\,\text{m}$ **c** $2.76\,\text{m}, 7.24\,\text{m}$

2 a $\frac{1}{2}b^2 + 1\frac{1}{2}b\,\text{m}^2$ **b** $4\,\text{m}$ **c** $2.77\,\text{m}$

3 a $l - 2\,\text{mm}$ **b** $l^2 - 2l\,\text{mm}^2$ **c** $6\,\text{mm}$

d 4.74 mm **e** 1.83 mm **4 a** $\dfrac{12}{v}$h **b** $\dfrac{12}{v+2}$h

c 10 km/h **d** 6.81 km/h **5 a i** $\dfrac{200}{v}$h

ii $\dfrac{200}{v+10}$h **b** 40 km/h

Exercises 12.14

1 a i $R^2 - 2R + 1 = 0$ **ii** $R = 1$
b i $R^2 - 13R + 36 = 0$ **ii** $R = 4, 9$
c i $R^2 - 15R + 25 = 0$ **ii** $R = 1.91, 13.1$
2 a $R = 2$ **b** $R = 1, 4$ **c** $R = 8$ **d** $R = 4, 9$
e $R = 1.37, 46.6$ **f** $R = 1.61, 22.4$
3 a i $V = 1$ **ii** $r = 0.333$ **iii** $I = 1$
b i $V = 3$ **ii** $r = 3$ **iii** $I = 0.333$ **c i** $V = 6$
ii $r = 3$ **iii** $I = 3$ **d i** $V = 1.80$ **ii** $r = 0.599$
iii $I = 0.898$ **4** 11.5 Ω, 78.5 Ω

Exercises 12.15

1 a 1 s, 4 s **b** 0.628 s, 6.37 s

2 a $\dfrac{5}{p}$h $\left(=\dfrac{18\,000}{p}\text{s}\right)$ **b** $\dfrac{5}{p+0.5}$h $\left(\dfrac{18\,000}{p+0.5}\text{s}\right)$

c 1.5 W for 3 h 20 min (= 12 000 s)
3 a $Q^2 + CEQ - 2CW = 0$ **b** 2.50 C
  c 1200 V

Self-test

1 a $x^5 - x^2 - 9x + 3$ **b** $2x^5 - x^4 + 6x^3 + 9$
2 -9 **3 a** $3x(1 + 4a - 3ax)$ **b** $6q(2pq - 3)$
c $(k + t)(a + 3)$ **d** $(t - 3)(1 - b)$
e $4e^{-3x}(3e^{2x} + 2)$ **f** $(x - 3y)(2a - b)$
g $(2a - b)(p - 1)$ **h** $2(a - b)(x - y)$
4 a $(5a - b)(3a + 7b)$ **b** $(1 + 6R)(1 - 6R)$
5 a $(R + 1)(R + 32)$ **b** $(t + 4)(t + 6)$
c $(Z + 1)(Z + 36)$ **d** $(Q + 3)(Q + 6)$
e $(4k + 3)(k - 2)$ **f** $(3n - 5)(3n - 2)$
g $(2x - 3)(2x + 1)$ **h** $(10t - 1)(t + 6)$
i $(3a - 4)(2a - 3)$ **6 a** $2(V + 18)(V - 1)$
b $7(1 + 2R)(1 - 2R)$ **c** $a(K - 6)(K - 1)$
d $2k(Q - 9)(Q + 2)$ **e** $3a(b + aC)(b - aC)$
f $2x(2x - 3y)(2x + 3y)$ **7 a** $\dfrac{C + 2}{C + 1}$
b $-3(2y + 1)$ **c** -1 **8 a** $R = 0, 5$ **b** $d = 0$
c $x = 0, 9$ **d** $R = -4, 3$ **e** $V = -2, 0, 2$
f $I = -2, \frac{5}{6}$ **g** $E = -4, -\frac{1}{8}$ **h** $x = \frac{2}{3}, \frac{7}{8}$

9 a $F = -1.72, 0.387$ **b** $d = -0.740, 0.540$
c $V = 0.349, 2.15$ **d** $F = -0.539, 1.65$
e $k = -0.221, 1.94$ **f** $C = -8.34, 0.839$
10 2.16 m, 1.25

13 Functions and their graphs

Exercises 13.1

1 a $4x - 3$ **b** $2x - 1$ **c** $-x^2 - 3x + 3$
d $2x - x^2$ **2 a** -1 **b** -13 **c** 2 **d** 8

3 a $\dfrac{1}{a + 1}$ **b** $\frac{1}{2}$ **c** $\dfrac{x}{x + 1}$ **d** 2 **4 a** 0

b $\dfrac{1 + x}{1 - x}$ **c** -1 **d** $\dfrac{2x + 1}{2x - 1}$ **5 a** $3y - 7$

b -13 **c** $2 - 3x$ **d** -7 **e** $3x^2 - 10$
f $6x + 2$ **6 a** 1 **b** $1\frac{1}{3}$ **c** 3 **d** 1 **7 a** $x = -1$
b $x = 1\frac{1}{2}$ **c** $x = 4$ **8 a** $x = \frac{1}{2}$ **b** $x = 0$
c $x = \frac{1}{2}$ **9** $y = 4$ **10** $x = -1, 1\frac{2}{3}$

Exercises 13.2

1 a i $x = -3$ **iv**
   ii $(-3, -5)$
   iii 4

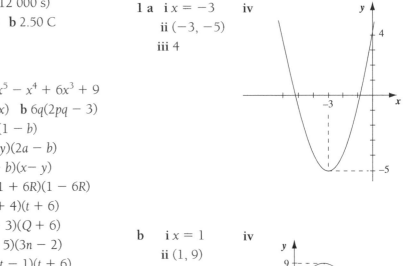

b   i $x = 1$ **iv**
   ii $(1, 9)$
   iii 7

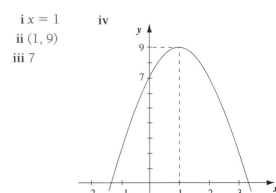

c **i** $x = 10$ **iv**
   ii $(10, -150)$
   iii -50

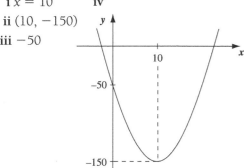

d **i** $x = 4$ **v**
   ii $(4, -8)$
   iii 24
   iv 2, 6

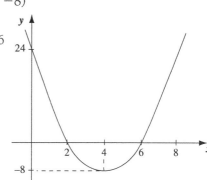

2 a **i** $x = -2$ **v**
   ii $(-2, -1)$
   iii 3
   iv $-1, -3$

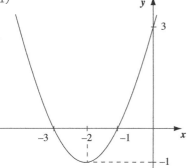

3 a **i** 0, 2   **iv**
   ii $x = 1$
   iii $(1, -3)$

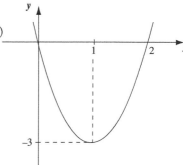

b **i** $x = -4$ **iv**
   ii $(-4, 8)$
   iii -24
   iv $-2, -6$

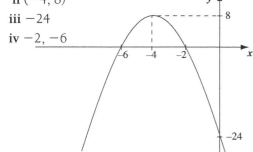

b **i** 0, 5   **iv**
   ii $x = 2.5$
   iii $(2.5, 62.5)$

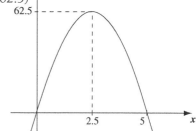

c **i** $x = 3$ **v**
   ii $(3, 12)$
   iii -15
   iv 1, 5

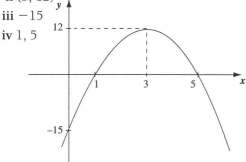

c **i** 0, -13 **iv**
   ii $x = -6.5$
   iii $(-6.5, 42.25)$

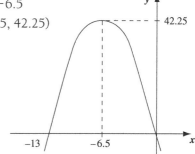

4

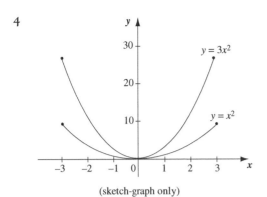

(sketch-graph only)

5

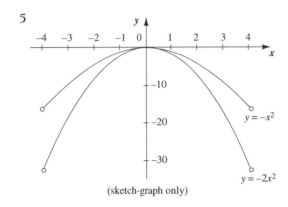

(sketch-graph only)

Exercises 13.3

1 a $y = 2x^2 + 4x - 6$ **b** $y = -3x^2 + 6x + 24$
c $y = 2x^2 - 6x - 8$ **2 a** $y = -x^2 + x + 6$
b $y = -3x^2 + 12x - 9$ **c** $y = 4x^2 + 16x + 12$
3 a $y = 2x^2 - 6x + 4$ **b** $y = -3x^2 - 6x + 9$
c $y = -2x^2 - 14x - 12$ **d** $y = 5x^2 - 45$
4 a $y = -x^2 + 4x - 5$ **b** $y = x^2 + 2x + 3$

Exercises 13.4

1 a 1.47 m   **b** 5.06 m   **2** 703 mm   **3** 62.5 cm
4 67 cm

Exercises 13.5

1 a circle with centre at
(0, 0) and radius 3 units

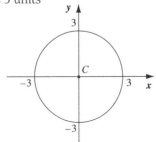

b circle with centre at $(-2, 3)$
and radius 1 unit

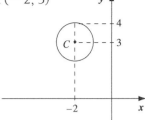

c circle with centre at $(-4, -1)$
and radius 1 unit

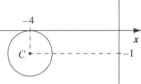

d circle with centre at (0, 0)
and radius 1 unit

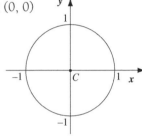

2 a $x^2 + y^2 = 64$ **b** $(x - 7)^2 + (y - 3)^2 = 1$
c $(x + 2)^2 + (y - 6)^2 = 9$ **3 a** $x^2 + y^2 = 16$
b $(x - 3)^2 + (y - 3)^2 = 9$ **c** $x^2 + y^2 = 25$
d $(x - 2)^2 + (y - 4)^2 = 16$ **e** $x^2 + y^2 = 169$
f $(x - 3)^2 + y^2 = 25$

Exercises 13.6

1   **2**

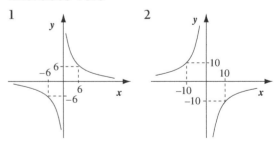

3   **4**

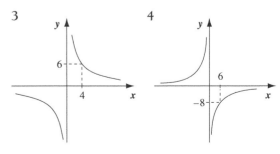

Exercises 13.7

1 $x = 2, y = 7; x = 11, y = 16$
2 $x = -3, y = -9\frac{1}{3}; x = 3, y = 9\frac{1}{3}$
3 $x = -2, y = 4\frac{1}{2}; x = 6, y = 9\frac{1}{6}$
4 $k = -6, m = 177; k = 2, m = 9$
5 $n = -1, g = -11; n = 2.5, g = 17.7$

Exercises 13.8

1 **a** to **f:** $x = 1, 4$ (*Note:* all these equations are reducible to the same equation and hence all have the same solutions.)
g $x = 0.7, 4.3$ **h** $x = 0.4, 4.6$ **i** $x = 1.4, 3.6$
2 **a** $x = -3, y = 0; x = 1, y = 0$ **b** $x = -2.4, y = 2; x = 0.41, y = 2$ **c** $x = -1, y = 4$ **d** no solutions 3 **a** $x = -3.7, y = 16; x = 4.7, y = -9.0$ **b** $x = -0.64, y = -1.3; x = 3.1, y = 0.57$ 4 $x = 0, y = 0; x = 5, y = -5$
5 $x = -2.4, y = -1.9; x = 1.4, y = 1.9$
6 $x = -3.5, y = 2.5; x = 1.5, y = 2.5$

Exercises 13.9

1 **a** $W = 6 - 2h$ **b** $A = 2h(3 - h)$
d $W = 3$ m, $h = 1.5$ m, $A = 4.5$ m^2
2 **b** **i** 1.8 m **ii** 5 m **iii** 18 m **iv** If this trajectory had originated from a point at ground level, this point would have been 2 m behind its actual projection point. **v** If this trajectory had been allowed to continue, instead of the ball striking the ground, it would have been 2.2 m below ground level when its horizontal displacement was 20 m.

Self-test

1 **a** −1 **b** −4 **c** 0 **d** 1 **e** $8t^3 - 4t$ 2 **a** −2
b 4 **c** 2 **d** 6 3 **a** $t = -2, 1$ **b** $k = -2, 3$
c $m = -1, 1.5$
4 **a**

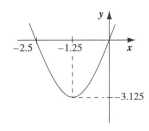

b

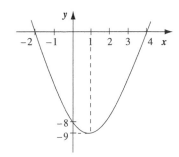

c

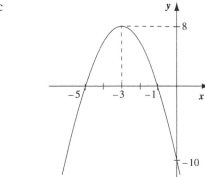

5 **a** $y = x^2 - 2x - 8$ **b** $y = x^2 + 3x - 10$
c $y = -\frac{1}{2}x^2 - 2x - 3$
6 **a** circle with centre at $(0, 3)$ and radius 2 units
b rectangular hyperbola with its branches in the second and fourth quadrants

14 Logarithms and exponential equations

Exercises 14.1

1 **a** $3^2 = 9$ **b** $5^{\frac{1}{2}} = \sqrt{5}$ **c** $7^0 = 1$ **d** $2^{-3} = \frac{1}{8}$
e $9^{-\frac{1}{2}} = \frac{1}{27}$ **f** $4^1 = 4$ 2 **a** $\log_5 25 = 2$
b $\log_4 \frac{1}{16} = -2$ **c** $\log_3 1 = 0$ **d** $\log_{16} 4 = \frac{1}{2}$
e $\log_9 9 = 1$ **f** $\log_4 \frac{1}{32} = -\frac{5}{2}$ 3 **a** $2^y = 7$
b $10^k = y$ **c** $e^t = M$ **d** $a^2 = Q$ **e** $10^y = x$
f $e^{10} = x$ 4 **a** $\log_t y = 2$ **b** $\log_e k = m$
c $\log_{10} Q = p$ **d** $\log_2 \frac{1}{8} = -3$

Exercises 14.2

1 **a** 4 **b** 2 **c** 1 **d** −2 **e** 1 **f** −1 **g** 4
h −3 **i** −8 2 **a** $\frac{1}{2}$ **b** 0 **c** $-\frac{1}{2}$ **d** −1 **e** $\frac{1}{3}$
f 0 **g** 1 **h** 7 **i** −15

Exercises 14.3

1 a 100, 115   b 0.1, 0.118   c 0.001, 0.001 35
d 2, 2.03   e 3, 2.99   f −2, −1.89   2 a 10.4
b 2.15   c 0.0141   d −2.75   e 29.9   f 6.32
g 27.2   h −15.1   3 a 94.2   b 16.0   c 2.45
d 21.2   e 9.39   f 1.34   g 2.13   h 10.9   i 22.3
j 46.0   k 5.82   l 0.361   4 a 0.434   b 0.809
c 2.29

Exercises 14.4

1 34 dB   2 1.6×10^6:1   3 3.2 mW   4 130:1
5 20 μ W/m^2 6 a 2.5 μ W/m^2 b 38 μ W/m^2
c 76 dB

Exercises 14.5

1 a $\log 2 + \log 15$, $\log 3 + \log 10$, $\log 5 + \log 6$
b $\log 6$ 2 a $\log 2 + \log 10$, $\log 4 + \log 5$
b $\log 5$ 3 a 2   b 1   c −3   d 3   e 1   f 0
4 a −1   b 1   c −1   d −1   e $\frac{1}{2}$ f $-\frac{1}{2}$
5 a −2   b −1   c 2   d 2   6 a $2.30x$ b $0.434\,x$
c x d x e $6.91x$ f $3.04x$ 7 a $k \log N$
b $\log M + \log N$ c $\log A - \log B$
d $\log k + n \log x$ e $\log m + \frac{1}{3} \log w$
f $\log 5 + 3 \log x$ g $\frac{1}{2} \log x - \log 3$
h $\log k + \frac{1}{2} \log t - \log m$
i $\log 5 + \frac{1}{2} \log x - \log y$ j $x \log e$
k $\log c + n \log x$ l $\log K + x \log e$

Exercises 14.6

1 a 1.85   b 6.64   c 8.58   d − 2.43   e 38.4
2 $\dfrac{1}{\log_a b}$ (or $-\log_a b$)

Exercises 14.7

1 a 1.3   b 3.8   c −4.6   d 8.8   e $2\frac{1}{6}$ f 0.7
2 a 2   b 2.3   c 6.9   d 10.58

Exercises 14.8

1 a $V = 48$ b $x = 0.364$ c $E = 0.1$
d $x = -13$ e $x = 1.75$ f $V = 0.1$
2 a $x = 12.2$ b $P = 0.903$ c $C = 2.45 \times 10^{-5}$
d $x = 3680$ e $P = 2.02 \times 10^{-3}$ f $k = 8.23$

Exercises 14.9

1 a $k = 4$ b $n = -5$ c $x = 1\frac{1}{2}$ d $t = 4$
2 a $m = 0$ b $x = 1$ c $t = 0$ d $x = -\frac{1}{2}$

3 a $n = 1\frac{1}{2}$ b $x = \frac{2}{3}$ c $x = -\frac{1}{6}$ d $t = \frac{1}{8}$
4 a $m = \frac{1}{2}$ b $k = 1\frac{1}{4}$ c $x = -\frac{1}{4}$ d $t = -\frac{1}{4}$
5 a $x = \frac{1}{3}$ b $t = 1$ c $m = 5$ 6 a $x = -\frac{1}{4}$
b $t = 0$ c $x = 0$ 7 a $x = 1.77$ b $x = 2.59$
c $n = 1.78$ d $x = -11.3$ 8 a $k = 1.71$
b $n = 0.783$ c $k = -0.244$ d $t = -3.19$
e $x = 1.25$ f $n = 2.54$

Exercises 14.10

1 a $t = \dfrac{\log F}{\log k}$ b $x = 10^y$ c $x = -\dfrac{\log y}{\log 3}$
d $a = \dfrac{b \log C}{\log x}$ e $b = (a \times 10^{-m}) + c$
f $x = y \times 10^n$ g $x = y$ h $x = 10^{-k}$
2 a $t = \dfrac{1}{k} \log_{10} \dfrac{y_0}{y}$ b $t = \dfrac{1}{k} \log_e \dfrac{Q_0}{Q}$
c $t = \dfrac{\log\left(1 - \dfrac{y}{y_0}\right)}{k \log A}$ d $t = \dfrac{1}{k} \log_e \left(1 - \dfrac{Q}{Q_0}\right)$

Exercises 14.11

1 a $b = \dfrac{\log S}{\log 2}$ b 9

3 $h = \dfrac{\log \dfrac{p}{p_s}}{\log 0.883} \left(= \dfrac{\log p - \log p_s}{\log 0.883} \right)$

5 $t = \dfrac{1}{k} \ln \dfrac{Q}{Q_0} = \dfrac{\log \dfrac{Q}{Q_0}}{k \log e} \left(= \dfrac{\log Q - \log Q_0}{k \log e} \right)$

6 $n = \dfrac{\log \dfrac{A}{P}}{\log R} \left(= \dfrac{\log A - \log P}{\log R} \right)$

7 a $D = 0.5d \times 10^{\frac{z}{276}}$ b 27.8 mm

Exercises 14.12

1
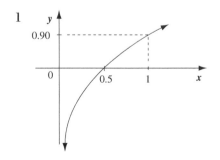

2

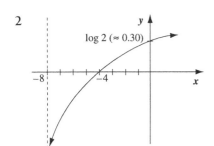

3, 4

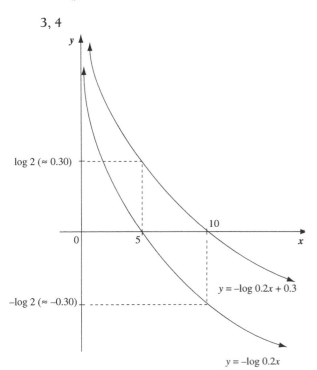

$y = -\log 0.2x + 0.3$

$y = -\log 0.2x$

Exercises 14.13
1 a 64   b 2   c 0.25   d 10   e 0.01   f 64
g 32   h 81   2 a $x = 2$ b $x = 4$ c $x = \pm 5$
d $x = -4, 1$ 3 a 1   b 2   c 1   d -2 e -3
f 0.5   g $-\frac{1}{2}$ h 1.5   4 a 9   b 0.25   c 0.125
d 0.05   e 5   f 4   g 8   h 3

Exercises 14.14
1 a 7   b ∞ c $-\infty$ d 0   e ∞ f 0   g 0   h 0
i 1   j $-\infty$ 2 a 0   b ∞ c $-\infty$ d ∞ e 0

Self-test
1 a -3 b 0   c $-\frac{2}{3}$ d $-\frac{1}{2}$ e $\frac{1}{2}$ f 7
2 a 5.70   b 3.13   c 47.5   d 83.2   e 0.165
f 1.06   3 a -1 b $-\frac{1}{2}$ c 1   d 1.5   4 a 2.32
b -0.631 c 0.605   5 a $x = 2\frac{1}{2}$ b $n = 0$

c $k = 6$ d $m = -1\frac{1}{2}$ e $t = -1$ 6 a $k = -6$
b $t = 7$ c $x = 3$ d $6x^2$ 7 a $t = 1.79$
b $n = 1.29$ c $k = 0.462$ d $x = 1.29$

8 a $t = \dfrac{k \log A}{\log b}$ b $x = a(1 - 10^{-k})$

c $t = \dfrac{\log_e k + b}{1 - a}$ d $y = \dfrac{x}{1 - e^{kt}}$

9 a $t = \dfrac{\log V - \log V_0}{\log 2}\left(= \dfrac{\log\left(\dfrac{V}{V_0}\right)}{\log 2}\right)$

b 11.7 ms   10 31.6 μ W/m^2

15 Non-linear empirical equations

Exercises 15.1
1 $K = 49.4 \pm 0.1, b = 1.5 \pm 0.1$ 2 a 16.8 ± 0.3,
b 954 ± 9 N   c 12.2 ± 0.1 m/s

Exercises 15.2
1 $K = 48 (\pm 3\%), a = 0.12 (\pm 3\%)$
2 $K = 2.9 (\pm 4\%), N = 1.5$
3 $C = 2.5 (\pm 4\%), n = -0.50 (\pm 4\%)$
4 a $K = 70, b = 0.48$ b $i = 70 \times e^{-0.744t}$

Self-test
1 $C = 5.6 (\pm 3\%), K = 1.2 (\pm 3\%)$
2 $C = 7.2 (\pm 3\%), n = 1.5 (\pm 3\%)$
3 $K = 33 (\pm 3\%), c = 0.012 (\pm 3\%)$

16 Compound interest: exponential growth and decay

Exercises 16.1
1 a   i 19.1%   ii 59.4%   b 12   2 a $2550 \times$
$(1.007)^5, \$2640.51$ b $[(1.007)^{12} - 1] \times$
100%, 8.73%   3 \$106.55

Exercises 16.2
1 a $\$P \times 1.08$ b $\$P \times (1.02)^4$ c $\$P \times e^{008}$
2 a \$10 400.00   b \$10 406.04   c \$10 407.42
d \$10 407.95   e \$10 408.08   f \$10 408.11

Exercises 16.3

1 a −0.181 per day **b** 4.40 days **2 a** 14 800
b 9.90 years **c** 4.29 years **3 a** 47 mA
b 9.6 mA **4 a i** 9.2 g **ii** 1.9 kg **b** 29.2 g
c 5.3 h **5** 13 kg

Exercises 16.4

1 (−0.6, 1.15), (− 0.3, 2.3), (0, 4.6), (0.3, 9.2),
(0.6, 18.4) **2** (−1.24, 16), (−0.62, 8), (0, 4),
(0.62, 2), (1.24, 1) **3** (−0.38, 176), (−0.19, 88),
(0, 44), (0.19, 22), (0.38, 11) **4** (−0.6, 7),
(− 0.3, 14), (0, 28), (0.3, 56), (0.6, 112)
5 (−0.24, 0.045), (−0.12, 0.09), (0, 0.18),
(0.12, 0.36), (0.24, 0.72)

Exercises 16.5

1 a, b

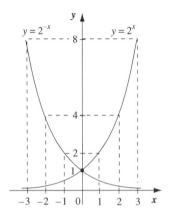

c

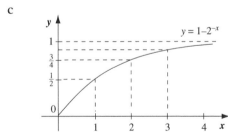

d Multiply the values on the y-axis scale for **c**
by 327.

2 (only sketch-graphs shown here)

a

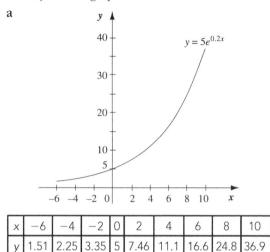

x	−6	−4	−2	0	2	4	6	8	10
y	1.51	2.25	3.35	5	7.46	11.1	16.6	24.8	36.9

b

x	0.1	0.2	0.3	0.4	0.5
y	13.2	19.1	21.8	23.0	23.6

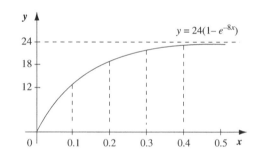

3 a

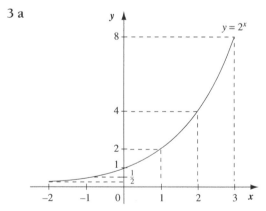

b i 0.5, 1, 1.5 **ii** 0.25, 0.5, 0.75 **iii** 0.69,
1.4, 2.1 **iv** 0.35, 0.69, 1.0 **c** 43, 86, 172, 344

4

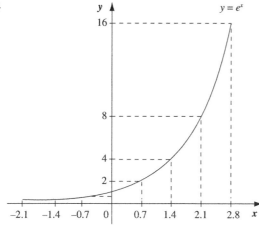

(sketch-graph only)

5 a 0.35 min (= 21 s)   **b** 0.7 min (= 42 s)
c 1.05 min (= 63 s)   **d** 54 s

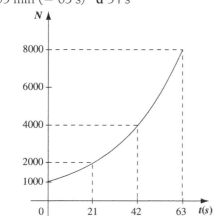

(sketch-graph only)

6 a 1.4 s, 2.8 s, 4.2 s
b

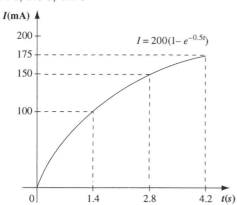

(sketch-graph only)

c 80 mA

Self-test

1 a \$8508.54   **b** \$8539.23   **c** \$8549.70
d \$8549.82   **2 a** 0.0715   **b** 17.0 g
3 a 85 V   **b** 5.9 s
4 a

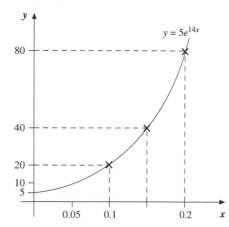

b

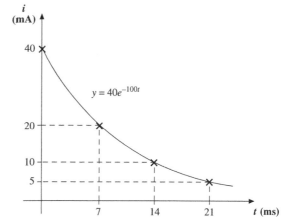

c

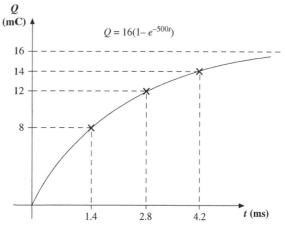

5 a i 7 years **ii** 14 years **iii** 21 years
iv 35 years **b** $m \approx 37$ g

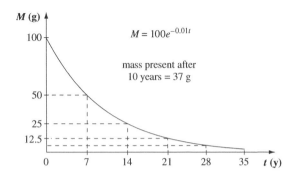

17 Circular functions

Exercises 17.1

1 a 0.2 **b** 0 **c** 0 **d** 0 **e** 1 **f** −0.2 **2 a** 0
b 0 **c** 2 **d** −2 **3 a** 0.3 **b** 0.3 **c** −0.3 **d** 0
e −0.3 **f** 0 **g** 0.3 **h** 1 **4 a** 2 **b** 1 **c** 0
d −1 **5 a** −1.2 **b** 1.2 **c** 0 **d** 1.2 **e** 0 **f** −1.2
6 a 0.98 **b** −0.98 **c** −6 **d** 0.98 **e** 0.2
f −0.2 **g** 6 **h** −0.2 **i** −6 **j** −0.98 **k** 0.2
l −0.98 **m** 0.98 **n** 6 **o** 0.98 **7 a** $\tan\theta$
b 1 **c** cot *i* **8 a** $\sin 150° = 0.5$, sin
$210° = -0.5$, $\sin 330° = -0.5$
9 $\tan 120° = -\sqrt{3}$, $\tan 240° = \sqrt{3}$,
$\tan 300° = -\sqrt{3}$ **10** $\sin 50° \approx 0.766$,
$\sin 130° \approx 0.766$, $\sin 310° \approx -0.766$
11 $\cos 70° \approx 0.342$, $\cos 110° \approx -0.342$,
$\cos 250° \approx -0.342$, $\cos 290° \approx 0.342$
12 a 90° **b** sin **c** 90° **d** cos **e** cos **f** 180°
g cos **h** 360° **i** 180° **j** cos **k** sin **l** 270°
m sin **n** 90° **o** 270° **p** 270° **13 a** θ, $180° - \theta$
b $90° + \theta$, $270° - \theta$ **c** $90° - \theta$, $90° + \theta$
d $270° - \theta$, $270° + \theta$ **e** 6, θ $360° - \theta$
f θ, $180° - \theta$ **14 a** T **b** F **c** F **d** T **e** F
f T **g** T **h** T **i** F **j** F **k** T **l** F **m** F **n** T
o F **p** T **q** T **r** T **s** F **t** F

Exercises 17.2

1 a 1.67 **b** 2.40 **c** 1.67 **d** 0.750 **e** 1.08
f 1.08 **g** 1.25 **h** 0.417 **i** 2.60 **j** 2.60
2 a 0.933 **b** 1.66 **c** 1.05 **d** 0.207 **e** 1.20
f 1.14 **3 a** 50.5 **b** 2.40 **c** 1.20 **d** 0.632
4 a 52.48° **b** 25°11′ **5 a** 2.571 **b** 0.4660

Exercises 17.3

1 (same answers in each case for **i** and **ii**) **a** $\dfrac{a}{c}$
b $\dfrac{b}{c}$ **c** $\dfrac{a}{b}$ **d** $\dfrac{b}{a}$ **e** $\dfrac{c}{b}$ **f** $\dfrac{c}{a}$ **2 a** 70°
b 50° **c** 57° **d** 41° **e** 42° **f** $(90 - A)°$
3 a sin **b** tan **c** cot

Exercises 17.4

1 a 45° **b** 180° **c** 90° **d** 18° **e** 75°
2 a $\dfrac{\pi}{2}$ **b** $\dfrac{\pi}{3}$ **c** $\dfrac{\pi}{4}$ **d** 2π **e** $\dfrac{\pi}{6}$ **3 a** 0.820 30 rad
b 0.017 453 rad **c** 5.9690 rad **d** 0.415 04 rad
e 3.3379 rad **f** 1.7453×10^{-4} rad
4 a 171.8873° **b** 57.2958° **c** 116.5797°
d 68.1818° **e** 330.4800° **f** 77.1429°
5 a 114°35′ **b** 79°21′ **c** 178°00′ **d** 27°42′
6 a 0 **b** 0 **c** 1 **d** 0 **e** −1 **f** −1 **g** 0 **h** 1
7 a 0.5720 **b** 0.8121 **c** −2.625 **d** 0.6613
8 a 0.743 **b** 0.891 **c** −0.825 **9 a** 4.87
b −10.6

Exercises 17.5

1 a 419 rad/s **b** 3.49 rad/s **c** 3140 rad/s
2 a 2480 rpm **b** 3820 rpm **3** 200°/s
4 a 1.44 s **b** 628 ms **c** 4.44 s **5 a** 37.2°
b 5.10°

Exercises 17.6

1 a i 4.19 m **ii** 16.8 m² **b i** 157 mm
ii 4.71×10^3 mm² **c i** 249 mm **ii** 15.3 ×
10^3 mm² **d i** 5.93 m **ii** 115 m² **2 a** 3.85 ×
10^3 mm² **b** 13.1×10^3 mm² **3 a** 2.69 m
b 1.51 m²

Self-test

1 a $\sin x$ **b** $-\cos\theta$ **2** $\cos 110° = -0.342$,
$\cos 250° = -0.342$, $\cos 290° = 0.342$
3 $\cos 38° = 0.788$, $\cos 142° = -0.788$,
$\cos 218° = -0.788$, $\cos 322° = 0.788$

4 $K = 30.35°, 120.35°, 210.35°, 300.35°$
5 $R = 165.15°$ 6 2.59 7 a $3.72°$ b $7.65°$
8 a 413 mm   b 30.1×10^3 mm^2
c 50.8×10^3 mm^2 d 139×10^3 mm^2

18 Trigonometric functions and phase angles

Exercises 18.1

1 a

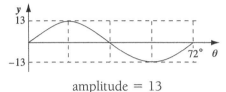

amplitude $= 13$

period $= \dfrac{360°}{5} = 72°$

b

amplitude $= 6$

period $= \dfrac{360°}{3} = 120°$

c

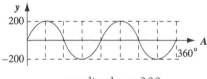

amplitude $= 200$
period $= 180°$

d

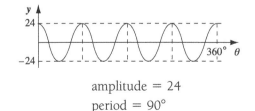

amplitude $= 24$
period $= 90°$

2 $0°, 45°, 90°, 135°, 180°$ 3 $30°, 90°, 150°$
4 7, when $\theta = 15°, 75°$ 5 -31, when $A = 36°,$
$108°, 180°, 252°, 324°$ 6 3 7 $2\frac{1}{2}$

Exercises 18.2

1   $a = 18$
    $T = 360°$

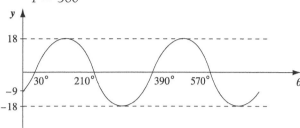

2   $a = 200$
    $T = 2\pi$

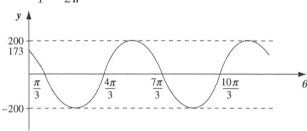

3   $a = 0.6$
    $T = 2\pi$

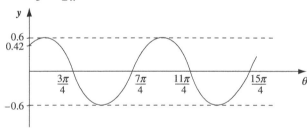

4   $a = 50$
    $T = 360°$

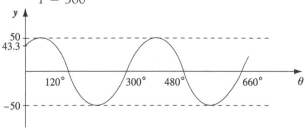

5     $a = 2.3$
       $T = 2\pi$

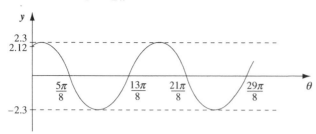

6     $a = 6$
       $T = 2\pi$

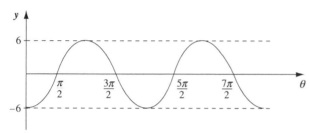

Exercises 18.3

1 a   **i** $\dfrac{2\pi}{3}(2.09)$ **ii** $120°$ **iii** 477×10^{-3}
cycles/rad   **iv** 8.33×10^{-3} cycles/degree   **v** 3
 vi

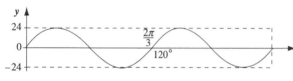

b   **i** $\dfrac{\pi}{2}(1.57)$ **ii** $90°$ **iii** 637×10^{-3} cycles/rad
iv 11.1×10^{-3} cycles/degree   **v** 4   **vi**

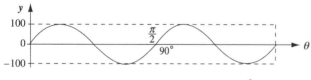

c   **i** π (3.14) **ii** $180°$ **iii** 318×10^{-3} cycles/rad
iv 5.56×10^{-3} cycles/degree   **v** 2
 vi

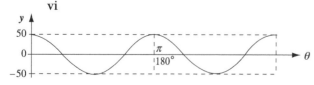

d   **i** $\dfrac{2\pi}{5}(1.26)$ **ii** $72°$ **iii** 796×10^{-3} cycles/rad
iv 13.9×10^{-3} cycles/degree   **v** 5
vi

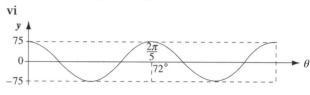

2 a $y = 12 \sin 6\theta$ **b** $\dfrac{7°}{6}$ **c** right

3 a $V = 40 \cos 5\theta$ **b** $\dfrac{29°}{5}$ **c** left

4 a $i = 17 \sin 4\theta$ **b** 0.3π **c** left

5 a $V = 35 \cos 3\theta$ **b** $40°$ **c** right

6 a $i = 23 \sin 8\theta$ **b** $\dfrac{\pi}{2}$ **c** right

7 a $V = 240 \cos 20\pi\theta$ **b** 0.01 **c** left

8 a   **i** $y = 50 \sin 2\theta$ **ii** $y = 50 \sin(2\theta - 30°)$

b   **i** $y = 18 \cos 32x$ **ii** $y = 18 \cos\left(32x + \dfrac{3\pi}{8}\right)$

9 a

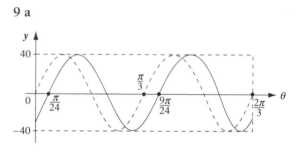

b

10 a

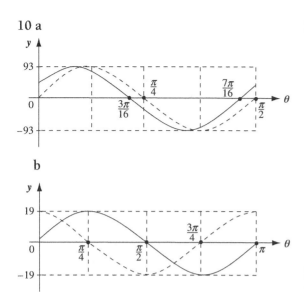

b

Exercises 18.4

1 The first leads by $30°$

2 The first leads by $40°$

3 The second leads by $100°$

4 The second leads by $80°$

5 The second leads by $160°$

6 The first leads by $\dfrac{\pi}{2}$

7 They are in phase

8 The first leads by $\dfrac{5\pi}{12}$

9 The first leads by $\dfrac{\pi}{2}$

10 The second leads by $\dfrac{\pi}{2}$

Exercises 18.5

1

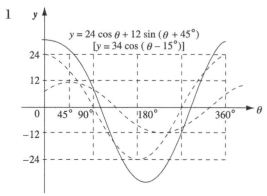

$y = 24 \cos \theta + 12 \sin (\theta + 45°)$
$[y = 34 \cos (\theta - 15°)]$

2

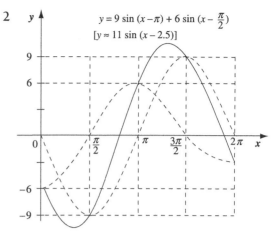

$y = 9 \sin (x - \pi) + 6 \sin (x - \frac{\pi}{2})$
$[y \approx 11 \sin (x - 2.5)]$

Self-test

1 25

2 a

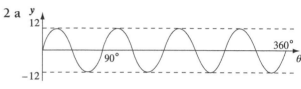

amplitude = 12
period = $90°$

b

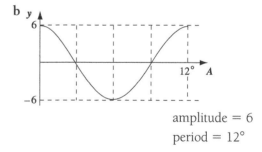

amplitude = 6
period = $12°$

3 17, when $\theta = 0°, 120°, 240°, 360°$

4 a

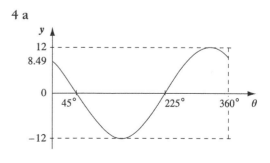

b

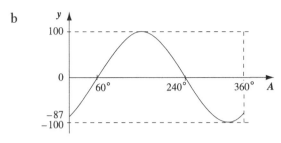

c

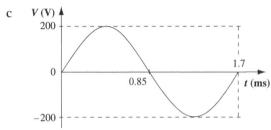

d

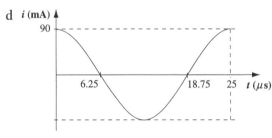

5 a

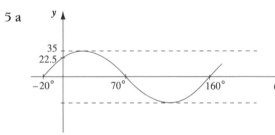

b

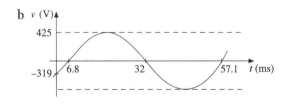

19 Trigonometry of oblique triangles

Exercises 19.1

1 a 7.6 m   b 590 mm   c 2.86 m   d 41.5 mm
e 1.90 m   f 32.9 mm   2 a 50°15′   b 42°43′
c 22°41′   d 28°24′   3 a 117°27′   b 134°17′,
34.0 mm

Exercises 19.2

1 a R, no   b A, yes   c H, yes   d T, yes   e K, no
f N, no   2 a 52°49′   b 92°45′   c 272 mm or
a 127°11′   b 18°23′   c 85.9 mm   3 a 41.34°
b 112.23°   c 45.5 m or a 138.66°   b 14.91°
c 12.7 m

Exercises 19.3

1 a $m^2 = k^2 + t^2 − 2kt \cos \theta$
b   i $t^2 = p^2 + q^2 − 2 pq \cos T$
ii $p^2 = q^2 + t^2 − 2qt \cos P$
c   i $x^2 = n^2 + v^2 − 2nv \cos X$
ii $v^2 = n^2 + x^2 − 2nx \cos V$ 2 a 11.9m
b 30°52′   c 1.70 m   d 3.01 m   e 119°28′
f 25.3 m   g 222 mm   h 35°26′   i 95°10′

Exercises 19.4

1 28°57′, 46°34′, 104°29′   2 23.3 cm   3 1.49 m,
2.39m   4 a 893 mm   b 14.5°   5 115.5
6 18°35′   7 a 836 km   b 46.4°   8 199 mm,
319 mm   9 193 m 10 S69.2°E 11 97.7°
12 19.7 km

Self-test

1 a 30.8 mm   b 1.14 m   2 75.1°   3 180 m
4 36°49′ or 143°11′   5 a 20.5 mm   b 39°35′
c 571 mm   d 134°03′   6 17.4 km from P and
11.5 km from Q   7 69.7 km on bearing 123.8°

20 Trigonometric identities

Exercises 20.1

1 a $x = −1, 0$ b $x = 0$ c $x =$ any number
d $x = 0$. Equation c is an identity.

Exercises 20.2

1 a 0.2191   b 0.2051   c 0.6545   d 0.9231

Exercises 20.3

1 a $\tan 47°$ b $\cos 23°$ c $\cot 38°$ d $\cot \theta$
e $\cot A$ f $\operatorname{cosec} x$ g $\cos \theta$ h $\sec \theta$ i $\cot \theta$
2 a $\tan$ b $\tan$ c $\sec$ d $\sec$ e $\tan$ f cosec
3 a $\cot \theta$ b $\sin A$ c $\tan x$ d $\cos \theta$ 4 a $\sin^2 \theta$
b 1   c $\cos^2 A$ d $−1$ e 1   f $−\sin^2 A$ g $−1$
h $3 \sin^2 K$ 7 a $\cos^2 \theta$ b $− 1$ c $\cot \theta$ d $\sin x$
e $\tan^2 \theta$ f $− \tan^2 \theta$

Exercises 20.4

1 b $F_2 \approx 646$ N 2 b $R \approx 540$ N 3 b 9.88 kN
c 34.2° 4 b $R \approx 3.46$ kΩ c $\phi = -16.7°$
5 a 0.6 b 30.96° c $A \approx 5.83$

Exercises 20.5

1 29.8 N 2 28.4 N 3 5.44 kN 4 a 8.06
b 60.3°

Exercises 20.6

1 $\left(\dfrac{\sqrt{3}}{2}, \dfrac{1}{2}\right)$ 2 $(\cos \theta, \sin \theta)$ 3 a $P(\cos \beta, \sin \beta)$,
$Q(\cos \alpha, \sin \alpha)$ b $PQ^2 = 2 - 2 \cos (\alpha - \beta)$
c $PQ^2 = (\cos \alpha - \cos \beta)^2 + (\sin \alpha - \sin \beta)^2$
d $\cos (\alpha - \beta) = \cos \alpha \cos \beta + \sin \alpha \sin \beta$
4 $\sin 2x = 2 \sin x \cos x$ 5 $\cos 2x = 2 \cos^2 x - 1$
6 $\cos 3x$ 7 $\sin y$ 8 $\cos x$ 9 $\tan y (\tan x + 1)$

10 $\dfrac{\tan A + \tan B}{1 + \tan A \tan B}$ 11 $\sin^2 A - \sin^2 B$

12 a $\tan y = \dfrac{7A - 1}{A + 7}$ b 0 13 $\sin x$ 14 $\sin 2A$

15 a $\tan^2 \theta$ b $2 \sec^2 \theta$ 16 a $\dfrac{\sqrt{3} - 1}{2\sqrt{2}}$

b i $\dfrac{\sqrt{3} + 1}{2\sqrt{2}}$ ii $-\dfrac{\sqrt{3} + 1}{2\sqrt{2}}$ iii $\dfrac{\sqrt{3} + 1}{\sqrt{3} - 1}$

iv $\dfrac{1 - \sqrt{3}}{2\sqrt{2}}$

Self-test

1 a $\cos^2 A$ b $\sec^2 \theta$ c $\csc^2 x$ d $\sec A$
e 1 2 a 1 b $-\tan^2 x$ c 1 d $\cot^2 x$

4 a and b $A = 18.61, \theta = 127.71°$

5 a $\frac{1}{2}\cos \theta - \dfrac{\sqrt{3}}{2}\sin \theta$ b $\dfrac{\tan x + 1}{\tan x - 1}$ 6 a $\cos \theta$

b $\tan \theta$ 9 $4.96 \sin (\theta + 56.4°)$

21 Introduction to vectors

Exercises 21.1

1 5.00 km from starting point in direction
bearing 53.1° 2 1.30 kN upwards making an
angle of 22.6° with the horizontal 3 2.44 kN

in direction S61.7°W 4 608 km in direction
N54°28′E 5 a downwards making angle 28.0°
with the vertical b 11.1 m s^{-2} 6 9.8 m s^{-1}
making an angle 18.4° with direction of car's
motion

Exercises 21.2

1 a 103 m s^{-1} b 292 m s^{-1} 2 a 5 h 00 min
b 5 h 19 min c 6 h 32 min d 10 h 00 min
e 28 h 48 min 3 a 8.61 s b 6.75 s c 14.4 s
d 23.8 s 4 F_1: 6.13 N to the right, F_2: 2.39 N
to the right, F_3: 9.40 N to the left, net horizontal
force = 874 mN to the left 5 1.98 N upwards
6 a 75.1 N b 7.51 m s^{-2} 7 a 177 N down the
plane b 157 N down the plane
8 a 0 b 7.0×10^{-12} N c 12×10^{-12}N
d 14×10^{-12}N

Exercises 21.3

1 2.6 kN in direction N5°E (i.e. in direction
bearing 005°) 2 134 km from P in direction
bearing 123° from P 3 447 kN in direction
N7°E 4 a 122 N directed between the
directions of the two given forces, making an
angle of 32° with the 83 N force (i.e. making
an angle of 40° with the 67 N force) b 5.9 kN
directed between the directions of the two given
forces, making an angle of 88° with the 4 kN
force (i.e. making an angle of 35° with the 7 kN
force) 5 25.6 km

Exercises 21.4

1 a $R = 43.5$ N, $F = 20.4$ N b $R = 39.6$ N,
$\theta = 41.3°$ c $T = 3.13$ kN, $P = 2.81$ kN
d $T = 389$ N, $\theta = 55.4°$ 2 a 3.51 kN b 7.2:1
3 1.43 kN

Self-test

1 a 22.5 m s^{-1} upwards at angle 27.5° with
horizontal b 27.6 m s^{-1} downwards at angle
43.5° with horizontal 2 0.652 kN in direction
between F_1 and F_4 making an angle of 19.9°
with force F_1 3 $T_1 = 320$ N, $T_2 = 240$
N 4 291°

22 Rotational equilibrium and frame analysis

Exercises 22.1

1 a i $R_{Av} + T\sin 50° = 196 + (50 \times 9.81)$
ii $R_{Ah} = T\cos 50°$ **iii** $(196 \times 2) + (50 \times 9.81 \times 4) = (T\sin 50° \times 3)$ **iv** $(R_{Av} \times 3) + (50 \times 9.81 \times 1) = 196 \times 1$ **v** $(R_{Av} \times 4) + (T\sin 50° \times 1) = 196 \times 2$ **b** 98.2 N downwards **c** 1.02 kN **d** 658 N **e** 686 N **f** 392 N **g** $R_A = 666$ N @ 351.5° **h** 1.02 kN @ 310° **2 a** 68.7 N **b** 208 N @ 109.3° **3 a** 235 N @ 0° **b** 847 N @ 106.1° **4 a** 839 N **b** 1.22 kN @ 130.0° **5 a** 226 N **b** 386 N @ 63.4° **6 a** 948 N **b** 3.06 kN @ 101.4°

Exercises 22.2

1 $R_A = 3.42$ kN @ 60.0°; $R_B = 2.66$ kN @ 130.0°; $F_1 = 0$; $F_2 = 3.42$ kN, compression; $F_3 = 2.66$ kN, compression. With roller at B: $F_1 = 1.71$ kN, tension; $R_A = 2.96$ kN @ 90.0°; $R_B = 2.04$ kN @ 90.0°
2 $R_A = 5.03$ kN @ 142.7°; $R_B = 2.95$ kN @ 90.0°; $F_1 = 5.11$ kN, tension; $F_2 = 3.24$ kN, compression; $F_3 = 5.90$ kN, compression
3 a $R_A = 8.70$ kN @ 20.0°; $R_D = 8.26$ kN @ 180.0°; $T = 8.26$ kN, tension; $F_1 = 5.42$ kN, compression; $F_2 = 4.13$ kN, compression; $F_3 = 6.20$ kN, tension. **b** $R_A = 11.4$ kN @ 7.6°; $R_D = 13.0$ kN @ 150.0°; $T = 13.0$ kN, tension; $F_1 = 8.66$ kN, compression; $F_2 = 3.00$ kN, compression; $F_3 = 10.0$ kN, tension.

23 Determinants and matrices

Exercises 23.1

1 7 **2** 2 **3** −24 **4** 22 **5** −10 **6** 15.1

Exercises 23.2

1 a 5 **b** −13 **c** 19 **2 a** −6 **b** 7 **c** 5
3 a −17.16 **b** −1.02 **c** −9.72 **4** $E \approx 1.03$, $V \approx 0.883$ **5** $L \approx 22.4$, $W \approx 13.6$ **6** $x \approx 0.984$, $y \approx -0.530$ **7** $I_1 \approx -3.61$, $I_2 \approx -5.01$

Exercises 23.3

1 a 2×3 **b** 3×2 **c** 3×1 **d** 2×3 **e** 1×3
2 a A, D **b** G **c** B, E, H, K **d** F **e** C

Exercises 23.4

1 a $x = 7, y = 3$ **b** $x = 3, y = 9$ **c** $x = 2, y = 6$ **d** $x = 4, y = -1$

2 $\begin{cases} 3x + 2y - 5z = 8 \\ 4x - 3y + 2z = 7 \\ 5x + 5y - 3z = 9 \end{cases}$

3 a A and D, B and H, E and K

b $\begin{pmatrix} 4 & 4 & 6 \\ 2 & 7 & 5 \\ 5 & 1 & 5 \end{pmatrix}$ **c** $\begin{pmatrix} 0 & 0 & 0 \\ 0 & 0 & 0 \end{pmatrix}$

d $\begin{pmatrix} 0 & 0 & 0 \\ 0 & 0 & 0 \\ 0 & 0 & 0 \end{pmatrix}$ **e** $\begin{pmatrix} 3 & 0 & 6 \\ 9 & 6 & 3 \end{pmatrix}$

f $\begin{pmatrix} 11 & 9 & 14 \\ 4 & 16 & 13 \\ 11 & 2 & 12 \end{pmatrix}$

Exercises 23.5

1 a i $\begin{pmatrix} 18 & 5 \\ 22 & 10 \end{pmatrix}$ **ii** $\begin{pmatrix} 17 & 9 \\ 18 & 16 \end{pmatrix}$

iii $\begin{pmatrix} 17 & 10 \\ 18 & 10 \end{pmatrix}$ **iv** $\begin{pmatrix} 6 & 3 \\ 4 & 12 \end{pmatrix}$ **v** $\begin{pmatrix} 9 & 13 \\ 11 & 12 \end{pmatrix}$

vi $\begin{pmatrix} 27 & 16 \\ 19 & 11 \end{pmatrix}$ **vii** $\begin{pmatrix} 10 & 3 \\ 6 & 6 \end{pmatrix}$ **viii** $\begin{pmatrix} 9 & 5 \\ 26 & 15 \end{pmatrix}$

ix $\begin{pmatrix} 2 & 6 \\ 8 & 9 \end{pmatrix}$ **x** $\begin{pmatrix} 10 & 9 \\ 4 & 3 \end{pmatrix}$ **b i** $\begin{pmatrix} 15 & -10 \\ 18 & -12 \end{pmatrix}$

ii $\begin{pmatrix} 2 & 0 \\ 0 & 2 \end{pmatrix}$ **iii** $\begin{pmatrix} -7 & 0 \\ -8 & 0 \end{pmatrix}$ **iv** $\begin{pmatrix} -14 & 7 \\ -16 & 8 \end{pmatrix}$

v $\begin{pmatrix} 8 & -7 \\ 16 & -14 \end{pmatrix}$ **vi** $\begin{pmatrix} 0 & 0 \\ 0 & 0 \end{pmatrix}$ **vii** $\begin{pmatrix} 0 & 0 \\ 0 & 0 \end{pmatrix}$

viii $\begin{pmatrix} 1 & 0 \\ -1 & 0 \end{pmatrix}$ **ix** $\begin{pmatrix} 2 & -1 \\ -2 & 1 \end{pmatrix}$

x $\begin{pmatrix} 8 & -4 \\ 12 & -6 \end{pmatrix}$ **2 a** $\begin{pmatrix} 7 & 2 & 9 \\ 4 & 3 & 5 \\ 2 & 13 & 9 \end{pmatrix}$

$$\mathbf{b}\begin{pmatrix} 5 & -5 & 1 \\ 3 & 0 & -3 \\ 5 & 2 & -1 \end{pmatrix} \quad \mathbf{c}\begin{pmatrix} 1 & 3 & 4 \\ 6 & 10 & 5 \\ 3 & 7 & 7 \end{pmatrix}$$

$$\mathbf{d}\begin{pmatrix} 3 & -2 \\ -1 & 12 \end{pmatrix} \quad \mathbf{e}\begin{pmatrix} 5 & 4 & 8 \\ 27 & 26 & 42 \end{pmatrix}$$

$$\mathbf{f}\begin{pmatrix} 4 & 2 & 5 \\ 14 & 13 & -16 \end{pmatrix} \quad \mathbf{g}\begin{pmatrix} 7 & 2 & 6 \\ 9 & 0 & 3 \end{pmatrix}$$

$$\mathbf{h}\begin{pmatrix} 9 & 15 \\ 2 & 4 \end{pmatrix} \quad \mathbf{i}\,(5) \quad \mathbf{j}\begin{pmatrix} -4 & 6 & -2 \\ 2 & -3 & 1 \\ -6 & 9 & -3 \end{pmatrix}$$

$$\mathbf{k}\begin{pmatrix} 16 \\ 38 \end{pmatrix} \quad \mathbf{l}\,(86) \quad \mathbf{m}\begin{pmatrix} -6 \\ 3 \\ 2 \end{pmatrix} \quad \mathbf{n}\,(-7)$$

$$\mathbf{o}\begin{pmatrix} 0 & -7 \\ -5 & -6 \end{pmatrix} \quad \mathbf{p}\begin{pmatrix} -20 & -1 \\ 8 & -6 \\ 6 & 2 \end{pmatrix}$$

$$\mathbf{3\,a}\begin{pmatrix} 11 & 1 & 5 \\ 1 & 7 & 2 \\ 3 & 5 & 5 \end{pmatrix} \quad \mathbf{b}\begin{pmatrix} -9 & 10 \\ 10 & -4 \end{pmatrix}$$

Exercises 23.6

1 a no b yes, 2×3 c yes, 2×1 d no
e yes, 1×3 f no g yes, 3×3 h yes, 2×3
i yes, 2×3 j no k yes, 2×1 l no

Exercises 23.7

$$\mathbf{1\,a}\begin{pmatrix} a & b \\ c & d \end{pmatrix} \quad \mathbf{b}\begin{pmatrix} a & b \\ c & d \end{pmatrix} \quad \mathbf{2\,a}\begin{pmatrix} a & b & c \\ d & e & f \\ g & h & i \end{pmatrix}$$

$$\mathbf{b}\begin{pmatrix} a & b & c \\ d & e & f \\ g & h & i \end{pmatrix} \quad \mathbf{3\,a}\begin{pmatrix} 1 & 2 & 3 \\ 4 & 5 & 6 \end{pmatrix}$$

$$\mathbf{b}\text{ incompatible } \mathbf{c}\begin{pmatrix} 1 & 2 & 3 \\ 4 & 5 & 6 \end{pmatrix}$$

$$\mathbf{d}\text{ incompatible } \mathbf{4\,a}\begin{pmatrix} 1 & 0 & 0 \\ 0 & 1 & 0 \\ 0 & 0 & 1 \end{pmatrix}$$

$$\mathbf{b}\begin{pmatrix} 1 & 0 \\ 0 & 1 \end{pmatrix} \quad \mathbf{c}\text{ no, because } \mathbf{A} \neq \mathbf{B}$$

$$\mathbf{5\,a}\text{ yes, }\begin{pmatrix} 1 & 0 \\ 0 & 1 \end{pmatrix} \quad \mathbf{b}\text{ no}\quad \mathbf{c}\text{ no}$$

$$\mathbf{d}\text{ yes, }\begin{pmatrix} 1 & 0 & 0 \\ 0 & 1 & 0 \\ 0 & 0 & 1 \end{pmatrix}$$

Exercises 23.8

$$\mathbf{1\,a}\begin{pmatrix} 1 & 0 \\ 0 & 1 \end{pmatrix}\text{ i.e. }\mathbf{I_2}\quad \mathbf{b}\begin{pmatrix} 1 & 0 \\ 0 & 1 \end{pmatrix}\text{ i.e. }\mathbf{I_2}$$

$$\mathbf{2\,a}\begin{pmatrix} 1 & 0 & 0 \\ 0 & 1 & 0 \\ 0 & 0 & 1 \end{pmatrix}\text{ i.e. }\mathbf{I_3}\quad \mathbf{b}\begin{pmatrix} 1 & 0 & 0 \\ 0 & 1 & 0 \\ 0 & 0 & 1 \end{pmatrix}\text{ i.e. }\mathbf{I_3}$$

c $\mathbf{CD} = \mathbf{DC} = \mathbf{I}$, $\mathbf{D} = \mathbf{C}^{-1}$ and $\mathbf{C} = \mathbf{D}^{-1}$.
$\mathbf{C}$ and $\mathbf{D}$ are inverses. 3 a $p \times s$ b $r \times q$
c $p = q = r = s = n$ d both are *square* matrices

$$\mathbf{4\,a}\ \mathbf{i}\begin{pmatrix} 7 & 16 & -13 \\ 6 & 14 & -11 \\ 9 & 21 & -17 \end{pmatrix}\quad \mathbf{ii}\begin{pmatrix} 1 & 0 & 0 \\ 0 & 1 & 0 \\ 0 & 0 & 1 \end{pmatrix}$$

$$\mathbf{iii}\begin{pmatrix} 3 & 5 & 4 \\ 2 & 3 & 4 \\ 3 & 6 & 6 \end{pmatrix}\quad \mathbf{iv}\begin{pmatrix} 3 & -8 & 5 \\ 0 & -4 & 3 \\ 1 & -7 & 5 \end{pmatrix}$$

$$\mathbf{v}\begin{pmatrix} -1 & -5 & 4 \\ -3 & -10 & 9 \\ -4 & -13 & 12 \end{pmatrix}\quad \mathbf{vi}\begin{pmatrix} -1 & 4 & 0 \\ -2 & 9 & 2 \\ -2 & 13 & 2 \end{pmatrix}$$

$$\mathbf{vii}\begin{pmatrix} 1 & 0 & 0 \\ 0 & 1 & 0 \\ 0 & 0 & 1 \end{pmatrix}\quad \mathbf{b}\ \mathbf{i}\begin{pmatrix} 1 & -1 & 0 \\ -1 & 0 & 1 \\ 0 & 2 & -1 \end{pmatrix}$$

$$\mathbf{ii}\begin{pmatrix} 3 & -4 & 2 \\ -2 & 1 & 0 \\ -1 & -1 & 1 \end{pmatrix}\quad \mathbf{5\,a}\ 3\begin{pmatrix} 1 & 0 & 0 \\ 0 & 1 & 0 \\ 0 & 0 & 1 \end{pmatrix}\text{ or}$$

$$\begin{pmatrix} 3 & 0 & 0 \\ 0 & 3 & 0 \\ 0 & 0 & 3 \end{pmatrix}\text{ i.e. }3\ \mathbf{I_3}\quad \mathbf{b}\ \frac{1}{3}\begin{pmatrix} 3 & -2 & 1 \\ 0 & 1 & -2 \\ 0 & 0 & -3 \end{pmatrix}$$

$$\text{or}\begin{pmatrix} 1 & -\frac{2}{3} & \frac{1}{3} \\ 0 & \frac{1}{3} & -\frac{2}{3} \\ 0 & 0 & -1 \end{pmatrix}\quad \mathbf{6\,a}\ 10\begin{pmatrix} 1 & 0 & 0 \\ 0 & 1 & 0 \\ 0 & 0 & 1 \end{pmatrix}\text{ or}$$

$$\begin{pmatrix} 10 & 0 & 0 \\ 0 & 10 & 0 \\ 0 & 0 & 10 \end{pmatrix} \text{ i.e. } 10\ \mathbf{I_3} \quad \mathbf{b}\ \frac{1}{10}\begin{pmatrix} 1 & 2 & -2 \\ 1 & 3 & -1 \\ 3 & 2 & 0 \end{pmatrix}$$

or $\begin{pmatrix} 0.1 & 0.2 & -0.2 \\ 0.1 & 0.3 & -0.1 \\ 0.3 & 0.2 & 0 \end{pmatrix}$

Exercises 23.9

1 a B **b** I **c** $ABA^{-1} + B$ **d** A $BA^{-1}B + A^2$
e I **f** 2I **g** $A^2 + 2A + I$ **h** $AB + A^2 + B + A$
i A **j** O **2 a** $X = \frac{1}{2}(A + B)$ **b** $X = \frac{1}{3}(C - D)$
c $X = -\frac{3}{4}A$ **d** $X = 2A + 9B$ **3 a** $X = A^{-1}B$
b $X = BA^{-1}$ **c** $X = A^{-1}(B - 2A)$ or $A^{-1}B -$
$2I$ **d** $X = A^{-1}(B - I)$ **e** $X = A^{-1}$ **f** $X =$
$(3I - A)^{-1}B$ **g** $X = (A + I)^{-1}B$
h $X = (A + I)^{-1}(C - B)$ **i** $X = B(A + I)^{-1}$
j $X = \frac{1}{2}(C - B)(A + I)^{-1}$ **4 a** yes **b** $B = C$
5 a no **b** $B = A^{-1}CA$ **6 a** $X = (C - B)$
$(2A + 3I)^{-1}$ **b** $X = (A - B)(3C - 2I)^{-1}$ or
$(B - A)(2I - 3C)^{-1}$

Exercises 23.10

1 a $\begin{pmatrix} 2x + 3y \\ 5x + 4 \end{pmatrix}$, order 2×1

b $\begin{pmatrix} 3x + 5y - 2z \\ x + 4y + 3z \\ 2x - 3y + 4z \end{pmatrix}$, order 3×1

2 a $\begin{pmatrix} 3 & 4 \\ 2 & -5 \end{pmatrix}\begin{pmatrix} x \\ y \end{pmatrix}$ **b** $\begin{pmatrix} 2 & -3 & 4 \\ 1 & 2 & -5 \\ 3 & -1 & 2 \end{pmatrix}\begin{pmatrix} x \\ y \\ z \end{pmatrix}$

c $\begin{pmatrix} 3 & -5 & 2 \\ 2 & 4 & 0 \\ 5 & 0 & -7 \end{pmatrix}\begin{pmatrix} a \\ b \\ c \end{pmatrix}$ **3 a** $x = 4, y = -3$

b $x = 2, y = 3$ **4 a** $\begin{cases} 2x + 3y = 3 \\ 5x - 2y = 4 \end{cases}$

b $\begin{cases} 3x + 5y = 6 \\ 2x - 4y = 7 \end{cases}$ **c** $\begin{cases} 3x - 2y + 4z = 6 \\ 2x + 3y - 5z = 7 \\ x - 4y + 2z = 8 \end{cases}$

d $\begin{cases} x + 2y + 3z = 4 \\ y + 2z = 5 \\ x - 2y = 6 \end{cases}$

5 a $\begin{pmatrix} 5 & 7 & -3 \\ 7 & -2 & -8 \\ 3 & 5 & 2 \end{pmatrix}\begin{pmatrix} a \\ b \\ c \end{pmatrix} = \begin{pmatrix} 17 \\ 13 \\ 19 \end{pmatrix}$

b $\begin{pmatrix} 4 & -5 & 6 \\ 7 & 3 & 0 \\ 3 & 0 & -8 \end{pmatrix}\begin{pmatrix} x \\ y \\ z \end{pmatrix} = \begin{pmatrix} 9 \\ 5 \\ 7 \end{pmatrix}$

Exercises 23.11

1 a $\begin{pmatrix} 2 & -3 \\ 5 & -7 \end{pmatrix}\begin{pmatrix} x \\ y \end{pmatrix} = \begin{pmatrix} 9 \\ 22 \end{pmatrix}$ **b** $x = 3, y = -1$

2 a i 2 **ii** $\frac{1}{2}\begin{pmatrix} 4 & -2 \\ -5 & 3 \end{pmatrix}$ **b** $x = -2, y = 5$

3 a 11 $\begin{pmatrix} 1 & 0 \\ 0 & 1 \end{pmatrix}$ **b** $\frac{1}{11}\begin{pmatrix} 2 & 3 \\ 1 & -4 \end{pmatrix}$ **c** $x = 4, y = -3$

4 a i $\begin{pmatrix} 1 & -3 \\ 4 & -11 \end{pmatrix}\begin{pmatrix} a \\ b \end{pmatrix} = \begin{pmatrix} -1 \\ -2 \end{pmatrix}$ **ii** 1

iii $\begin{pmatrix} -11 & 3 \\ -4 & 1 \end{pmatrix}$ **iv** $a = 5, b = 2$

b i $\begin{pmatrix} 1 & -1 \\ 2 & 3 \end{pmatrix}\begin{pmatrix} a \\ b \end{pmatrix} = \begin{pmatrix} 6 \\ 2 \end{pmatrix}$ **ii** 5

iii $\frac{1}{5}\begin{pmatrix} 3 & 1 \\ -2 & 1 \end{pmatrix}$ **iv** $a = 4, b = -2$

c i $\begin{pmatrix} 3 & -1 \\ -2 & 1 \end{pmatrix}\begin{pmatrix} x \\ y \end{pmatrix} = \begin{pmatrix} 6 \\ 2 \end{pmatrix}$ **ii** 1 **iii** $\begin{pmatrix} 1 & 1 \\ 2 & 3 \end{pmatrix}$

iv $x = 8, y = 18$

d i $\begin{pmatrix} 4 & 4 \\ 5 & -3 \end{pmatrix}\begin{pmatrix} p \\ q \end{pmatrix} = \begin{pmatrix} 8 \\ 2 \end{pmatrix}$ **ii** -32

iii $\frac{1}{-32} \times \begin{pmatrix} -3 & -4 \\ -5 & 4 \end{pmatrix}$ **iv** $p = 1, q = 1$

e i $\begin{pmatrix} 1 & -3 \\ 4 & 8 \end{pmatrix}\begin{pmatrix} x \\ y \end{pmatrix} = \begin{pmatrix} 4 \\ 6 \end{pmatrix}$ **ii** 20

iii $\frac{1}{20}\begin{pmatrix} 8 & 3 \\ -4 & 1 \end{pmatrix}$ **iv** $x = 2.5$
$y = -0.5$

5 a $a = -2, b = -14, c = 9$

6 a $\begin{pmatrix} 1 & 0 & 0 \\ 0 & 1 & 0 \\ 0 & 0 & 1 \end{pmatrix}$

$$\mathbf{b}\begin{pmatrix} 11 & -5 & -3 \\ -8 & 4 & 2 \\ -7 & 3 & 2 \end{pmatrix}\begin{pmatrix} x \\ y \\ z \end{pmatrix} = \begin{pmatrix} -12 \\ 10 \\ 10 \end{pmatrix}$$

$\mathbf{c}\,x = -1, y = 2, z = -3$

$$7\,\mathbf{a}\,\mathbf{i}\,PQ = \begin{pmatrix} 1 & 0 & 0 \\ 0 & 1 & 0 \\ 1 & 3 & -2 \end{pmatrix}$$

$$\mathbf{ii}\,PR = \begin{pmatrix} 15 & -18 & 8 \\ -8 & 9 & -4 \\ -5 & 5 & -2 \end{pmatrix}$$

$$\mathbf{iii}\,QR = \begin{pmatrix} 1 & 0 & 0 \\ 0 & 1 & 0 \\ 0 & 0 & 1 \end{pmatrix}$$

$\mathbf{b}$ Q and R $\;\mathbf{c}\begin{pmatrix} 1 & 2 & -2 \\ 2 & 5 & -4 \\ 3 & 7 & -5 \end{pmatrix}\begin{pmatrix} a \\ b \\ c \end{pmatrix} = \begin{pmatrix} 3 \\ 7 \\ 8 \end{pmatrix}$

$\mathbf{d}\,a = -3, b = 1, c = -2$
$8\,\mathbf{a}\,x = 1, y = 2, z = 5$ $\;\mathbf{b}\,a = 3, b = 2, c = -2$
$\mathbf{c}\,p = 2, q = 2, r = -4$ $\;\mathbf{d}\,x = 2, y = 1, z = -3$
$\mathbf{e}\,x = 1, y = 2, z = 1$ $\;\mathbf{f}\,a = 3, b = 2, c = -3$
$\mathbf{g}\,a = 3, b = -4, c = -1$ $\;\mathbf{h}\,x = 2, y = 1,$
$z = -1$ $\mathbf{i}\,p = 2, q = 1, r = 3$ $\mathbf{j}\,k = 2, n = -24,$
$t = -27$ $\;\mathbf{k}\,x = 5, y = -2, z = 3$ $\;\mathbf{l}\,a = 3, b = 2,$
$c = -6$ $\;\mathbf{m}\,x = 1, y = -1, z = 0$ $\;\mathbf{n}\,p = 2,$
$q = -3, r = 4$ $\;\mathbf{o}\,x = 1, y = 3, z = -5$

$$9\,\mathbf{a}\,\mathbf{i}\begin{pmatrix} 2 & -2 & -1 \\ 1 & -5 & -3 \\ 1 & 6 & 4 \end{pmatrix}\begin{pmatrix} x \\ y \\ z \end{pmatrix} = \begin{pmatrix} 1 \\ 2 \\ -3 \end{pmatrix}\,\mathbf{ii}\,-1$$

$$\mathbf{iii}\,\frac{1}{-1}\begin{pmatrix} -2 & 2 & 1 \\ -7 & 9 & 5 \\ 11 & -14 & -8 \end{pmatrix}$$

$\mathbf{iv}\,x = 1, y = 4, z = -7$

$$9\,\mathbf{b}\,\mathbf{i}\begin{pmatrix} 2 & 3 & 2 \\ 2 & 2 & 2 \\ 5 & 1 & 4 \end{pmatrix}\begin{pmatrix} a \\ b \\ c \end{pmatrix} = \begin{pmatrix} -4 \\ -2 \\ -6 \end{pmatrix}\,\mathbf{ii}\,2$$

$$\mathbf{iii}\,\frac{1}{2}\begin{pmatrix} -6 & -10 & -2 \\ 2 & -2 & 0 \\ -8 & 13 & -2 \end{pmatrix}$$

$\mathbf{iv}\,a = -8, b = -2, c = 9$

$$9\,\mathbf{c}\,\mathbf{i}\begin{pmatrix} 1 & 1 & 1 \\ 2 & -2 & 1 \\ 2 & 2 & 4 \end{pmatrix}\begin{pmatrix} p \\ q \\ r \end{pmatrix} = \begin{pmatrix} 4 \\ 5 \\ 6 \end{pmatrix}\,\mathbf{ii}\,-8$$

$$\mathbf{iii}\,\frac{1}{-8}\begin{pmatrix} -10 & -2 & 3 \\ -6 & 2 & 1 \\ 8 & 0 & -4 \end{pmatrix}$$

$\mathbf{iv}\,p = 4, q = 1, r = -1$

$$9\,\mathbf{d}\,\mathbf{i}\begin{pmatrix} 2 & 2 & -2 \\ -2 & 2 & -1 \\ -4 & -3 & 3 \end{pmatrix}\begin{pmatrix} x \\ y \\ z \end{pmatrix} = \begin{pmatrix} -2 \\ 1 \\ 0 \end{pmatrix}$$

$$\mathbf{ii}\,-2\;\;\mathbf{iii}\,\frac{1}{-2}\begin{pmatrix} 3 & 0 & 2 \\ 10 & -2 & 6 \\ 14 & -2 & 8 \end{pmatrix}$$

$\mathbf{iv}\,x = 3, y = 11, z = 15$

$$9\,\mathbf{e}\,\mathbf{i}\begin{pmatrix} 4 & 2 & -4 \\ 4 & -5 & 2 \\ 5 & -2 & -1 \end{pmatrix}\begin{pmatrix} a \\ b \\ c \end{pmatrix} = \begin{pmatrix} -4 \\ -2 \\ -5 \end{pmatrix}$$

$$\mathbf{ii}\,-4\;\;\mathbf{iii}\,\frac{1}{-4}\begin{pmatrix} 9 & 10 & -16 \\ 14 & 16 & -24 \\ 17 & 18 & -28 \end{pmatrix}$$

$\mathbf{iv}\,a = -6, b = -8, c = -9$

Exercises 23.12

$1\,\mathbf{a}\,-8$ $\;\mathbf{b}\,0$ $\;\mathbf{c}\,-8$ $\;\mathbf{d}\,0$ $\;\mathbf{e}\,1$ $\;\mathbf{f}\,0$ $\;\mathbf{g}\,-10\,080$
$\mathbf{h}\,-19\,608$ $\;2\,\mathbf{a}\,x = 3$ $\;\mathbf{b}\,x = -1, 2$ $\;\mathbf{c}\,n = -5$
$\mathbf{d}\,t = \frac{1}{3}, 1$ $\;\mathbf{e}\,k = -2, -4$

Exercises 23.13

$1\,\mathbf{a}\,x = 2, y = -3, z = 1$ $\;\mathbf{b}\,p = -3, q = -1,$
$r = -2$ $\;\mathbf{c}\,x = -1, y = 2, z = -2$ $\;\mathbf{d}\,n = -2,$
$p = 0, t = 1$ $\;2\,\mathbf{a}\,x = 0, y = 2, z = 0$
$\mathbf{b}\,x = 3.04, y = 0.896, z = 1.23$ $\;\mathbf{c}\,k = 0.81,$
$t = -0.75, w = -0.91$ $\;3\,F_1 = -0.583$ kN,
$F_2 = 1.67$ kN, $F_3 = 1.25$ kN. F_1 acts in the
opposite direction to that shown in the diagram.
$4\,I_1 = 3.66$ A, $I_2 = 4.97$ A, $I_3 = 3.84$ A $\;5\,10$
of 2 ha, 20 of 0.5 ha, 70 of 0.2 ha $\;6$ A: 40 kg,
B: 10 kg, C: 50 kg

24 Statistics and probability

Exercises 24.1

$\mathbf{a}$ Mean 6.0 Mode 3.0 Range 6.0 Median 6.5
$\mathbf{b}$ Mean 6.0 Mode 7.0 Range 7.0 Median 6.5

c Mean 21.0 Mode 21.0 Range 8.0 Median 21.0
d Mean 18.3 Mode 12.0 Range 53.0 Median 12.0
e Mean 9.5 Mode 5.0 Range 32.0 Median 5.5

Exercises 24.2

1 Mean 101.50 Range 53.00 Standard deviation 15.04

2 a Mean 5.00 Variance 2.40 Standard deviation 1.55 **b** Mean 6.00 Variance 13.33 Standard deviation 3.65 **c** Mean 8.00 Variance 10.67 Standard deviation 3.27 **d** Mean 4.67 Variance 3.56 Standard deviation 1.89 **e** Mean 5.40 Variance 18.64 Standard deviation 4.32

3 a Variance 3.00 Standard deviation 1.73
b Variance 16.00 Standard deviation 4.00
c Variance 16.00 Standard deviation 4.00
d Variance 4.27 Standard deviation 2.07
e Variance 23.30 Standard deviation 4.83

Exercises 24.3

1 a Mean 72.46 Range 31 **b** Variance 60.71 Standard deviation 7.79 **c** Joshua : A Alice : C Jessica : B **d** Yes, Charles **e** No

2 a 0.65 **b** 0.82 **c** 1.22 **d** 1.24 **e** 0.83

Exercises 24.4

1 a 0.08, (8%) **b** 0.46, (46%) **c** 0.46, (46%)
2 a 0.127, (12.7%) **b** 0.436, (43.6%) **c** 0.436, (43.6%) **d** 0.928, (92.8%) **e** 0.999, ≈ 1
3 a 0.709, (70.9%) **b** 0.051, (5.1%) **c** 0.181, (18.1%)

Self-test

1 a Mean 20.8 Mode 22.0 Range 2.9 Median 21.0 Variance 1.6 Standard deviation 1.3 **b** Mean 8.4 Mode 2.0 Range 18.0 Median 7.0 Variance 36.5 Standard deviation 6.0 **c** Mean 9.2 Mode 8.6 Range 1.7 Median 8.9 Variance 0.5 Standard deviation 0.7

2 a Variance 1.9 Standard deviation 1.4
b Variance 42.6 Standard deviation 6.5
c Variance 0.6 Standard deviation 0.8

3 a Mean 87.8 **b** Standard deviation 10.8
c Wetter than a sock in a puddle **4 a** 9.8%
b 48.6% **5 a** 10.2% **b** 50% **6 a** 24.2%
b 27.4% **c** 34.7% **d** 35.9% **e** 13.5%, 95.9%, 11.3%, 93.6%

INDEX